Central and Eastern European Development Studies (CEEDES)

For further volumes:
http://www.springer.com/series/3862

Jan Harff · Svante Björck · Peer Hoth
Editors

The Baltic Sea Basin

With 174 Figures and 16 Tables

Springer

Editors
Prof. Dr. Jan Harff
Leibniz Institute for Baltic Sea Research
Warnemünde
Seestr. 15
18119 Rostock
Germany
jan.harff@io-warnemuende.de

Dr. Peer Hoth
Federal Institute for Geosciences and
Natural Resources
Berlin Office
Wilhelmstrasse 25-30
13593 Berlin
peer.hoth@freenet.de

Prof. Svante Björck
Department of Earth and Ecosystem
Sciences
Division of Geology, Quaternary Sciences
Lund University
Sölveg. 12
SE-223 62 Lund
Sweden
svante.bjorck@geol.lu.se

ISSN 1614-032X
ISBN 978-3-642-17219-9 e-ISBN 978-3-642-17220-5
DOI 10.1007/978-3-642-17220-5
Springer Heidelberg Dordrecht London New York

Library of Congress Control Number: 2011921542

Cover design: deblik, Berlin

Printed on acid-free paper

Springer is part of Springer Science+Business Media (www.springer.com)

Contents

Part I Introduction

1 **The Baltic Sea Basin: Introduction** 3
Jan Harff, Svante Björck, and Peer Hoth

Part II Geological and Tectonical Evolution

2 **Geological Evolution and Resources of the Baltic Sea Area from the Precambrian to the Quaternary** 13
Saulius Šliaupa and Peer Hoth

3 **Glacial Erosion/Sedimentation of the Baltic Region and the Effect on the Postglacial Uplift** 53
Aleksey Amantov, Willy Fjeldskaar, and Lawrence Cathles

Part III The Basin Fill as a Climate and Sea Level Record

4 **The Development of the Baltic Sea Basin During the Last 130 ka** 75
Thomas Andrén, Svante Björck, Elinor Andrén, Daniel Conley, Lovisa Zillén, and Johanna Anjar

5 **Late Quaternary Climate Variations Reflected in Baltic Sea Sediments** 99
Jan Harff, Rudolf Endler, Emel Emelyanov, Sergey Kotov, Thomas Leipe, Matthias Moros, Ricardo Olea, Michal Tomczak, and Andrzej Witkowski

6 **Geological Structure of the Quaternary Sedimentary Sequence in the Klaipėda Strait, Southeastern Baltic** 133
Albertas Bitinas, Aldona Damušytė, and Anatoly Molodkov

Part IV Coastline Changes

7 Coastlines of the Baltic Sea – Zones of Competition Between Geological Processes and a Changing Climate: Examples from the Southern Baltic 149
Jan Harff and Michael Meyer

8 Palaeogeographic Model for the SW Estonian Coastal Zone of the Baltic Sea 165
Alar Rosentau, Siim Veski, Aivar Kriiska, Raivo Aunap, Jüri Vassiljev, Leili Saarse, Tiit Hang, Atko Heinsalu, and Tõnis Oja

9 Palaeoreconstruction of the Baltic Ice Lake in the Eastern Baltic . 189
Jüri Vassiljev, Leili Saarse, and Alar Rosentau

10 Submerged Holocene Wave-Cut Cliffs in the South-eastern Part of the Baltic Sea: Reinterpretation Based on Recent Bathymetrical Data 203
Vadim Sivkov, Dimitry Dorokhov, and Marina Ulyanova

11 Drowned Forests in the Gulf of Gdańsk (Southern Baltic) as an Indicator of the Holocene Shoreline Changes 219
Szymon Uścinowicz, Grażyna Miotk-Szpiganowicz, Marek Krąpiec, Małgorzata Witak, Jan Harff, Harald Lübke, and Franz Tauber

12 Holocene Evolution of the Southern Baltic Sea Coast and Interplay of Sea-Level Variation, Isostasy, Accommodation and Sediment Supply 233
Reinhard Lampe, Michael Naumann, Hinrich Meyer, Wolfgang Janke, and Regine Ziekur

Part V Sediment Dynamics

13 On the Dynamics of "Almost Equilibrium" Beaches in Semi-sheltered Bays Along the Southern Coast of the Gulf of Finland 255
Tarmo Soomere and Terry Healy

14 Modelling Coastline Change of the Darss-Zingst Peninsula with Sedsim 281
Michael Meyer, Jan Harff, and Chris Dyt

Part VI Interactions Between a Changing Environment and Society

15 Settlement Development in the Shadow of Coastal Changes – Case Studies from the Baltic Rim 301
Hauke Jöns

16 Geological Hazard Potential at the Baltic Sea and Its Coastal Zone: Examples from the Eastern Gulf of Finland and the Kaliningrad Area 337
Mikhail Spiridonov, Daria Ryabchuk, Vladimir Zhamoida, Alexandr Sergeev, Vadim Sivkov, and Vadim Boldyrev

17 Seafloor Desertification – A Future Scenario for the Gulf of Finland? 365
Henry Vallius, Vladimir Zhamoida, Aarno Kotilainen, and Daria Ryabchuk

18 Sources, Dynamics and Management of Phosphorus in a Southern Baltic Estuary 373
Gerald Schernewski, Thomas Neumann, and Horst Behrendt

Part VII Hydrogeological Modeling

19 Potential Change in Groundwater Discharge as Response to Varying Climatic Conditions – An Experimental Model Study at Catchment Scale 391
Maria-Theresia Schafmeister and Andreas Darsow

Part VIII Monitoring

20 Monitoring the Bio-optical State of the Baltic Sea Ecosystem with Remote Sensing and Autonomous In Situ Techniques 407
Susanne Kratzer, Kerstin Ebert, and Kai Sørensen

Index 437

Contributors

Aleksey Amantov VSEGEI, St. Petersburg, Russia, 4448470@mail.wplus.net

Thomas Andrén School of Life Sciences, Södertörn University, SE-141 89 Huddinge, Sweden, thomas.andren@sh.se

Elinor Andrén School of Life Sciences, Södertörn University, SE-141 89 Huddinge, Sweden, Elinor.Andren@sh.se

Johanna Anjar Department of Earth and Ecosystem Sciences, Quaternary Sciences, Lund University, SE-223 62 Lund, Sweden, Johanna.Anjar@geol.lu.se

Raivo Aunap Department of Geography, University of Tartu, 51014 Tartu, Estonia, raivo.aunap@ut.ee

Horst Behrendt Leibniz Institute of Freshwater Ecology and Inland Fisheries, Berlin, Germany

Albertas Bitinas Coastal Research and Planning Institute, Klaipėda University, LT-92294 Klaipėda, Lithuania; Department of Geology and Mineralogy, Faculty of Natural Sciences, Vilnius University, LT-03101 Vilnius, Lithuania, albertas.bitinas@corpi.ku.lt; albertas.bitinas@gmail.com

Svante Björck Department of Earth and Ecosystem Sciences, Division of Geology, Quaternary Sciences, Lund University, Sölveg. 12, SE-223 62 Lund, Sweden, svante.bjorck@geol.lu.se

Vadim Boldyrev Atlantic Branch, P.P. Shirshov Institute of Oceanology, Russian Academy of Sciences (ABIORAS), Kaliningrad, Russia, Vadim_Boldyrev@mail.ru

Lawrence Cathles Cornell University, Ithaca, NY, USA, lmc19@cornell.edu

Daniel Conley Department of Earth and Ecosystem Sciences, Quaternary Sciences, Lund University, SE-223 62 Lund, Sweden, Daniel.Conley@geol.lu.se

Aldona Damušytė Lithuanian Geological Survey, LT-03123 Vilnius, Lithuania, aldona.damusyte@lgt.lt

Andreas Darsow Department of Environmental Geosciences, University of Vienna, 1090 Vienna, Austria, andreas.darsow@univie.ac.at

Dimitry Dorokhov Atlantic Branch, P.P. Shirshov Institute of Oceanology, Russian Academy of Sciences, Kaliningrad, Russia, d_dorokhov@mail.ru

Chris Dyt CSIRO Petroleum Resources, Bentley, WA 6102, Australia, chris.dyt@csiro.au

Kerstin Ebert Laboratoire d'Océanographie de Villefranche (LOV), Universite Pierre et Marie Curie, UMR CNRS 7093, Quai de la Darse, 06230 Villefranche-sur-Mer Cedex, France, kerstin.ebert@obs-vlfr.fr

Emel Emelyanov Atlantic Branch, P.P. Shirshov Institute of Oceanology, Russian Academy of Sciences (ABIORAS), Kaliningrad, Russia, abio@atlas.baltnet.ru

Rudolf Endler Leibniz Institute for Baltic Sea Research Warnemünde, D-18119 Rostock, Germany, rudolf.endler@io-warnemuende.de

Willy Fjeldskaar IRIS, Stavanger, Norway, WF@tectonor.com

Tiit Hang Department of Geology, University of Tartu, 51014 Tartu, Estonia, tiit.hang@ut.ee

Jan Harff Leibniz Institute for Baltic Sea Research Warnemünde, D-18119 Rostock, Germany; presently at Institute of Marine and Coastal Sciences, University of Szczecin, PL-70-383 Szczecin, Poland, jan.harff@io-warnemuende.de

Terry Healy† **(28.11.1944–20.07.2010)** Coastal Marine Group, Earth and Ocean Sciences, University of Waikato, Hamilton 3240, New Zealand

Atko Heinsalu Institute of Geology, Tallinn University of Technology, 19086 Tallinn, Estonia, heinsalu@gi.ee

Peer Hoth Federal Institute for Geosciences and Natural Resources, Berlin Office, 13593 Berlin (presently at: Federal Ministry of Economics and Technology, Energy Department), peer.hoth@freenet.de

Wolfgang Janke 17489 Greifswald, Germany, wofajanke@web.de

Hauke Jöns Lower Saxony Institute for Historical Coastal Research, D-26382 Wilhelmshaven, Germany, joens@nihk.de

Aarno Kotilainen Geological Survey of Finland, FIN-02151 Espoo, Finland, aarno.kotilainen@gtk.fi

Sergey Kotov St. Petersburg State University, St. Petersburg, Russia, kotov_s@yahoo.co.uk

Marek Krąpiec University of Science and Technology, Kraków, Poland, mkrapiec@agh.edu.pl

Susanne Kratzer Department of Systems Ecology, Stockholm University, 106 91 Stockholm, Sweden, Suse@ecology.su.se

Aivar Kriiska Institute of History and Archaeology, University of Tartu, Tartu, Estonia, aivar.kriiska@ut.ee

Reinhard Lampe Institut für Geographie und Geologie, Ernst-Moritz-Arndt-Universität Greifswald, D-17487 Greifswald, Germany, lampe@uni-greifswald.de

Thomas Leipe Leibniz Institute for Baltic Sea Research Warnemünde, D-18119 Rostock, Germany, thomas.leipe@io-warnemuende.de

Harald Lübke Roman-Germanic Commission, German Archaeological Institute, 60325 Frankfurt a.M, Germany, harald.luebke@schloss-gottorf.de

Hinrich Meyer Institut für Geographie und Geologie, Ernst-Moritz-Arndt-Universität Greifswald, D-17487 Greifswald, Germany, himeyer@uni-greifswald.de

Michael Meyer Leibniz Institute for Baltic Sea Research Warnemünde, D-18119 Rostock, Germany; Institute for Environmental Engineering, University Rostock, 18057 Rostock, Germany, michael.meyer@uni-rostock.de

Grażyna Miotk-Szpiganowicz Polish Geological Institute, National Research Institute, Gdańsk, Poland, grazyna.miotk-szpiganowicz@pgi.gov.pl

Anatoly Molodkov Research Laboratory for Quaternary Geochronology, Institute of Geology, Tallinn University of Technology, 19086 Tallinn, Estonia, molodkov@gi.ee

Matthias Moros Leibniz Institute for Baltic Sea Research Warnemünde, D-18119 Rostock, Germany, Matthias.moros@io-warnemuende.de

Michael Naumann Leibniz Institute for Baltic Sea Research Warnemünde, D-18119 Rostock, Germany; presently at Landesamt für Bergbau, Energie und Geologie, 30655 Hannover, Germany, michael.naumann@lbeg.niedersachsen.de

Thomas Neumann Leibniz Institute for Baltic Sea Research Warnemünde, Rostock, Germany, thomas.neumann@io-warnemuende.de

Tõnis Oja Department of Physics, Tallinn University of Technology, 19086 Tallinn, Estonia, tonis.oja@maaamet.ee

Ricardo Olea Leibniz Institute for Baltic Sea Research Warnemünde, D-18119 Rostock, Germany; presently at US Geological Survey, Reston, VA, USA, olea@usgs.gov

Alar Rosentau Department of Geology, University of Tartu, 51014 Tartu, Estonia; Institute of History and Archaeology, University of Tartu, Tartu, Estonia, alar.rosentau@ut.ee

Daria Ryabchuk A.P. Karpinsky Russian Research Geological Institute (VSEGEI), St. Petersburg 199106, Russia, Daria_Ryabchuk@vsegei.ru

Leili Saarse Institute of Geology, Tallinn University of Technology, 19086 Tallinn, Estonia, saarse@gi.ee

Maria-Theresia Schafmeister Institute for Geography and Geology, University of Greifswald, 17489 Greifswald, Germany, schaf@uni-greifswald.de

Gerald Schernewski Leibniz Institute for Baltic Sea Research Warnemünde, Rostock, Germany, gerald.schernewski@io-warnemuende.de

Alexandr Sergeev A.P. Karpinsky Russian Research Geological Institute (VSEGEI), St. Petersburg, 199106, Russia, sergeevau@yandex.ru

Vadim Sivkov Atlantic Branch, P.P. Shirshov Institute of Oceanology, Russian Academy of Sciences (ABIORAS), Kaliningrad, Russia, sivkov@kaliningrad.ru

Saulius Šliaupa Institute of Geology and Geography, Vilnius University, Vilnius 01013, Lithuania, sliaupa@geo.lt

Tarmo Soomere Institute of Cybernetics, Tallinn University of Technology, 12618 Tallinn, Estonia, soomere@cs.ioc.ee

Kai Sørensen Norwegian Institute for Water Research (NIVA), Gaustadalléen 21, NO-0349 OSLO, Norway, kai.sorensen@niva.no

Mikhail Spiridonov A.P. Karpinsky Russian Research Geological Institute (VSEGEI), St. Petersburg 199106, Russia, Michail_Spiridonov@vsegei.ru

Franz Tauber Leibniz Institute for Baltic Sea Research Warnemünde, D-18119 Rostock, Germany, franz.tauber@io-warnemuende.de

Michal Tomczak Institute of Marine and Coastal Sciences, University of Szczecin, Szczecin, Poland, tomcz.michal@gmail.com

Marina Ulyanova Atlantic Branch, P.P. Shirshov Institute of Oceanology, Russian Academy of Sciences, Kaliningrad, Russia, marioches@rambler.ru

Szymon Uścinowicz Polish Geological Institute, National Research Institute, Gdańsk, Poland, szymon.uscinowicz@pgi.gov.pl

Henry Vallius Geological Survey of Finland, FIN-02151 Espoo, Finland, henry.vallius@gtk.fi

Jüri Vassiljev Institute of Geology, Tallinn University of Technology, 19086 Tallinn, Estonia, vassilje@gi.ee

Siim Veski Institute of Geology, Tallinn University of Technology, 19086 Tallinn, Estonia, veski@gi.ee

Małgorzata Witak Institute of Oceanography, University of Gdańsk, Gdańsk, Poland, ocemaw@univ.gda.pl

Andrzej Witkowski Institute of Marine and Coastal Sciences, University of Szczecin, Szczecin, Poland, witkowsk@univ.szczecin.pl

Vladimir Zhamoida A.P. Karpinsky Russian Research Geological Institute (VSEGEI), St. Petersburg, 199106, Russia, Vladimir_Zhamoida@vsegei.ru

Regine Ziekur Leibniz-Institut für Angewandte Geophysik, D-30655 Hannover, Germany, Regine.Ziekur@liag-hannover.de

Lovisa Zillén Department of Earth and Ecosystem Sciences, Quaternary Sciences, Lund University, SE-223 62 Lund, Sweden, Lovisa.Zillen@geol.lu.se

Part I
Introduction

Chapter 1
The Baltic Sea Basin: Introduction

Jan Harff, Svante Björck, and Peer Hoth

Abstract The Baltic Sea Basin serves as an example of a region where the use of natural resources and the need of environmental protection require a comprehensive and holistic approach in terms of geosciences, environmental sciences, and socio-economics. In this book, authors from countries around the Baltic Sea and overseas shed light on the Baltic Sea Basin with respect to (1) the formation of the Baltic Basin and its geological resources, (2) the stratigraphic record – mirror of climatic changes during the last glacial cycle, (3) coastal processes and sediment dynamics including aspects of coastal engineering, (4) interaction between socio-economic driving forces and the natural environment since the prehistoric colonization, (5) management of the marine ecosystem, and (6) monitoring strategies, respectively remote sensing. The editors intend not only to provide a record of the current state of the art in the investigation of the Baltic Sea Basin, but also to initiate innovative interdisciplinary and international research activities.

Keywords Baltic basin · Geology · Tectonics · Climate history · Sea level change · Coastal dynamics · Socio-economy · Archaeology · Coastal zone management · Anthropogenic impact · Monitoring · Remote sensing

The Baltic Sea, connected to the North sea and the North Atlantic via the Danish straits, is the largest brackish water basin in the world. Geologically, the basin is confined to the northwest by the highlands of the Scandinavian Caledonides, situated between two major tectonic regional units: the eastern and the western European platforms. The Baltic basin serves as a natural laboratory for a variety of geological structures and key processes crucial in the exploration of mineral resources and engineering, the formation of intra-continental sedimentary basins, and the interaction of hydrosphere, geosphere, and biosphere in basinal and coastal environments.

J. Harff (✉)
Leibniz Institute for Baltic Sea Research Warnemünde, D-18119 Rostock, Germany; presently at Institute of Marine and Coastal Sciences, University of Szczecin, PL-70-383 Szczecin, Poland
e-mail: jan.harff@io-warnemuende.de

J. Harff et al. (eds.), *The Baltic Sea Basin*, Central and Eastern European Development Studies (CEEDES), DOI 10.1007/978-3-642-17220-5_1,

Additionally, Baltic Sea sediments provide high-resolution records of climate and environmental changes during the Quaternary for the eastern North Atlantic realm. That record allows tracing back not only the change in the natural environment for the last 130,000 years but also the human impact and therefore socio-economic developments for at least the last 10,000 years.

The densely populated Baltic drainage area and the exploitation of the Baltic Sea resources cause permanent conflicts between economic interests and the protection of the unique ecological environment of the Baltic Sea. Therefore, the design of an effective interface between the different stakeholders is of vital importance for the community in the Baltic area and of great methodological interest for scientists, managers, and politicians not only in Europe but also worldwide.

The 33rd International Geological Congress (IGC) did provide the unique opportunity to discuss questions related to the points listed above in a very general way with the international geological scientific community. Therefore, a special symposium "The Baltic Sea Basin" was held on August 11, 2008, within the frame of the 33rd IGC at Oslo, Norway, in order to foster the understanding of the Baltic basin as a unit in terms of genesis, structure, ongoing processes and utilization. At the symposium, geoscientists, climate researchers, biologists, archaeologists, and computer scientists discussed questions regarding

- the formation of the Baltic basin and geological resources,
- the stratigraphic record – mirror of climatic changes during the last glacial/interglacial cycle,
- coastal processes and sediment dynamics,
- the feedback between socio-economic driving forces and the natural environment since the prehistoric colonization,
- the management of the marine ecosystem, and
- monitoring strategies and technical device design, including satellite observation methods.

In this book we report the results of the symposium. It is the first time that in a joint publication, scientists from different disciplines give a comprehensive and general overview about the Baltic Sea basin.

After this introduction, Part II is devoted to the *geological and tectonic evolution* of the Baltic basin. Sliaupa and Hoth give an overview about the geological history of the Baltic Sea basin from the Precambrian to the Quaternary, including the genesis of geological resources. The chapter gives a summary of the evolution and the known resources of the Baltic sedimentary basin focusing on its central part. According to new evidence for the origin of the Baltic Sea, the basin was formed during Late Ediacaran–Early Cambrian time caused by the reactivation of the weakest lithosphere part of the East European craton. All the following stages of basin subsidence were dominated by extensional tectonics. However, the crust was most intensively structured in NW–SE-directed compression during Late Silurian and Early Devonian time due to the collision of Laurentia and Baltica. The Permo-Carboniferous period is mainly marked by magmatic intrusions in the southern part of the Baltic Sea, in northern Poland, and in the area of the Rügen Island. After small amplitude faulting in the Mesozoic, the tectonic activities increased during the Cretaceous inversion in the south-western part of the basin. The bottom morphology of the Baltic Sea mirrors large-scale ancient structures, but glacial erosional processes contributed undoubtedly to the shape and the depth of the Baltic Sea. Oil, gas, geothermal energy, and reservoir formations which can be used as storage sites (natural gas, CO_2, compressed air) are the major resources of the deeper underground of the Baltic basin. Amantov et al. assume that Plio-Pleistocene erosion and sedimentation significantly impact post-glacial uplift of the basin. The authors estimate that in the last glacial cycle, sedimentation could produce up to 155 m of subsidence, and erosion 32 m of uplift. The analysis is based on the changes in surface load caused by glacial and post-glacial erosion and sedimentation over 1,000 year time intervals (coarser intervals before 50,000 years) utilizing a largely automated interpretation of regional geological and geomorphological observations. The analysis suggests that the first glaciations probably shaped the major over-deepened troughs, and younger glaciations mainly removed sediments left by their predecessors, decreasing the thickness of the Quaternary succession and only locally incising and changing the dip of the bedrock surface. The basin fill provides in particular for the last glacial cycle (LGC) valuable records for the reconstruction of the changing climate of the northern Europe.

The Quaternary sedimentary fill of the Baltic basin provides the records for the *reconstruction of the climate and sea level history* (Part III) of the border area between the northeast Atlantic and Eurasia. Despite the erosional effects of the Weichselian ice sheet, sediments displaying the whole LGC are at least fragmentarily preserved, and Late Pleistocene to Holocene sediments display the environmental change continuously by high-resolution proxy-data records. This topic of climate history is approached here by three articles. Andrén et al. describe the environmental change within the Baltic area for the last 130,000 years. First, the authors compare the conditions of the Eemian interglacial with the modern warm period and conclude that both salinity and sea surface temperature of the Baltic Sea were significantly higher during at least parts of the last interglacial, 130–115 ka BP. Also, the hydrology of the Baltic Sea was significantly different from the Holocene because of a pathway between the Baltic basin and the Barents Sea through Karelia that existed during the first ca. 2.5 ka of the interglacial. A first early Weichselian Scandinavian ice advance is recorded in NW Finland during marine isotope stage (MIS) 4 and the first Baltic ice lobe advance into SE Denmark is dated to 55–50 ka BP. After the last glacial maximum (LGM), ca. 22 ka BP, the ice sheet retreated northwards with a few still stands and re-advances, and by ca. 10 ka BP the entire basin was deglaciated. After different freshwater stages, full brackish marine conditions were reached at ca. 8 ka BP. The present Baltic Sea is characterized by a marked halocline, preventing vertical water exchange and resulting in hypoxic bottom conditions in the deeper part of the basin. Harff et al. have investigated sediment echosounder data and sediment cores from the eastern Gotland basin in order to reconstruct Holocene hydrographic and climatic conditions for the Baltic Proper. The down-hole physical facies variations from the eastern Gotland have been correlated basin-wide. Thickness maps of the freshwater and the brackish sediments ascribe the general change in the hydrographic circulation from a coast-to-basin to a basin-to-basin system along with the Littorina transgression. Variations in the salinity of the brackish (Littorina Baltic basin) are attributed to changes in the North Atlantic oscillation (NAO), ascribing the wind forces and driving the inflow of marine water into the Baltic basin. Time series analysis of facies variations reveals distinct periodicities of 900 and 1,500 years. These periods identify global climate change effects in Baltic basin sediments.

A main prerequisite for palaeo-environmental reconstructions based on sediment proxies is the establishment of correct-age models. For dating Holocene sediments the radiocarbon method is the most common one, but problems emerge for glacial and coastal sediments poor in organic matter. In these cases, optical-stimulated luminescence (OSL) dating has become more common. Bitinas et al. used this method to date lacustrine inter-till sandy sediments of the Klaipėda strait. The dating and detailed geological investigations imply that the sediments are allochthonous, having formed during marine isotope stages (MIS) 4. This conclusion sheds new light on the genesis of the till beds beneath the bottom of the Klaipėda strait.

Controlled by climate change, but also by the glacial isostatic adjustment (GIA), the relative sea level changes serve as the most important steering factor for long-termed *coastline change* (Part IV) in the Baltic Sea. Harff and Meyer describe a model that is applied to reconstruct the palaeogeographic development of a coastal area and that generates future projections as coastline scenarios. For the hind-cast, relative sea level, curves are compared with recent digital elevation models. For future projections, data of vertical crustal displacement, measured from gauge measurements, are superimposed with eustatic changes based on climate scenarios. The authors classify the Baltic coasts in those influenced by crustal uplifting and another type determined by subsidence and eustatically controlled sea level rise. For the first type, Rosentau et al. combined geological, geodetic, and archaeological shore displacement evidences to create a temporal and spatial water-level change model for the SW Estonian coast of the Baltic Sea since 13.3 ka BP. A water-level change model was applied together with a digital terrain model in order to reconstruct coastline change in the region and to examine the relationships between coastline change and displacement of the Stone Age human settlements that moved in connection with transgressions and regressions on the shifting coastline of the Baltic Sea.

Vassiljev et al. show in a GIS-based palaeogeographic reconstruction the development of the Baltic ice lake (BIL) in the eastern Baltic during the deglaciation of the Scandinavian ice sheet. The study shows that at about 13.3 ka BP the BIL extended to the ice-free areas of Latvia, Estonia, and NW Russia, represented by the highest shoreline in this region. Reconstructions demonstrate a detailed palaeogeographic history of BIL and glacial lakes Peipsi and Võrtsjärv, which is determined by the glacio-isostatic uplift.

At the transition to sea level rise controlled coasts along the Sambian Peninsula, erosional processes outweigh sediment accumulation. Sivkov et al. investigated the bottom relief along the coast. The authors derived digital bathymetric and slope angle maps from the modern 1:25,000, 1:50,000, and 1:100,000 nautical charts. A total of five axial isolines of post-glacial, wave-undercut cliffs were identified: two dated to the Yoldia Sea (58–45 and 52–40 m), one assigned to the Ancylus lake (38 m), and two dated to the Littorina Sea (29 and 21 m).

In the southern Baltic the Littorina transgression leads to inundations of the coastal lowlands. Due to the sheltered position in the Gulf of Gdańsk, the terrestrial ecosystems have been preserved, forming a unique inventory of palaeo-ecological proxies. Uścinowicz et al. have investigated the nature of the plant communities, and also tree stumps position in relation to the palaeo-sea level. Tree stumps occurring in situ on the sea floor along with peat deposits are the most reliable indicators of sea level changes. The characteristic forest composition of that time was the broad deciduous forest with oak (*Quercus*), elm (*Ulmus*), and lime (*Tilia*). The climate was characterized by good thermal and moisture conditions, which is confirmed by the presence of pollen grains of mistletoe (*Viscum*) and ivy (*Hedera*). The obtained data from the time of accumulation of the investigated sediments indicate that the sea level at that time was about 19–20 m lower than is at present. At open coasts, a slowly rising sea level in the Late Holocene, together with storm-induced wave action, has lead to amplified cliff erosion. In the southern and south-eastern lowlands, the accumulation of eroded sediments leads to the formation of sandy barriers and spits. Lampe et al. have studied the factors influencing the formation of sandy spits, with the Darss-Zingst Peninsula as an example. These are among others, the eustatic sea level rise, the rates of land uplift and subsidence, the inclination of the pre-transgressional bottom relief, and the amount and type of supplied sediments. In a final synopsis the authors assess the interplay of all factors, explaining the distribution, volume, and stability of the barriers along the German Baltic coast.

For future projections of coastal processes and the protection of coasts the numerical modelling of *sediment dynamics* (Part V) plays a key role. Soomere and Healy use the concept of the equilibrium beach profile as an adequate tool for their analysis of Estonian beaches. As an example, beach parameters and long-shore transport patterns are evaluated for Pirita beach based on a granulometric survey and long-term simulation of wave climate . It is demonstrated that net sand changes for such beaches can be estimated directly from the properties of the equilibrium profile, land uplift rate, and loss or gain of the dry beach area. Meyer et al. use the southern coast of the Baltic Sea as a notable example for the impact of erosion, transport, and accumulation of sediments on coastline change during the Holocene. Since the end of the Littorina transgression the coastline morphology has been shaped mainly by long-shore sediment transport controlled by the geological situation and glacio-isostatic influence. The long-shore sediment transport is driven by wind and consequently waves shaping young Holocene structures like the Darss-Zingst Peninsula. In order to model these processes, SEDSIM (sedimentary basin simulation), a stratigraphic forward-modelling software, has been applied for the Darss-Zingst Peninsula on a centennial timescale. The results of the numerical experiments show possible implications to the area of investigation and may serve as a basis for the elaboration of strategies for the coastal protection against erosion.

Coastal protection strategies require concepts for the sustainable development of the utilization, i.e. *interaction between a changing environment and society* (Part VI) of the Baltic Sea. This utilization has already a long history. Jöns describes that the maritime (coastal) zone of the Baltic basin was, during all the phases of its settlement history, of special importance to the people living there because of resources, transportation, and communication. This is especially true in areas with high rates of shore displacement, where the data and models can be used to reconstruct environmental conditions and to date prehistoric coastal sites. Conversely, well-excavated and dated archaeological sites that were originally located on the shore can provide detailed information about the sea level at the time of their occupation and can be used as sea level index points. In his paper, Jöns discusses the opportunities and problems arising from the use of shore displacement models for the interpretation of archaeological sites. Both models and sites are introduced in case studies that represent not only the different areas and localities but also the different stages in the development of the Baltic Sea. One of the current requirements is an integrated management of the coastal zone. Spiridonv et al. claim mapping and assessment of the geological hazard potential to be the main objectives for the protection of coastal zones. Ecological hazards may threaten human life, result in serious property damage, and may significantly influence normal development of biota. They are caused by natural endogenic and exogenic driving forces or generated by anthropogenic activities. An interaction of geological processes and intense anthropogenic activities, e.g. construction of buildings, harbours, oil and gas pipelines, hydro-engineering facilities, and land reclamation, has resulted in hazard potential, especially for the densely populated areas of the Russian Baltic coastal zone. These hazards may in addition be harmful for the sensitive ecosystem of the Baltic Sea. Vallius et al. mention seafloor desertification as a possible future scenario in parts of the Baltic Sea environment as the result of its utilization. During its whole post-glacial history the seafloor of the gulf has been periodically anoxic, and anoxia below halocline can thus be seen as a natural phenomenon. During the last decades, however, this has been accompanied by an annually repeated seasonal anoxia in the shallower basins triggered by substantial eutrophication of the sea, and is a clear signal of anthropogenic pressure. Phosphorus, which is bound to oxic seafloor sediments, is easily released from sediments during episodes of anoxia, which further intensifies eutrophication. Schernewski et al. mention that phosphorus is today regarded as the key nutrient for Baltic Sea eutrophication management. Major sources are large rivers like the Oder, Vistula, and Daugava in the southern Baltic region. Taking the Oder/Odra estuary as an example, the authors analyse the long-term pollution history and the major sources for phosphorus and calculate a phosphorus budget, with special focus on anoxic phosphorus release from sediments. A phosphorus emission reduction scenario is presented. Phosphorus load reductions have only limited effect on the eutrophic state of the lagoon. The lagoon is more sensitive to nitrogen load reductions. Therefore, the authors mention that both elements have to be taken into account in measures to reduce eutrophication.

For the assessment of interrelation between the Baltic Sea basin and terrestrial areas, subsurface water exchange has to be considered in *hydrogeological modelling* (Part VII). Schafmeister and Barsow have analysed the possible change in groundwater discharge from a medium-scale catchment to the Baltic by means of a numerical groundwater flow model. The test areas for groundwater recharge north-east of Wismar (Mecklenburg-Vorpommern, Germany) is calculated to 24% of the recent average annual precipitation of 600 mm in the test area, and its submarine groundwater discharge is modelled to 14% of the precipitation. Based on climate scenarios calculated by the Swedish Meteorological and Hydrological Institute (SMHI) and the Hadley Centre (HC), three sea level scenarios in combination with four precipitation scenarios are modelled for steady-state groundwater conditions in order to assess potential response in discharge. The modelled scenarios indicate that changes in groundwater recharge as a consequence of climate-induced changes in precipitation lead to notable variations of submarine groundwater discharge.

The base for modelling and management is a continuous *monitoring* (Part VIII) of the marine system, and during the last decade, remote sensing methods have been developed successfully. Kratzer et al. focus on recent advances in water quality monitoring of the Baltic Sea using remote sensing techniques in combination with optical in situ measurements. Here the Baltic Sea ecosystem is observed through its bio-optical properties, which are defined by the concentration of optical in-water constituents governing the spectral attenuation of light. The authors explain differences in the investigation of the marine and the coastal environment. An overview of existing monitoring approaches is given, and operational online systems are discussed that combine remote sensing and autonomous in situ measurements.

Acknowledgements At least two peer reviewers have reviewed each paper. Here, we express our thanks to their valuable critics and advise for revisions to the authors. The reviewers who have agreed to be identified are Ole Bennike, Mikael Berglund, Reinhard Dietrich, Martin Ekman, Berit Eriksen, Rimante Guobyte, Algimantas Grigelis, Matthias Hauff, William W. Hay, Heiko Hüneke, Antoon Kuijpers, Thomas Leipe, Robert Mokrik, Ralf Otto Niedermeyer, Renate Pilkaityte, Werner Stackebrandt, Szymon Uscinowicz, Boris Winterhalter, Andrzej Witkowski, and Lovisa Zillen.

We thank Dr. Teresa Radziejewska for her help in linguistic improvement of some of the papers.

Michal Tomczak provided valuable assistance in the production of this volume; we are greatly indebted to him for his efforts.

We also acknowledge the support of the Springer Publishing House in the production of this book.

This book is addressed to professionals and students in the geosciences, the social sciences, economics, and coastal engineering, and decision makers in management of marine systems. The book shall not only summarize the state of the art in the investigation of the Baltic Sea basin but also raise the community's awareness of new interdisciplinary challenges and initiate discussion about innovative research projects, establishment of international research laboratories, and monitoring strategies including technical devise design.

During the work on this book, one of the authors, Prof. Dr. Terry Healy, Hamilton, New Zealand, has passed away on July 20, 2010. Born on November 28, 1944, he has left the international stage of science much too early. We, his colleagues and friends, will keep the remembrance of an outstanding scientist and above all a wonderful person.

Part II
Geological and Tectonical Evolution

Chapter 2
Geological Evolution and Resources of the Baltic Sea Area from the Precambrian to the Quaternary

Saulius Šliaupa and Peer Hoth

Abstract The Baltic Sea is a young geomorphologic feature that formed during Quaternary time. It covers the western and the central part of the Baltic sedimentary basin. The origin of the Baltic Sea and of the corresponding morphological low is still controversial, considered by some as an erosional structure and as a tectonic depression by others. The chapter gives a summary of the evolution and the known resources of the Baltic sedimentary basin focussing on its central part and thus tries to present new evidence for the origin of the Baltic Sea. The Baltic sedimentary basin was formed during Late Ediacaran–Early Cambrian time. Its formation was caused by the reactivation of the weakest lithospheric part of the East European craton. All the following stages of pronounced basin subsidence (major subsidence phase during Late Ordovician–Middle Silurian), including the recent tectonic stage, were dominated by extensional tectonics. However, the most intense structuring of the crust in the region took place in a compressional setting during Late Silurian and Early Devonian time. The NW–SE-directed compression was caused by the collision of Laurentia and Baltica. It caused the formation of an Early Palaeozoic thrust and fold belt at the margin of the East European craton and led to the formation of E–W and NE–SW striking faults in the Baltic basin northeast of the Danish–North German–Polish Caledonides during that time. Typical for the Permocarboniferous period are magmatic intrusions in the southern part of the Baltic Sea, in northern Poland, and in the area of the Rügen Island. Tectonic activities ceased within the Permian and only small amplitude faulting is detected in the Mesozoic. Later on, tectonic activities increased during the Cretaceous inversion in the southwestern part of the basin. The typical wrench-dominated faulting is related to the reactivation of Pre-Permian fault systems by Late Cretaceous inversions of the Mesozoic Danish and Polish basins. Large-scale ancient structures of the Baltic basin are reflected in the sea bottom morphology. Detailed analysis indicates that those morphological structures are mainly passive features related to selective glacial erosion,

S. Šliaupa (✉)
Institute of Geology and Geography, Vilnius University, Vilnius 01013, Lithuania
e-mail: sliaupa@geo.lt

J. Harff et al. (eds.), *The Baltic Sea Basin*, Central and Eastern European Development Studies (CEEDES), DOI 10.1007/978-3-642-17220-5_2,

but some hints for neotectonic activities do also exist. Glacial erosional processes undoubtedly contributed to the shape and depth of the Baltic Sea. Evidences available today, however, suggest the existence of a pre-existing tectonic depression. Major resources of the deeper underground of the Baltic basin are oil, gas, geothermal energy and reservoir formations which can be used as storage sites (natural gas, CO_2, compressed air). Location of known oil and gas fields shows a strong relation to the major fault zones.

Keywords Baltic Sea · Baltic basin · Geodynamic evolution · Structure · Resources · Hydrocarbons

2.1 Introduction

Although the Baltic Sea was formed during Quaternary time, a more detailed look at the origin of the corresponding morphological low implies that a reactivated ancient tectonic structure could have been a major factor for its development. The origin of the Baltic Sea and of the corresponding morphological low is still a matter of controversy, considering it on the one side as an erosional structure and on the other as a tectonic depression. The different views are summarized by Šliaupa et al. (1995b) and Schwab et al. (1997). The debate is not purely academic. It is important to understand the processes of the formation of the Baltic Sea as a basic frame and data input for the prognosis of the future development of the Baltic Sea in a changing environment and for the search of mineral resources.

The Baltic Sea covers the western and the central part of the Baltic sedimentary basin and is therefore intimately connected to the development of the underlying basin. The chapter gives a synthesis of the sedimentation and the structural history of the basin from the Proterozoic to the Cenozoic, focusing on its central part. It describes important tectonic mechanisms of the basin development and the subsidence history and points out links between the development of both the Baltic Sea and the underlying sedimentary basin. Within this context the resource potential and consequences for further exploration are discussed.

2.2 Geological Framework and History of Sedimentation

2.2.1 The Baltic Basin

The Baltic basin (Fig. 2.1) is located above the margin of the East European craton, which was consolidated during the Early Proterozoic (Linnemann et al. 2008), except for the westernmost part which was formed during the Mesoproterozoic (Bingen et al. 2002, Obst et al. 2004). The thickness of the sedimentary section is less than 100 m in northern Estonia, increasing to around 1,900 m in southwest

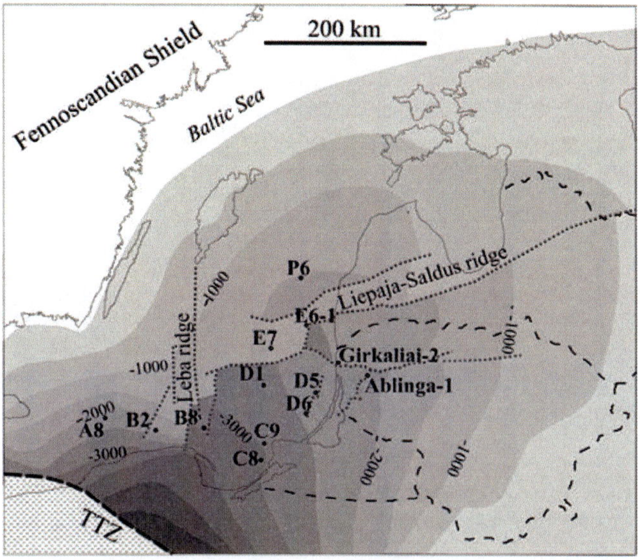

Fig. 2.1 Depths of the Early Precambrian crystalline basement below the sedimentary section of the Baltic basin (nearly equivalent to sediment thickness). Wells referred in the text are shown. *Dotted lines* indicate major faults. TTZ indicates Teisseyre–Tornquist zone

Latvia and 2,300 m in western Lithuania. The maximum thickness of sediments is reached in the western part of the basin (central north Poland) where the depth of the Early Precambrian crystalline basement exceeds 4,000 m. The basin extended further to the southeast and the northwest prior to the Caledonian deformation phase and its extent had been limited to the subsidence and sedimentation area northeast of the Danish–North German–Polish Caledonides thereafter. Today, the Baltic basin borders on the North German basin, the Polish basin and the Danish basin (Fig. 2.2). The western boundary of the basin is formed by the Teisseyre–Tornquist zone.

The oldest non-metamorphosed sediments, infilling local depressions in the Baltic basin area, are of Mesoproterozoic age. The corresponding Hogland Series is locally distributed in the Gulf of Finland, on Saaremaa Island and on the Kurzeme Peninsula. The series is represented by quartz sandstones and conglomerates intercalating with mafic and felsic volcanic rocks. The isotopic age of the volcanic rocks was dated from 1580 to 1670 million years using the K–Ar dating method (Puura et al. 1983). The 130-m-thick stratotype section is located on the Hoghland Island in the Gulf of Finland, where the layers are tilted at an angle between 5 and 30°. These rocks are spatially associated with cratonic granitoids of Middle Proterozoic age (Vyborg, Riga massif). Contemporaneous Sub-Jotnian sediments are mapped in the Gotska Sandön Island (north of Gotland). They are composed of quartz sandstones, which fill local half-graben structures with up to 400–500-m-thick sections. Ages of these sandstones are dated to a time period between 1490 and 1540 Ma (Gorbatchev 1962).

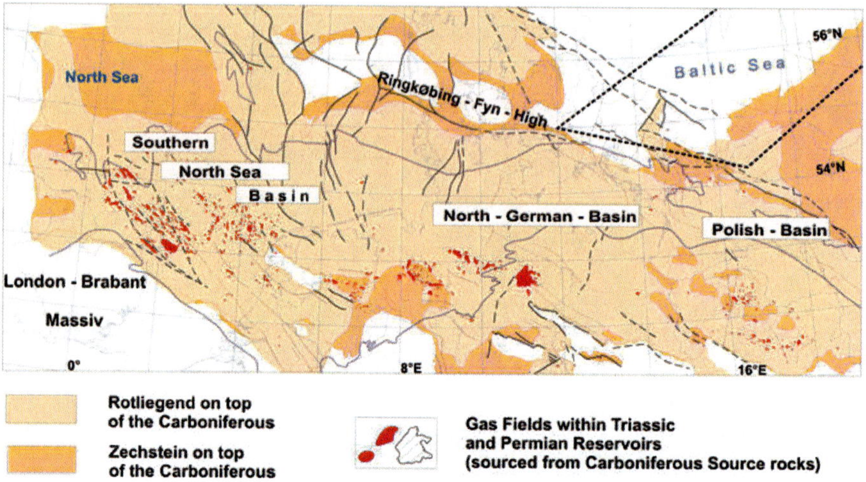

Fig. 2.2 Main sedimentary basins in the vicinity of the Baltic Sea area (after Bandlowa 1998, Ziegler 1990, Hoffmann et al. 2001)

The younger Jotnian quartzites, siltstones and conglomerates are reported from the Gotska Sandön area and the southern periphery of the Åland rapakivi massif in the north of the Baltic Sea. These sediments have also accumulated in graben depressions, reaching 900 m of maximum thickness. According to Gorbatchev (1962), the Gotska Sandön sandstones were accumulated between 1300 and 1400 Ma. They were deposited in fluvial, tidal or aeolian environments and are generally not affected by folding or other deformation.

After a long break in sedimentation, the deposition was re-established in Early Ediacaran time. Corresponding sediments are preserved in local areas of western Latvia and the adjacent offshore (well P6-1, see Fig. 2.1 for location). The Early Ediacaran sediments are defined as the Zura formation and composed of 2–30-m-thick partly tuffitic sandstones and conglomerates, siltstones and shales.

The first wide transgression in the Baltic region took place in Late Ediacaran–earliest Cambrian time (Figs. 2.2 and 2.3). Sea transgressions occurred from the east and from the west. Therefore a typical western facies is distributed in the south-west of the Baltic Sea and in the adjacent onshore area which is attributed to the Żarnowiec formation (wells A8-1, B16-1) and the Nexø formation (Bornholm area). Sandstones and conglomerates exceeding 100 m thickness were mainly deposited in a floodplain environment (Jaworowski and Sikorska 2003).

The Late Ediacaran transgression in the east was of a much wider extent and related to the gradual widening of the Moscow marine basin in the east. Arkosic conglomerates and sandstones of up to 200 m thickness are the dominating sediments there. The succession is bounded by the lowermost Cambrian blue clays of the Moscow basin (Jankauskas and Lendzion 1994). They are up to 120 m thick and crop out along the northern coast of Estonia.

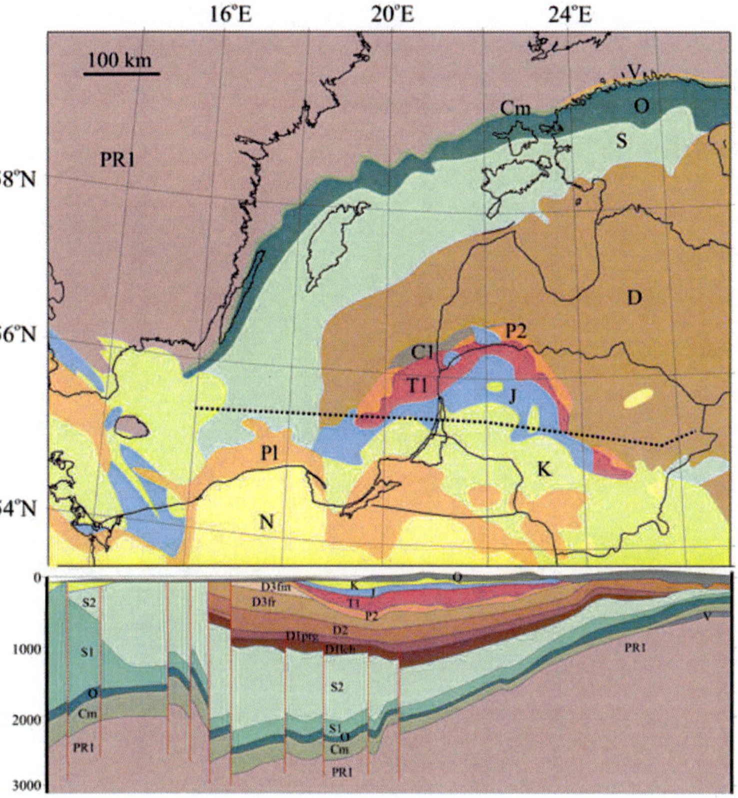

Fig. 2.3 Geological subcrop map (Pre-Quaternary level) of the Baltic region and geological cross section (*dotted line* shows location of the profile)

A drastic rearrangement of the sedimentation pattern took place in the middle of Early Cambrian time. While sedimentation ceased in the Moscow basin, a vast marine transgression from the west took place and resulted in the deposition of quartz sandstones, siltstones and shales. The thickness of the Cambrian section attains 250 m in the central part of the Baltic Sea and more than 500 m in the area of central north Poland. Figure 2.4 shows thickness maps of sub-stages of the Middle Cambrian. The distribution of those sediments is nearly consistent to the recent outline of the Baltic Sea. Although later erosion also plays a role in the distribution of these formations, the outline impressively supports that this time period is defined as the nucleation stage of the basin.

The Cambrian is overlain by a shaly carbonaceous succession of Ordovician age, which is between 60 and 160 m thick in the offshore part and reaches up to 250 m thickness onshore. The sediment pattern shows a split between a carbonate-dominated facies in the east and a deeper marine facies with graptolitic shales in

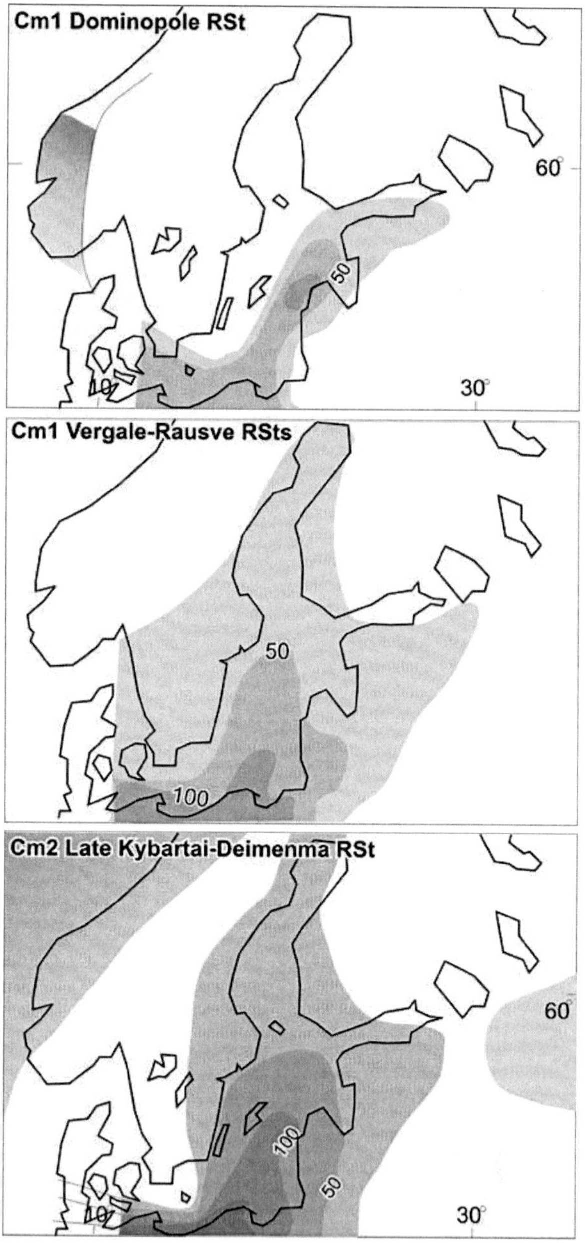

Fig. 2.4 Initiation of the Baltic sedimentary basin in Early–Middle Cambrian time – thickness maps for the Dominopole, Vergale-Rausve and Kybartai-Deimena regional stages

the west (Laskovas 2000). Deposition occurred nearly continuously throughout the Ordovician. The sedimentation rate considerably accelerated during the Silurian. Thickness patterns show an increase to the west and a maximum thickness of around 3,500 m in the southwestern part of the Baltic Sea (and north Poland). This maximum thickness was even larger in the past since parts of the section have been eroded. The Silurian is composed of graptolitic shales with some marlstones and limestone interlayer in the deeper marine central basin parts, while carbonates predominate in the shallow periphery of the basin (Lapinskas 2000).

Sedimentation shifted to the central part of the Baltic basin during the Devonian (Fig. 2.3). Shallow marine and lagoon carbonates and marlstones alternate with sandstones and shales which were deposited in shallow marine and continental environments. The maximum thickness is reached in the Klaipeda area (west Lithuania and adjacent offshore) with up to 1,050 m. Lowermost Carboniferous sediments (sandstones and carbonaceous shales with a thickness of up to 110 m) are of limited extent and so far only known from northwest Lithuania, southwest Latvia and the adjacent offshore areas. Numerous Carboniferous diabase sills are also identified in the central part of southern Baltic Sea (Šliaupa et al. 2004).

The Permian, Mesozoic and Cenozoic deposits show a shift of sedimentation to the southwest (Fig. 2.3). By contrast to the Palaeozoic period, which was mainly marked by rather continuous sedimentation and persistent subsidence, the Mesozoic and Cenozoic periods were dominated by non-deposition which was only partly interrupted by recurrent marine transgressions from the west (Fig. 2.5). The Upper Permian consists of carbonates and evaporates with a maximum thickness of up to 350 m in the southern part of the Gdańsk depression. The Lower Triassic reaches its maximum thickness in the same area and is composed of red coloured lacustrine mudstones with subordinate fine-grained arkosic sandstones (Suveizdis and Katinas 1990). The Jurassic succession shows a typical development from lacustrine sediments in the lower part to marine sediments in the upper part. It is composed predominantly of fine-grained sandstones, siltstones and shales and shows limestone interlayer in the upper part. The thickness attains 200 m in the southern Baltic Sea area. Two distinct facies can be defined in the Cretaceous section. While sediments of Albian age are composed of glauconitic sandstones and siltstones, chalk, marlstones and siltstones are typical for the Upper Cretaceous. The total thickness of the Cretaceous section reaches 400 m along the southern coast of the Baltic Sea.

Cenozoic terrigenous sediments are mapped only in the southernmost part of the Baltic Sea and further south onshore. The thickness of the Palaeogene attains 80 m. It is composed of shallow marine shales, sandstones and siltstones. A large deltaic complex with amber deposits developed in the western part of the Kaliningrad district. Sediments of Neogene age are distributed south of the Baltic Sea. They were deposited in lacustrine–alluvial environments and are composed of fine-grained sandstones, siltstones and shales of grey and dark-grey colours.

Figures 2.3 and 2.5 give a summarizing picture on the geological setting and the development of sedimentation in the Baltic basin and the surrounding areas.

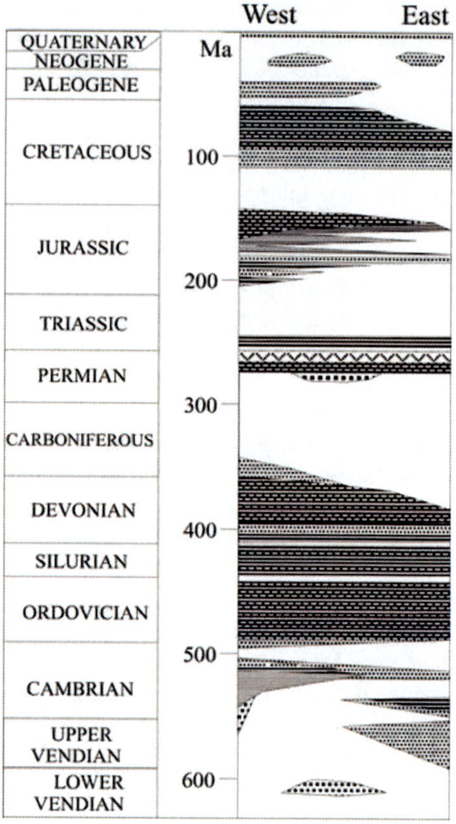

Fig. 2.5 Ediacaran–Phanerozoic chronostratigraphic chart of the Baltic basin

2.2.2 *The Southwestern Basin Rim*

The southwesternmost part of the Baltic basin, west of the Tornquist zone (southern Møn, Falster, Lolland, Rügen and Usedom islands), is located in a different tectonic setting and forms the bordering area between the North German basin and the Baltic basin. The area is characterized by an Early Palaeozoic thrust and fold belt which forms part of the Danish–North German–Polish Caledonides (Piske and Neumann 1993, Meissner et al. 1994, Hoffmann and Franke 1997). The Palaeozoic section of the area is known only from the results of a few oil and gas exploration wells (see Fig. 2.6). Corresponding borehole information is summarized by Hoth et al. (1993), Piske et al. (1994), Hoth and Leonhardt (2001) and Doornenbal and Stevenson (2010).

The borehole G14 which is located 36 km east–northeast of Arkona (northern tip of Rügen Island) and north of the Caledonian deformation front drilled into intensely altered granite which was dated to around 1450 Ma using the U/Pb dating method of the zircon fraction (Tschernoster et al. 1997). This age is similar to the

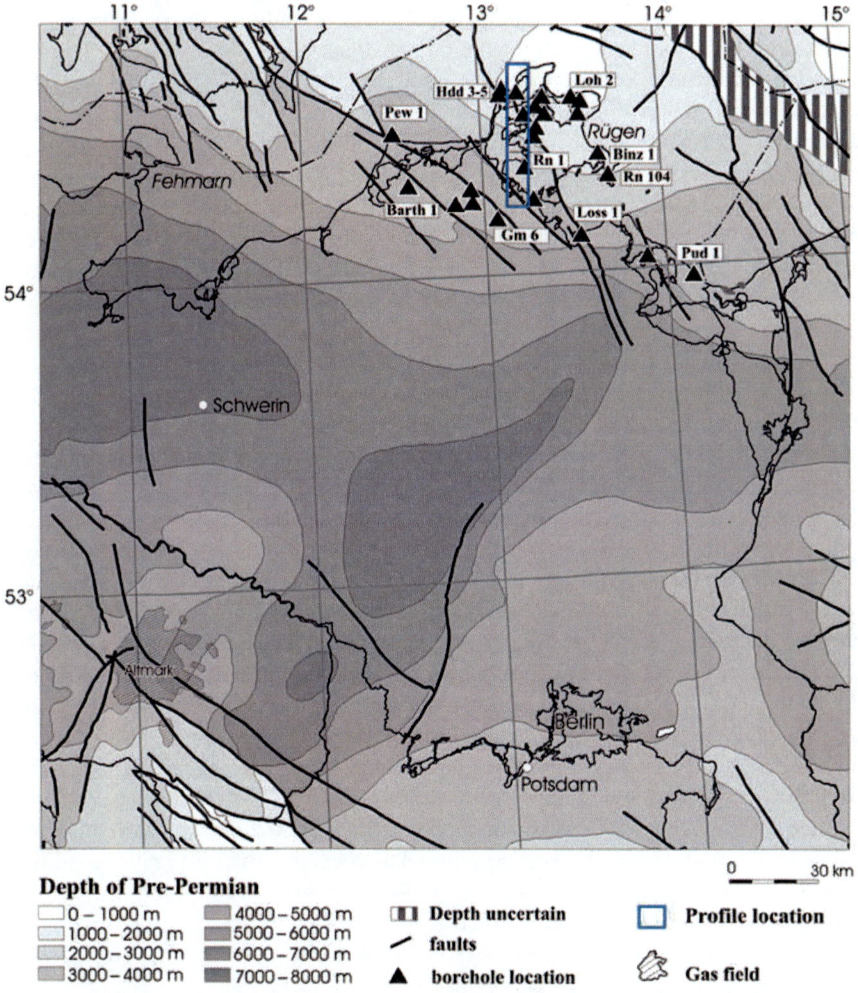

Fig. 2.6 Borehole location and depth of the Pre-Permian for the Rügen and neighbouring areas

Bornholm basement rocks (Jørgart 2000). The crystalline basement is overlain by around 150–160-m-thick Cambrian sandstone and a roughly 30-m-thick alum shale formation (Piske and Neumann 1990). Ordovician sediments are around 60 m thick in the well G14 and consist mainly of black to grey shales with additional siltstone and carbonate layers. High coalification and sonic velocity values as well as the tectonic deformation of the shales hint of a significant burial depth of the section and a severe later erosion of the overburden.

Several boreholes have encountered the Lower Palaeozoic south of the Caledonian deformation front (for instance, H2-1/1990, K5-1/1988, Rügen 3/1965, Rügen 5/1966, Dranske 1/1968, Lohme 2/1970, Binz 1/1973, Loissin 1/1970) and both the Loissin 1 and the Rügen 5 borehole have additionally reached presumably Neoproterozoic sediments at their final depth (Beier et al. 2001). The most complete Ordovician section was encountered by the Rügen 5 borehole with more than 3,000 m intensely deformed Ordovician sediments (Hoth et al. 1993, Franke and Illers 1994). The lower and around 120–150-m-thick part (Late Tremadoc) is mainly characterized by fine-grained sandstones. It is overlain by a more than 1,000-m-thick black shale formation (Llanvirn) and a formation formed by greywackes, black shales and siltstones. The upper formation is of Caradocian age and shows a thickness of more than 2,000 m. The whole Ordovician section shows a multiphase Caledonian deformation history and is interpreted to belong to an Avalonian (peri-Gondwanan) sedimentation area (Beier and Katzung 2001). Most of the other boreholes in the area have drilled only some 100–200 m into the Ordovician.

All the boreholes of the area show a severe erosion unconformity on top of the Ordovician and the corresponding sedimentation gap decreases from the north to the south. Triassic sediments are located on top of the Ordovician in the northern part of Rügen, whereas Middle to Upper Devonian or Lower Carboniferous sediments form part of the overburden in the southern part and on Usedom Island.

The Middle Devonian of the area is mainly characterized by a clastic sedimentation which was followed by a marine transgression during the Upper Devonian and the deposition of marine shales and carbonates. The boreholes have encountered a Devonian section thickness between some hundred and up to 2,300 m in the Binz 1 borehole (Hoth et al. 1993). Marine conditions were also typical for the Visean and the Dinantian. Two main sedimentary facies are described by Hoffmann et al. (2006): a carbonate-dominated shelf facies and a facies which is typically for graben structures and dominated by shales, siltstones and marlstones. Sediment thickness varies between some hundred and up to around 2,000 m. During the transition to the Namurian, the sea became shallower and finally paralic conditions prevailed and led to the deposition of 100–700-m-thick clastic sections. The following Lower Westphalian sedimentation occurred within deltas, flood plains and swamps. Several coal seams are therefore typical for the Westphalian A and B. During the Westphalian C and D, fluvial and limnic sedimentation was more important and Stephanian sedimentation occurred partly even under semiarid conditions. The thickness of the whole Westphalian to Stephanian section reaches around 2,100 m in the Rügen/Vorpommern area (Hoth et al. 2005).

Figure 2.6 shows the location of the boreholes which have drilled to the Carboniferous and the depth of the Pre-Permian for the Rügen area and also for the neighbouring parts of the North German basin. The profile in Fig. 2.7 highlights the importance of granites and magmatic dykes mainly of Variscan age for the described area. These magmatic rocks have caused severe coalification anomalies within Carboniferous and older Palaeozoic sections (Hoth 1997) and are partly connected to thick Permocarboniferous volcanic rocks within the North German basin and the Polish basin. Thickness of clastic Rotliegend sequences is also increasing to the south into the central basin part of the North German basin. Both figures show the Rügen area forming the northeastern boundary of the North German basin to the

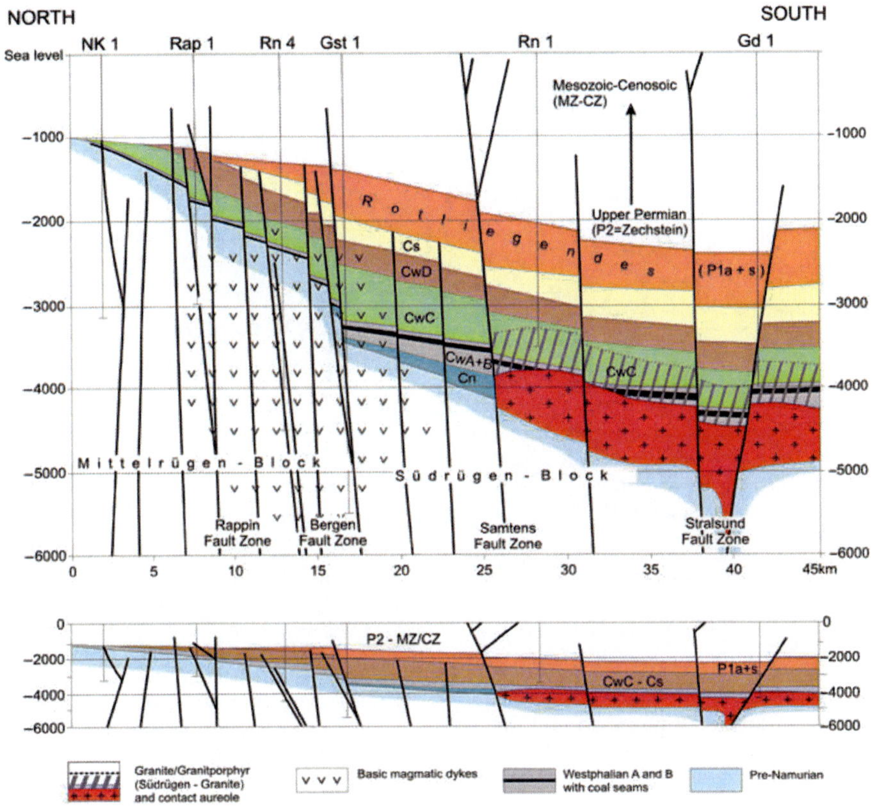

Fig. 2.7 N–S cross section through the island of Rügen (after Hoth et al. 2005)

Baltic basin. The subsidence history from the Permian to the Mesozoic is described for that part of the North German basin by Hoth (1997).

2.3 Basin Subsidence and Geodynamic Evolution

The continental crust of the Baltic region formed during the Palaeoproterozoic (Bogdanova et al. 2006). It was reactivated later on by extensive intrusion of rapakivi granites and associated igneous rocks during the time period between 1.67 and 1.45 Ga (Haapala and Rämö 1992, Puura and Flodén 2000, Åhall et al. 2000).

Volcanic and sedimentary rocks mainly filling graben structures are spatially associated with Mesoproterozoic intrusions. The largest feature of this type of extensional depressions is the Bothnian Sea depression. It has many characteristic features of a palaeo-rift such as a topographic low, a thin crust, large crustal thickness gradients and a voluminous bimodal magmatism (Korja et al. 2001). The Bothnian aborted rift is probably a part of a honeycomb-like wide rift area that

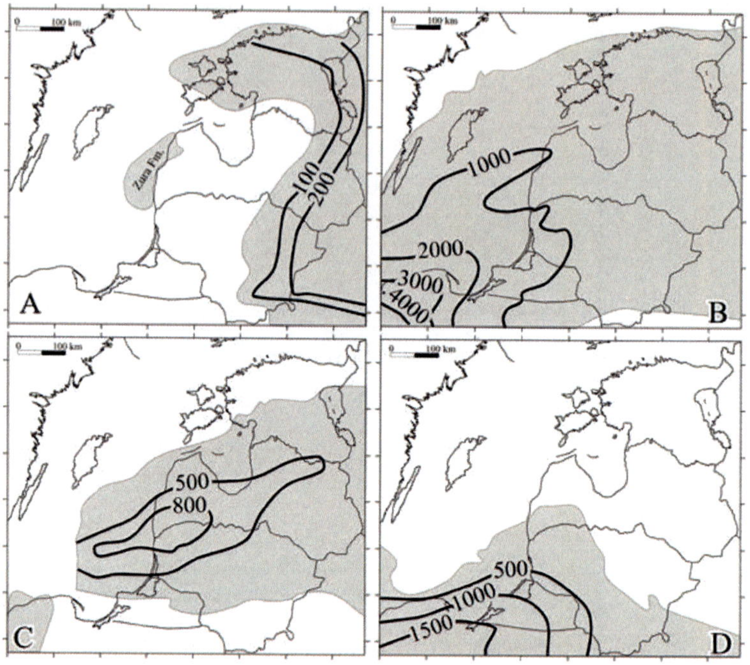

Fig. 2.8 Sediment thickness of major structural complexes of the Baltic basin. **a** Baikalian (Ediacaran–lowermost Cambrian); **b** Caledonian (Cambrian to *lower part* of Lower Devonian); **c** Hercynian (*upper part* of Lower Devonian–Carboniferous); **d** Alpine (Permian–Cenozoic)

extends from Lake Ladoga to the Caledonides and has seeds of many localized narrow rifts. The Jotnian sediments were intruded by Post-Jotnian diabase sills and dykes (e.g. diabases in the Kvarken area dated 1268 ± 13 Ma by Suominen 1991). Sediments of Riphean age are not known from the Baltic Sea area.

The Baltic sedimentary basin was initiated on this type of continental crust during Late Ediacaran–Early Cambrian time. It is a special tectonic structure because of its long-lasting subsidence history reaching from Late Precambrian to Quaternary. Subsidence rates and patterns varied considerably throughout the Phanerozoic (Figs. 2.5, 2.8 and 2.9). This is related to changing geodynamic mechanism driving the basin evolution. The following main geodynamic stages can be distinguished.

2.3.1 Failed Rift Stage

The Baltic Sea area was affected by intense magmatic activities during the Mesoproterozoic. The ages of the rapakivi granites and associated igneous rocks are in the range of 1.67–1.45 Ga (Haapala and Rämö 1992, Puura and Flodén 2000, Åhall et al. 2000). Age data reflect a general trend with southward younger ages

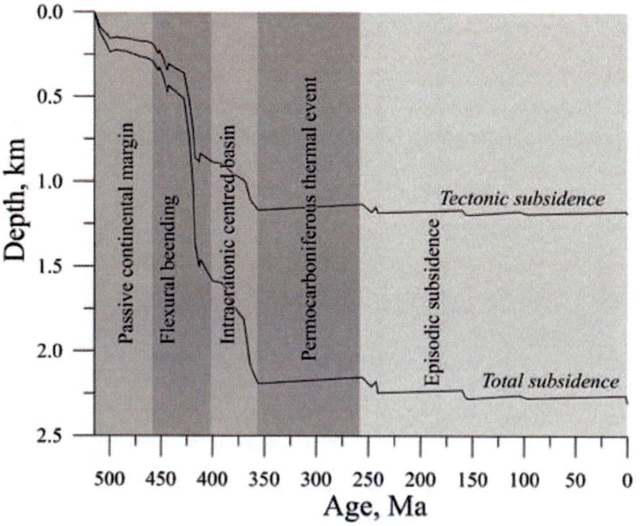

Fig. 2.9 Total and tectonic subsidence curves for the whole basin history – example of the central basin part (well Ablinga-1, west Lithuania). The major geodynamic stages are marked

of intrusions. This might be an evidence of mantle plume migration to the south. During the thermal relaxation phase (1500–1200 Ma), thermal domes associated with the rapakivi complexes were eroded and the Baltic basin area was provided with clastic sediments. The extensional regime was likely related to the opening of the Grenvillian Sea; intrusion of dykes and sills are hints for it. A reactivation of the rifting processes coupled with the intrusion of diabase dyke S-type granites might have taken place around 950–850 Ma after the Sveconorwegian orogeny (Wilson 1982). No geological records are so far known from the region for the period between 850 and 600 Ma, which points to low-rate geodynamic processes.

2.3.2 *Passive Continental Margin Stage*

The Baltic basin was established in Late Ediacaran–Early Cambrian time (Fig. 2.8) as a passive continental margin basin in response to the break apart of the Rodinia supercontinent (Šliaupa et al. 2006). This is reflected in the typical concave-shaped subsidence curves for the Cambrian–Ordovician times (Figs. 2.9 and 2.10). The incipient stage of continent fragmentation and activation of tectonic processes in the Baltic Sea area is marked by the localized deposition of the Zura tuffitic sandstones and conglomerates of the Lower Ediacaran in Latvia (Paskevičius 1997; see Fig. 2.8a).

Sedimentation progressed in Late Ediacaran–earliest Cambrian time as evidenced by the deposition of the Żarnowiec and the Nexø sandstones in the south-westernmost part of the Baltic basin. This is interpreted to be caused by the

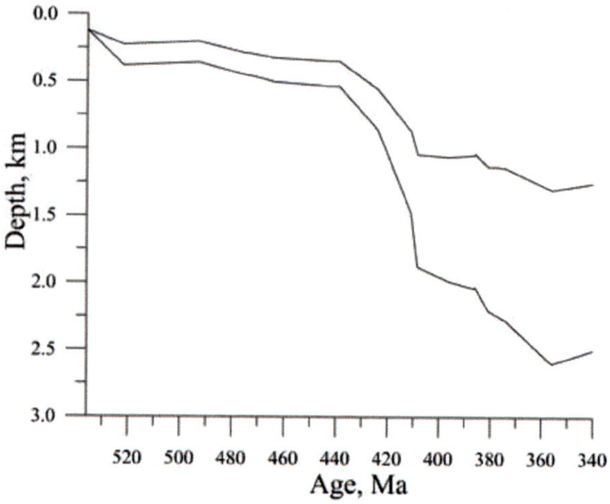

Fig. 2.10 Total and tectonic subsidence curves for the time period of Middle Cambrian to Lower Carboniferous – example of the central basin part (well D1-1, Baltic Sea)

break apart of continental landmasses and opening of the Tornquist Sea in the west (Poprawa et al. 1999). With progressing continental separation, the marine basin expanded to the east, first within the area of what is known now as the Baltic Sea and later into the present-day onshore regions. Lithosphere extension and sedimentary and thermal loads of adjacent rift system, presumably situated west of the Tornquist–Teisseyre zone, are accounted for the subsidence of the Baltic basin during that time (Šliaupa 2002).

The passive continental margin subsidence of the Baltic basin gradually decelerated during the Ordovician causing further but slower basin subsidence. By contrast to the Cambrian, subsidence in the western part of the basin was not compensated by conformable sedimentation rates which imply cessation of the terrigenous source in the west due to widening of the Tornquist Sea. The Ordovician is characterized by a nearly continuous sedimentation in a basinal facies and in a shallow marine environment. Minor thickness variations hint of considerable decreasing tectonic activities.

2.3.3 Foreland Stage

Subsidence intensity started to accelerate in the Late Ordovician again and increased drastically during the Silurian (Figs. 2.8 and 2.9). This change in subsidence was due to the flexural bending of the western margin of the Baltica plate because of the docking of the East Avalonian plate in the west (Poprawa et al. 1999). The progressing advancement of the North German–Polish orogenic build-up in the west is reflected by the compensation of the subsidence by the sedimentary load during Late Silurian time. The basin was finally completely filled by Early Devonian Old Red deposits. By contrast to the hard coupling of Laurentia and Baltica, which caused intense faulting in the Baltic region, the soft docking of East Avalonia and Baltica did not result in any significant faulting of the Baltic region.

2.3.4 *Intracratonic Basin Stage*

The foreland subsidence stage with high subsidence rates was followed by more stable tectonic conditions and continuous but much slower subsidence during the Devonian. The subsidence pattern changed considerably. Maximum subsidence rates occurred in the central part of the Baltic region, where the thickness of the Devonian succession reaches up to 1.1 km. Results of subsidence backstripping (Figs. 2.6 and 2.7) show that the calculated tectonic subsidence is accountable only for roughly 300 m of the total subsidence during that time. Driving mechanisms for the basin subsidence are so far not completely understood. It is presumed that larger scale processes influencing the whole East European platform have also triggered the subsidence (e.g. Ismail-Zadeh 1998). This hypothesis is mainly justified by the similarities in sedimentation and subsidence trends between the Baltic and the Moscow basins (McCann et al. 1997). Nevertheless, the Baltic basin was also influenced by compressive tectonic forces related to the Variscan deformation processes in the western part of Europe (Šliaupa 2004).

2.3.5 *Thermal Doming and Thermal Sag Stage*

The subsidence ceased at the beginning of the Carboniferous and the subsequent period of basin development was characterized by a break in sedimentation until the Middle/Upper Permian. Furthermore, the basin flanks were considerably uplifted and eroded.

Numerous diabase sills and dykes of Permocarboniferous age are known from the southern and the central part of the Baltic Sea basin (Motuza et al. 1994) as well as from Scania, Bornholm and the Rügen area in the west (Obst 2000). The chemical composition of the diabases has an affinity to rift-related intrusions. This hints of a tectonic reactivation of the Baltic Sea area. The corresponding lithosphere heating is accounted for the uplift of the basin during the Permocarboniferous thermal doming stage.

The following thermal relaxation led to the re-establishment of the subsidence regime and sedimentation during Late Permian time (Fig. 2.5). Subsidence took place especially in those areas which were uplifted before, in particular the Mazury High. This is a clear hint for the involvement of thermal sag processes as major subsidence-driving mechanism. The thermal sag is coupled with some wrench fault movements along the craton margin. These mechanisms were most active in Late Permian and Early Triassic time and gradually ceased throughout the Mesozoic and the Cenozoic. Therefore only episodic sedimentation related to global sea level changes occurred in the Baltic basin during that time. Only in the southwestern part of the basin, tectonic activities increased during a Cretaceous inversion phase (e.g. Krzywiec et al. 2003).

2.4 Major Tectonic Phases and Basin Structures

As described above, five defined geodynamic stages were important for the Baltic basin development. Moreover, several tectonic phases and multiphase fault zones and other structural features can be distinguished (Figs. 2.11, 2.12, 2.13, 2.14 and 2.15).

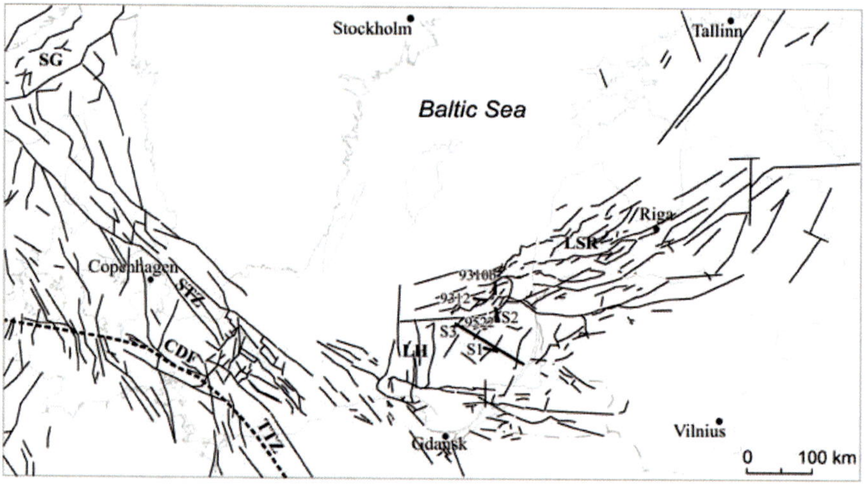

Fig. 2.11 Faults detected in the sedimentary cover. Locations of seismic and seismo-acoustic profiles referred in the text are marked. Major tectonic zones are marked: SG, Skagerrak graben; STZ, Sorgenfrei–Tornquist zone; TTZ, Teisseyre–Tornquist zone; LH, Leba high; LSR, Liepaja-Saldus ridge. Caledonian deformation front (CDF) is indicated

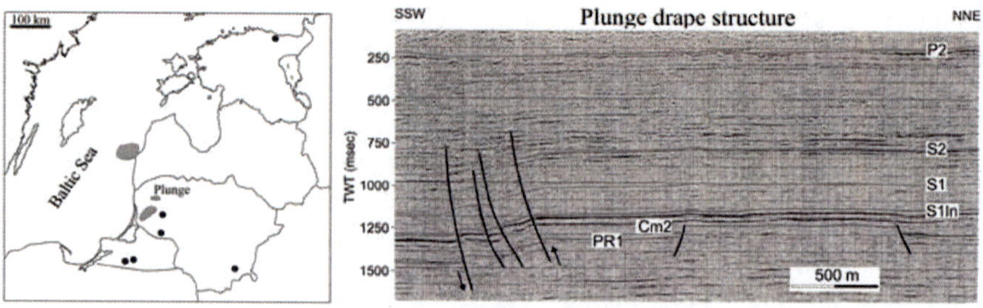

Fig. 2.12 **a** Distribution of Ediacaran drape structures (*black dots*) and clusters (*grey polygons*) in the Baltic region. **b** Seismic profile across the plunge drape structure (west Lithuania); a set of transpressional Caledonian faults (Telsiai zone) are seen in the *left part* of the profile

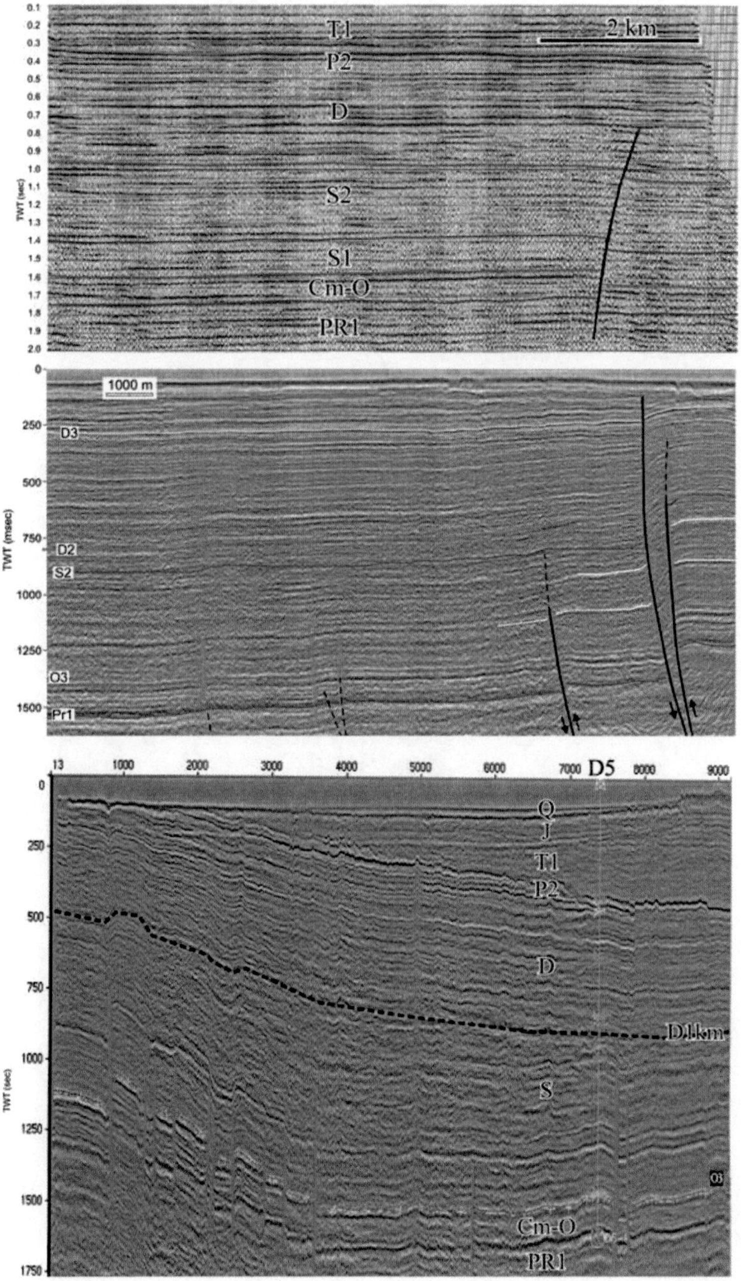

Fig. 2.13 Seismic profiles S1, S2 and S3 (see Fig. 2.11 for location). The *upper* profile crosses the compressional NNE–SSW-trending west Nida fault of Caledonian age. The *middle* profile crosses the transpressional fault zone bordering the Liepaja-Saldus ridge in the south. Faulting took place there during Late Silurian to Early Devonian; the zone was reactivated during the Permocarboniferous phase. The *lower* seismic profile crosses the axial part of the Baltic basin

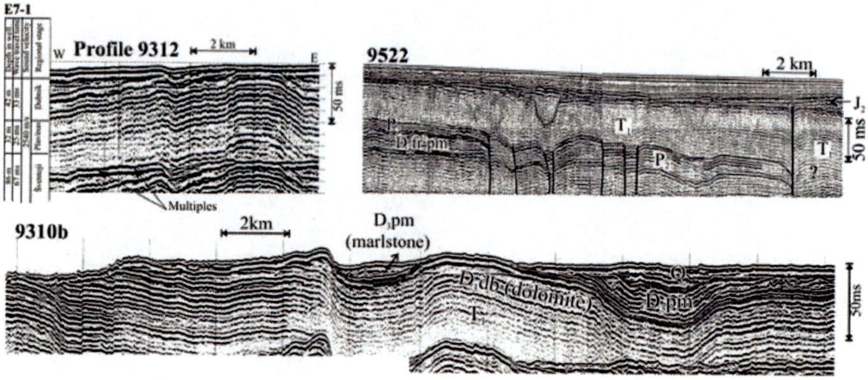

Fig. 2.14 Seismo-acoustic profiles 9312, 9522 and 9310b (see Fig. 2.8 for location of profiles). Profile 9312 shows several steep flexures spaced roughly at the distance of 2 km in an Upper Devonian succession (Liepaja-Saldus ridge). Profile 9522 indicates a set of steep Permian faults spaced in a 0.5–2-km range and cutting Upper Devonian sediments. Profile 9310b crosses the southern part of the Liepaja-Saldus ridge. The Upper Devonian succession shows strong deformations; some are well reflected in the sea bottom morphology

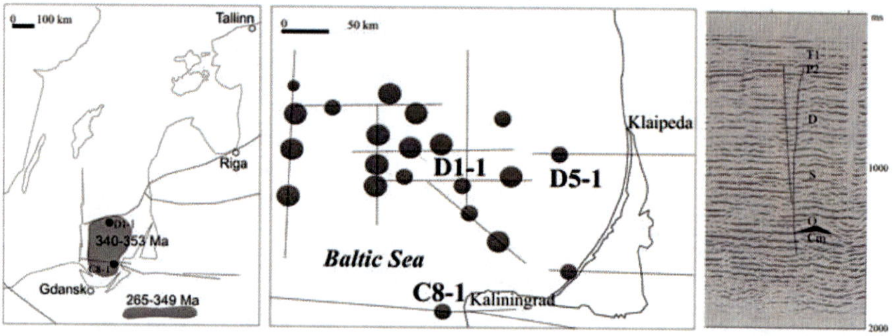

Fig. 2.15 *Left part* shows clusters (*grey polygons*) of Permocarboniferous intrusions. Major faults and wells penetrating intrusions are indicated. *Middle part* shows magnetic anomalies related to diabase intrusions in the Baltic Sea. Controlling faults were defined from gravity and magnetic fields. *Right part* shows seismic profile across a diabase intrusion north of the well D5-1

2.4.1 Early Ediacaran Tectonic–Igneous Phase

As mentioned above, the break apart of the Rodinia continent was at first reflected in the Early Ediacaran with the deposition of the Zura formation in western Latvia and adjacent offshore areas. A dense cluster of drape structures (basement blocks covered by the Zura formation and Lower Cambrian sediments) occurs in the Zura depression (Brangulis and Kanevs 2002). Similar drape structure clusters were identified in westernmost Lithuania, Estonia and the western Kaliningrad area (Fig. 2.9). Seismic data hint of an extensional kinematic type of those fault blocks (Fig. 2.9). The fault vertical displacement reaches 170 m. These structures suggest an intense tectonic extensional regime that mainly affected the eastern Baltic Sea and neighbouring areas.

The tectonic activity ceased during the Late Ediacaran and Cambrian with the onset of wide marine transgression. The tectonic strain accumulation shifted to the Teisseyre–Tornquist zone in the west. This is reflected, for example, in the formation of funnel grabens with extensions between 1 and up to 200 m and depth ranges between 1 and over 50 m. They formed mainly during the Early Cambrian and partly also during Middle and Late Cambrian and are documented from south Sweden/Scania (Lindström 1967, Scholz et al. 2009). In general, no evidences of significant major faulting are recognized during the Cambrian, Ordovician and Early Silurian times, suggesting low tectonic stresses affecting the Baltic basin. Fractures filled with Lower Cambrian sandstones are mapped in the northern Baltic Sea region. According to Drake et al. (2009), they might have formed in relation to far-field extensional effects of the opening of the Iapetus Ocean. Cambrian sandstone fractures in the coastal region around Simpevarp generally follow the orientation of the basement fracture sets with dominant NNE–ENE directions (Drake 2008). In Bornholm, a well-exposed sandstone dyke swarm strikes NW–SE. The opening and filling of the fissures were caused by normal extension movements in NNE–SSW direction in several steps, probably during the Early Cambrian (Katzung and Obst 1997). Funnel structures and clastic dykes are also reported from the coast of the Baltic Sea south of Vik (Scania). Their formation is also related to extensional tectonics (Scholz et al. 2009).

In some seismic profiles, evidences of Late Ordovician faulting were reported from the Lithuanian and the Latvian offshore areas. Although fault amplitudes reach only a few dozens of metres, some of them controlled the growth of Ordovician reefs (Kanev and Peregudov 2000). They mainly show reverse kinematic features implying compressional tectonic activity during Late Ordovician when the lithosphere flexuring was initiated due to East Avalonia docking.

2.4.2 Late Silurian–Early Devonian Phase

The main structuring phase of the Baltic Sea basin took place during the time period between the latest Silurian and the earliest Devonian. A detailed structural analysis revealed that the region was exposed to NW–SE-directed horizontal compression in relation to the collision of Laurentia and Baltica (Šliaupa 1999). Two dominating groups of E–W (ENE–WSW) and NE–SW (NNE–SSW) striking reverse faults have been formed. Typical for the first group are transpressional geometries, while the second fault group belongs mainly to a compressional type.

The faulting was focussed on special areas (Fig. 2.8). Main tectonic strain accumulated in the Liepaja-Saldus ridge zone and the Telsiai fault zone in the central part of the Baltic basin. Such a selective faulting can be explained in terms of structural inheritance. The Liepaja-Saldus ridge is an ancient fault zone that is marked by sharp changes in the Moho depths, thus pointing to a first-order structure. The Telsiai fault zone is also a reactivated Palaeoproterozoic zone (Šliaupa et al. 2002b).

The amplitude of the Liepaja-Saldus ridge reaches 600 m. Bounding faults dip at angles between 50 and 80°. They show very complex geometries. Flower structures are typical implying strike-slip-type faulting.

To the south of the Liepaja-Saldus-Telsiai zone, the prevailing direction of Caledonian faults is NE–SW. The amplitudes are in the range of 50–200 m. These faults are rather regularly spaced at a distance of about 30 km and show quite simple compressional geometries. They dip to the west at high angles of 70–80°. The Leba ridge faults were also probably established during the Caledonian stage, but their main activity happened during the Permocarboniferous. The onset of this large-scale feature during the Late Silurian is supported by the presence of associating gas fields in the Polish offshore area. As it is shown below, the gas was generated during Silurian, while source rocks were already overmature by the beginning of the Devonian.

The faulting north of Liepaja-Saldus ridge is only of minor intensity, which is somehow surprising as the stress source is located in the northwest (Scandinavian Caledonides). Several faults trending NE–SW are reported from Estonia. The amplitudes are in the range of 10–30 m only. A network of smaller faults striking NW–SE is mapped in northeast Estonia (Sokman et al. 2008). Here too, amplitudes reach only a few metres. The faults are dipping mainly to the northwest at predominating angles of 60–70° and show a compressional style.

Detailed seismo-acoustic surveys of the northern Baltic Sea area revealed a cluster of linear disturbance zones with 1–4-km-wide spacing. These zones strike several tens of kilometres north–south and show offsets of several tens of metres. The seismic profiles revealed a weak flexure-like bending of the layers in the zones; locally they are intersected by small-scale faults (Tuuling and Flodén, 2001). There is so far no stratigraphic control to estimate the time of this faulting.

Small-scale faulting associated with the migration of hydrothermal fluids is known from the Early Devonian in the northern part of the Baltic basin. This hydrothermal activity is about 10–15 Ma younger than the corresponding ones in Sweden and Finland (Alm et al. 2005). Fluid inclusion investigations of the fluorite–calcite–galena veins in the Baltic basin indicate depositional temperatures of 100–150°C (Alm and Sundblad 2002).

2.4.3 Permocarboniferous Phase

During the Permocarboniferous, tectonic processes were reactivated. Most intense tectonic deformation took place in the southwesternmost Baltic Sea area along the Bornholm–Darlowo fault zone which is a part of the Teisseyre–Tornquist zone. This NW–SE striking fault system forms an up to 100 km broad zone of horsts and grabens (Vejbæk 1985, Krzywiec et al. 2003). These structures were mainly formed in Late Carboniferous–Early Permian (Wikman 1986) and are related to post-orogenic destruction of the Variscan foreland initiated by wrenching and strike-slip movements (Brochwicz-Lewinski et al. 1984, Ziegler 1990).

In the southern part of the Baltic basin, a set of large E–W striking faults were established. This faulting is associated with the intense doming of the lithosphere that also leads to the erosion of Devonian and older sediments. The largest fault of this group is the Kaliningrad fault striking across the Gdańsk Bay and further onshore to the east (Fig. 2.8). The amplitudes of those faults are in the range of 30–50 m.

The Leba ridge is built of a wide set of N–S striking faults, the activity of which led to the truncation of more than 1 km of Devonian and uppermost Silurian sediments (Domżalski et al. 2004). The seismic profiles reveal the compressional nature of the Leba faults. The other Caledonian faults were also reactivated in a compressional regime during the Permocarboniferous. The most intense fault reactivation is reported from the Liepaja-Saldus ridge (Figs. 2.13 and 2.14).

A peculiar feature of the Permocarboniferous phase is the activation of igneous processes in the southern part of the Baltic Sea and in northern Poland (Fig. 2.12). The intrusions were dated to 340–355 Ma (Birkis and Kanev 1991, Šliaupa et al. 2002c) which is contemporaneous to the Chmielno volcanic formation of the Pomeranian basin in Poland. So far 21 intrusions have been identified by characteristic magnetic anomalies (Šliaupa et al. 2004). They are connected mainly to N–S and E–W trending faults (Fig. 2.12). Well C8-1 drilled, for example, a 6-m-thick intrusion hosted by Silurian shales, which is connected to the Kaliningrad fault. Well D1-1 penetrated a 25-m-thick sill also hosted by Silurian shales and connected to an E–W striking fault. This fault also hosts another intrusion located close to the well D5-1 (Fig. 2.12). This fault is very well traced by 30 m offset of Upper Permian layers in the onshore area. It is noticeable that this fault shows inverse relationship offset of Devonian sediments pointing to tectonic inversion. The chemical composition of D1-1 diabases is close to the continental rift basalts (Motuza et al. 1994). Diabases are of sub-alkaline composition with modal olivine and nepheline, of porphyritic texture (3–5% of plagioclase phenocrysts). D1-1 sill intruded in two phases – the early phase is represented by fine-grained diabase, while very fine-grained diabase intruded in the second phase. The chemical composition suggests a formation of the magma chamber at 150–120 km depth. The magnetic source depth modelling of the magnetic field data indicates that diabase sills are mainly hosted by Cambrian and Silurian sediments and only partly by the crystalline basement.

Contemporaneous igneous activities are documented from northern Poland, the region that experienced the most intense uplift during the Permocarboniferous. Rb–Sr ages of the Elk massif are around 355 Ma and thus similar to K–Ar ages of the Pish gabbro intrusion (Depciuch et al. 1975). Several smaller ultramafic and mafic intrusions were identified in the area and show age dates between 347 and 344 Ma (Depciuch et al. 1975). A second phase of igneous activity took place in northern Poland between 295 and 265 Ma. This phase corresponds to the phase of intense magmatism in the North German basin (Benek et al. 1996).

Furthermore, a 355-Ma Rb–Sr age of Ordovician K-bentonites was identified in Estonia (Kirsimae et al. 2002). All these data point to basin-scale thermal processes during the Permocarboniferous.

2.4.4 *Late Cretaceous Inversion Phase*

Tectonic processes ceased during the Middle Permian and were of minor activity throughout the main part of the Mesozoic. In the southwesternmost part of the Baltic basin, tectonic activity strongly increased during Late Cretaceous. Strike-slip and reverse faults were established within the Bornholm–Darlowo fault zone (Krzywiec et al. 2003) that is a part of a larger Fennoscandian border zone. The island of Bornholm is a composite fault block. The island and associating Palaeozoic fault blocks are bounded by WNW–ESE and NNW–SSE trending faults. The faults of the Mesozoic blocks follow the same trends in part, but the fault orientations have a wider scatter and an additional NW–SE trending segment. This wrench-dominated Mesozoic faulting was related to the reactivation of the Pre-Permian fault system. The Sorgenfrei–Tornquist zone continued to experience tectonic activity in Triassic and Middle Jurassic times. The zone has experienced an uplift of up to 1,700–2,000 m during the Late Cretaceous to Early Tertiary inversion tectonic phase and the Late Tertiary regional uplift of Fennoscandia.

An intense Late Cretaceous inversion tectonics is also documented from southern Lithuania and the Kaliningrad district. Amplitudes of inverted structures reach 200 m there (Šliaupa 2004).

2.5 Tectonic Evolution of the Southwestern Basin Rim During the Early Palaeozoic

The Baltic basin extended further to the west and the southwest during its passive continental margin stage. The situation changed with the build-up of the Danish–North German–Polish Caledonides during the foreland stage of the main basin. An Early Palaeozoic thrust and fold belt formed the southwestern basin rim since that time (Meissner et al. 1994, Hoffmann and Franke 1997, McCann 1998, Katzung 2001). Detailed biostratigraphical analysis of Lower Palaeozoic sediments from Maletz (1997) gave evidence for a Llanvirnian age of the first Caledonian deformation phase. Frost et al. (1981) dated the low-grade metamorphism of the Ringkobing-Fyn High to 440 Ma. This metamorphic age seems to mark the peak of the orogenic processes within the area. Isotope studies from Lower Palaeozoic sediments of the boreholes from Rügen Island point to a major deformation in the Early Silurian (Giese et al. 1995). Beier and Katzung (2001) reconstructed three to five Caledonian deformation phases and interpreted those as deformation signs in an accretionary wedge in the forefield of Avalonia that was subsequently thrusted over the passive margin of Baltica.

Provenance studies of sediments from the Rügen and the Danish area indicate a sediment transport from a Gondwana-type western provenance and a mixing with a Baltica source during Late Ordovician to Silurian times (Vecoli and Samuelsson 2001a, b), thus reflecting the docking of Eastern Avalonia to the margin of Baltica. Furthermore, in Bornholm Silurian tuffaceous sandstones deposited in front of the

Caledonides on the East European craton were dated to 420–430 Ma, reflecting volcanic activity in the adjacent orogen (Hansen 1995, Obst et al. 2002). Giese et al. (1997), Torsvik et al. (1996) and the MONA LISA Working Group (1997) assume a closure of the Tornquist Ocean between Avalonia and Baltica for the Late Ordovician. Intense erosion of the uplifted orogen occurred afterwards. Within the Rügen area the erosion period lasted until the Middle to Upper Devonian.

The Caledonian deformation front represents the northern boundary of the thrust and fold belt. Its southern extension is not known because no single well has reached the Lower Palaeozoic in the central part of the North German basin and in the basin part south of Rügen Island. Both areas are characterized by a very thick overlying sedimentary succession of Carboniferous to Cenozoic age (Fig. 2.6, Hoth 1997). According to Hoffmann et al. (2001) the gently south-dipping Cambro-Ordovician alum shales form the basal decollement on which the orogen wedge was thrusted onto Baltica. This horizon is assumed to be located deeper than 10 km in the mainland south of Darß and in the area south of the wells Greifswald 1, Loissin 1 and Grimmen 6 (Gd1, Loss1, Gm6 in Figs. 2.6 and 2.7). Fault blocks downthrusted southwest were identified in offshore seismic profiles close to the Rügen area (Schlueter et al., 1997). They cut the basement, as well as Cambrian and the lowermost Ordovician sediments, and thus document an intense rifting during the earliest Palaeozoic.

For a long time, the boundary between Baltica and Eastern Avalonia was considered to be confined to the Caledonian deformation front (CDF) in the southwestern Baltic Sea (Cocks and Fortey 1982). However, since the EUGENO-S deep seismic survey in the 1980s, it was realized that the major tectonic boundary between the two plates is located further to the south and west. This was supported by deep seismic sounding studies BABEL (Meissner et al. 1994), DEKORP (Meissner and Krawczyk 1999) and MONA LIZA (MONA LIZA Working Group 1998). Some authors even concluded that the major suture is related to the Elbe zone (Abramovitz et al. 1998), which is located 200–300 km west and south of the CDF. A sporadically strong sub-Moho structure dipping 20–30° to the NE was observed in the Bornholm area. It is interpreted as the subducted slab of East Avalonia (McCann and Krawczyk 2001, Krishna et al. 2007). Furthermore, a series of profiles of the Baltic'96 experiment show north-dipping reflections at Moho level in the southwestern Baltic Sea area. Hence, northward subduction in the uppermost mantle is indicated by the available information (Thybo 2000). During the Variscan stage, the southwestern Baltic Sea area represented the northeastern margin of the foreland basin predominantly controlled by flexure induced by the Variscan orogen to the south.

2.6 Present Morphology of the Baltic Sea Depression

The present Baltic Sea depression was formed during Cenozoic time. There is still no consensus with respect to the role of erosional and tectonic processes for the formation of the depression. A group of researchers suggest that the main forms are of pre-glacial tectonic origin and that glacial erosion and deposition had only a very limited role (e.g. Voipio 1981, Šliaupa et al. 1995b), while the others stress the essential role of erosional processes (e.g. Marks 2004).

Due to the absence of Palaeogene and Neogene marine sediments, except for the southernmost part of the Baltic region, the reconstruction of the neotectonic movements is very uncertain. The smoothed sub-Quaternary surface is often considered to mainly reflect vertical tectonic movements (Šliaupa et al. 1995a). However, it does not completely remove and exclude erosional components either. Figure 2.16 shows the altitudes of this smoothed surface; it ranges from +100 m in northeast Lithuania to –280 m in the Gotland low and generally reflects the shape of the Baltic sedimentary basin. The pattern of morphological highs and lows is dominated by N–S trends. Pandevere, Vidzeme and south Lithuanian structures compose the eastern system of highs separated by Riga and Kaunas lows from Saaremaa and Kurzeme-Zemaitija highs. The west Gotland, Gotland and Gdańsk lows are the main sub-Quaternary features in the sea area. This dominating N–S pattern is superimposed by lower order features striking generally ENE–WSW.

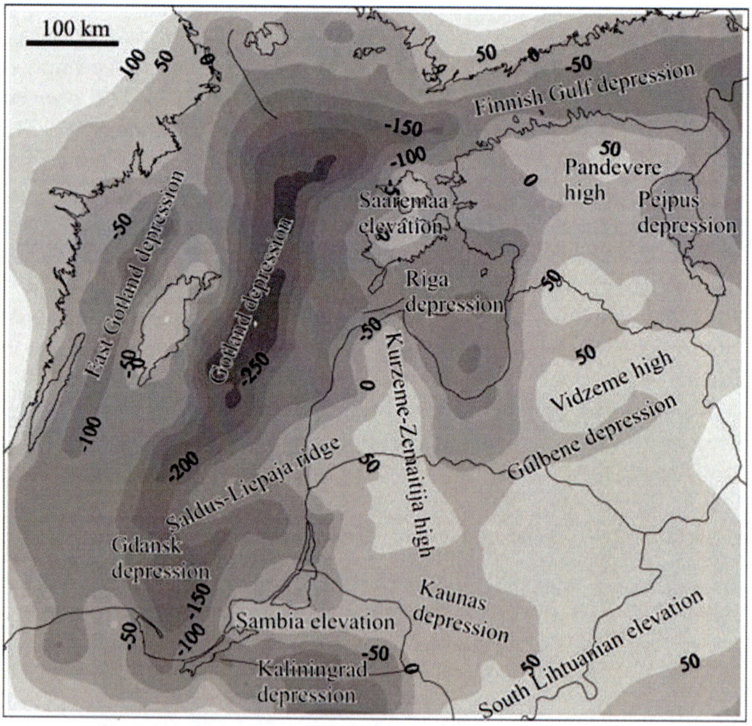

Fig. 2.16 Smoothed depth map of the sub-Quaternary surface in the Baltic area (after Šliaupa et al. 1995b, with modifications)

Intensive glacial erosion led to thickness reduction of the Quaternary cover in the Baltic Sea area. Therefore it is difficult to reconstruct the successive events during the Pleistocene. At the beginning of the Pleistocene, the Baltic Sea low was occupied by the so-called Baltic stream, flowing from northeast towards southwest (Gibbard 1988). The knowledge about the Baltic stream is very limited, because the Elsterian ice sheet removed all older sediments.

The first ubiquitous evidences for the existence of the Baltic Sea low are marine sediments of the Holstein interglacial that are distributed in the Baltic Sea area and adjacent regions (Marks and Pavlovskaya 2003). The second marine event took place in the Eemian interglacial and the limits of the marine basin roughly coincided with the present-day Baltic Sea shoreline.

2.7 Geological Resources

Besides drinking waters, sand and gravel deposits (Harff et al. 2004, Kramarska et al. 2004) are the main resources of the shallow subsurface. The mining of amber is of additional importance. It is exploited in the Sambia Peninsula (Kharin et al. 2004) and prospects are considered in the Polish coastal zone (Kosmowska-Ceranowicz 2004). A small amber exploitation was performed in the Kursiai lagoon during the previous century.

Important resources of the deeper underground of the Baltic basin are related to reservoir horizons and hydrocarbon fields. Oil exploitation was initiated in the area at Kinnekulleverken on Gotland in the 1940s of the previous century (Johansson et al. 1943). The offshore hydrocarbon exploitation started in the Polish economic zone in the 1980s and a decade earlier in the onshore area of Lithuania and the Kaliningrad district.

Reservoir horizons are of importance for gas storage and for geothermal energy recovery. Additional future utilization of reservoir rocks might be connected to the issues of CO_2 storage (Šliaupa et al. 2008) and the storage of compressed air as an energy storage option for wind power stations. Major reservoir horizons for all these utilizations are sandstone layers within the Devonian and the Cambrian. The recovery of geothermal energy from the corresponding formation waters of the reservoir horizons requires certain temperature levels. The 40°C level is only reached in certain areas of the Baltic basin, where the reservoirs are located in a depth below 1,000 m. Perspective areas exist particularly near the Lithuanian coast because of the heat flow anomaly in this area. So far only the station in Klaipeda produces geothermal energy on a larger scale in the area (Radeckas and Lukosevicius 2000).

2.7.1 Hydrocarbon Fields

The Baltic basin represents a proven hydrocarbon province (Fig. 2.17). In total about 40 hydrocarbon accumulations have been discovered (Brangulis et al. 1993,

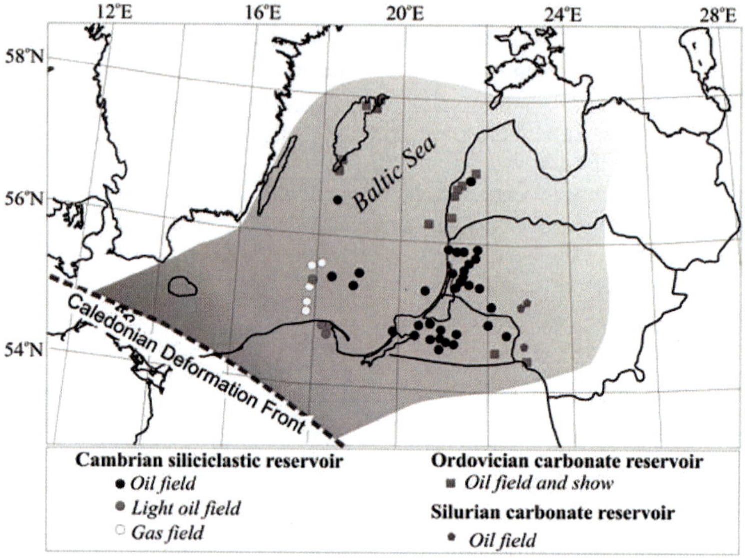

Fig. 2.17 Distribution of oil and gas fields and shows in the Baltic basin

Freimanis et al. 1993, Kanev et al. 1994, Dobrova et al. 2003, Šliaupa et al. 2004). Most of them are oil accumulations, but offshore Poland gas accumulations also do occur.

In the Kaliningrad district, oil production began in 1975. Currently 5–6 Mbbl per year are produced from the onshore fields. Production from the offshore D6 oil field started in the second half of 2004. The Lithuanian onshore oil production started after the restoration of independence in 1991. It reached the production peak in 2004 with 2.8 Mbbl. There is light oil and gas production in the Polish sector of the Baltic Sea. In the northern part of the basin, there is a small-scale oil production in Gotland. In Latvia, several small oil accumulations were discovered. Only very minor, short time oil production took place in 1990.

2.7.2 Major Reservoirs

The major hydrocarbon reservoirs are sandstone horizons of Middle Cambrian age. They are underlain by Middle Cambrian shales and are capped by shales of Ordovician–Silurian age. The total thickness of these sandstone reservoirs is between 50 and 70 m. They are represented by shallow marine quartz sandstones with subordinate shale and siltstone layers. The mineral composition of the sandstones is dominated by quartz that composes 96–99.8% of the rock volume. The clay content varies between 0.5 and 3.5%. Illite commonly dominates the clay admixture of the lower part and kaolinite predominates in the upper part of the reservoir sections. This is related to either a regression phase or a more intense percolation of the upper part of the reservoir sections by meteoric waters during early burial stages. The reservoir properties of the sandstones are mainly controlled by authigenic quartz cement which ranges from 10 to 32%. The best reservoirs are identified in the Liepaja-Saldus ridge and the Leba ridge, where average porosities are between 14 and 18%. The good reservoir properties are mainly related to the shallow (1,100–1,600 m) burial of sandstones. Southeast of the ridges, porosities and permeabilities of the sandstones decrease dramatically to values between 1 and 9% (average 5%) and <0.01 and 25 mD, respectively.

The Ordovician carbonates show, in general, poor reservoir properties; the porosity is mainly 2–5% only. However, some oil shows and oil inflows were reported from western Latvia. The corresponding reservoir layers are related to the "Porkuni regional stage" of Upper Ordovician age. They are composed of oolitic and bioclastic limestones (Laskovas 1994). The open porosity of the Porkuni carbonate reservoirs of the wells E6-1 and E7-1 varies in the range of 3–24%. Best permeabilities are around 40 mD. Oil shows were reported from well E7-1 and an oil inflow of 2.7 m^3/day was reported from well E6-1. This Upper Ordovician reservoir belt is confined to the Liepaja-Saldus ridge. Prospective resources of this area were assessed to around 8.8 million tons (Laskovas and Jacyna 1998).

In the western part of the Baltic Sea, oil is produced from Upper Ordovician carbonate mounds at the Gotland Island (Sivhed et al. 2004). Between 1974 and 1992, total oil production amounted to 100,000 m^3. The mounds contain large numbers of vugs and moulds which communicate through dissolution fractures and surfaces. Sediments represent sub-mound, intra-mound, cap and flank, and supra-mound facies. Algae and stromatolites dominate the intra-mound facies, providing an organic framework for the entire structure. A large field of Ordovician reefs was identified between Gotland and Latvia, but so far no drilling has been carried out to prove its hydrocarbon potential (Kanev and Peregudov 2000).

The Silurian consists mainly of black shales and clayey marlstones, representing a 1-km-thick source rock package. An oil show was reported from the well Nida-44 in the Curonian Spit. It is confined to the uppermost part of the Silurian section containing dolomite interlayers of around 7.5 m thickness with porosities from 12 to 14%. However, this is the only discovery so far. Lower and Upper Silurian reefs are reported from Gotland and the area east of the island (Manten 1971, Kershaw 1990, Flodén et al. 2001), but no evidences of hydrocarbons were reported. Still those reef build-ups have some potential, as several oil accumulations were discovered in Upper Silurian reefs in central Lithuania (Lapinskas 2000).

2.7.3 Source Rocks

Major source rocks of the Baltic basin are Cambrian, Ordovician and Silurian shales. The TOC content of the Lower–Middle Cambrian shales is rather low and varies between 0.03 and 2%. The lowest values are typical for Lithuania. The well-known alum shales (middle part of Upper Cambrian–Tremadocian) are distributed in the western part of the basin (Buchardt and Lewan 1990, they contain 11–12% TOC).

The Ordovician carbonate deposits contain generally low amounts of organic matter; TOC is commonly less than 0.2%. Other source rocks of the Ordovician are black shales of the Mossen formation (Middle Ordovician age) and of the Fjacka formation (Upper Ordovician). The thickness of these black shales varies between 2.0 and 4.5 m. They have been deposited in deeper shelf zones and are characterized by a high content of sapropelic organic matter and TOC contents of up to 14.9% (Kaduniene 1978; Kaduniene et al. 1978).

The Silurian is represented by a 750–1,150-m-thick succession of mainly dark grey graptolite shales. Two parts are typical for the succession. While the lower part with 300-m-thick shales of Llandovery–Lower Ludlow age contains up to 11.2–16.5% TOC (Kaduniene et al. 1978), the upper part contains significantly lower amounts of organic matter. However, the distribution of Silurian source rocks is still poorly understood in the Baltic Sea area because it is based there just on extrapolation of data of a few offshore wells.

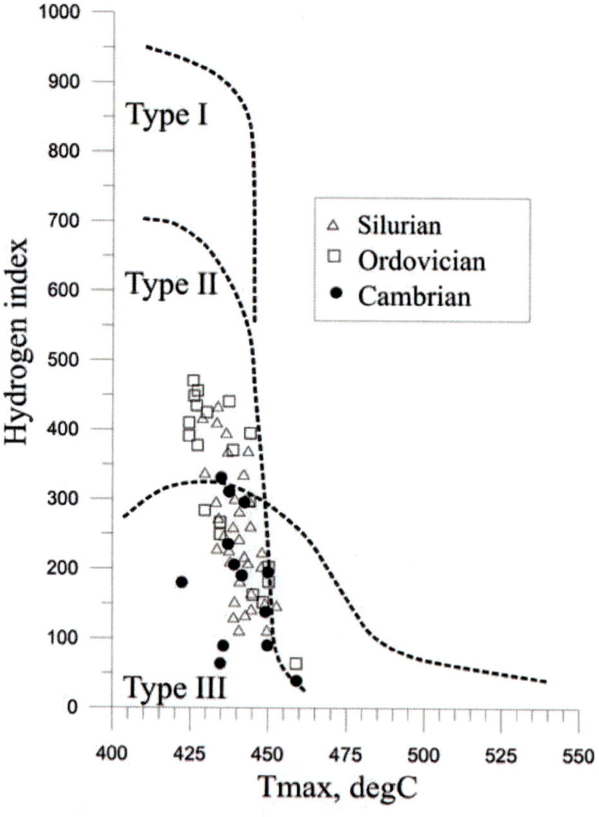

Fig. 2.18 Kerogen type of Cambrian, Ordovician and Silurian source rocks of the Baltic basin (after Kanev et al. 1994, Zdanaviciute and Sakalauskas 2001)

The kerogen of all mentioned shales can be mainly classified as type II. Irrespective of source rock age, kerogens show a rather uniform trend on the T_{max}–hydrogen index plot (Fig. 2.18). Organic maturity increases from the east to the west, exceeding reflectance values of 2.0% Ro in the western part of the Polish offshore and 4.5–5.0% Ro in the Rügen area (Hoth 1997, Hoffmann et al. 2001).

2.7.4 Oil and Gas Generation

One-dimensional modelling of HC generation was carried out for selected offshore wells. Thereby burial history was calibrated with sonic and density well log data (Šliaupa et al. 2002a) and organic maturity data (Brangulis et al. 1993, Buchardt et al. 1997).

The burial reconstruction indicates that maximum burial depth exceeded 4.5 km in the southwestern Baltic basin by the end of the Devonian (Fig. 2.19, well A8-1). Hydrocarbon generation started during Late Silurian time, the period with the maximum subsidence rate. Cambrian and Ordovician source rocks lost their hydrocarbon generation potential by the end of the Silurian. Modelling results show that the hydrocarbon generation from Silurian shales lasted up to the beginning of the Carboniferous. It is therefore inferred that the structures of the Leba ridge were filled by migrating hydrocarbons from the west during the latest Silurian and Devonian times.

Further west, the Lower Cambrian sandstones from Bornholm, and in particular the Hardeberga sandstone, contain a substance that has been interpreted to be pyrobitumen. It causes the dark colour seen at many outcrops (Møller and Friis 1999). The presence of pyrobitumen indicates the former presence of migrating hydrocarbons. Petrographic observations show, even though the sandstones are now extensively compacted, that only low amount of diagenetic cement was formed during hydrocarbon generation and migration (probably during the Silurian).

The modelling results of the well B2-1, located on the Leba ridge, show that the oil generated there during Devonian–earliest Carboniferous time. In this area, only between 7 and 17% of the HC potential of the Silurian, Ordovician and Cambrian shales was realized.

Intensity of oil generation was also rather low in the eastern part of the Baltic basin due to both low burial and heat flow (40–50 mW/m^2). The oil generation started in latest Devonian–earliest Carboniferous time. In the area of well B8-1, only Cambrian and Ordovician shales entered the oil window (Fig. 2.19, well B8-1), but only 9 and 6% of the oil generation potential was realized. In the D6 area, oil generation started not before Mesozoic time (Fig. 2.19, well D6-1).

West Lithuania was and is characterized by an anomalous heat flow reaching 70–90 mW/m^2 today. This has caused a more intense hydrocarbon generation compared to the eastern part of the Baltic Sea. Cambrian, Ordovician and Silurian source rocks entered the oil window in western Lithuania during the Middle Devonian time. Maximum hydrocarbon generation took place in the Early Carboniferous.

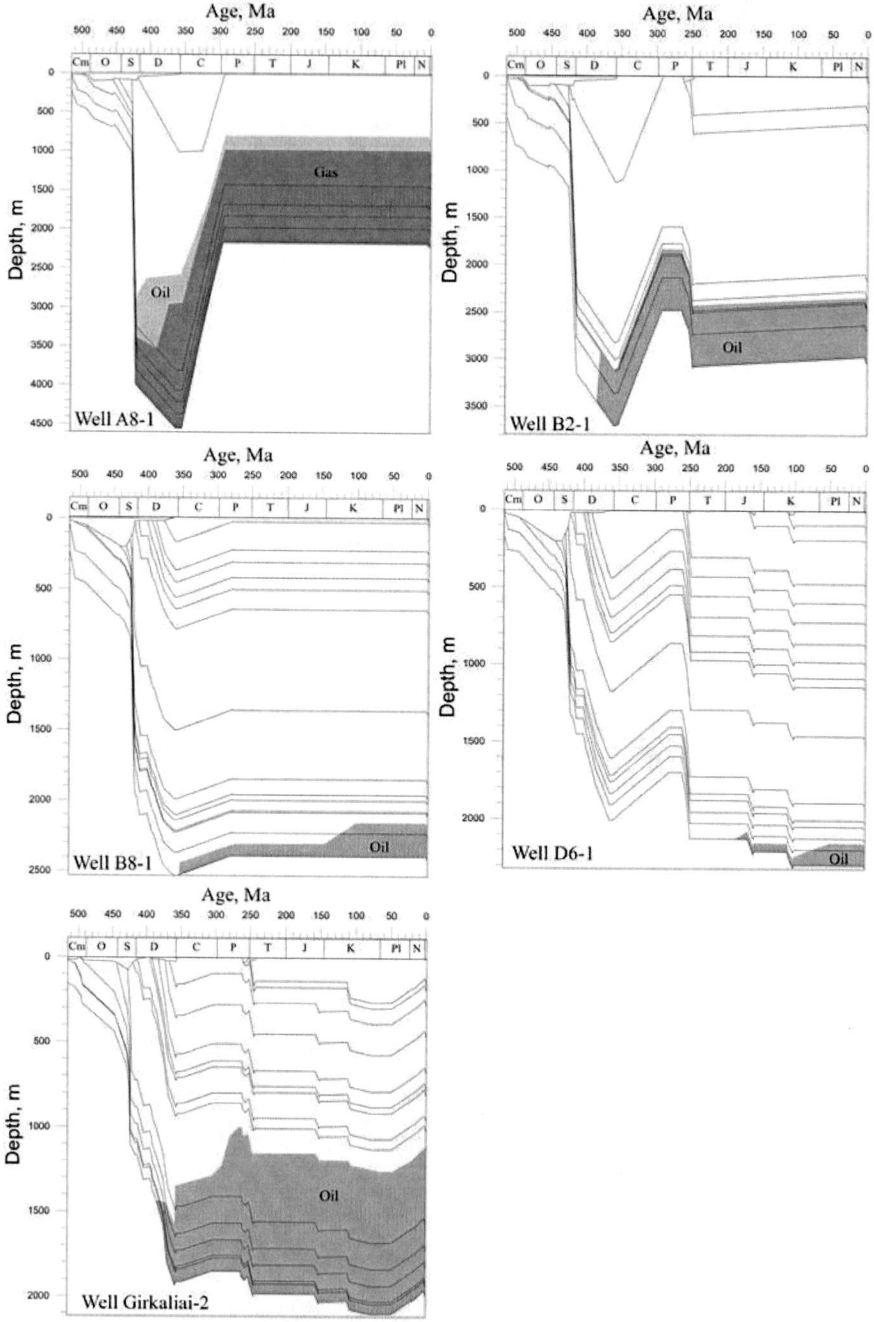

Fig. 2.19 Burial graphs and modelling results for oil/gas generation for the wells A8-1, B2-1, B8-1, D6-1 (offshore) and Girkaliai-2 (western Lithuania)

2.8 Discussion and Conclusions

The Baltic Sea is a young geomorphologic feature that was established in Cenozoic time, most probably during the Quaternary period as evidenced by Middle Pleistocene marine sediments. However, some authors present evidences of Neogene ages of the N–S striking features, which dominate the Baltic Sea depression morphology (Grigelis 1991). Glacial erosional processes undoubtedly contributed to the shaping and deepening of the depression. But, even assuming the essential role of erosion, it is rather difficult to explain the exceptional ice sheet and melt-water activity in the area without a pre-existing tectonic depression.

A strong evidence for the tectonic nature of the Baltic Sea depression is the coincidence of the outline of the Cambrian marine basin and the recent Baltic Sea (Fig. 2.8). The Cambrian marks the onset stage of the Baltic basin that was initially established in response to the continent break-up, thus implying a strong extensional regime during the Cambrian, as supported by structural studies. Furthermore, the Mesoproterozoic time was marked by voluminous intrusions of rapakivi granitoids and related igneous rocks, which all concentrated in the Baltic Sea area. Early Ediacaran tectonic extension and Permocarboniferous magmatism also anomalously affected the Baltic Sea area.

Rheological modelling of the lithosphere, based on a rather dense network of deep seismic sounding profiles both onshore and offshore (e.g. Baltic Sea, Babel), proved that the Baltic Sea depression is characterized by the weakest lithosphere in the Baltic region (Ershov and Šliaupa 2000). The effective elastic thickness (EET) of the lithosphere is in general between 20.5 and 21.5 km in the Baltic sea area (28 km in the Gulf of Finland), while it is in the range of 30–40 km in surrounding territories and more than 40 km outside the Baltic basin (Fig. 2.20). Variations in mechanical properties are mainly due to different lithologies and temperatures. It is noticeable that those variations are discordant to crustal thickness variations which are dominated by E–W and NW–SE trends, most likely reflecting the Palaeoproterozoic accretionary system (Fig. 2.20). These lithosphere strength variations are mirrored in the sub-Quaternary surface of the Baltic region, reflecting the general shape of the Baltic basin, and the Baltic Sea depression in particular.

If in-plane tectonic extension is strong enough, it can result in subsidence of a weak lithosphere (Artyushkov et al. 2000). The 2D dynamic modelling of the Baltic lithosphere indicates that extensional tectonic forces, typical for cratonic areas, may have induced 150–200 m of subsidence. Taking into consideration the deepening effects of glacial erosion, this is in good agreement with the sub-Quaternary subsurface in the Baltic Sea area. Thus, the presented model implies an extensional tectonic regime affecting the Baltic Sea area during the Quaternary time. The extensional nature of the Baltic Sea depression comprising smaller scale graben-like structures was also suggested by some previous studies (e.g. Schwab et al. 1997).

There are only a few breakout stress field measurements in deep wells in the southernmost Baltic Sea (Jarosinski 1994) that indicate N–S maximum stress orientation, which is possibly related to the impact of the Alpine chain in the south. On the other hand, the crust of Fennoscandia is believed to be affected by the ridge push

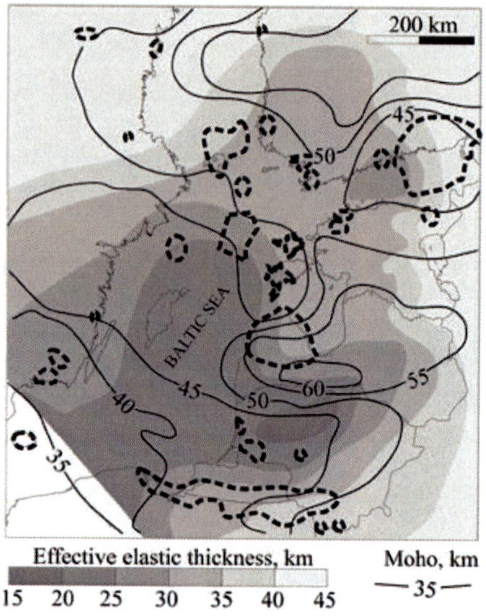

Fig. 2.20 Effective elastic thickness of the lithosphere (Ershov and Šliaupa 2000) and Moho depths (Rapakivi granites and related rocks are indicated by the *dashed lines*)

from the Mid-Atlantic ridge as indicated by the general NW–SE-orientated maximum horizontal stress (Olsson 2002). However, this model alone cannot explain the recent seismic activity in the region. Isostatic glacial rebound movements strongly influence the tectonic stresses of the shield (Muir Wood 1993). GPS measurements indicate a doming of the crust centred in the Bothnian Bay. The eccentric shift of the GPS sites is coherent with a vertical doming (Scherneck et al. 2001). Examination of the strain field of Fennoscandia by means of a glacial isostatic adjustment model suggests that elastic extension is the dominant style of deformation, controlled by horizontal displacement (Scherneck et al. 2003, Marrota and Sabadini 2004).

The Baltic Sea is located on the western flank of this Fennoscandian dome and thus may be a part of this geodynamic system. This suggestion is supported by recent GPS measurements in the Baltic countries (Pan and Sjöberg 1999). The modelling of the stress field distribution from those GPS data revealed two major stress provinces. For the western parts of Lithuania and Latvia and most of the Estonian area, uniaxial and diaxial tectonic extensions are shown, while the eastern part of the Baltic region is exposed to compression; the strain rate is in the order of 10^{-8}–10^{-9} year^{-1} (Zakarevičius et al. 2008). Therefore, it is hypothesized that the western part of the Baltic region and the Baltic Sea area are affected by the same geodynamic mechanism as the Fennoscandian dome.

It is established that the Fennoscandian doming preceded the Quaternary glaciation and it is thus obvious that the Baltic Sea area was exposed to an extensional regime before Quaternary time. Moreover, comparison of data of GPS sites around the Baltic Sea (BIFROST) carried some authors to the conclusion of the existence of a dextral strike-slip fault with a relative velocity of about 1.5±0.5 mm/year along a N–S direction in the middle of the Baltic Sea.

Taking all evidences together, we conclude that although glacial erosional processes undoubtedly contributed to the shape and depth of the Baltic Sea depression, it formed primarily as a tectonic depression before glaciation.

Acknowledgements The study was supported by the Lithuanian Science and Study Foundation (V–05/2009). We are thankful for suggestions and critical comments from Ricardo Olea (United States Geological Survey), Heiko Hünecke (University of Greifswald) and Werner Stackebrandt (Geological Survey of Brandenburg) which helped to improve the manuscript substantially.

References

Abramovitz T, Thybo H, MONA LIZA Working Group (1998) Seismic structure across the Caledonian deformation front along MONA LIZA profile 1 in the southeastern North Sea. Tectonophysics 288:153–176

Åhall K-I, Connely JN, Brewer TS (2000) Episodic rapakivi magmatism due to distal orogenesis? Correlation of 1.69–1.50 Ga orogenic and inboard, orogenic events in the Baltic Shield. Geology 28:823–826

Alm E, Sundblad K (2002) Fluorite–calcite–galena-bearing fractures in the counties of Kalmar and Blekinge, Sweden. SKB R-02-42, Svensk Kärnbränslehantering AB, 116pp

Alm E, Sundblad K, Huhma H (2005) Sm–Nd isotope determinations of low-temperature fluorite–calcite–galena mineralization in the margins of the Fennoscandian Shield. Report of activities carried out during 2004. SKB Rapport R-05-66, 58pp

Artyushkov EV, Lindstrom M, Popov LE (2000) Relative sea-level changes in Baltoscandia in the Cambrian and early Ordovician: the predominance of tectonic factors and the absence of large scale eustatic fluctuations. Tectonophysics 320:375–407

Bandlowa T (1998) Erdgasführung im Karbon-Perm-Trias-Komplex der Mitteleuropäischen Senke. Geologisches Jahrbuch, A 151:3–65

Beier H, Katzung G (2001) The deformation history of the Rügen Caledonides (NE-Germany) – implications from the structural inventory of the Rügen 5 borehole. Neues Jahrbuch für Geologie und Paläontologie, Abhandlungen 222(1–2):269–300

Beier H, Katzung G, Maletz J, Schallreuter R (2001) Vorpommern und angrenzende Ostsee (Rügen Kaledoniden). In: Hoth K, Leonhardt D (eds) Stratigraphie von Deutschland II – Ordovizium, Kambrium, Vendium, Riphäikum, Teil III, Stuttgart, 186pp

Benek R, Kramer W, McCann T, Scheck M, Negendank JFW, Korich D, Huebscher H-D, Bayer U (1996) Permo-Carboniferous magmatism of the northeast German basin. Tectonophysics 266(1–4):379–404

Bingen B, Mansfeld J, Sigmond EMO, Stein H (2002) Baltica-Laurentia link during the Mesoproterozoic: 1.27 Ga development of continental basins in the Sweconorwegian orogen, southern Norway. Canadian Journal of Earth Sciences 39:1425–1440

Birkis A, Kanev S (1991) Diabase intrusions in Baltic Sea. Bulletin of the Geological Survey of Estonia, Tallin

Bogdanova S, Gorbatchev R, Grad M, Guterch A, Janik T, Kozlovskaja E, Motuza G, Skridlaite G, Starostenko V, Taran L, EUROBRIDGE and POLONAISE Working Groups (2006) EUROBRIDGE: new insight into the geodynamic evolution of the East European Craton. In: Gee DG, Stephenson RA (eds) European lithosphere dynamics. Geological Society Memoirs no. 32, pp 599–625

Brangulis AP, Kanev SV, Margulis LS, Pomerantseva RA (1993) Geology and hydrocarbon prospects of the Palaeozoic in the Baltic region. In: Parker JR (ed) Petroleum geology of NW Europe. Proceedings of the 4th conference, Geological Society, London, March–April 1992

Brangulis A, Kanevs S (2002) Latvijas Tektonika – Riga, 50pp

Brochwicz-Lewinski W, Pozaryski W, Tomczyk H (1984) Sinistral strike-slip movements in central Europe in the Paleozoic. Publications of the Institute of Geophysics – Polish Academy of Sciences A-13(160):3–13

Buchardt B, Lewan D (1990) Reflectance of vitrinite-like macerals as a thermal maturity indicator for Cambrian–Ordovician alum shale, southern Scandinavia. AAPG Bulletin 74:394–406

Buchardt B, Nielsen AT, Schovsbo NH (1997) Alun skiferen i Skandinavia. Geologisk Tidsskrift 3:1–30

Cocks LR, Fortey RA (1982) Faunal evidence for ocean separations in the Palaeozoic of Britain. Journal of the Geological Society, London 150:465–478

Depciuch T, Lis J, Sylwestrzak H (1975) Wiek izotopowy K–Ar skal podloza krystalicznego polnocno – wschodniej Polski. Kwartalnik Geologiczny 19(4):759–777

Dobrova H, Kolly E, Schmitz U (2003) E&P ventures in the eastern–central Europe. Transformation states after 1989 – a review of expectations and results. Oil Gas European Magazine, 4/2003

Domżalski J, Górecki W, Mazurek A, Myoeko A, Strzetelski W, Szamałek K (2004) The prospects for petroleum exploration in the eastern sector of southern Baltic as revealed by sea bottom geochemical survey correlated with seismic data. Przegląd Geologiczny 52(8/2): 792–799

Doornenbal H, Stevenson A (eds) (2010) Petroleum geological atlas of the southern Permian Basin Area. EAGE, 342pp

Drake H (2008) Proterozoic to Quaternary events of fracture mineralisation and oxidation in SE Sweden. Doctoral thesis A120, University of Gothenburg, 21pp

Drake H, Tullborg E-L, Page L (2009) Distinguished multiple events of fracture mineralisation related to far-field orogenic effects in Paleoproterozoic crystalline rocks, Simpevarp area, SE Sweden. Lithos 110(1–4):37–49

Ershov A, Šliaupa S (2000) Lithosphere strength variations in the Baltic basin. Geophysical Journal 22(4):86

Flodén T, Bjerkéus M, Tuuling I, Eriksson M (2001) A Silurian reefal succession in the Gotland area, Baltic Sea. GFF 123:137–152

Franke D, Illers K (1994) The Caledonian deformational fabric of the Ordovician of Rügen island (NE-Germany). Zeitschrift für Geologische Wissenschaften 22:359–370

Freimanis A, Margulis LS, Brangulis A, Kanev S, Pomerantseva R (1993) Geology and hydrocarbon prospects of Latvia. Oil & Gas Journal 6(12):71–74

Frost RTC, Fitch FJ, Miller JA (1981) The age and nature of the crystalline basement of the North Sea basin. In: Illing LV, Hobson GD (eds) Petroleum geology of North West Europe. Institute of Petroleum, London, pp 43–57

Gibbard PL (1988) The history of the great north-west European rivers during the past three million years. Philosophical Transactions of the Royal Society of London, B318:559–602

Giese U, Dallmeyer RD, Kramm RD, Mingram B (1995) First isotope investigations (U/Pb, Ar/Ar) of Early Palaeozoic rock samples from Rugen boreholes. Nachrichten Deutsche Geologische Gesellschaft 54:71–72

Giese U, Katzung G, Walter R, Weber J (1997) The Caledonian deformation of the Brabant Massif and the Early Palaeozoic in northeast Germany: compared. Geological Magazine 134(5): 637–652

Gorbatchev R (1962) The Precambrian sandstone of the Gotska Sandön boring core. Bulletin of the Geological Institution, Uppsala 39(2):1–30

Grigelis (ed) (1991) Geology and geomorphology of the Baltic Sea. Leningrad "Nedra", 419pp

Haapala I, Rämö T (1992) Tectonic setting and origin of the Proterozoic rapakivi granites of southeastern Fennoscandia. Transactions of the Royal Society of Edinburgh. Earth Sciences 83:165–171

Hansen K (1995) Fennoscandian border zone: thermal and tectonic history of a tuffaceous sandstone and granite from fission track analysis, Bornholm, Denmark. Tectonophysics 244(1–3):153–160

Harff J, Bobertz B, Granitzki K, Lemke W, Wehner K (2004) Sand and gravel deposits in the South-Western Baltic Sea, their utilization and sustainable development. Zeitschrift für Angewandte Geologie, Sonderheft 2:111–124

Hoffmann N, Franke D (1997) The Avalonia–Baltica suture in NE Germany – new constraints and alternative interpretations. Zeitschrift für Geologische Wissenschaften 25:3–14

Hoffmann N, Joedicke H, Gerling P (2001) The distribution of Pre-Westphalian source rocks in the North German Basin – evidence from magnetotelluric and geochemical data. Netherlands Journal of Geosciences 80:71–84

Hoffmann N, Lindert W, Weyer D (2006) Unterkarbon – Südwestrand der Osteuropäischen Plattform (Inseln Rügen, Hiddensee, Usedom; Mecklenburg-Vorpommern). In: Amler M, Stoppel D (eds) Stratigraphie von Deutschland VI – Unterkarbon (Mississippium). Schriftenreihe der Deutschen Gesellschaft für Geowissenschaften 41:480–493

Hoth K, Rusbült J, Zagora K, Beer H, Hartmann O (1993) Die tiefen Bohrungen im Zentralabschnitt der Mitteleuropäischen Senke – Dokumentation für den Zeitabschnitt 1962–1990. Schriftenreihe für Geowissenschaften, Heft 2:145

Hoth K, Leonhardt D (2001) Stratigraphie von Deutschland II – Ordovizium, Kambrium, Vendium, Riphäikum. Courier Forschungsinstitut Senckenberg, Frankfurt, vol 235, 186pp

Hoth K, Lindert W, Hoth P, Weyer D (2005) Das Oberkarbon des Nordrandes der Mitteleuropäischen Senke im Bereich Vorpommern, Rügen, Pommersche Bucht 254:355–386

Hoth P (1997) Fazies und Diagenese von Präperm-Sedimenten der Geotraverse Harz-Rügen. Schriftenreihe für Geowissenschaften, Heft 4, 140pp

Ismail-Zadeh IJ (1998) The Devonian to Permian subsidence mechanisms in basins of the East-European platform. Journal of Geodynamics 26(1):69–83

Jankauskas T, Lendzion K (1994) Biostratigraphic correlation of Lower and Middle Cambrian sections in the Baltic Syneclise and adjacent areas. Przegląd Geologiczny 5:365–370

Jarosinski M (1994) Stress field break-out measurements in Poland. Pomiery kierunow wspoczesnych naprężen skorupy ziemskiej w Polsce na podstawie analizy breakouts. Pr Geol 142(12):996–1003 (in Polish)

Jaworowski K, Sikorska M (2003) Composition and provenance of clastic material in the Ediacaran – Lowermost Cambrian from northern Poland: geotectonic implications. Polish Geological Institute Special Papers 8:1–59

Johansson S, Sundius N, Andwestergärd AH (1943) Beskrivning till kartbladet Lidköping. Sveriges Geologiska ündersökn.Aal 82:197

Jørgart T (2000) The basement geology of Bornholm. An excursion guide. Paper for The field conference "TransBaltic Precambrian Correlations", the Visby programme/Swedish Institute and EUROBRIDGE/ESF. Bornholm-Blekinge, 17–30 July 2000

Kaduniene E (1978) Distribution of the organic matter in the deposits of Caledonian geotectonic stage in the Baltic syneclise. Achievements and perspectives of the geological researches of the Lithuanian SSR, Vilnius, pp 135–138

Kaduniene E, Lapinskas P, Laskovas J, Laskova L, Sakalauskas K, Suveizdis P (1978) Occurrence of the oil in the sedimentary formations (geogenerations) of the Lower and Middle Palaeozoic on the western margin of the East European platform. Sedimentary formations and occurrence of oil and gas, Moscow, pp 248–249 (in Russian)

Kanev S, Margulis L, Bojesen-Koefoed JA, Weil WA, Merta H, Zdanaviciute O (1994) Oil and hydrocarbon source rocks of the Baltic Syneclise. Oil and Gas Journal 92(8): 69–73

Kanev S, Kushik E, Peregudov Y (2000) A large Ordovician carbonate buildup offshore Latvia. Latvijas Geologijas Vestis Journal 8/00

Kanev S, Peregudov Y (2000) A large carbonate buildup offshore Latvia. Latvija Geologija Viestis 8:10–14

Katzung G, Obst K (1997) The sandstone dyke swarm of Vang, Bornholm (Denmark). Bulletin of the Geological Society of Denmark 44:161–171

Katzung G (2001) The Caledonides at the southern margin of the East European Craton. Neues Jahrbuch für Geologie und Paläontologie, Abhand- lungen 222:3–53

Kershaw S (1990) Stromatoporoid palaeobiology and taphonomy in a Silurian biostrome on Gotland, Sweden. Palaeontology 33:681–705

Kharin G, Emelyanov EM, Zagorodnich YA (2004) Paleogene mineral resources of the SE Baltic Sea and Sambian Peninsula. Zeitschrift für Angewandte Geologie, Sonderheft 2:63–72

Kirsimae K, Gorokhov I, Kallaste T, Kiiply T, Kikas R (2002) Illite–smectite diagenesis of Ordovician K-bentonites of the Baltic basin: implications for basin development. Proceedings of the 5th Baltic stratigraphical conference, Vilnius, pp 84–85

Korja A, Heikkinena P, Aarob S (2001) Crustal structure of the northern Baltic Sea palaeorift. Tectonophysics 331(4):341–358

Kosmowska-Ceranowicz B (2004) Quaternary amber-bearing deposits on the Polish Coast. Zeitschrift für Angewandte Geologie, Sonderheft 2:73–86

Kramarska R, Maslowska M, Uscinowicz S, Zachowicz J (2004) Review of marine sand and gravel resources in the polish exclusive economic zone of the Baltic Sea. Zeitschrift für Angewandte Geologie, Sonderheft 2:125–133

Krishna VG, Meissner R, Thomas SA (2007) Unusual sub-Moho events near Bornholm, Baltic Sea: modelling of Babel seismic wide-angle data and a tectonic interpretation. Geophysical Journal International 125(1):193–198

Krzywiec P, Kramarska R, Zientara P (2003) Strike-slip tectonics within the SW Baltic Sea and its relationship to the inversion of the Mid-Polish trough—evidence from high-resolution seismic data. Tectonophysics 373:93–105

Lapinskas P (2000) Structure and petroleum potential of the Silurian in Lithuania. Institute of Geology, Vilnius, 203pp

Laskovas E (2000) The sedimentation environments of the Ordovician basin in the SW margin of the East European platform and lithogenesis of deposits, Vilnius, 314p

Laskovas J (1994) The composition and reservoirs of the Upper Ordovician Porkuni oil-bearing stage in Lithuanian aquatory of the Baltic Sea. Problems of entrails research, utilisation and protection in Lithuania, Vilnius, pp 63–64

Laskovas J, Jacyna J (1998) Oil potential prospectivity of the Ordovician carbonate complex in Lithuania. Prospectives of petroleum exploration in the Baltic region, Vilnius, pp 24–29

Lindström M (1967) "Funnel grabens" and Early Paleozoic tectonism in south Sweden. GSA Bulletin 78(9):1137–1154

Linnemann U, Romer RL, Pin C, Aleksandrowski P, Buła Z, Geisler T, Kachlik V, Krzemińska E, Mazur S, Motuza G, Murphy JB, Nance RD, Pisarevsky SA, Schulz B, Ulrich J, Wiszniewska J, Zaba J, Zeh A (2008) Precambrian. In: McCann T (ed) The geology of Central Europe, Volume I: Precambrian and Palaeozoic. Geological Society, London, pp 21–101

Maletz J (1997) Ordovician and Silurian strata of the G-14 well (Baltic Sea): graptolite faunas and biostratigraphy. Zeitschrift für Geologische Wissenschaften 25(1/2):29–39

Manten AA (1971) Silurian reefs of Gotland. Developments in Sedimentology 13. Elsevier, Amsterdam, 539pp

Marks L, Pavlovskaya IE (2003) The Holsteinian interglacial river network of mid-eastern Poland and western Belarus. Boreas 32:337–346

Marks L (2004) Quaternary research in Poland: selected achievements and prospects. Przeglad Geologiczny 52(8/2):800–805

Marrota AM, Sabadini R (2004) The signatures of tectonics and glacial isostatic adjustment revealed by the strain rate in Europe. Geophysical Journal International 157:865–870

McCann T, Šliaupa S, Poprawa P, Narkiewich M, Matyja H, Lipiec M, Lazauskiene J (1997) The Devonian–Carboniferous tectonic evolution of selected intracratonic and passive margin basins from central and eastern Europe: constraints from subsidence analysis. Terra Nostra 11:85–89

McCann T (1998) Lower Palaeozoic evolution of the NE German basin/Baltica borderland. Geological Magazine 135:129–142

McCann T, Krawczyk CM (2001) The trans-European fault: a critical reassessment. Geological Magazine 138(1):19–29

Meissner R, Sadowiak P, Thomas SA, BABEL Working Group (1994) East Avalonia, the third partner in the Caledonian collisions: evidence from deep seismic reflection data. Geol Rdsch 83:186–196

Meissner R, Krawczyk CH (1999) Caledonian and Proterozoic terrane accretion in the southwest Baltic Sea. Tectonophysics 314(1–3):255–267

Møller LN, Friis H (1999) Petrographic evidence for hydrocarbon migration in Lower Cambrian sandstones, Bornholm, Denmark. Bulletin of the Geological Society of Denmark 45:117–127

MONA LISA Working Group (1997) Closure of the Tornquist Sea: constraints from MONA LISA deep seismic reflection data. Geology 25(12):1071–1074

Motuza G, Kepezinskas P, Šliaupa S (1994) Diabases from the well D-1 in the Baltic Sea. Geologija 16:16–20

Muir Wood R (1993) A review of the seismotectonics of Sweden. Swedish Nuclear Fuel and Waste Management Co., Stockholm, Technical Report, 93-13, 225pp

Obst K (2000) Permo-Carboniferous dyke magmatism on the Danish Island Bornholm. Neues Jahrbuch für Geologische und Paläontologisch Abhandlungen 218:243–266

Obst K, Böhnke A, Katzung G, Maletz J (2002) Pb–Pb zircon dating of tuff horizons in the Cyrtograptus shale (Wenlock, Silurian) of Bornholm, Denmark. Bulletin of the Geological Society of Denmark 49:1–8

Obst K, Hammer J, Katzung G, Korich D (2004) The Mesoproterozoic basement in the southern Baltic Sea: insights from the G 14-1 offshore borehole. International Journal of Earth Science 93:1–12

Olsson R (2002) A conceivable explanation of the Fennoscandian seismicity. Journal of Geodynamics 33(3):345–351

Pan M, Sjöberg LE (1999) Estimating present-day postglacial rebound and horizontal movements in Fennoscandia by repeated GPS Campaigns in 1993 and 1997. Geophysical Research Letters 26(6):771–774

Paskevičius J (1997) The geology of the Baltic Republics. Geological Survey of Lithuania, Vilnius, 388p

Piske J, Neumann E (1990) Zur Ausbildung der Ablagerungen des Altpaläozoikums im Gebiet der südwestlichen Ostsee. Zeitschrift für Angewandte Geologie 36:342–346

Piske J, Neumann E (1993) Tektonische Gliederung des prävariszischen Untergrundes in der südwestlichen Ostsee. Geologisches Jahrbuch (Reihe A) 131:361–388

Piske J, Rasch H-J, Neumann E, Zagora K (1994) Geologischer Bau und Entwicklung des Präperms der Insel Rügen und des angrenzenden Seegebietes. Zeitschrift für Geologische Wissenschaften 22:211–226

Poprawa P, Šliaupa S, Stephenson RA, Lazauskiene J (1999) Late Ediacaran–Early Palaeozoic tectonic evolution of the Baltic basin: regional implications from subsidence analysis. Tectonophysics 314:219–239

Puura V, Vacher R, Klein V, et al. (1983) Crystalline basement of Estonia. Nauka, Moscow, 208pp (in Russian)

Puura V, Flodén T (2000) Rapakivi-related basement structures in the Baltic Sea area – a regional approach. GFF 122(3):257–272

Radeckas B, Lukosevicius V (2000) Klaipeda geothermal demonstration project. Proceedings of the word geothermal congress 2000, Kyushu-Tohoku, Japan, May 28–June 10, pp 3547–3550

Scherneck H-G, Johansson JM, Vermeer M, Davis JL, Milne GA, Mitrovica JX (2001) BIFROST project: 3-D crustal deformation rates derived from GPS confirm postglacial rebound in Fennoscandia. Earth Planets Space 53:703–708

Scherneck H-G, Haas R, Arvidsson R, Bergstrand S, Johansson J, Milne GA (2003) Contemporary strain rates in Fennoscandia. EGS–AGU–EUG Joint Assembly, Abstracts from the meeting held in Nice, France, 6–11 April 2003, Abstract #6743

Schlueter H-U, Best G, Juergens U, Binot F (1997) Interpretation reflexionsseimischer Profile zwischen baltischer Kontinentalplatte und Kaledonischem Becken in der suedlichen Ostsee – erste Ergebnisse. Zeitschrift der Deutschen Geologischen Gesellschaft Band 148(1):1–32

Scholz H, Frieling D, Obst K (2009) Funnel structures and clastic dykes in Cambrian sandstones of southern Sweden – indications for tensional tectonics and seismic events in a shallow marine environment. Neues Jahrbuch für Geologie und Palaontologie, Abhandlungen 251:355–380

Schwab G, Karabanov A, Aizberg R, Garbar D, Kockel F, Ludwig AO, Lukke-Andersen H, Ostaficzuk S, Palienko V, Sim L, Sliaupa A, Sokolowski I, Stackebrandt W (1997) The neogeodynamics of Central and Western regions of Europe. Geological Journal (Kiev, Ukraine) 3–4:32–41

Sivhed U, Erlström M, Bojesen-Koefbed JA, Löfgren A (2004) Upper Ordovician carbonate mounds on Gotland, central Baltic Sea: distribution, composition and reservoir characteristics. Journal of Petroleum Geology 27(2):115–140

Šliaupa A, Gelumbauskaite Z, Straume J, Šliaupa S (1995a) Methods of neotectonic investigation of middle part of Baltic Sea. Technika Poszukiwan Geologicznych, Krakow, pp 59–61

Šliaupa A, Gelumbauskaite Z, Straume J, Šliaupa S (1995b) Neotectonic structure of eastern part of Baltic Sea and adjacent land area. Technika Poszukiwan Geologicznych, Krakow, pp 63–65

Sliaupa S (1999) Far-field stress transmission indications in Early Palaeozoic structural evolution of the Baltic basin. Romanian Journal of Tectonics and Regional Geology 77:59

Šliaupa S (2002) Origin and geodynamic evolution of the Baltic Cambrian basin. Geologija 37: 31–43

Šliaupa S, Hoth P, Piske J, Laskova L, Bleschert K-H (2002a) Burial history and maturation of Palaeozoic sediments between Rügen and Lithuania – conclusions for hydrocarbon exploration. Proceedings of the 7th marine geological conference “Baltic-7” abstracts, Kaliningrad, pp 119–120

Šliaupa S, Katinas V, Vosylius G, Šliaupienė R, Vėjelytė I (2002b) Reconstruction of palaeostress of Telsiai fault in west Lithuania. Geologija 38:12–23

Šliaupa S, Motuza G, Timermann M, Korabliova L (2002c) Age and distribution of the diabase intrusions of the Baltic Sea. Proceedings of the 7th marine geological conference “Baltic-7” abstracts, Kaliningrad, p 121

Šliaupa S (2004) Geodynamic evolution of the Baltic sedimentary basin. Synopsis of Dr. Habil dissertation, Geological Institute, Vilnius, 45pp

Šliaupa S, Laškovas E, Lazauskienė J, Laškova L, Sidorov V (2004) The petroleum system of the Lithuanian offshore region. Zeitschrift für Angewandte Geologie, Sonderheft 2:41–59

Šliaupa S, Fokin P, Lazauskienė J, Stephenson R (2006) The Vendian–Early Palaeozoic sedimentary basins of the East European craton. In: Gee DG, Stephenson RA (eds) European lithosphere dynamics. Geological Society Memoir no. 32, pp 449–462

Šliaupa S, Shogenova A, Shogenov K, Sliaupiene R, Zabele A, Vaher R (2008) Industrial carbon dioxide emissions and potential geological sinks in the Baltic States. Oil Shale 25(4):409–429

Sokman K, Kattai V, Vaher R, Systra YJ (2008) Influence of tectonic dislocations on oil shale mining in the Estonia deposit. Oil Shale 25(2):175–187

Suominen V (1991) The chronostratigraphy of southwestern Finland with special reference to Postjotnian and Subjotnian diabases. Geological Survey of Finland Bulletin 356:100

Suveizdis PI, Katinas V (1990) Atlas of the lithologic-paleogeographical maps of the Soviet Baltic and adjacent areas: Permian – Neogene. Leningradskaia kartograficheskaia fabrika VSEGEI, Vilnius, 45p

Torsvik TH, Smethurst MA, Meert JG, Van der Voo R, McKerrow WS, Brasier MD, Sturt BA, Walderhaug HJ (1996) Continental break-up and collision in the Neoproterozoic and Palaeozoic – a tale of Baltica and Laurentia. Earth Science Reviews 40:229–258

Tschernoster R, Kramm U, Giese U, Glodny J (1997) The evolution of the Baltica–Gondwana suture along the TESZ during Lower Palaeozoic times – implications from detritus analysis an isotope studies. Terra Nostra 11:148–152

Thybo H (2000) Crustal structure and tectonic evolution of the Tornquist Fan region as revealed by geophysical methods. Bulletin of the Geological Society of Denmark 46:145–160

Tuuling I, Flodén T (2001) The structure and relief of the bedrock sequence in the Gotland-Hiiumaa area, northern Baltic Sea. GFF 123(Part 1):35–49

Vecoli M, Samuelsson J (2001a) Quantitative evaluation of microplankton palaeobiogeography in the Ordovician–Early Silurian of the northern trans European suture zone: implications for the timing of the Avalonia–Baltica collision. Review of Palaeobotany and Palynology 115:43–68

Vecoli M, Samuelsson J (2001b) Reworked acritarchs as provenance indicators in the Lower Palaeozoic of Denmark. C. R. Acad Sci Paris, Sciences de la Terre et des planètes 332:465–471

Vejbæk OW (1985) Seismic stratigraphy and tectonics of sedimentary basins around Bornholm. Geological Survey of Denmark, Series A8, 30pp

Voipio A (ed) (1981) The Baltic Sea. Elsevier Oceanography Series. Elsevier, Amsterdam, 418pp

Wikman H (1986) Precambrian basement and Permocarboniferous diabases Den prekambriska berggrunden och de permo-karboniska diabaserna. In: Sivhed U, Wikman H (eds) Beskrivning till berggrundskartan Helsingborg SV. Sveriges geologiska undersökning. Serie Af 149, Uppsala (in Danish)

Wilson MR (1982) Magma types and the tectonic evolution of the Swedish Proterozoic. Geologische Rundschau 71(1):120–129

Zakarevičius A, Parseliunas E, Šliaupa S, Stanionis A, Stephenson R (2008) Horizontal deformations of the earth's crust in the Baltic region from GPS data. Proceedings of the 7th international conference "Environmental engineering": selected papers, 3. Technika, Vilnius, pp 1503–1507

Zdanaviciute O, Sakalauskas K (eds) (2001) Petroleum geology of Lithuania and southeastern Baltic. GI Publications, Vilnius, 204pp

Ziegler PA (1990) Geological Atlas of western and central Europe. Shell Internationale Petroleum Maatschappij. Geological Society, The Hague, 239pp

Chapter 3
Glacial Erosion/Sedimentation of the Baltic Region and the Effect on the Postglacial Uplift

Aleksey Amantov, Willy Fjeldskaar, and Lawrence Cathles

Abstract Plio-Pleistocene erosion and sedimentation significantly impact postglacial uplift. We estimate in the last glacial cycle sedimentation could produce up to 155 m of subsidence and erosion 32 m of uplift. To show this we determine the changes in surface load caused by glacial and postglacial erosion and sedimentation over 1,000 year time intervals (coarser intervals before 50,000 years) utilizing a largely automated interpretation of regional geological and geomorphological observations that is constrained by plausible bounds on the rate of erosion of various lithologies and the known general pattern and behavior of glacial ice (ice boundaries over time, the dendritic pattern of ice movement, geometry of fast-flowing ice streams, plausible changes in frozen-bed conditions, etc.). Mass balance between erosion and deposition is enforced at all times. The analysis is regional and obliged to agree with all known geological constraints. Although the focus is on the last glacial cycle, all previous cycles are considered. The analysis suggests that the first glaciations probably shaped the major overdeepened troughs, although it is possible that the deepening was distributed evenly over all the cycles. Younger glaciations mainly removed sediments left by their predecessors, decreasing the thickness of the Quaternary succession and only locally incising and changing the dip of the bedrock surface. Over the last glacial cycle, ~20–90 m of sediments (and locally more) was removed in the zones of most active erosion.

Keywords Pleistocene · Glaciation · Erosion · Sedimentation · Isostasy · Fennoscandia · Baltic · Ice-stream · Uplift · Bedrock

3.1 Introduction

The role of glacial erosion and sedimention in creating the modern landscape of the Baltic Sea basin has been appreciated for a long time. Glacial and fluvioglacial erosion had a decisive influence in shaping the Baltic–White Sea lowland on the

A. Amantov (✉)
VSEGEI, St. Petersburg, Russia
e-mail: 4448470@mail.wplus.net

J. Harff et al. (eds.), *The Baltic Sea Basin*, Central and Eastern European Development Studies (CEEDES), DOI 10.1007/978-3-642-17220-5_3,

margin of the Fennoscandian (Baltic) shield (Amantov 1992, 1995), for example. The Atlantic margin shows increased Late Pliocene and Pleistocene deposition rates (Riis and Fjeldskaar 1992). Worldwide, erosion of exposed unconsolidated clastic shelf sediments and consequent isostatic compensation has resulted in large masses of sediment being offloaded from the continental shelves onto deep-sea fans and abyssal plains by turbidity currents (Hay 1994). But opinions differ on the intensity of the glacial erosion. To some the glaciations were crucial in changing the landscape. These authors emphasize that glacial erosion can be much greater than fluvial erosion (White 1972, 1988, Bell and Laine 1982, 1985, Clague 1986, Braun 1989, Harbor and Warburton 1992, 1993, Clayton 1996, Hallet et al. 1996, Montgomery 2002, James 2003). In mountain glaciers, the erosion rate is greatest near the equilibrium line altitude (ELA) where ice accumulation changes to melting. Here the glaciers are often considered "buzz saws" (Brozovic et al. 1997, Meigs and Sauber 2000, Montgomery et al. 2001, Mitchell and Montgomery 2006). Glaciers increase topographic relief through a combination of focused erosion in valleys and the regional isostatic rebound the incision induces (Small and Anderson 1998), and this, in turn, increases erosion.

Other researches point to the moderate transformation of preglacial landscapes and find evidence for low rates of glacial erosion and little difference between fluvial and glacial erosion rates (e.g., Gravenor 1975, Sugden 1976, 1978, Lindström 1988, Hebdon et al. 1997). In this view, the glaciers merely polished the northern shields, and the erosion they caused (although sometimes highly variable; Lidmar-Bergström 1997) was generally less than tens of meters in magnitude.

Glacial erosion is intriguing because on the local scale it is highly irregular but at the large scale it is regular. We would like to understand it quantitatively. For example, we would like to assess whether most of the sediment redistribution took place during the first or last glacial cycles. The shifts of sediment loading could be enough to affect subsurface temperature and cause isostatic tilting. But local spatial variations, the wealth of data that must be assembled and integrated, and the large spatial scales involved make analysis difficult.

Our approach is to apply computer software adept at creating and manipulating surfaces to infer glacial erosion and sedimentation rates across Europe in a locally detailed but regionally coherent way. At every instant of time and across the Quaternary, our method requires that erosion and sedimentation are balanced, locally and across all of Europe. Our analysis honors bounds on what erosion and sedimentation rates are reasonable, and a great many local geological constraints. The redistribution is process-driven. We develop algorithms that honor the pattern of glacial flow suggested by geological evidence and the locations of the ice margins as the glaciers grew and retreated. From this we build erosion and sedimentation modules that redistribute the sediments. We calibrate these tools to current glaciers and to the observed present-day sediment pattern, and this assures they are reasonable (but not necessarily correct). In this manner, we infer how the sediments may have been created and redistributed across Quaternary time and tentatively conclude that most of the major bedrock landscape changes were probably produced by the earliest glaciations. Even so, the erosion and sedimentation that occurred over the last glaciation were still sufficient to induce isostatic movements comparable to those caused by glacial loading. The analysis suggests interesting phenomenological connections.

The purpose of this chapter is to present and describe the results of this new kind of analysis. We motivate the methods used with a fairly extensive review of geological observations that provide insight into the processes that are occurring and the parameters that appear to be important and then give a fairly brief discussion of the methods themselves. The results of our analysis are then given in reasonable detail.

Although physically based, our methods remain largely empirical algorithms. As such they are difficult to fully describe in any reasonable space, and, in any case, their validity rests largely in their predictions. We will describe the methods in full detail in subsequent publications. Here we hope mainly to show that the sediment redistributions that result from the analysis we describe are reasonable and interesting.

3.2 Glacial Erosion and Sedimentation

Rates of glacial erosion have been estimated between 0.1 and 10 mm/year. Erosion of glaciated catchments of fjords of southern Alaska exceeds 10 mm/year (Hallet et al. 1996). Long-term averaged exhumation rates are 3 mm/year in the Chugach–St. Elias Range, Alaska, where the maximum rates of denudation are thought to be limited by rates of tectonic uplift (Spotila et al. 2004). In Western Nunavut, 6–20 m of rock is believed to have been eroded during the last glacial cycle (Kaszycki and Shilts 1980).

In Northeast Scotland, where both glacial and preglacial landforms exist in close proximity, the expansion of ice sheets across the area in the middle Quaternary was associated with a sharp increase in the rates of erosion (>0.13 mm/year), but the last (Late Devensian) ice sheet in the area was less erosive (<0.095 mm/year) (Glasser and Hall 1997).

On the assumption that the erosional work was achieved over 10,000–20,000 years of each 100,000 year glacial cycle, the rates of surface lowering during glaciations in Britain fall in the range of 11–23 mm/year (Clayton 1996). The average erosion rate over the full glacial cycle is comparable to the 1 mm/year figure regarded as "typical" by Boulton et al. (1991) for glacial erosion of resistant rocks. Erosion in Britain is several times faster for weaker rocks flooring major lowlands and much of the shelf (Clayton 1996). Average erosional rates of the sedimentary bedrock of the Barents Sea during the last ~2.5 million years were estimated to be between 0.1 and 1.1 mm/year (Faleide et al. 1996) or 0.2–0.6 mm/year (Solheim et al. 1996).

Assuming glacial erosion for 1 million years over the past 2.57 million years, the average rate of glacial erosion in the Sognefjord drainage basin, western Norway, was ~0.4 mm/year by subtracting the present topography from a reconstructed preglacial (paleic) surface. Considering the selective nature of glacial erosion along ice streams, the annual erosion rate for ice streams is most likely 2 ± 0.5 mm/year (Nesje et al. 1992, Nesje and Sulebak 1994). Comparable mean rates were reported for Isfjorden region of Svalbard over the glacial cycle (Elverhøi et al. 1995). In the Antarctic, average erosion rates are considered to be three times higher beneath ice-stream tributaries which are underlain by deep subglacial troughs (0.6 mm/year) than beneath ice-stream trunks (0.2 mm/year) (Bougamont and Tulaczyk 2003).

Remnants of marine Quaternary sedimentary sequences indicating high glacial/fluvioglacial erosion rates in the Baltic–White Sea lowland are a cornerstone in the validation of our erosion–accumulation modeling. The sporadic distribution of the youngest marine interglacial strata in the form of remnants around the Gulf of Finland attests to strong erosion during even the last glaciations (which was the smallest of all in this area). Sediments from previous glacial cycles are very rare in the axial part of the lowland, but in rare isolated locations remnants can be 40 m thick (Malakhovsky and Amantov 1991). Surface reconstruction suggests that, in addition to thick marine strata, at least 10–20 m of the underlying sediments were removed. In ice-stream zones like Lake Ladoga, remnants of older Quaternary beds survived the deep erosion in protected positions, indicating more than 60–70 m of erosion during the last glaciation. This suggests that in zones of active erosion the present cover belongs nearly entirely to the last glaciations (moraine cover and late-postglacial sediments).

Where soft sedimentary sequences have been glaciated, buried channels and hollows of several generations suggest local linear erosion of 100–200 m (Amantov 1992). Rarely, older channels can be seen to be entrenched at shallower depths than the younger channels that cross them (Amantov 1992). The nature of these channels depends on whether they are radial or parallel to the glacial front, affected by sedimentary infilling, deformed by ice or melting waters, etc. Lithology and structure are also dominant factors. The channels may often have nearly parallel orientation, sometimes with arc shape that roughly coincides with the boundaries of retreated glacial tongues. The depth of the channels decreases in the direction to the modern shield, so that the base of the channels tends to parallel the relief of the basal platform sediments, mostly entrenching only into the weathered top of the resistant crystalline basement. A similar rapid decrease in channel depth occurs toward resistant lithologies such as carbonate rocks forming prominent scarp-like features on the bedrock topography.

The depth of both glacial and fluvioglacial erosions strongly depends on lithology (Amantov 1992, 1995). In the Baltic–White Sea, depressions in bedrock topography suggesting maximum long-term erosion are evident in zones with pliable sediments. Here, glacial erosion rates inferred geologically and in our analysis reached 2 mm/year, with local short-term rates up to 8 mm/year. The thickness of erodable sediments should be taken into account. The rate of erosion should decrease if a pliable sedimentary unit is completely removed in an area with exhumation of resistant surface. The Landsort Deep illustrates how removal of a thickness of pliable sediment can create a strongly overdeepened ice-proximal negative form.

Another key factor controlling glacial erosion is the ice sliding velocity at the ice-bed contact (Humphrey and Raymond 1994). Our analysis addresses ice-stream zonation and accounts for the radial increase in ice velocity outward from the central zone of ice accumulation to the abrasion maximum near the ice terminus. In our models, abrasion increases up to a point and then possibly decreases due to overwhelming of the abrasive content that reduces basal sliding velocity by increased basal friction. Ice boundaries thus control concentric changes of the erosion rates. This broad pattern provides a regional context for further refinements. The main refinement in the erosion pattern is caused by fast-flowing ice streams near the glacial margins that have an enhanced capacity for erosion (Fig. 3.2). Ice streams move at high velocities under low driving stresses in a basal zone environment mostly because their base is lubricated (see discussion in Marshall et al. 1996, Tulaczyk et al. 2000, Stokes and Clark 2001, Kamb 2001, Bougamont and Tulaczyk, 2003, Hall and Glasser 2003).

The bedrock surface determines the topography of ice streams with profound erosion capacity. The location of bedrock troughs or elongated lowlands was initially controlled or at least influenced by the bedrock topography. Domination of elongated landforms of smaller scale is taken to indicate zones of faster ice flow. The elongation ratio of bedrock forms and megascale lineations are known to be a useful proxy for ice velocity (Anderson and Shipp 2001). Long subglacial bedforms (length:width ratios 10:1) are indicative of fast ice flows (Stokes and Clark 2002). The geological–geomorphological impact of ice streams cannot be underestimated, since modern ones literally control ice discharge. For example, over 90% of ice discharging from the West Antarctic Ice Sheet into the Ross Ice Shelf (Joughin and Tulaczyk 2002) is carried by ice streams.

Bedrock surface forms may also suggest very low ice velocities and erosion. Areas with abundant distribution of relict landforms indicate slow ice. Special grid filtering to emphasize outliers with a relevant search window can identify these areas best. In zones adjacent to weathered bedrock, possible frozen-bed conditions and weak erosional capacity can be manually input as constraints.

3.3 Methods

The preceding section suggests what must be taken into account by any glacial erosion analysis. Not discussed thus far is that the mass of glacial sediments must equal the mass of material eroded. We compile a huge quantity of published seismic and sedimentological data and make our best estimates of the total sediments deposited across the Quaternary. This provides a bound on the total Quaternary erosion. We use denudation surfaces to estimate the erosion directly. This stage of analysis is essentially an automation of traditional methods (Riis and Jensen 1992). Surfaces capture stages of Tertiary uplift and erosion (Amantov 2007). The surfaces connect isolated summit outcrops, patches of exhumed peneplains, and etchplains. Surfaces emerging from under sedimentary cover can be extrapolated and correlated with onshore saprolites and (or) remnants of cover so that the grids measure missing volumes. The surfaces can also illustrate past geological conditions. Regional compilations always have some uncertainties due to gaps in confirmation of seismic stratigraphy, different estimations of drainage provinces, and possible input of eroded material from irrelevant provinces to depocenters, etc. We estimate that the amount of material eroded in the Baltic region during Plio-Pleistocene is about 90,000 km^3 (Amantov 1995).

We estimate both the erosion and the sedimentation over specific intervals of time and require that erosion equal sediment accumulation over these periods. We use 1 ka timesteps over the last 50,000 years and longer 5–10 ka steps for early Weichselian stadials and across earlier glacial cycles. For the early Weichselian we assume two interstadials with ice-free conditions following Lundqvist (1992) and Lokrantz and Sohlenius (2006) as corrected by Svendsen et al. (2004) and Sarala's (2005) interpretation for southern Finnish Lapland.

The margins of the glacial ice sheets are the starting point for our analysis. The ice margins at the LGM are shown in Fig. 3.1. We use a number of tools to simulate erosion under the ice cover and sedimentation under, at the margin, and outside the ice. The tools are computation modules that allow useful geological analysis procedures to be repeated easily. The procedures might include sampling of gridded data (sub-ice lithology, for example), connecting sparse kinds of data with a best fitting surface, inferring velocity fields from the distance to an ice depocenter and topography, subtracting surfaces to determine the material removed, visualizing the geology in particular ways, etc.

Erosion under the ice sheets is estimated using such tools by requiring that the long-term glacial erosion rates are reasonable and the pattern of erosion conforms to the concentric (radial) changes in erosion observed as well as the "spider's web" pattern of grounded ice sheet's movement (ice streams). This is illustrated in Fig. 3.2. Figure 3.2a shows the erosion and sedimentation that might occur if only the ice velocity were considered. The concentric pattern results from the low ice velocity under the center of the continental glaciers and the more rapid basal ice velocity near the margins. Figure 3.2b shows how this simple pattern is modified if the likely effect of the spider-web pattern of ice flow with the enhanced erosional capacity of ice streams is taken into account. Figure 3.2c illustrates the effect of different erodability of sedimentary bedrock and basement lithologies. The glacial erosion module contains adjustable parameters that allow the sediment redistribution it "predicts" to be controlled by only concentric factors (Fig. 3.2a) or increasingly influenced by lithology, dendritic ice flow, and ice streams (Fig. 3.2b, c).

An important control is sub-ice topography which helps control the spider-web flow with "topographic" ice streams and erosion paths. The drainage pattern is determined from the paths raindrops runoff would follow in reaching the sea. Submodules refine the interpretation. For example, overdeepening of bedrock surface is imposed where slopes are >10–20° and oriented such as to cause rotational ice flow that could locally maximize basal sliding (Lewis 1949). The modules create grids that capture erosion surfaces over time and show the exhumation of sedimentary rocks, the boundaries of the sedimentary cover, expansion of the crystalline shield exposure, etc.

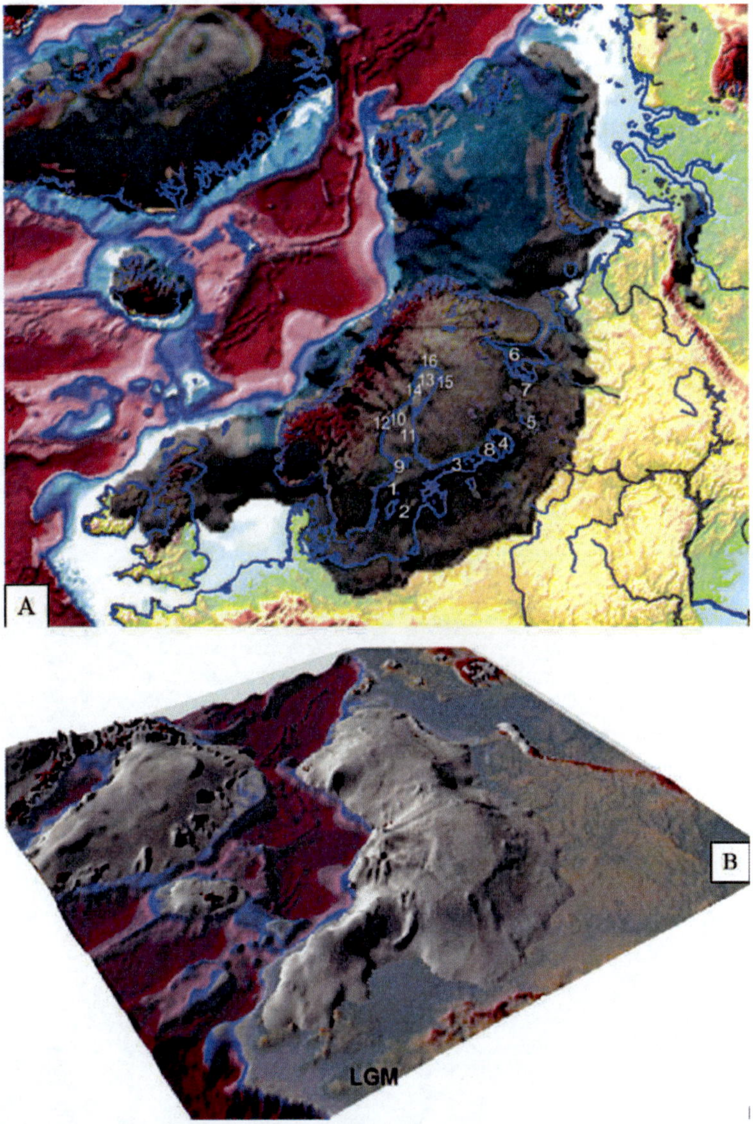

Fig. 3.1 Sample output of ice thickness module (LGM): **a** Orthographic view, ice is shown in half-transparent mode. Present-day shorelines are shown in *blue color*. Figures illustrate localities mentioned in the text. Central Baltic Proper: *1* – Landsort Deep, *2* – Gotland Deep; *3* – Gulf of Finland; *4* – Lake Ladoga; *5* – Lake Onega; *6* – White Sea; *7* – Vetryany Poyas; *8* – Karelic peninsula; *9* – Åland Deep; Bothnian Sea: *10* – Hörnösand Deep; *11* – Aranda Rift; *12* – Sundsvall; *13* – Bothnian Bay; *14* – Shellefteo; *15* – Ouly; *16* – Nordkalott. **b** 3D view

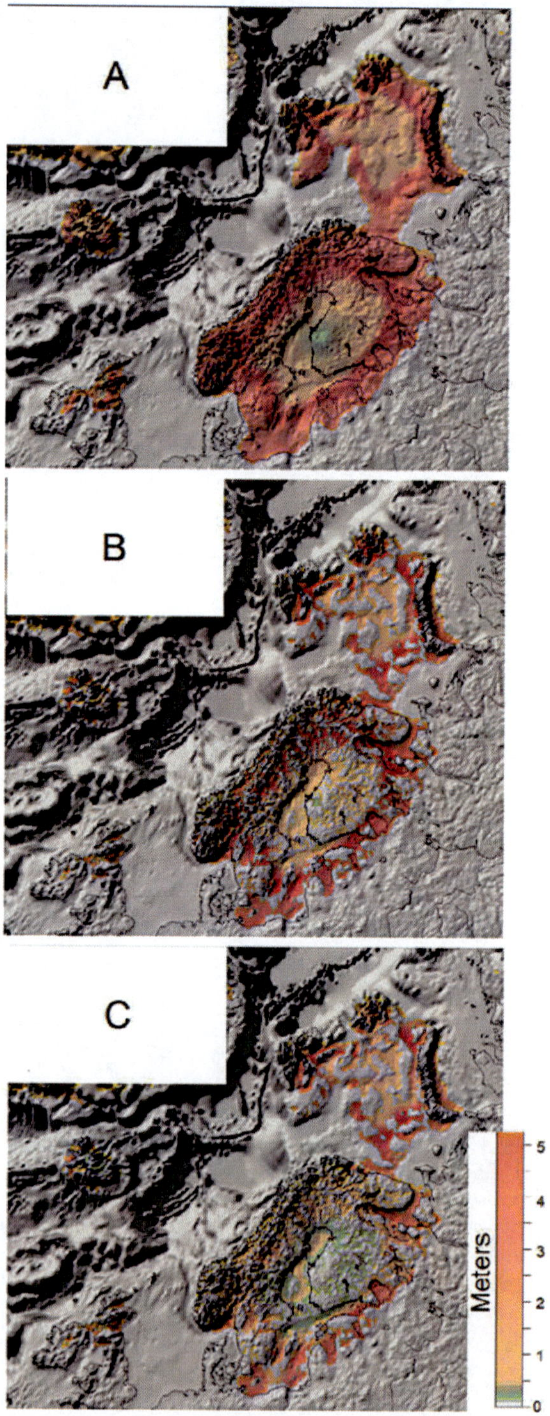

Fig. 3.2 Sample output of glacial erosion module: routine transformation from general simplified concentric pattern (**a**) to ice-stream flow (**b**) and further account of different lithology and erosion resistance (**c**)

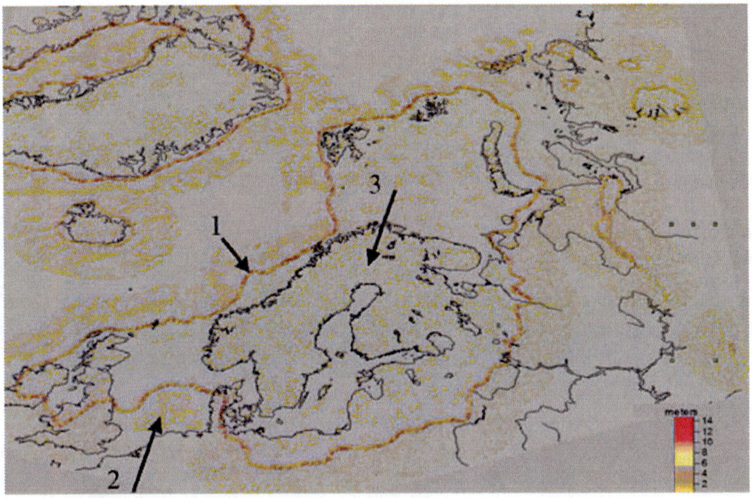

Fig. 3.3 Sample output of glacial sedimentation module: *1* – end-moraine ridges; *2* – peripheral sediments; *3* – products of subglacial sedimentation

The complement to erosion is sediment accumulation. This can occur under or outside the ice. We distinguish the glacial, interglacial, and postglacial sediment deposition patterns. For example, a *glacial sedimentation module* simulates the formation of end-moraine ridges, subglacial (i.e., drumlins, flutes, eskers), and peripheral deposition that deals with meltwater redeposition of a material (Fig. 3.3). The thickness and width of end-moraine ridges are approximated as random within defined bounds that are controlled by presumed sediment supply to the ice margin, the mobility of the ice front, and the ice-stream pattern. Time-dependent grids specify the lithology at the base of ice. An *interglacial–postglacial deposition module* forecasts thickness of debris accumulated between and after Weichselian erosion episodes, when additional automated time-slice modules estimate the possible thickness of interglacial sediments. This module was calibrated against Holocene offshore and onshore data. Figure 3.3 shows the pattern of sediment accumulation. Any sediment pockets could be individually resolved, depending on input grid resolution.

The results of this kind of analysis are illustrated in a corridor that runs from the northern Gulf of Bothnia across Finland and the Gulf of Finland into the Russian Plain in Fig. 3.4. The northern shore of the Gulf of Finland marks the approximate northern border of the Baltic–White Sea lowland – the area that contains the most erodable material that was particularly affected by glacial erosion. Figure 3.4a shows how we believe this transect looked at the end of the Tertiary before it was affected by any continental glaciations. Figure 3.4b shows the erosion that was accomplished in the first glacial cycles, showing the situation just after one

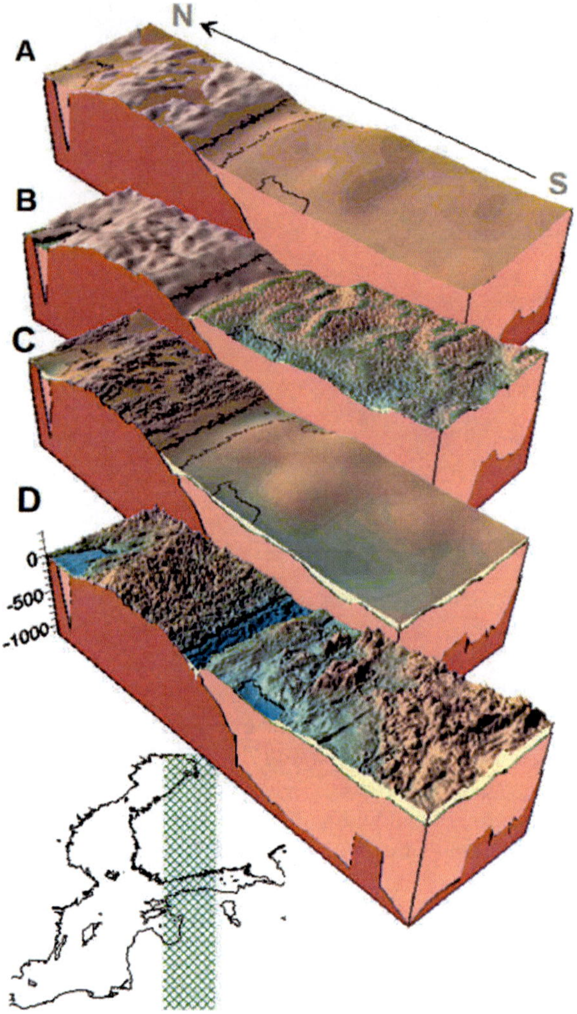

Fig. 3.4 3D slices showing simplified principal development of the Baltic region: **a** preglacial stage; **b** bedrock erosion of first major glacial impact; **c** later preglacial stage, illustrating input of interglacial sedimentation; **d** present. Color indications: basement – *dark red*, cover – *pink*, Quaternary cover – *yellow*

of these early cycles. The surface is rough and sculpted, and significant material has been completely removed from the Baltic–White Sea lowland area particularly. Figure 3.4c shows the situation at the end of the interglacial period that followed Fig. 3.4b. A layer of interglacial sediment has been laid over the rough lowland surface, and as a result the surface is smoothed in numerous areas. Finally, Fig. 3.4d shows the present situation that reflects intensive glacial erosion of mostly glacial and interglacial sediments, with resulting cumulative effects of all the Quaternary glacial cycles.

3.4 Results and Discussion

The analysis of sediment redistribution in the Baltic area using the methods sketched above is clearly a complex task, and to a considerable degree the validity of the methods used must be assessed by how geologically reasonable the product is perceived to be. The results of our analysis are described below, first in the areas peripheral to the Baltic–White Sea lowland and then in the lowland itself.

The Baltic–White Sea lowland (lowland for short) exhibits a regional first-order bedform that was to a significant degree created by strong diverse and multiphase glacial and fluvioglacial erosion of pliable sedimentary rocks covering the slope of the Baltic (Fennoscandian) shield (Amantov 1992, 1995). Its approximate shape is shown in Fig. 3.5. The lowland can be traced from the Baltic Proper with Gulf of Finland to the lakes Ladoga and Onega and then to the White Sea. It seems to have formed during rapid erosional lowering of wide Tertiary plains with the progressive removal of saprolites and less stable sediments. Basement features such as the sub-Cambrian or sub-Upper Vendian peneplains were exhumed around the present margin of the shield (Amantov 1995).

A narrow zone of eroded post-Late Vendian cover and Riphean–Jotnian formations is traced by the deepest indentations of the bedrock surface where hundreds of meters of unmetamorphosed rocks had been eroded. The deepest parts of the lowland usually coincide with areas where the sedimentary cover is truncated or, more rarely, with zones where the most friable sedimentary units thin. Major aquifers are often involved in the detachment of huge masses of cover. For example, the Gdov aquifer at the base of Late Vendian cover probably facilitated stripping along zones of disintegrated sandstone cementation and in areas with deep dissection by tunnel valleys or glacial hollows.

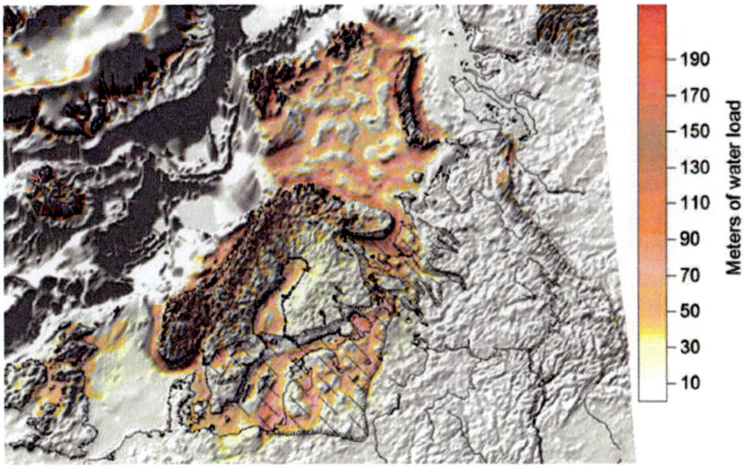

Fig. 3.5 Weischselian net erosion recalculated in meters of water load using averaged rock density. Baltic–White Sea erosion lowland is marked by slash pattern with half-ticks outline

The sub-Late Vendian or sub-Cambrian (basement) peneplain is an important reference for erosion up to where it is covered by sediments that are not penetrated by glacial erosion (Amantov 1992). It can be reconstructed on the adjacent shield area by interpolating preserved fragments of exposed peneplain under sedimentary cover and connecting summit highs of the Archean–Early Proterozoic crystalline bedrock. The slope of the stripped and slightly dissected peneplain presents one flange of the lowland onshore slope of parts of Sweden and Finland (Lidmar-Bergström 1992). The exhumed surface usually has shallow dip relative to its cover and is not significantly affected by faults. This contrasts strongly with the rugged (30–80 m) bedrock topography on crystalline rocks on the periphery of glaciated areas. There the topography is controlled by crystalline rock properties and structural peculiarities, like faults and fracture zones.

Bedrock depressions are often localized in the more erodable formations. They are separated by minor asymmetric basement highs whose steeper side faces the shield. The shallowest and narrowest lowland of this kind occurs in the Lake Onega–Vetryany Poyas region and on the Karelic peninsula, where elevation of the bedrock roof is 20–40 mbsl (meters below sea level), and locally overdeepened troughs with erosible lithology or structures can extend to 300–400 mbsl. On average the basement lies 50–200 mbsl. The depth to basement gradually increases from 55 mbsl in the eastern part of the Gulf of Finland to more than 200 mbsl in the Central Baltic Proper, where a paleo ice stream could have been located in the Gotland Deep.

Smoothed onshore scarps and slopes often bound the lowland. They are considered to be products of selective glacial denudation. Scarps and slopes usually face the shield, and their outline roughly corresponds to the outline of the ice at a particular glacial stage. The bounding is not distinct in areas where bedrock seems to be worn down and smoothed by ice streams. The more resistant strata control the plains between scarps and slopes. Evidence of their origin is provided by escarpments that can be traced in overdeepened locations like the >100 m scarps in the zone of maximal erosion in northern Lake Ladoga. These scarps can be seen to be localized by Riphean gabbroic sills that penetrated the sedimentary sequence (Amantov 1992, Amantov et al. 1995).

A significant percentage of the glacial erosion occurs in negative structures filled by more erodable, usually Riphean–Jotnian sequences. Examples are the Landsort, Åland, and Lake Ladoga deeps where the bedrock surface has been overdeepened by hundreds of meters (Amantov 1995). Rare thick Quaternary remnants in protected positions indicate the decisive role of first glaciations in excavating the troughs (Amantov 1992) and suggest that the subsequent glaciations mainly removed Quaternary sediments left by their predecessors and affected the bedrock surface in only a minor fashion. As a result, in zones of deepest erosion the tills now present belong mainly to the last glaciation and are overlain by the late-postglacial mantle. The last glacier could have removed 20–50 m of rocks of different density over wide zones of maximum erosion. Locally, in narrow overdeepenings, hollows, and glacial valleys, this figure increases to 70–90 m.

Tills, fluvioglacial, and other relevant sediments cover the peripheral accumulation belt. Late Pleistocene–Holocene uncompacted sediments starting with varved clays cover the entire Baltic Sea floor. Local sediment transfer is common. Numerous local overdeepenings of the late-glacial surface rapidly accumulated sediments immediately after glacial retreat. As a result, a thick (tens of meters over wide areas) veneer of sediments has been deposited. The impact on sediment loading is less than might be expected, however, because these postglacial clays are relatively uncompacted and have low density.

The central area of the Gulf of Bothnia is not a zone of low erosion as often expected from its position in the central zone of maximum ice accumulation. Lower erosion is expected in the northeastern part of the Bothnian Bay with adjacent onshore areas. In the western offshore part of the Bothnian Sea area, along the Swedish coast, erosion could be of the same order as in the Baltic Proper–White Sea lowland. According to our time-slice computations, much of the erosion occurred in the Early-Mid Weichselian, when ice marginal fluctuations occurred around the modern northwestern coast of the Bothnian Sea. Erosion was also strong during the piedmont phase and during glacial retreat.

In some ways the Swedish coast of the Bothnian Sea is comparable with the northwestern rim of the Lake Ladoga basin and other areas where unmetamorphosed Riphean–Jotnian sediments subcrop (Amantov 1992, Amantov et al. 1995, 1996). These areas are zones of deep glacial erosion. A west–east seismic profile from the Sundsvaal area (Axberg 1980, fig. 18) is similar to profiles crossing the coastal slope of the Northern Ladoga basin. Trends of bedrock topography are similar; even comparable scarps are observed in the Bothnian Sea in connection with resistant dolerite intrusions, but here they rise 25–30 m above the bedrock surface instead of 60–100 m in Lake Ladoga. The most intensive erosion resulted in the large negative relief form that is today the Hörnösand Deep. The present day bottom depths here range between 150 and 260 m, and the bedrock topography is slightly deeper. Such values are similar to those in the deepest northern part of the Lake Ladoga basin and to deeps in the steep coastal zone. Climate, the duration of ice activity, and ice streams can account for lateral changes of bedrock overdeepening along the contact zone between the crystalline rocks and the Riphean–Jotnian sediments.

Topographic similarities are connected with geological ones. In the authors' opinion, the deeps have been formed by the selective erosion of Riphean–Jotnian sandstones that fill tectonic basins. In the Hörnösand area erosion-resistant Ordovician limestones, which armor the bedrock surface to the south, are absent, thinning out at the southern slope. Here the erosion of the Cambrian–Ordovician platform produced a composite 100–120 m scarp-like slope that faces to the north. It is similar in form, magnitude, and lithology to the Cambrian–Ordovician slopes and escarpments in the Baltic Proper. The axial part of the Hörnösand Deep has sublongitudinal strike, joining to the south with a 100 m deep buried tunnel valley called the Aranda Rift (Winterhalter 1972). At some time-slices, an ice stream is expected southeast of the Hörnösand Deep and further toward the south, following an elongated bedform with depths between 110 and 160 m below sea level.

Locally, especially around the northern slope, Quaternary deposits up to 100–150 m thick occur in the Hörnösand Deep. A distinct acoustic appearance (Axberg 1980) may indicate that they belong to different glacial and interglacial events and are remnants that survived in shadow position, as in the Lake Ladoga basin. If so, this supports a scenario of excavating and shaping the major bedforms by the first glaciations, with subsequent oscillation between sedimentation and "cleaning out" of outlet zones.

In spite of the presence of pliable presumably Lower Cambrian–Upper Vendian sedimentary formations, the erosion of the northern part of the Baltic, the Bothnian Bay, is mild to moderate. This is supported by both bedrock topography and the pattern of glacial accumulation. The bedrock surface is rarely deeper than 130–170 m below sea level, and somewhat steeper along the Swedish coast. The southwestern area seems to have eroded, especially to the southeast from Shellefteo, but the erosion is mild compared to the Hörnösand Deep area. In the northeastern part of the Bothnian Bay, the bedrock surface on the Riphean sediments is 50–120 m lower than the surrounding crystalline rocks in the coastal area of Finland southeast of Ouly, where the Riphean–Upper Vendian Muhos formation comprises the half-graben appendage of the major Riphean–Jotnian basin. The bedrock is overlain by 50–80 m Quaternary sediments (Tynni and Uutela 1984). Thus, the bedrock surface is relatively deepened, as is noted everywhere where Riphean sediments are surrounded by harder crystalline basement, but to a lesser degree. The Quaternary sequence suggests moderate erosion prior to Weichselian. The total Quaternary section attains great thickness, frequently 50–100 m, and pre-Weichselian till deposits may be expected in the southwestern parts of the basin (Floden et al. 1979). Survival of the thick and complicated Quaternary succession in the subbottom area is in agreement with onshore observations. In the continuation of the major lowland in the Nordkalott area, north of the Bothnian Bay, the cover is comprised of two or more till beds, Eemian sediments are common, and even Saalian and older deposits occur (Hamborg et al. 1986). The survival of these remnants is compatible with their location in a complicated zone of ice divide, where the flow of ice was slow and its direction complicated with a dominance to the southeast (Hirvas and Tynni 1976). The first glaciations significantly transformed the region, by strongly eroding pliable terrigenous formations, which, together with the consequent isostatic adjustment, separated central sedimentary outliers of the Bothnian Sea and Bothnian Bay from each other and from the sedimentary domain of the Baltic Sea Proper.

3.4.1 Sediment accumulation and mass balance

Sediments accumulated around the areas glaciated as shown in Fig. 3.3. This sediment mass must, of course, match the mass of material eroded, taking into account the redistribution of material over a wider area. Our analysis assures that this is the case, not only today but also for every increment of erosion that occurred over the entire Quaternary (e.g., all the glacial cycles, including the last). There is great uncertainty regarding how the erosion is distributed between the glacial cycles, but we make an attempt to apportion it in a reasonable fashion.

The history of ice sheet development is relatively well known for the last 25,000 years, but uncertainties of earlier ice sheet oscillations are an important factor in possible model variations. In spite of uncertainties of imprecise estimations of erosion and accumulation rates in different areas, time-scale reconstructions provide a good picture of the regional loading–unloading cycles.

Our modeling also assumes variability of erosion and accumulation rates in time and space. For the Baltic area the largest short-term erosion rates are expected in the case when sediments are incorporated into the ice or pushed in front of glacier on initial advances in areas where intensive interglacial accumulation created unconsolidated extra-soft beds. Even on relatively hard argillaceous Late Vendian clays in eastern Gulf of Finland, the zone of very intensive dislocation has a normal thickness of 4–8 m with common thick slabs in overlaying tills. Increasing erosion rates during rapid deglaciation are related to highly dynamic ice masses, fluvioglacial processes, and outbursts from glacial lakes.

Modeling shows that the deepest sedimentary bedrock erosion is related to soft formations in depressions, i.e., graben-like structures, proximal to ice-flow contact zones between rocks of highly contrasting erodability. In such cases, hard abrasive material comes to the ice–bedrock contact zone, while the contact zone usually forms a relatively steep slope, possibly providing rotational flow with a sufficient supply of fresh firm abrasive. Major aquifers may serve as an additional factor in bedrock removed by other mechanisms.

Knowledge of bedrock topography and measure of its overdeepening and lowering from reconstructions of older geomorphic facets serve as important validation steps in the determination of the erosion magnitude. However, it cannot be used to judge erosion rates. In many cases, glacially shaped topography, with elongated basins alternating with conformal ridges and riegels, produced multiple local depocenters for interglacial (postglacial) sedimentation, partly being inherited. For such basins, erosion and later sedimentation could be compared with a pendulum, when the nature "masked its wounds." Local zones of deep erosion appeared as zones of thick sedimentation with maximum rates immediately after glacial retreat, but roles reversed again on the next advance. The initial glaciation(s) affected the bedrock, but later ones eroded glacial and interglacial deposits over wide areas (Fig. 3.4). We think that further development of joint simulation of different processes could be productive, in spite of the multiple assumptions and imperfection of our current simple tools.

The load redistribution caused by erosion and sedimentation is compensated isostatically. To assess this, sediment thicknesses must be converted to mass. Where the conditions are submarine, the load is the equivalent buoyant load. Whether on land or submerged, the porosity of the sediments must be taken into account. The algorithms we have designed take these matters into account. Figure 3.6 (right) shows the isostatic uplift and subsidence pattern that would be produced by the sediment redistribution that we estimate occurred over the last glacial cycle. Full isostatic equilibrium is assumed and the load is filtered by a lithosphere of flexural rigidity 10^{23} Nm (effective elastic thickness of 20 km) (Fjeldskaar et al. 2000). The modeling shows that the isostatic response to erosion and sediment loading (Fig. 3.6 (right)) is significant compared to that caused by deglaciation and sea level changes. The rise of sea level caused ca. 40 m of hydro isostatic subsidence under

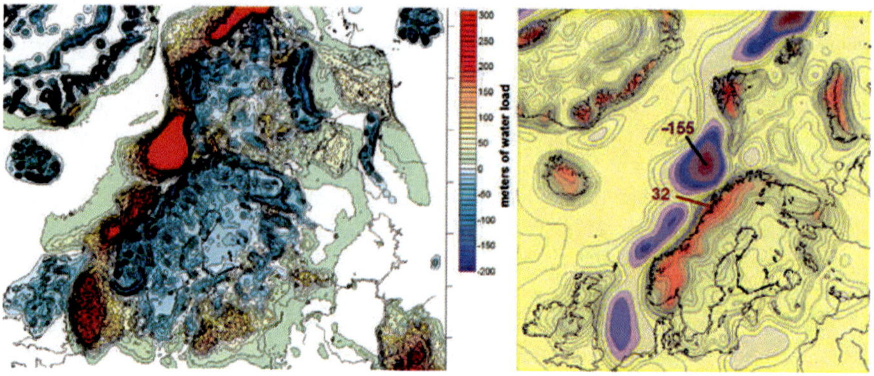

Fig. 3.6 Weischselian erosion and accumulation redistribution load recalculated in meters of water load using averaged rock density (*left*) and its total isostatic uplift–subsidence effect in meters (*right*)

the ocean load. Figure 3.6 (right) shows that the sediment loading of marine areas can cause isostatic subsidence five times greater than the loading by glacial meltwater. The uplift associated with erosion is smaller (<10% of the glacial isostasy) for the Baltic area, but for some areas of coastal Norway it could be a significant part of the observed postglacial uplift.

3.5 Conclusions

Although it is possible that bedrock erosion was evenly distributed between all the glacial cycles, most of the modification of the bedrock surface and shaping major overdeepened troughs was probably accomplished by the first glaciations of the Quaternary. Younger glaciations mainly removed sediments deposited by previous glacial cycles, reducing the thickness of the Quaternary succession and locally incising the bedrock surface. The isostatic effect of the glacial erosion and sedimentation significantly impact the total postglacial rebound. Subsidence in submarine areas adjacent to the continental glaciers can be much larger than that induced by the postglacial rise in sea level. Isostatic uplift caused by erosion is minor for the Baltic area, but could be a significant part of the observed postglacial uplift in coastal areas of Norway.

Acknowledgments This study was funded by the Research Council of Norway and StatoilHydro, as part of the project “Ice Ages – Subsidence, Uplift and Tilting of Traps – The Influence on Petroleum Systems” (Petromaks 169291; “GlaciPet”). The authors wish to express their gratitude for the support. We also want to thank William W. Hay for constructive comments on an earlier version of this chapter. We are grateful to Patrick Madison and Golden Software team for the development of Surfer, MapViewer and other products that were involved in investigations. Thanks also to M. Amantova who digitized numerous data used in the research.

References

Amantov A (1992) Geological structure of the sedimentary cover of the basins in Northwestern Russia (in Russian). Sedimentary cover of the glacial shelves of the Northwestern seas of Russia, St. Petersburg, pp 25–47

Amantov A (1995) Plio-Pleistocene erosion of Fennoscandia and its implication for the Baltic Area. PPIG CXLIX, Warszawa. Proceedings of the 3rd Marine Geological Conference "The Baltic", pp 47–56

Amantov A, Sederberg P, Hagenfeldt S (1995) The Mesoproterozoic to Lower Palaeozoic sedimentary bedrock sequence in the Northern Baltic Proper, Aland Sea, Gulf of Finland and Lake Ladoga, Prace Panstwow. Institute of Geology, CXLIX, Warsawa, pp 19–25

Amantov A, Laitakari I, Poroshin Ye (1996) Jotnian and Postjotnian: sandstones and diabases in the surroundings of the Gulf of Finland. Geological Survey of Finland, Special Paper 21:99–113

Amantov A (2007) Research process automation systems with smart multidimensional analysis as a tool of geological-geomorphological solutions. Regional Geology and Metallogeny, No. 30–31, pp 85–92

Anderson J, Shipp S (2001) Evolution of the West Antarctic Ice Sheet. Antarctic Research Series 77:45–57

Axberg S (1980) Seismic stratigraphy and bedrock geology of the Bothnian sea, Northern Baltic. Stockholms Contributions in Geology XXXVI:153–213

Bell M, Laine E (1982) New evidence from beneath the Western North Atlantic for the depth of glacial erosion in Greenland and North America: reply to Andrew's comment. Quaternary Research 17:125–127

Bell M, Laine E (1985) Erosion of the laurentide region of North America by glacial and glaciofluvial processes. Quaternary Research 23:154–174

Bougamont M, Tulaczyk S (2003) Glacial erosion beneath ice streams and ice-stream tributaries: constraints on temporal and spatial distribution of erosion from numerical simulations of a West Antarctic ice stream. Boreas 32:178–190

Boulton GS, Peacock JD, Sutherland DG (1991) Quaternary. In: Craig GY (ed) Geology of Scotland, 3rd edn. Geological Society, London, pp 503–543

Braun DD (1989) Glacial and periglacial erosion of the Appalachians. Geomorphology 2:233–256

Brozovic N, Burbank DW, Meigs AJ (1997) Climatic limits on landscape development in the northwestern Himalaya. Science 276:571–574

Clague JJ (1986) The Quaternary stratigraphic record of British Columbia—evidence for episodic sedimentation and erosion controlled by glaciation. Canadian Journal of Earth Sciences 23:885–894

Clayton K (1996) Quantification of the impact of glacial erosion on the British Isles. Institute of British Geographers Transactions 21:124–140

Elverhøi A, Svendsen JI, Solheim A, Andersen ES, Milliman J, Mangerud J, Hooke RLeB (1995) Late Quaternary sediment yield from the high Arctic Svalbard area. Journal of Geology 103: 1–17

Faleide JI, Solheim A, Fiedler A, Hjelstuen BO, Andersen ES, Vanneste K (1996) Late Cenozoic evolution of the western Barents Sea-Svalbard continental margin. Global and Planetary Change 12:53–74

Fjeldskaar W, Lindholm C, Dehls JF, Fjeldskaar I (2000) Post-glacial uplift, neotectonics and seismicity in Fennoscandia. Quaternary Science Reviews 19:1413–1422

Floden T, Jacobsson R, Kumpas MG, Wadstein P, Wannes K (1979) Geophysical investigations of the western Bothnian Bay. Geologiska Föreningens i Stockholm. Förhandlingar 101:321–327

Glasser NF, Hall AM (1997) Calculating Quaternary glacial erosion rates in North East Scotland. Geomorphology 20:29–48

Gravenor CP (1975) Erosion by continental ice sheets. American Journal of Science 275:594–604

Hall AM, Glasser NF (2003) Reconstructing the basal thermal regime of an ice stream in a landscape of selective linear erosion: Glen Avon, Cairngorm Mountains, Scotland. Boreas 32:191–207

Hallet B, Hunter L, Bogen J (1996) Rates of erosion and sediment yield by glaciers: a review of field data and their implications. Global and Planetary Change 12:213–235
Hamborg M, Hirvas H, Lagerbäck R, Mäkinen K, Nenonen K, Olsen L, Rodhe L, Sutinen R, Thoresen M (1986) Map of Quaternary geology, sheet 1: Map of Quaternary stratigraphy, Scale 1: 1 000 000. Nordkalott project. Geological Survey of Finland, Geological Survey of Norway, Geological Survey of Sweden
Harbor J, Warburton J (1992) Glaciation and denudation rates. Nature 356:751
Harbor J, Warburton J (1993) Relative rates of glacial and nonglacial erosion in alpine environments. Arctic and Alpine Research 25:1–7
Hay WW (1994) Pleistocene-Holocene fluxes are not the earth's norm. In: Hay WW, Usselmann T (eds) Material fluxes on the surface of the earth. Studies in geophysics. National Academy Press, Washington, DC, pp 15–27
Hebdon NJ, Atkinson TC, Lawson TJ, Young IR (1997) Rate of glacial valley deepening during the late Quaternary in Assynt. Scotland. Earth Surface Processes and Landforms 22:307–315
Hirvas H, Tynni R (1976) Tertiary clay deposit at Savykoski, Finnish Lapland, and observations of tertiary microfossils, preliminary report (In Finnish). Geologi 28:33–40
Humphrey NF, Raymond CF (1994) Hydrology, erosion and sediment production in a surging glacier; the Variegated Glacier surge, 1982–83. Journal of Glaciology 40(136):539–552
James AL (2003) Glacial erosion and geomorphology in the northwest Sierra Nevada, CA. Geomorphology 55:283–303
Joughin I, Tulaczyk S (2002) Positive mass balance of the Ross Ice Streams, West Antarctica. Science 295:476–480
Kamb B (2001) Basal zone of the West Antarctic ice streams and its role in lubrication of their rapid motion. The West Antarctic Ice Sheet. Behavior and Environment, Antarctic Research Series 77:157–201
Kaszycki CA, Shilts WW (1980) Glacial erosion of the Canadian Shield-calculation of average depths. Atomic Energy of Canada, Technical Record, TR-106
Lewis WV (1949) Glacial movement by rotational slipping. Geografiska Annaler 31:146–158
Lidmar-Bergström K (1992) Denudation surfaces and tectonics in the southernmost part of the Baltic Shield. Precambrian Research, 64. Elsevier Science Publication, Amsterdam, pp 337–345
Lidmar-Bergström K (1997) A long-term perspective on glacial erosion. Earth Surface Processes and Landforms 22:297–306
Lindström E (1988) Are roches mountoneś mainly preglacial forms? Geografiska Annaler 70A:323–332
Lokrantz H, Sohlenius G (2006) Ice marginal fluctuations during the Weichselian glaciation in Fennoscandia, a literature review. Geological Survey of Sweden (SGU) Technical Report TR-06-36. http://www.skb.se/upload/publications/pdf/TR-06-36webb.pdf
Lundqvist J (1992) Glacial stratigraphy in Sweden. Geological Survey of Finland, Special Paper 15:43–59
Malakhovsky D, Amantov A (1991) Geologic and geomorphic anomalies in the North of Europe (in Russian). Geomorphologya 1:85–95
Marshall SJ, Clarke GKC, Dyke AS, Fisher DA (1996) Geologic and topographic controls on fast flow in the Laurentide and Cordilleran ice sheets. Journal of Geophysical Research 101(B8):17827–17839
Meigs A, Sauber J (2000) Southern Alaska as an example of the long-term consequences of mountain building under the influence of glaciers. Quaternary Science Reviews 19:1543–1562
Mitchell SG, Montgomery DR (2006) Influence of a glacial buzzsaw on the height and morphology of the Washington Cascade Range, Washington State, USA. Quaternary Research 65(1):96–107
Montgomery DR, Balco G, Willett SD (2001) Climate, tectonics, and the morphology of the Andes. Geology 29:579–582
Montgomery DR (2002) Valley formation by fluvial and glacial erosion. Geology 30:1047–1050
Nesje A, Dahl SO, Valen V, Øvstedal J (1992) Quaternary erosion in the Sognefjord drainage basin, western Norway. Geomorphology 5:511–520

Nesje A, Sulebak JR (1994) Quantification of late Cenozoic erosion and denudation in the Sognefjord drainage basin, western Norway. Norsk geografisk Tidsskrift 48:85–92
Riis F, Fjeldskaar W (1992) On the magnitude of the Late Tertiary and Quaternary erosion and its significance for the uplift of Scandinavia and the Barents Sea. Structural and tectonic modelling and its application to petroleum geology, NPF Special Publication 1, pp 163–185
Riis F, Jensen LN (1992) Introduction: measuring uplift and erosion – proposal for a terminology. Norsk Geologisk Tidsskrift 72:223–228
Sarala P (2005) Glacial morphology and dynamics with till geochemical exploration in the ribbed moraine area of Peräpohjola, Finnish Lapland. Geological Survey of Finland, Espoo, 17p
Small EE, Anderson RS (1998) Pleistocene relief production in Laramide Mountain ranges, western U.S. Geology 26:123–126
Solheim A, Riis F, Elverhøi A, Faleide JI, Jensen LN, Cloetingh S (1996) Impact of glaciations on basin evolution: data and models from the Norwegian Margin and adjacent areas. Global and Planetary Change 12:1–9
Spotila JA, Buscher JT, Meigs AJ, Reiners PW (2004) Long-term glacial erosion of active mountain belts: example of the Chugach–St. Elias Range, Alaska. Geology 32(6):501–504
Stokes CR, Clark CD (2001) Palaeo-ice streams. Quaternary Science Reviews 20:1437–1457
Stokes CR, Clark CD (2002) Are long subglacial bedforms indicative of fast ice flow? Boreas 31:239–249
Sugden DE (1976) A case against deep erosion of shields by ice sheets. Geology 4:580–582
Sugden DE (1978) Glacial erosion by the Laurentide ice sheet. Journal of Glaciology 20:367–391
Svendsen J, Alexanderson H, Astakhov V, Demidov I, Dowdeswell J, Funder S, Gataullin V, Henriksen M, Hjort C, Houmark-Nielsen M, Hubberten H, Ingolfsson O, Jakobsson M, Kjær K, Larsen E, Lokrantz H, Lunkka J-P, Lyså A, Mangerud J, Matiouchkov A, Murray A, Möller P, Niessen F, Nikolskaya O, Polyak L, Saarnisto M, Siegert C, Siegert M, Spielhagen R, Stein R (2004) Late Quaternary ice sheet history of northern Eurasia. Quaternary Science Reviews 23:1229–1271
Tulaczyk S, Kamb WB, Engelhardt HF (2000) Basal mechanics of Ice Stream B, West Antarctica. 2. Undrained plastic bed model. Journal of Geophysical Research 105:483–494
Tynni R, Uutela A (1984) Microfossils from the Precambrian Muhos formation in Western Finland. Geological Survey of Finland Bulletin 330:38
Winterhalter B (1972) On the geology of the Bothnian Sea, an epeiric sea that has undergone Pleistocene glaciation. Geological Survey of Finland Bulletin 258:1–66
White WA (1972) Deep erosion by continental ice sheets. Geological Survey of Finland Bulletin 83:1037–1056
White WA (1988) More on deep glacial erosion by continental ice sheets and their tongues of distributary ice. Quaternary Research 30:137–150

Part III
The Basin Fill as a Climate and Sea Level Record

Chapter 4
The Development of the Baltic Sea Basin During the Last 130 ka

Thomas Andrén, Svante Björck, Elinor Andrén, Daniel Conley, Lovisa Zillén, and Johanna Anjar

Abstract During the Eemian interglacial 130–115 ka BP, the hydrology of the Baltic Sea was significantly different from the Holocene. A pathway between the Baltic basin and the Barents Sea through Karelia existed during the first ca. 2.5 ka of the interglacial. Both sea surface temperature and salinity of the SW Eemian Baltic Sea were much higher, ca. 6°C and 15‰, respectively, than at present. A first early Weichselian Scandinavian ice advance is recorded in NW Finland during marine isotope stage (MIS) 4 and the first Baltic ice lobe advance into SE Denmark is dated to 55–50 ka BP. From the last glacial maximum that was reached ca. 22 ka BP, the ice sheet retreated northward with a few still-stands and readvances; however, by ca. 10 ka BP the entire basin was deglaciated. Weak inflows of saline water were registered in the southern and central Baltic Sea ca. 9.8 ka BP with full brackish marine conditions reached at ca. 8 ka BP and the maximum Holocene salinity was recorded between 6 and 4 ka BP. The present Baltic Sea is characterized by a marked halocline preventing the vertical water exchange resulting in hypoxic bottom conditions in the deeper part of the basin.

Keywords Baltic Sea · Eemian · Scandinavian ice sheet · Weichselian · Baltic Ice Lake · Yoldia Sea · Ancylus Lake · Littorina Sea · Hypoxia

4.1 Introduction

During the last decade, significant efforts have been focused on the recent development of the Baltic Sea. This has resulted in different explanatory mechanisms for its present state and different possible remedies to change its present eutrophication status (Conley et al. 2009). The increased knowledge about the array of processes influencing the Baltic has meant that it has gradually become more common to place

T. Andrén (✉)
School of Life Sciences, Södertörn University, SE-141 89 Huddinge, Sweden
e-mail: thomas.andren@sh.se

J. Harff et al. (eds.), *The Baltic Sea Basin*, Central and Eastern European Development Studies (CEEDES), DOI 10.1007/978-3-642-17220-5_4,

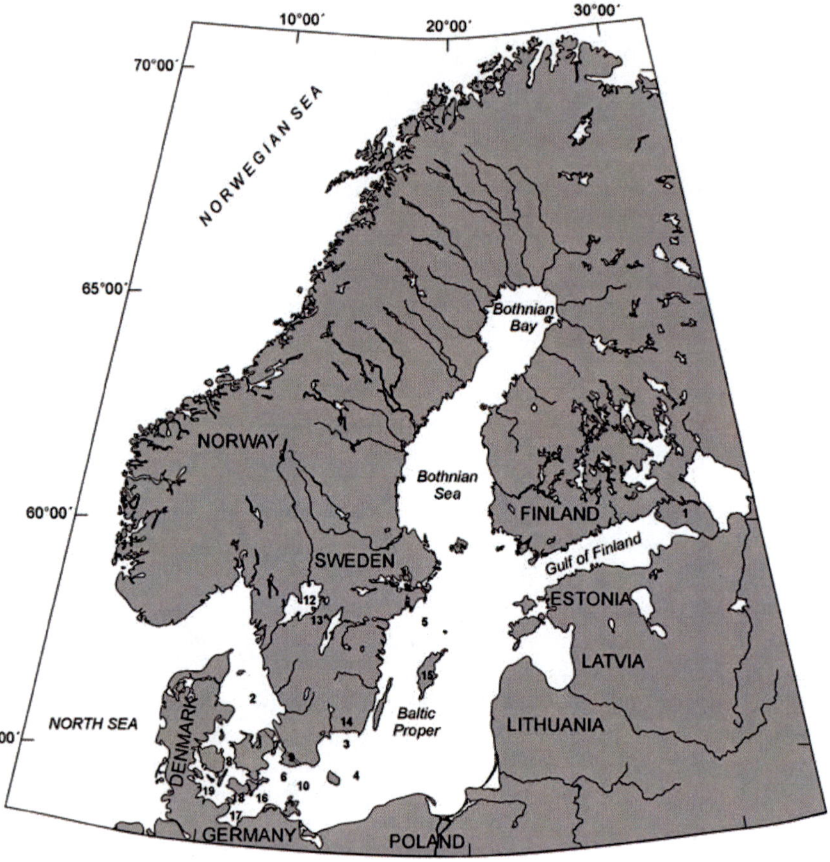

Fig. 4.1 Map of the Baltic Sea and its surroundings. The *numbers* refer to sites mentioned in the text. *1* = Karelia; *2* = Kattegat; *3* = Hanö Bay; *4* = Bornholm Deep; *5* = Landsort Deep; *6* = Kreigers Flak; *7* = Öresund Strait; *8* = Store Belt; *9* = Esrum/Alnarp bedrock valley; *10* = Arkona Basin; *11* = Lake Vättern; *12* = Lake Vänern; *13* = Mt. Billingen; *14* = Blekinge; *15* = Gotland; *16* = Darss Sill; *17* = Mecklenburg Bay; *18* = Fehmarn Belt

the Baltic Sea and its huge drainage area into a long-term perspective to gain a better understanding of the natural internal and external dynamics influencing the basin.

The Baltic Sea basin is located in a glaciation-sensitive high northern latitude, which has resulted in a very dynamic development during its young geological history (Fig. 4.1). This owes to the fact that the recurring Quaternary glaciations over northern Europe have repeatedly covered parts of or the whole basin, and that the subsequent deglaciations have resulted in largely differential uplift in the region of the Baltic Sea and its drainage area. The last interglacial/glacial cycle is a good example of the variety of processes that the basin also has undergone during previous glacial cycles, and the differences between the last and the present interglacial exemplify the variety of potential processes that can influence the basin.

The geologic deposits in the Baltic basin, as well as in the surrounding region, are thus archives of its history. If these can be retrieved and "read" by geologists, we will understand the background on which we shall base our interpretations of the most recent history on, as well as plan for the future in a continuously changing world. Therefore, we think it is appropriate to summarize some important, and for the Baltic basin history decisive, aspects of its youngest geologic history. It should be noted that this summary is from a slightly south Scandinavian perspective and from the wealth of papers published on the subject we have chosen those considered the most appropriate, i.e., with high quality of data and with a proper chronological control.

The interglacial/glacial cycles and their recurring glaciations have had different types of impacts on the Baltic basin and can be summarized into some main categories:

- Glacial and glaciofluvial erosion of the basin and its catchment, resulting in displacement of clastic sediments (ground bedrock) from the surrounding land areas to the basin floor.
- Repeated cycles of downwarping of the lithosphere, as an effect of the glacial expansion and loading, and uplift/unloading during phases of deglaciation or thinning of the ice sheet.
- Varying ice thicknesses during more or less extensive glaciations of the Scandinavian ice sheet have resulted in highly different uplift rates (high in the north and low in the south) during subsequent deglaciations.
- The combination of glacially forced global sea level changes and regional isostatic movements has resulted in changing water levels (depths) – in both time and space – of the basin and of the critical threshold areas.
- The above-mentioned processes have been the main salinity regulator for the Baltic basin, allowing more or less saline water to enter the basin through more or less broad and deep straits.
- The setting of the Baltic Sea basin at the rim of the northeastern Atlantic means that it is sensitive to changes in atmospheric and marine circulation patterns of the region. These have caused large changes in both temperature and precipitation/evaporation ratios. These have had a direct and also indirect impact on the Baltic Sea; the latter through changing river and surface run-off from the huge catchment area, four times as large as the basin itself.

4.2 History of the Baltic Sea Prior to the Last Glacial Maximum (LGM)

4.2.1 130–70 ka BP

Deposits from the Last Interglacial, the Eemian (basically corresponding to Marine Oxygen Isotope Substage (MIS) 5e) ca. 130–115 ka BP, have been described from a number of marine and terrestrial sites in the North Atlantic region, but is only partly documented in the NGRIP (North Greenland Ice Core Project members 2004) ice core. A delay between the beginning of MIS 5e and that of the European terrestrial Eemian was demonstrated for the first time by Sánchez Goñi et al. (1999) based on a land–sea correlation between the European pollen zones and the marine isotope stages (discussed by Kukla et al. 2002). High-resolution Eemian marine shelf records (here correlated with MIS 5e) from northern Europe are, however, very scarce and usually only contain fragmentary paleoenvironmental information. The same is valid for the early Weichselian stadials and interstadials (MIS 5d to 5a), which were, however, fully recovered in the, e.g., NorthGRIP ice core. Data from marine sediments in the Nordic Seas show three substantial sea surface temperature fluctuations during MIS 5e (Fronval and Jansen 1996). These results imply that the Last Interglacial at high northern latitudes was characterized by rapid changes in the polar front movement, ocean circulation, and oceanic heat fluxes. This may have resulted in noticeable temperature changes in neighboring land areas, which is different from the Holocene climate development, with only minor fluctuations on a general cooling trend.

Records from Eemian lacustrine and marine sediments (MIS 5e), presently situated on shore, show that the Eemian in the Baltic Sea Basin (BSB) began with a lacustrine phase during ca. 300 years before marine conditions were established (Kristensen and Knudsen 2006). A pathway existed between the BSB and the Barents Sea through Karelia during the first ca. 2–2.5 ka of the interglacial due to the large isostatic depression as a result of the extensive Saalian ice sheet which probably was much thicker than the Weichselian ice sheet (Fig. 4.2). It is debatable as to what degree this pathway was of importance for the general circulation in the BSB and the climate of north Europe (Funder et al. 2002). It did, however, have significant effect on oceanography during the first ca. 4 ka of the Eemian Baltic Sea, and possibly also on the surrounding terrestrial climate (Björck et al. 2000), resulting in a strong W–E temperature and salinity gradient with a winter sea surface water temperature ca. 6° higher and a salinity ca. 15‰ higher than today in the Belt Sea and western BSB. At the same time, lower salinity and colder bottom water (ca. 2.5°C) conditions prevailed in the eastern BSB. This circulation pattern with high salinities may have created strong salinity stratification and the development of a permanent halocline resulting in hypoxic bottom conditions during a great part of the Eemian. These conditions resemble in many ways the development of the Baltic Sea during the last 8,000 years and today's situation. Also, the difference between the warm and well-ventilated southwestern Eemian BSB and the cold, stagnant conditions of its easternmost parts implies that the ocean–continental climate gradient from the west to the east in N Europe was steeper than during the Holocene (Funder et al. 2002). After ca. 6 ka into the interglacial, the Eemian Baltic Sea was characterized by a falling sea level and decreased salinity seen in diatom and foraminifera records (Eiríksson et al. 2006, Kristensen and Knudsen 2006), but its

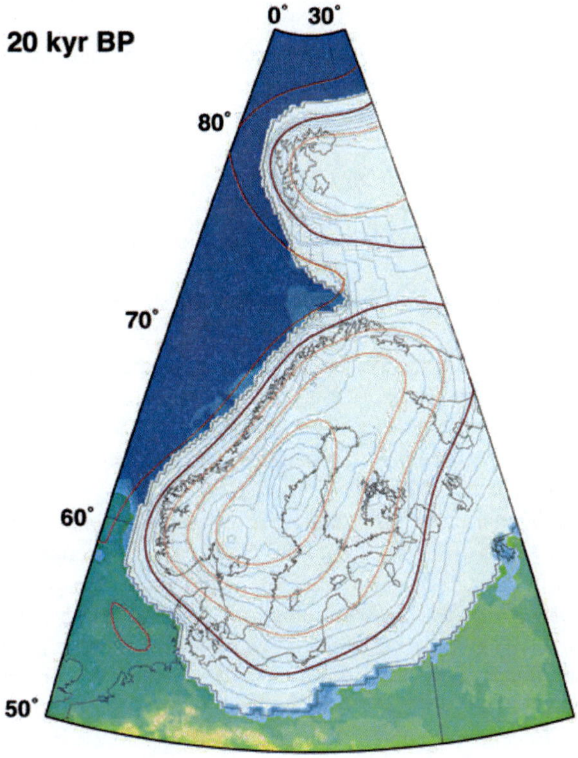

Fig. 4.2 Paleoreconstructions for the LGM at ca. 20 ka BP. The paleotopography and water depths are shown by the color coding. Ice thickness contours are 200 m. The positive relative sea-level contours are indicated in *orange*, and negative contours in *red*, with contour intervals of 150 m (Lambeck et al. 2010)

further development during the subsequent MIS 5 stadials and interstadials is largely unknown. It is indicated, however, that two early Weichselian glacial advances (MIS 5d and MIS 5b) may only have reached as far south as ca. 60.5°N and thus did not affect the central and southern BSB (Robertsson et al. 2005).

Several paleoclimatic records, both terrestrial and marine, from the north Atlantic margin (e.g., Rasmussen et al. 1997, Dickson et al. 2008, Grimm et al. 2006, Wohlfarth et al. 2008) display the same high climate variability during the last glacial (MIS 4–MIS 2) as recorded in Greenland ice cores from, e.g., GRIP and GISP 2 (Johnsen et al. 1992, Grootes et al. 1993). The Weichselian ice sheet, which covered the Baltic basin, was the largest ice sheet in Eurasia and together with the Wisconsinan ice sheet in North America contributed to this high degree of variability. It can be assumed that by advances and retreats, releases of icebergs and freshwater, and shifting sea ice conditions, these ice sheets recurrently impacted the North Atlantic thermohaline circulation and thereby also the climate of NW Europe.

The Baltic glacial history is only fragmentarily known, but it appears that a first Baltic glacial event occurred during MIS 4 as recorded in sediments from

NW Finland at ca. 64°N (Salonen et al. 2007), the first advance into northern Denmark at 65–60 ka BP blocking any Baltic outlets through Kattegat (Larsen et al. 2009), while the first Baltic ice lobe advance into southeastern Denmark is dated to ca. 55–50 ka BP (Houmark-Nielsen 2007). It is likely that freshwater lakes covered the deeper subbasins of the central and southern BSB until at least 60 ka BP when sea level was >50 m lower than today (Lambeck and Chappell 2001, Siddall et al. 2003). In the Hanö Bay, Bornholm and Landsort Deep basins were probably sediments deposited during several tens of millennia through the first half of the last glacial.

Based on detailed correlations and dating of the southwestern Baltic glacial stratigraphies, Houmark-Nielsen and Kjær (2003) and Houmark-Nielsen (2007, 2008) conclude that the SW Baltic may have experienced two major ice advances during MIS 3, at ca. 50 and 30 ka BP. The latter is being vividly discussed (Wohlfarth 2010) as well as the general asynchroneity of MIS 3 ice advances at the western margin of the Scandinavian ice sheet (SIS) (Mangerud 2004) compared to the ice margins in the south (Houmark-Nielsen et al. 2005). This less well-known period between ca. 50 and 25 ka BP with its partly incompatible records is followed by a complex glaciation in the southern BSB (Houmark-Nielsen and Kjær 2003) leading up to the LGM. Previous off-shore studies in the southern Baltic have documented the presence of marine brackish sediments, dated to MIS 3 or older, that were overridden by a glacier (Klingberg 1998) at Kriegers Flak and, e.g., two varved clay sequences – the upper one dates from the last deglaciation – separated by an organic-rich layer dated to >35 ka ^{14}C BP (bulk date) in Hanö Bay (Björck et al. 1990).

Recently, 40 cores were obtained from drillings for the planning of the Kriegers Flak wind-mill park, of which 9 indicate that complex yet incomplete stratigraphies occur in this shallow part of the BSB. The shallow Kriegers Flak area shows a surprisingly complex stratigraphy with a variety of lithologic units, gravel–sand–silt, clays of glaciolacustrine and brackish origin, interstadial lacustrine, and terrestrial organic sediments with five ^{14}C dates between 36 and 41 ka BP, sandwiched between several glacial diamicts (Fig. 4.3) (Anjar et al. 2010).

The geographic location and altitude (in relation to sea level) of the critical threshold, or "gateway," between the open ocean and the BSB are a key factor for the BSB history, as it controls if, and how much, water can flow in or out of the BSB. Presently, the two main thresholds are the Öresund Strait (–7 m) and the Store Belt (ca. –20 m).

However, during earlier stages in the history of the BSB, a bedrock threshold situated 60 m below sea level, the buried Esrum/Alnarp bedrock valley running through SW Skåne in Sweden and northernmost Sjaelland in Denmark, 120 km long and 6 km wide, has been suggested as a possible connection to the oceans (Lagerlund 1987, Andrén and Wannäs 1988). Deep corings in the 1970s of this main aquifer recovered fluvial and lacustrine sediment units with an organic carbon content that made radiocarbon dating possible. The ages presented by Miller (1977) indicate that the valley was sediment-filled during the later part of MIS 3. The valley may thus have served as the outlet route for the entire BSB until it later was filled up by sediments.

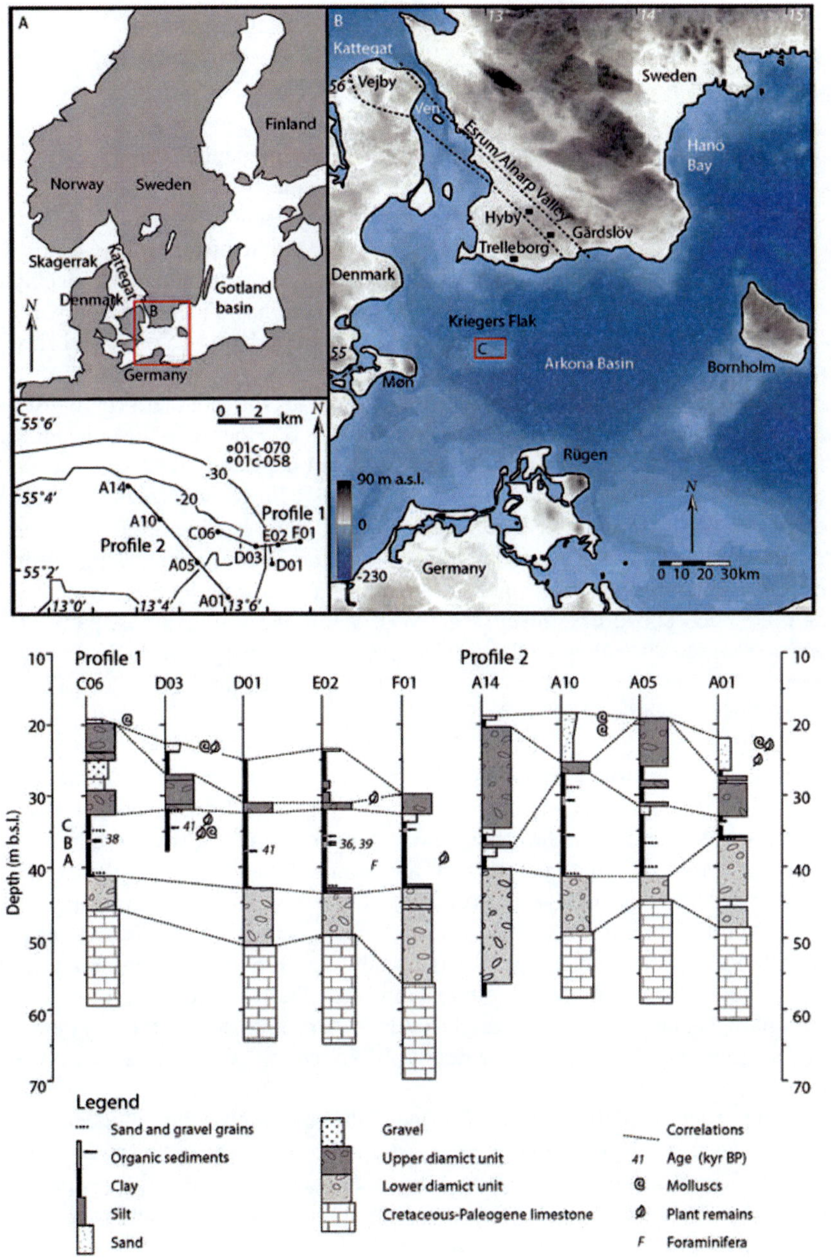

Fig. 4.3 *Upper panel*: **a** Overview map of the Baltic Sea region. **b** Localities mentioned in the text. Bathymetry from Seifert et al. (2001). **c** Locations of the cores from Kriegers Flak investigated in this study and of the cores from Klingberg (1998). *Lower panel*: Lithostratigraphic logs of the sediment cores from Kriegers Flak. Subunits **a–c** with clays, organic sediments were recorded between two diamict units. From Anjar et al. (2010)

4.3 Late and Postglacial History of the Baltic Sea

4.3.1 16,0–11,7 ka BP

While the development of the Baltic Sea during the last glacial period is only fragmentarily known, its history since the last deglaciation is better understood. It is based on studies of numerous sediment cores from different parts of the basin as well as on analyses of the mechanisms behind the geodynamic history of the Baltic Sea (Björck 1995, 2008). The latter can be evaluated from the many curves displaying the water level changes in different parts of the Baltic basin and with comparisons to the relative sea level changes seen in parts of Denmark and on the Swedish west coast.

The complexities of the post-LGM history of the southern parts of the Scandinavian ice sheet (SIS) have been summarized by Houmark-Nielsen and Kjær (2003). They conclude that a first embryo of a final deglacial lake basin, the Baltic Ice Lake (BIL), should have an age of ca. 16 ka BP. Lower valleys in Denmark as well as the Öresund area between Sweden and Denmark were possibly the main drainage pathways for the glacial melt water, and the Store Belt and Öresund straits were most likely formed as a consequence of gradual erosion by these rivers as the area rebounded above base level (sea level). While the southernmost parts of the basin were filled up with glacial deposits formed at the margin of the ice sheet, the deeper parts, such as the Arkona Basin and Bornholm Basin, later on constituted a glacial lake as the deglaciation continued.

During the initial stage of the BIL, it was most likely at level with the sea. However, as the isostatic rebound of the outlet in the Öresund threshold area between Copenhagen and Malmö – made up by glacial deposits on top of chalk bedrock – was greater than the sea level rise, the Öresund outlet river eroded its bed in pace with the emerging land. In fact, the island of Ven with its complex glacial stratigraphy (Adrielsson 1984) is a remnant of this eroded glacial landscape. When the fluvial downcutting reached the flint-rich chalk bedrock, the erosion must have ceased more or less completely. This is possibly an important turning point in the BIL development: the uplift of the threshold lifted the BIL above sea level and the updamming (ponding) of this large glacial lake started. Based on the apparent sudden changing rate of shore displacement in Blekinge this seems to have occurred at ca. 14 ka BP.

The deglaciation of the central Baltic led to the formation of the so-called highest shoreline since the deglaciation of the coastal areas was followed by rapid isostatic rebound. Because of the deglaciation, the sedimentation in the BIL was predominantly of a glaciolacustrine character resulting in either glacial varved clay or more homogenous glacial clay: as the ice sheet retreated north the BIL grew in size with varved clay forming in proximal areas of the ice sheet, while homogeneous clay was deposited in more distal areas. Organic productivity was very low and even diatoms were rare.

Due to the fact that the isostatic uplift of the outlet area in Öresund was more rapid than the eustatic sea level rise, the altitudinal difference between the level of the BIL and the sea gradually rose. It has been estimated that this difference at 13 ka BP was in the order of 10 m (Björck 2008), and around this time there are strong indications that a first drainage of the BIL occurred (Björck 1981, 1995). This is thought to have been the consequence of ice recession north of the south Swedish highlands and Mt. Billingen, situated between Lake Vättern and Lake Vänern (Björck and Digerfeldt 1984). This deglaciation uncovered parts of the middle Swedish lowlands and opened up a contact between the sea in the west, occupying, e.g., Lake Vänern, and the BIL. Due to a later ice readvance and erosion of the deglaciated terrain, the proofs for this drainage are more of circumstantial character, though the circumstantial evidences are many (cf. Björck 1995), than concrete drainage deposits. It may have been recorded in the Arkona basin as basin-wide sandy layer (Moros et al. 2002). There is, however, no evidence that marine water entered the Baltic basin.

When the Younger Dryas cooling set in at ca. 12.8 ka BP, the ice sheet advanced south over the previously deglaciated areas and once again blocked the northern drainage of the BIL at Mt. Billingen. This ponding effect might have been a gradual process but must have led to a more or less rapid transgression, depending on how long the updamming took, until the Öresund outlet functioned again. Complex Younger Dryas sediment lithologies in lakes in Blekinge (Björck 1981), in more or less contact with the BIL during Younger Dryas, imply that the BIL experienced a complex water level history during this time period. It has been shown that during this phase the BIL reached as far southwest as into the Kiel Bay (Jensen et al. 1997, 2002).

Fig. 4.4 Paleogeographic map showing the Baltic Ice Lake just prior to the maximum extension and final drainage at ca. 11.7 ka BP

The uplift of Öresund continued to be greater than the sea level rise, which meant that the altitudinal difference increased further between the closed-in BIL and the open sea (Fig. 4.4). At the end of Younger Dryas, we have ample evidence for milder conditions in NW Europe (e.g., Bakke et al. 2009), which triggered a retreat of the ice sheet even before the Younger Dryas cold period ceased (Björck and Digerfeldt 1984, Johnson and Ståhl 2010). Although we do not know the details about the final drainage of the BIL, we know from many independent evidence that there was a sudden lowering of the Baltic level of ca. 25 m down to sea level, and it occurred over a time period of 1–2 years just prior to the onset of the Holocene (Björck et al. 1996, Jakobsson et al. 2007), which dates it to ca. 11.7 ka BP (Walker et al. 2009). The effects both inside and outside the Baltic basin have been described in more detail by Björck (1995), but it must have had a huge impact on the whole circum-Baltic environment, with large coastal areas suddenly subaerially exposed, large changes in fluvial systems, considerable reworking of previously laid down sediments as well as the establishment of a large land-bridge between Denmark and Sweden.

4.3.2 11.7–10.7 ka BP

Hence, the onset of the next Baltic Sea stage, the Yoldia Sea (YS), coincides more or less exactly with the base/start of the Holocene Series/Epoch (Walker et al. 2009) and the rapid warming connected with that. In fact, varved clay thicknesses in northwestern Baltic Proper and $\delta^{18}O$ values in the GRIP ice core display a strikingly similar pattern over a 150 year long Younger Dryas–Preboreal transition period (Andrén et al. 1999, 2002), showing a distinct increase in sedimentation rate as the ice sheet began to melt and rapidly retreat (Fig. 4.5). Apart from being characterized by the rapid deglaciation of the Scandinavian ice sheet, the relative sea level changes of the YS played an important role and were a combination of rapid regression in the recently deglaciated regions and normal regression rates in southern Sweden (1.5–2 m/100 years).

Although the YS were at level with the sea, it would take ca. 300 years before saline water could enter through the fairly narrow straits of the southcentral Swedish lowland. This brackish phase has been documented by the varve lithology, geochemistry, and marine/brackish fossils, such as *Portlandia (Yoldia) arctica*. Occasionally this phase shows up as sulfide banding, implying a halocline, and the maximum duration of this brackish phase has been estimated to 350 years (Andrén et al. 2007), although some records indicate it only lasted some 70–120 years (Andrén and Sohlenius 1995, Wastegård et al. 1995). Due to the high uplift rate in southcentral Sweden, the strait area shallowed up rapidly, which together with the large outflow prevented saline water to enter the Baltic (Fig. 4.6). This turned the Yoldia Sea into a freshwater basin again although there was an open contact with the sea in the west through Lake Vänern and valley systems further west.

At the end of this stage the ice sheet had receded far north and most of today's Baltic Sea basin was deglaciated, with the exception of the Bothnian Bay. This

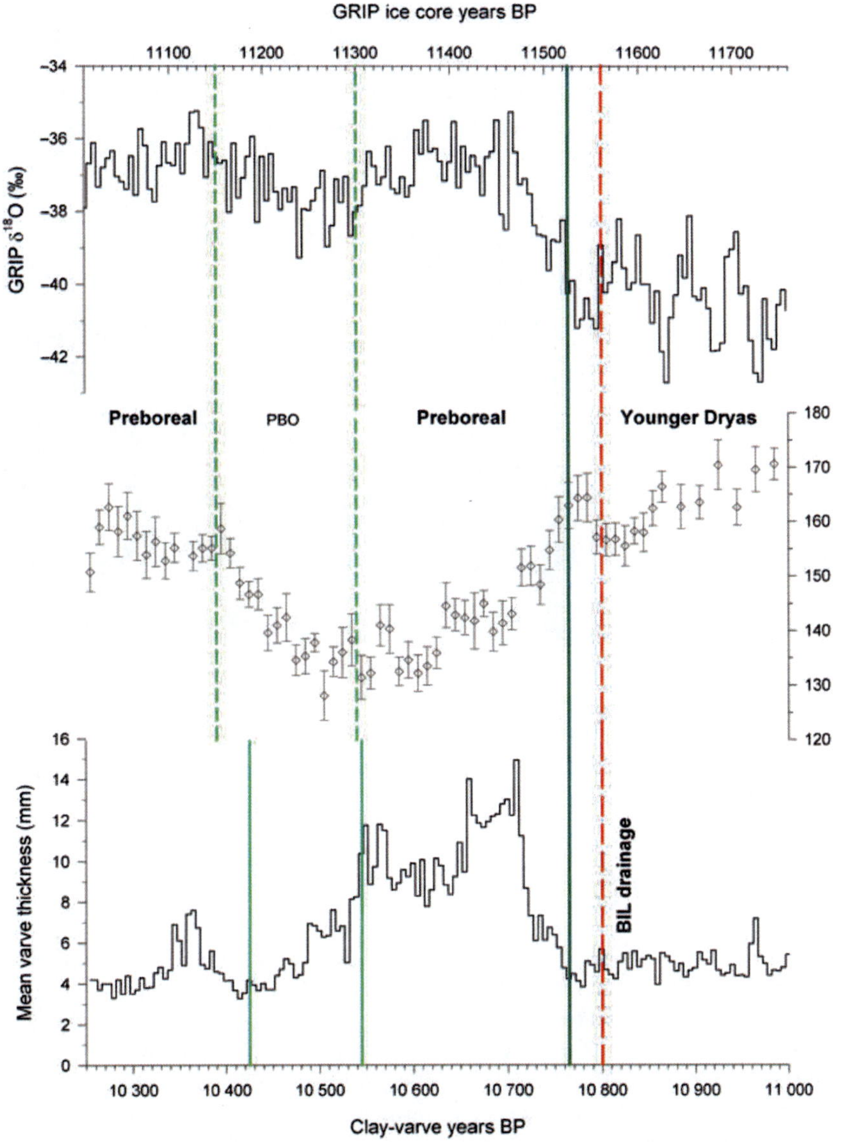

Fig. 4.5 Correlation between the thickness of the glacial clay varves from the Baltic Sea, the $\delta^{18}O$ from the Greenland GRIP ice core, and atmospheric $\delta^{14}C$ variation (included to show the climate variations over the Younger Dryas/Preboreal transition). GRIP ice core years and chronostratigraphy as defined by Björck et al. (1998). Redrawn from Andrén et al. (2002)

resulted in sedimentation in the BSB where varved glacial clay was deposited in the Bothnian Bay and during the same time postglacial sedimentation in the central and southern part of the basin (Ignatius et al. 1981). The isostatic rebound of the

Fig. 4.6 Paleogeographic map showing the Yoldia Sea at the end of the brackish phase ca. 11.1 ka BP

areas around Lake Vänern led to a situation where the outlets west of Vänern shallowed up and could not "swallow" all outflowing water from the Baltic. This marks the end of the Yoldia Sea.

4.3.3 10.7–9.8 ka BP

When the shallowing up of the outlets west of Lake Vänern forced the water level inside the Baltic to rise, the next stage, the Ancylus Lake (AL), began. The sediments of this large freshwater lake are usually poor in organic material, which is partly a consequence of the melt water input to the Baltic from the final deglaciation of the Scandinavian ice sheet and the fairly pristine soils of the mainly recently deglaciated drainage area. Together, this resulted in an aquatic environment with low nutrient input and hence low productivity. The absence of a halocline in the AL led to a well-mixed oxygenated water body and the fairly common sulfide-banded sediments of this stage can probably be explained by H_2S diffusion from younger sediments (Sohlenius et al. 2001).

The onset of the AL is displayed by a simultaneous switch in relative water level change in the areas south of Stockholm–Helsinki (corresponding to the mean isobase of the outlets west of Lake Vänern), which had previously experienced regression since deglaciation: transgression now took over. This is documented not only by an array of shore displacement curves, but also by simultaneous flooding

Fig. 4.7 Paleogeographic map showing the Ancylus Lake during the maximum transgression at ca. 10.5 ka BP

of pine forests around the coasts of the southern Baltic basin. In the very southern Baltic is a transgression recorded as evident from submerged pine trees and peat deposits dated between 11.0 and 10.5 ka BP (Andrén et al. 2007). This gradual updamming – the outlet areas were rising faster than sea level – had varying impacts. While areas to the north experienced a more or less slowed-down regression, the extent of the transgression in the south varied largely, altitudinally and aerially, depending on if areas were isostatically rising or submerging (Fig. 4.7). The end of the transgression often shows up as a beautiful raised beach along the Swedish, Latvian, and Estonian coasts as well as on the island of Gotland. By a large number of ^{14}C dates of underlying peat as well as tree remains (mainly pine) in the beaches, the time span for this so-called Ancylus transgression can be estimated to ca. 500 years. The pattern of the isobases over S Sweden for the time of the Ancylus Lake/transgression shows that the level of Baltic was higher than the sea in the west, showing that the Ancylus Lake was updammed. The final and total updamming effect has been estimated to have raised the Baltic ca. 10 m above sea level (Björck et al. 2008), which means that (isostatically) submerging areas in the southernmost Baltic experienced a larger transgression than that.

The transgression and flooding in the south as a consequence of a "tipping bathtub effect" would inevitably result in a new outlet in the south. Since Öresund had been uplifted more than potential outlet/sill areas further south, these southern areas were now lower than Öresund. What now might have followed is described in detail by Björck et al. (2008), but available data indicate that the Darss Sill area, between

Darss and Møn, was inundated by Ancylus waters. Through Mecklenburg Bay and Fehmarn Belt, along the eastern side of Langeland and out through the Great Belt to Kattegat, a large river was established through Denmark, i.e., the Dana river (von Post 1929). The idea of a sudden and large drainage of AL was proven impossible by Lemke et al. (2001) but due to initial erosion of the riverbed of soft Quaternary deposits this might have caused an initial lowering of the AL in the order of 5 m, followed by a period of rising base level, i.e., the sea in the north. This resulted in a complex river system through Denmark with river channels, levées, and lakes (Bennike et al. 2004) with a gradually smaller fall-gradient as the sea level was rising. When sea level in Kattegat had reached the level of the AL inside the Baltic basin we think it is possible that saltwater could penetrate all the way through the long river system into the Baltic, at least during periods when the Baltic region had been under influence of a long-lasting high-pressure suddenly followed by deep low pressure systems and strong westerly winds. We therefore place the end of the Ancylus Lake when it was at approximate level with the sea and we see the first signs of marine influence since the YS.

4.3.4 9.8–8.5 (8) ka BP

According to independent evidence in the Blekinge archipelago (Berglund et al. 2005) and from the Bornholm basin (Andrén et al. 2000b), the first, though weak, signs of saline water entering the Baltic basin after the AL have been ^{14}C dated to 9.8 ka BP. Also in the eastern Gotland basin is an increase in brackish freshwater diatom taxa recorded at ca. 9.8 ka BP (Andrén et al. 2000a). This period with very low saline influence has been named the Initial Littorina Sea (Andrén et al. 2000b), which is very appropriate considering the fact that the Baltic was at level with the sea. The Scandinavian ice sheet finally melted during the early part of this stage, and although most of the Baltic Sea coastline still experienced regression due to the rebound, the "0-line," i.e., the areas where eustasy and isostatic uplift balanced, moved north. The 0-line during the first part of this stage was possibly along a line from SE Sweden to Estonia, i.e., all areas south of such a line would have experienced a transgression.

In comparison with the AL sediments the organic content often rises gradually throughout this stage. In the coastal zone, a diverse brackish diatom flora, the so-called Mastogloia flora, was established (Miettinen 2002). In the open basin, however, a very low diatom abundance characterizes the Initial Littorina Sea (e.g., Andrén et al. 2000a, b, Paabo 1985, Sohlenius et al. 1996, 2001, Thulin et al. 1992, Westman and Sohlenius 1999). Pigment biomarkers indicate that cyanobacteria were abundant during the Initial Littorina Sea (Bianchi et al. 2000, Borgendahl and Westman 2007). Stable nitrogen isotopes which are indicative of the origin of nitrogen have been used to show that these early blooming cyanobacteria were actually nitrogen fixers (Borgendahl and Westman 2007, Voss et al. 2001).

How long this period of almost freshwater character lasted is not known in detail since the age of the onset of the next stage is not unequivocal, but it certainly lasted more than a millennium. The reasons behind the age uncertainty are many: different types of material have been dated with different types of dating methods, the ^{14}C reservoir effect in the Baltic is poorly known, and the simultaneousness in the Baltic of this shift is debated. Bulk sediments, shells, and terrestrial macrofossils have been ^{14}C dated and fine sand quartz samples have been OSL dated with the SAR procedure (Kortekaas et al. 2007).

4.3.5 8.5 (8) ka BP–Present

The onset of the next stage, the Littorina Sea, is seen as a marked lithological change in Baltic Sea cores. It shows up as a very distinct increase in organic content as well as increasing abundance of brackish marine diatoms (e.g., Sohlenius et al. 2001). It has been discussed if this sudden increase in organic carbon content is exclusively coupled to changes in primary production or if it is partly due to better preservation of carbon during anoxic conditions (Sohlenius et al. 1996). It has also been proposed that an increase in the secchi depth due to flocculation of clay particles and subsequent rapid sedimentation could attribute to an increased primary production (Winterhalter 1992). Distribution of trace elements in sediments, especially enrichment of barium and vanadium, is linked to the cycling of organic carbon and imply that increased productivity in the basin caused the rise in organic carbon content (Sternbeck et al. 2000).

Due to dating problems it has not been possible to absolutely date the transition from fresh to brackish water. In general, ^{14}C dates between 8.5 and 8 ka BP are very common for the onset of this important shift (Sohlenius et al. 1996, Sohlenius and Westman 1998, Andrén et al. 2000a), while the OSL-based age-depth model of Kortekaas et al. (2007) suggests an age of 6.5 ka BP for the same shift in the Arkona Basin. Both ^{14}C ages of bulk sediment and bivalves from the very same core give older ages than the OSL ages, but younger than the expected age of 8.5–8 ka BP. This discrepancy is difficult to explain unless the shift was not the same as determined in other studies; diatom analysis was not carried out by Kortekaas et al. (2007).

Rising sea level and flooding of the Öresund Strait is believed to be the main mechanism behind the onset of the Littorina Sea; melting of the Laurentide and Antarctic ice sheets over couple of millennia caused a 30-m rise in the absolute sea level (Lambeck and Chappell 2001). Episodic melting events of these ice sheets may explain the so-called Littorina transgressions in the Baltic Sea (e.g., Berglund et al. 2005), which are found in areas south of Stockholm. For example, the rapid sea level rise (4.5 m in a few hundred years) in Blekinge centered at 7.6 ka BP has been ascribed to the final decay of the Labrador sector of the Laurentide ice sheet (Yu et al. 2007).

4.3.6 Salinity

The outlets/inlets at Öresund and Great Belt widened and became deeper until ca. 6 ka BP, resulting in increasing and maximum postglacial salinities (e.g., Westman and Sohlenius 1999). Based on model calculations, Gustafsson and Westman (2002) suggest that changes in the morphology and depths of the sills in the inlet area only partly explain the salinity variations during the last ca. 8 ka BP. They found that a major cause of the salinity changes was variations in the freshwater input to the basin. The latter study demonstrated that the freshwater supply to the basin may have been 15–60% lower than at present during the phase of maximum salinity around 6 ka BP (Fig. 4.8). In addition, climate-driven long-term freshwater discharge variability may have been an important factor controlling the salinity and the stratification in the Baltic Sea during the last ca. 8 ka BP (Zillén et al. 2008).

The eustatic sea level rise ceased sometime between 6 and 5 ka BP. The remaining, though slow, rebound resulted in shallower Öresund and Great Belt straits and decreased salinities. An estimate of the Baltic basin paleosalinity was presented by Gustafsson and Westman (2002). They used presence or absence of mollusks and a silicoflagellate to infer different salinity intervals of the last 8 ka. Emeis et al. (2003) reconstructed Baltic salinity fluctuations throughout Holocene using stable carbon isotopes. A comparison of these two salinity reconstructions (Zillén et al. 2008) shows great discrepancies: Gustafsson and Westman (2002) show a decrease in salinity from the maximum value between 5 and 6 ka BP, while Emeis et al. (2003) infer increased salinity during the last 2 ka years. Another method to infer

Fig. 4.8 Paleogeographic map showing the Littorina Sea during the most saline phase at ca. 6.5 ka BP

paleosalinity of the Baltic Sea is to use the strontium isotopic ratio in carbonate mollusk shells (Widerlund and Andersson 2006) and quantify salinity with a precision better than ±5%. However, this method can only be used when carbonate shells are present/preserved in the sediments. Donner et al. (1999) suggest a salinity ca. 4‰ higher than present at the coastal areas of the Gulf of Finland and the Gulf of Bothnia between ca. 7.5 and 4.5 ka BP based on the $^{18}O/^{16}O$ ratio in mollusk shells.

4.3.7 Nutrient Conditions and Hypoxia

The inflowing marine water in the Baltic Sea 8–7 ka BP probably caused the release of phosphorus from sediments, enhancing the growth of cyanobacteria (Bianchi et al. 2000, Borgendahl and Westman 2007, Kunzendorf et al. 2001). The salinity stratification together with increased primary production initiated periods of deepwater hypoxia in the open Baltic basin, evident in the sediment record as extended periods of laminated sediments (Sohlenius and Westman 1998, Zillén et al. 2008). Increased upward transport of nutrients from the anoxic bottom water has been suggested as an explanation of the enhanced primary productivity at the Ancylus/Littorina transition (Sohlenius et al. 1996). Between 8 and 4 ka BP, the Littorina Sea experienced a long sustained period of hypoxia (Zillén et al. 2008). Oxygen conditions improved considerably after ca. 4 ka BP, also as salinity decreased (Gustafsson and Westman 2002). This coincided with the onset of the neoglaciation in N Europe with a more humid and cold climate (Snowball et al. 2004). The shift to colder and wetter conditions probably increased the net precipitation in the watershed leading to increased freshwater supplies to the basin and decreased salinities (Gustafsson and Westman 2002). Such a freshening, in combination with increased wind stress over the Baltic Sea, would result in a weakened halocline and enhanced vertical mixing allowing more efficient exchange of oxygen across the halocline (Conley et al. 2002). This scenario would promote more oxic bottom water conditions and explain the diminishing of the hypoxic zone around 4 ka BP (Zillén et al. 2008).

Hypoxia occurred again during the middle-late Littorina Sea (ca. 2–0.8 ka BP). In contrast to the period of oxygen deficiency during the early and more saline phase of the Littorina Sea, hypoxia during the late Littorina Sea does not show a relationship to any known changes in salinity. During this time, the surface salinity in the Baltic Proper probably ranged between 7 and 8‰, i.e., similar to the last ca. 4–3 ka (Gustafsson and Westman 2002). Hypoxia between 2 and 0.8 ka BP overlaps with a climate anomaly known as the Medieval Warm Period (Lamb 1965) when atmospheric temperatures in NW Europe were ca. 0.5–0.8° warmer than today (Snowball et al. 2004). However, temperatures have no proven effects on the oxygen conditions in the Baltic Sea and the relationship between primary production and climate change is not straightforward (Richardson and Schoeman 2004). Furthermore, the link between phytoplankton abundance and sea surface temperature is only indirectly coupled to temperature. The ecological response to NAO

(North Atlantic Oscillation) has been reviewed and several correlations between climate and ecological changes have been observed, although the mechanism is not understood (e.g., Ottersen et al. 2001). At the Swedish west coast, a strong correlation between phytoplankton biomass and NAO has been found, possibly caused by an increased stratification in Skagerrak (Belgrano et al. 1999). In the North Sea, there has been an increase in phytoplankton season length and abundance since the mid-1980s, interpreted as a response to climatic forcing (Reid et al. 1998). Although NAO is well known to influence climate conditions in the Baltic Sea, no direct links between NAO, hypoxia, and inflow of saline water have been established.

The causes of hypoxia during the middle-late Littorina Sea are not fully understood. An alternative trigging mechanism to widespread hypoxia during this time period is increased anthropogenic forcing via eutrophication. It has been proposed that hypoxia correlates with population growth and large-scale changes in land use that occurred in the Baltic Sea watershed during the early Medieval expansion between AD 750 and 1300 (Zillén et al. 2008, Zillén and Conley 2010). The large land use changes increased soil nutrient leakage significantly in the Baltic Sea watershed, leading to high nutrient variability in the basin and associated hypoxia (Zillén and Conley 2010). The late Littorina Sea record of hypoxia in the Baltic Sea may thus be due to multiple stressors, where both climate and human impacts may have interacted. It is known that human activities have affected the Baltic Sea already AD 200 which is recorded as a change in the lead composition in the sediments from the Eastern Gotland basin. This change coincides with a geographic shift in the Roman lead mining from the Iberian Peninsula to other areas, e.g., Germany and the British Isles during the first to third centuries AD (Bindler et al. 2009).

Hypoxia again appeared in the Baltic Sea around the turn of the last century with all sediments below 150 m in the Gotland Deep laminated (Hille et al. 2006). This period corresponds to a climate amelioration, which has lasted over most of the twentieth century as well as the onset of the Industrial Revolution when the European population increased rapidly (about six times since AD 1800) and technological advances in agriculture and forestry exploded (Zillén and Conley 2010). The eutrophication we now experience (e.g., Elmgren 2001) is caused by the increased discharge of nutrients with a growing population and the use of synthetic fertilizers on arable land after World War II (Elmgren 1989), but these effects are also superimposed on effects caused by the ongoing climate warming (Andrén et al. 2000a; Leipe et al. 2008). Revealing the relative importance between climate and anthropogenic forcing on the Baltic Sea ecosystem is one of the major scientific challenges for the future.

References

Adrielsson L (1984) Weichselian lithostratigraphy and glacial environments in the Ven-Glumslöv area, southern Sweden. LUNDQUA Thesis 16

Andrén E, Shimmield G, Brand T (1999) Changes in the environment during the last centuries on the basis of siliceous microfossil records from the southwestern Baltic Sea. The Holocene 9:25–38

Andrén E, Andrén T, Kunzendorf H (2000a) Holocene history of the Baltic Sea as a background for assessing records of human impact in the sediments of the Gotland Basin. The Holocene 10:687–702

Andrén E, Andrén T, Sohlenius G (2000b) The Holocene history of the southwestern Baltic Sea as reflected in a sediment core from the Bornholm Basin. Boreas 29:233–250

Andrén E, Clarke A, Telford R, Weckström K, Vilbaste S, Aigars J, Conley D, Johnsen T, Juggins S, Korhola A (2007) Defining reference conditions for coastal areas in the Baltic Sea. TemaNord 2007:583

Andrén T, Wannäs K (1988) Late Quaternary development of the Bornholm Gatt. Geological Survey of Finland, Special Paper 6:23–29

Andrén T, Sohlenius G (1995) Late Quaternary development in the north-western Baltic Proper – result from the clay-varve investigation. Quaternary International 27:5–10

Andrén T, Björck J, Johnsen S (1999) Correlation of Swedish glacial varves with the Greenland (GRIP) oxygen isotope record. Journal of Quaternary Science 14:361–371

Andrén T, Lindeberg G, Andrén E (2002) Evidence of the final drainage of the Baltic Ice Lake and the brackish phase of the Yoldia Sea in glacial varves from the Baltic Sea. Boreas 31:226–238

Anjar J, Larsen NK, Björck S, Adrielsson L, Filipsson HL (2010) MIS 3 marine and lacustrine sediments at Kriegers Flak, southwestern Baltic Sea. Boreas 10:360–366

Bakke J, Lie Ø, Heegaard E, Dokken T, Haug GH, BirksHH, Dulski P, Nilsen T (2009) Rapid oceanic and atmospheric changes during the Younger Dryas cold period. Nature Geoscience 2:202–205

Belgrano A, Lindahl O, Hernroth B (1999) North Atlantic Oscillation primary productivity and toxic phytoplankton in the Gullmar Fjord, Sweden (1985–1996). Proceedings Royal Society London B 266:425–430

Bennike O, Jensen JB, Lemke W, Kuijpers A, Lomholt S (2004) Late- and postglacial history of the Great Belt, Denmark. Boreas 33:18–33

Berglund BE, Sandgren P, Barnekow L, Hannon G, Jiang H, Skog G, Yu S (2005) Early Holocene history of the Baltic Sea, as reflected in coastal sediments in Blekinge, southeastern Sweden. Quaternary International 130:111–139

Bianchi TS, Westman P, Rolff C, Engelhaupt E, Andrén T, Elmgren R (2000) Cyanobacterial blooms in the Baltic Sea: natural or human-induced? Limnology and Oceanography 45:716–726

Bindler R, Renberg I, Rydberg J, Andrén T (2009) Widespread waterborne pollution in central Swedish lakes and the Baltic Sea from pre-industrial mining and metallurgy. Environmental Pollution 157:2132–2141

Björck S (1981) A stratigraphic study of Late Weichselian deglaciation, shore displacement and vegetation history in south-eastern Sweden. Fossils and Strata 14:1–93

Björck S, Digerfeldt G (1984) Climatic changes at Pleistocene/Holocene boundary in the Middle Swedish endmoraine zone, mainly inferred from stratigraphic indications. In: Mörner NA, Karlén W (eds) Climatic changes on a yearly to millennial basis. Reidel, Dordrecht

Björck S, Dennegård B, Sandgren P (1990) The marine stratigraphy of the Hanö Bay, SE Sweden, based on different sediment stratigraphic methods. GFF 112:265–280

Björck S (1995) A review of the history of the Baltic Sea, 13.0–8.0 ka BP. Quaternary International 27:19–40

Björck S, Kromer B, Jonsen S, Bennike O, Hammarlund D, Lemdahl G, Possnert G, Ramussen TL, Wohlfarth B, Hammer CU, Spurk M (1996) Synchronized terrestrial-atmospheric deglacial records around the North Atlantic. Science 274:1155–1160

Björck S, Walker MJC, Cwynar LC, Johnsen S, Knudsen KL, Lowe JJ, Wohlfarth B, INTIMATE Members (1998) An event stratigraphy for the Last Termination in the North Atlantic region based on the Greenland ice-core record: a proposal by the INTIMATE group. Journal of Quaternary Science 13:283–292

Björck S, Noe-Nygaard N, Wolin J, Houmark-Nielsen M, Hansen HJ, Snowball I (2000) Eemian lake development, hydrology and climate – a multistratigraphic study of the Hollerup site in Denmark. Quaternary Science Reviews 19:509–536

Björck S (2008) The late Quaternary development of the Baltic Sea basin. In: The BACC Author Team (eds) Assessment of climate change for the Baltic Sea Basin. Springer, Berlin, Heidelberg

Björck S, Andrén T, Jensen JB (2008) An attempt to resolve the partly conflicting data and ideas on the Ancylus-Litorina transition. Polish Geological Institute Special Papers 23:21–25

Borgendahl J, Westman P (2007) Cyanobacteria as a trigger for increased primary productivity during sapropel formation in the Baltic Sea – a study of the Ancylus/Litorina transition. Journal of Paleolimnology 38:1–12

Conley DJ, Humborg C, Rahm L, Savchuk OP, Wulff F (2002) Hypoxia in the Baltic Sea and basin-scale changes in phosphorus biogeochemistry. Environmental Science and Technology 36:5315–5320

Conley DJ, Bonsdorff E, Carstensen J, Destouni G, Gustafsson BG, Hansson LA, Rabalais NN, Voss M, Zillén L (2009) Viewpoint: tackling hypoxia in the Baltic Sea: is engineering a solution? Environmental Science and Technology 43:3407–3411

Dickson AJ, Austin WEN, Hall IR, Maslin M, Kucera M (2008) Centennial-scale evolution of Dansgaard-Oeschger events in the northeast Atlantic Ocean between 39.5 and 56.5 ka BP. Paleoceanography 23:PA3206

Donner J, Kankainen T, Karhu JA (1999) Radiocarbon ages and stable isotope composition of Holocene shells in Finland. Quaternaria Ser A 7:31–38

Eiríksson J, Kristensen PH, Lykke-Andersen H, Brooks K, Murray A, Knudsen KL, Glaister C (2006) A sedimentary record from a deep Quaternary valley in the southern Lillebælt area: Eemian and Early Weichselian lithology and chronology at Mommark. Boreas 35:320–331

Elmgren R (1989) Man's impact on the ecosystem of the Baltic Sea: energy flows today and at the turn of the century. Ambio 18:326–332

Elmgren R (2001) Understanding human impact on the Baltic ecosystem: changing views in recent decades. Ambio 30:222–231

Emeis KC, Struck U, Blanz T, Kohly A, Voss M (2003) Salinity changes in the central Baltic Sea (NW Europe) over the last 10 000 years. The Holocene 13:411–421

Fronval T, Jansen E (1996) Rapid changes in ocean circulation and heat flux in the Nordic seas during the last interglacial period. Nature 383:806–810

Funder S, Demidov I, Yelovicheva Y (2002) Hydrography and mollusc faunas of the Baltic and the White Sea-North Sea seaway in the Eemian. Palaeogeography, Palaeoclimatology, Palaeoecology 184:275–304

Grimm EC, Watts WA, Jacobson GL Jr, Hansen BCS, Almquist HR, Dieffenbacher-Krall AC (2006) Evidence for warm wet Heinrich events in Florida. Quaternary Science Reviews 25:2197–2211

Grootes PM, Stuvier M, White JWC, Johnsen S, Jouzel J (1993) Comparison of oxygen isotope records from the GISP2 and GRIP Greenland ice cores. Nature 366:552–554

Gustafsson BG, Westman P (2002) On the causes for salinity variations in the Baltic Sea for the last 8500 years. Paleoceanography 17:1–14

Hille S, Leipe T, Seifert T (2006) Spatial variability of recent sedimentation rates in the Eastern Gotland Basin (Baltic Sea). Oceanologia 48:297–317

Houmark-Nielsen M, Kjær KH (2003) Southwest Scandinavia, 40–15 ka BP: palaeogeography and environmental change. Journal of Quaternary Science 18:769–786

Houmark-Nielsen M, Krüger J, Kjær KH (2005) De senaste 150.000 år i Danmark. Istidsladskabet og naturens udvikling. Geoviden – Geologi og Geografi 2:2–19

Houmark-Nielsen M (2007) Extent and age of Middle and Late Pleistocene glaciations and periglacial episodes in southern Jylland, Denmark. Bulletin of the Geological Society of Denmark 55:9–35

Houmark-Nielsen M (2008) Testing OSL failures against a regional Weichselian glaciations chronology from southern Scandinavia. Boreas 37:660–677

Ignatius H, Axberg S, Niemistö L, Winterhalter B (1981) Quaternary geology of the Baltic Sea. In: Voipio A (ed) The Baltic Sea. Elsevier Oceanography Series 30, Elsevier, Amsterdam

Jakobsson M, Björck S, Alm G, Andrén T, Lindeberg G, Svensson NO (2007) Reconstructing the Younger Dryas ice dammed lake in the Baltic Basin: Bathymetry, area and volume. Global and Planetary Change 57:355–370

Jensen JB, Bennike O, Witkowski A, Lemke W, Kuijpers A (1997) The Baltic Ice Lake in the southwestern Baltic: sequence-, chrono- and biostratigraphy. Boreas 26:217–236

Jensen JB, Kuijpers A, Bennike O, Laier T, Werner F (2002) New geological aspects for freshwater seepage and formation in Eckernf.orde Bay, western Baltic. Continental Shelf Research 22:2159–2173

Johnsen SJ, Clausen HB, Dansgaard W, Fuhrer K, Gundestrup N, Hammer CU, Iversen P, Jouzel J, Stauffer B, Steffensen JP (1992) Irregular glacial interstadials recorded in a new Greenland ice core. Nature 359:311–313

Johnson MD, Ståhl Y (2010) Stratigraphy, sedimentology, age and paleoenvironment of marine varved clay in the Middle Swedish end-moraine zone. Boreas 39:199–214

Klingberg F (1998) A late Pleistocene marine clay succession at Kriegers Flak, westernmost Baltic, southern Scandinavia. Boreas 13:245–253

Kortekaas M, Murray AS, Sandgren P, Björck S (2007) OSL chronology for a sediment core from the southern Baltic Sea: a continuous sedimentation record since deglaciation. Quaternary Geochronology 2:95–101

Kristensen P, Knudsen KL (2006) Palaeoenvironments of the Eemian deposits at Mommark, South Denmark: foraminifera, ostracods and stable isotopes. Boreas 35:349–366

Kukla G, Bender MD, de Beaulieu JL, Bond G, Broecker WS, Cleveringa P, Gavin JE, Herbert TD, Imbrie J, Jouzel J, Keigwin LD, Knudsen KL, McManus J, Merkt J, Muhs DR, Müller H, Poore RZ, Porter SC, Seret G, Shackleton NJ, Turner C, Tzedakis PC, Winograd IJ (2002) Last interglacial climates. Quaternary Research 57:2–13

Kunzendorf H, Voss M, Brenner W, Andrén T, Vallius H (2001) Molybdenum in sediments of the central Baltic Sea as an indicator for algal blooms. Baltica 14:123–130

Lagerlund E (1987) An alternative Weichselian glaciation model, with special reference to the glacial history of Skåne, South Sweden. Boreas 16:433–459

Lamb HH (1965) Climate, history and the modern world, 2nd edn. Routledge, London

Lambeck K, Chappell J (2001) Sea level change through the Last Glacial cycle. Science 292:679–686

Lambeck K, Purcell A, Zhao J, Svensson NO (2010) The Scandinavian ice sheet: from MIS 4 to the end of the Last Glacial Maximum. Boreas 39:410–435

Larsen NK, Knudsen KL, Krohn CF, Kronborg C, Murray AS, Nielsen OB (2009) Late Quaternary ice sheet, lake and sea history of southwest Scandinavia – a synthesis. Boreas 38:732–759

Leipe T, Dippner JW, Hille S, Voss M, Christiansen C, Bartholdy J (2008) Environmental changes in the central Baltic Sea during the last 1000 years: inferences from sedimentary records, hydrography and climate. Oceanologia 50:23–41

Lemke W, Jensen JB, Bennike O, Endler R, Witkowski A, Kuijpers A (2001) Hydrographic thresholds in the western Baltic Sea: Quaternary geology and the Dana River concept. Marine Geology 176:191–201

Mangerud J (2004) Ice sheet limits on Norway and the Norwegian continental shelf. In: Ehlers J, Gibbard P (eds) Quaternary glaciations – extent and chronology. Elsevier, Amsterdam

Miettinen A (2002) Relative sea level changes in the Eastern part of the Gulf of Finland during the last 8000 years. Annales Academiae Scientarum Fennicae Geologica-Geographica 162

Miller U (1977) Pleistocene deposits of the Alnarp Valley, southern Sweden – microfossils and their stratigraphical applications. University of Lund, Department of Quaternary Geology, Thesis 4

Moros M, Lemke W, Kuijpers A, Endler R, Jensen JB, Bennike O, Gingele F (2002) Regressions and transgressions of the Baltic basin reflected by a new high-resolution deglacial and postglacial lithostratigraphy for Arkona Basin sediments (western Baltic Sea). Boreas 31: 151–162

North Greenland Ice Core Members (2004) High-resolution record of Northern Hemisphere climate extending into the last interglacial period. Nature 431:147–151

Ottersen G, Planque B, Belgrano A, Post E, Reid PC, Stenseth NC (2001) Ecological effects of the North Atlantic Oscillation. Oecologia 128:1–14

Paabo K (1985) Diatomological studies of two cores from the western Baltic. Striae 23:83–91

Rasmussen TL, VanWeering TCE, Labeyrie L (1997) Climatic instability, ice sheets and ocean dynamics at high northern latitudes during the last glacial period (58–10 ka BP). Quaternary Science Reviews 16:71–80

Reid PC, Edwards M, Hunt HG, Warner AJ (1998) Phytoplankton change in the North Atlantic. Nature 391:546

Richardson AJ, Schoeman DS (2004) Climate impact on plankton ecosystems in the Northeast Atlantic. Science 305:1609–1612

Robertsson A-M, Lundqvist J, Brunnberg L (2005) Dark clayey till in central and northern Sweden microfossil content and stratigraphical importance. GFF 127:169–178. http://www.gff-online.se/site/article.asp?articleID=833

Salonen VP, Kaakinen A, Kultti S, Miettinen A, Eskola KO, Lunkka JP (2007) Middle Weichselian glacial event in the central part of the Scandinavian Ice Sheet recorded in the Hitura pit, Ostrobothnia, Finland. Boreas 37:38–54

Sánchez Goñi MF, Eynaud F, Turon JL, Shackleton NJ (1999) High resolution palynological record off the Iberian margin: direct land-sea correlation for the Last Interglacial complex. Earth and Planetary Science Letters 17:123–137

Seifert T, Tauber F, Kayser B (2001) A high resolution spherical grid topography of the Baltic Sea. Proceedings of the Baltic Sea Science Congress 25–29 November 2001, poster 147. Stockholm, Sweden

Siddall M, Rohling EJ, Almogi-Labin A, Hemleben Ch, Meischner D, Schmelzer I, Smeed DA (2003) Sea-level fluctuations during the last glacial cycle. Nature 423:853–858

Snowball IF, Korhola A, Briffa KR, KoçN (2004) Holocene climate dynamics in Fennoscandia and the North Atlantic. In: Battarbee RW, Gasse F, Stickley CE (eds) Past climate variability through Europe and Africa. Springer, Dordrecht

Sohlenius G, Sternbeck J, Andrén E, Westman P (1996) Holocene history of the Baltic Sea as recorded in a sediment core from the Gotland Deep. Marine Geology 134:183–201

Sohlenius G, Westman P (1998) Salinity and redox alternations in the northwestern Baltic proper during the late Holocene. Boreas 27:101–114

Sohlenius G, Emeis KC, Andrén E, Andrén T, Kohly A (2001) Development of anoxia during the fresh–brackish water transition in the Baltic Sea. Marine Geology 177:221–242

Sternbeck J, Sohlenius G, Hallberg RO (2000) Sedimentary trace elements as proxies to depositional changes induced by a Holocene fresh-brackish water transition. Aquatic Geochemistry 6:325–345

Thulin B, Possnert G, Vuorela I (1992) Stratigraphy and age of two postglacial sediment cores from the Baltic Sea. Geologiska Föreningen i Stockholm Förhandlingar 114:165–179

von Post L (1929) Svea, Göta och Dana älvar. Ymer 49:1–33

Voss M, Kowalewska G, Brenner W (2001) Microfossil and biochemical indicators of environmental changes in the Gotland Deep during the last 10,000 years. Baltica 14:131–140

Walker M, Johnsen S, Rasmussen S, Popp T, Steffensen JP, Gibbard P, Hoek W, Lowe J, Andrews J, Björck S, Cwynar L, Hughen K, Kershaw P, Kromer B, Litt T, Lowe D, Nakagawa T, Newnham R, Schwander J (2009) Formal definition and dating of the GSSP (Global Stratotype Section and Point) for the base of the Holocene using the Greenland NGRIP ice core, and selected auxiliary records. Journal of Quaternary Science 24:3–17

Wastegård S, Andren T, Sohlenius G, Sandgren P (1995) Different phases of the Yoldia Sea in the north-western Baltic Proper. Quaternary International 27:121–129

Westman P, Sohlenius G (1999) Diatom stratigraphy in five off-shore sediment cores from the north-western Baltic implying large scale circulation changes during the last c. 8500 years. Journal of Paleolimnology 22:53–69

Widerlund A, Andersson PS (2006) Strontium isotopic composition of modern and Holocene mollusc shells as a palaeosalinity indicator for the Baltic Sea. Chemical Geology 232:54–66
Winterhalter B (1992) Late-Quaternary stratigraphy of Baltic Sea basins – a review. Bulletin of the Geological Society of Finland 64:189–194
Wohlfarth B, Veres D, Ampel L, Lacourse T, Blaauw M, Preusser F, Andrieu-Ponel V, Kéravis D, Lallier-Vergès E, Björck S, Davies S, de Beaulieu JL, Risberg J, Hormes A, Kasper HU, Possnert G, Reille M, Thouveny N, Zander A (2008) Rapid ecosystem response to abrupt climate changes during the last glacial period in Western Europe, 40–16 kyr BP. Geology 36:407–410
Wohlfarth B (2010) Simulated climate conditions in Fennoscandia during a MIS 3 stadial. Boreas 39:436–456
Yu S, Berglund B, Sandgren P, Lambeck K (2007) Evidence for a rapid sea-level rise 7600 yr ago. Geology 35:891–894
Zillén L, Conley DJ, Andrén T, Andrén E, Björck S (2008) Past occurrences of hypoxia in the Baltic Sea and the role of climate variability, environmental change and human impact. Earth Science Reviews 91:77–92
Zillén L, Conley DJ (2010) Hypoxia and cyanobacterial blooms are not natural features of the Baltic Sea. Biogeoscience 7:2567–2580

Chapter 5
Late Quaternary Climate Variations Reflected in Baltic Sea Sediments

Jan Harff, Rudolf Endler, Emel Emelyanov, Sergey Kotov, Thomas Leipe, Matthias Moros, Ricardo Olea, Michal Tomczak, and Andrzej Witkowski

Abstract Late Pleistocene to Holocene climate change of the Atlantic and the northern European realm is reflected by the facies of sediments in the Baltic Sea. The sedimentary sequence have been subdivided into zones reflecting the main postglacial stages of the Baltic Sea basin development according to sediment echosounder profiling and investigating sediment cores from the central Baltic. The changes in the environment of Baltic Sea bottom water is displayed by sediment physical, geochemical, and microfossil proxies. These proxies mark the main shift in the sedimentary facies of the Baltic Basin at 8.14 cal. years BP, from a freshwater to a brackish/marine environment due to the Littorina transgression of marine water masses from the North Sea. The downhole physical facies variation from the Eastern Gotland can be correlated basinwide. Thickness maps of the freshwater and the brackish sediments ascribe the general change in the hydrographic circulation from a coast-to-basin to a basin-to-basin system along with the Littorina transgression. Variations in the salinity of the brackish Littorina Baltic Basin are attributed to changes in the North Atlantic Oscillation (NAO) ascribing the wind forces driving the inflow of marine water into the Baltic Basin. Time series analysis of facies variation reveals distinct periodicities of 900 and 1,500 years. These periods can be compared with data from North Atlantic marine sediments and Greenland ice cores identifying global climate change effects in Baltic Basin sediments.

Keywords Eastern Gotland Basin · Holocene · Physico stratigraphical zona

---

J. Harff (✉)
Leibniz Institute for Baltic Sea Research Warnemünde, D-18119 Rostock, Germany; presently at Institute of Marine and Coastal Sciences, University of Szczecin, PL-70-383 Szczecin, Poland
e-mail: jan.harff@io-warnemuende.de

J. Harff et al. (eds.), *The Baltic Sea Basin*, Central and Eastern European Development Studies (CEEDES), DOI 10.1007/978-3-642-17220-5_5,

5.1 Introduction

The study of recent global climate change commonly involves the reconstruction of climate variation during the late Quaternary based on adequate proxy variables (Bond et al. 1997, 2001). Due to the high sedimentation rate, sediments from the Baltic Sea provide ideal climate archives for climate and environmental reconstructions. In this book, Andrén et al. (Chap. 4) give an overview about the environmental change for the Baltic area during the last glacial cycle (LGC). The postglacial climate and environmental change have been intensively studied based on sediment proxies from the Baltic Basin by Ignatius et al. (1981), Winterhalter et al. (1981), Emelyanov (1994), Björck (1995, 2008), Sohlenius et al. (1996), Winterhalter (2001a), Repecka (2001), Andrén et al. (2001, 2002), Harff et al. (2001a, b), Emeis and Dawson (2003), Dippner and Voss (2004) among others. Based on a multi-proxy approach, Harff et al. (1999, 2001a) subdivided the Late Pleistocene to Holocene sediments from the central Eastern Gotland Basin into physico-stratigraphic facies zones. Lower parts of the postglacial sediments (facies zones A1–A6) represent mainly freshwater sediments accumulated in an isolated basin. At about 8,000 cal. years BP the system changed rapidly to a brackish–marine environment resulting in the accumulation of sediments with changing intensity of lamination. Harff et al. (2001a) structured the brackish sediments into physico-stratigraphic facies zones B1–B6 and ascribe a change in lamination intensity to differences in ventilation of the bottom water during the deposition. Westman and Sohlenius (1999) and Sohlenius et al. (2001) showed, on the basis of diatom analysis and oxygen isotope measurements, that the changes from homogeneous to laminated layers coincide with variations in salinity. However, main findings are still the subject of discussion, and an important scientific question is any coupling of the depositional environment of the Baltic Basin to global climate driving forces. To contribute to this discussion an international research team of geoscientists studied sedimentary sequences from the Baltic Sea during 2004–2006. A main task was to interpret the facies variation as an environmental signal reflecting climatic change during the late Pleistocene and Holocene (Project GISEB: GIS for Time/Space Modeling of Sediment Distribution as a Function of Changing Environment in the Baltic Sea). During an expedition to the central Baltic in 2005 the German R/V "Poseidon" (Harff 2005) sampled Late Quaternary sediments for detailed studies and the comparison with earlier research results. Numerical methods have been applied for stratigraphic core zonation, correlation of sediment cores, development of 3D space models for stratigraphic units, and interpretation and time series analysis of proxy data. Here, we report about results achieved within the frame of this research project.

5.2 The Area of Investigation and the Geological Development as a Response to Climate Variability

The Baltic Sea is a semi-enclosed intra-continental sea surrounded by the landmasses of Scandinavia, northern central Europe, and northeastern Europe.

Quaternary glaciations created the morphology of the Baltic region. The relief of the northwest European Caledonides, with elevations up to 2,470 m, and the surficial topography on the crystalline Precambrian rocks of the Fennoscandian Shield were shaped by a combination of weathering and glacial erosion, and the lowlands of the Russian Plate and the west European Platform were covered by glacial sediments. Glaciers also excavated the Baltic Basin (which has an average water depth of 55 m) and formed a series of sub-basins (Mecklenburgian Bight, 25 m; Arkona Basin, 45 m; Bornholm Basin, 100 m; Gotland Basin, 250 m; Golf of Bothnia, 120 m) separated by shallower sills (Figs. 5.1 and 5.3).

The postglacial history of the Baltic Sea Basin is explained in detail by Andrén et al. (Chap. 4) in this book. The hydrology of the Baltic Sea can be described as a typical estuarine current system. One driving force is the positive water balance resulting from precipitation within the Baltic drainage basin, which belongs to the European humid climate belt. Westerly winds form the second driving force pushing the denser marine water from the North Sea into the Baltic close to the bottom. These winds are the result of atmospheric low-pressure systems tracking from the central North Atlantic to Europe.

The relation between the Icelandic low-pressure and the Eurasian high-pressure systems controls whether north-easterlies and a cold atmosphere or westerlies and relative warm air masses govern the climate in central and northern Europe.

The variation of the system follows a hierarchically superimposed cyclic pattern. The Arctic Oscillation (AO) is the dominant pattern of non-seasonal variations in the stratospheric air pressure of the Northern Hemisphere. The North Atlantic-European sector of the AO is represented through the well-known North Atlantic Oscillation (NAO) at sea level. The NAO describes fluctuations in the strength of geostrophic westerlies affecting predominantly winter climate in the Baltic area. Here, according to Alheit and Hagen (1997) a positive NAO causes a "maritime mode" with strengthened westerlies transporting warm humid air masses eastward and producing mild winters over the Baltic Sea. The opposite situation (negative NAO: continental mode) is determined by strengthened westward transport of cold and dry Siberian air towards Europe. This is accompanied by severe winters in the Baltic Sea area. The NAO fluctuates periodically on a decadal time scale (Hurrell, 1995, Hagen 2006, Hagen and Feistel 2008). In addition, Justino and Peltier (2005) report about a so-called Atlantic Multi-decadal Oscillation (AMO) of about 30 years. Hagen and Feistel (2005) showed that the decadal NAO/AMO periodicity is obviously superimposed on a century lasting trend. In this study we intend to show that this periodicity is reflected by the facies variation in the central Baltic Basin.

Figure 5.1 shows a digital terrain model of the Baltic area. Within the central Baltic Sea (Baltic Proper) the halocline prevents vertical water exchange and leads here to anoxic conditions below a permanent redoxcline (Fig. 5.2). The absence of higher benthic biota prevents the sediments from being bioturbated and causes laminated sequences that record environmental change with high age resolution (Sohlenius et al. 1996, Sohlenius and Westman 1998, Sohlenius et al. 2001). Within the shallow Belt Sea the water column is not stratified due to the mixing effect of

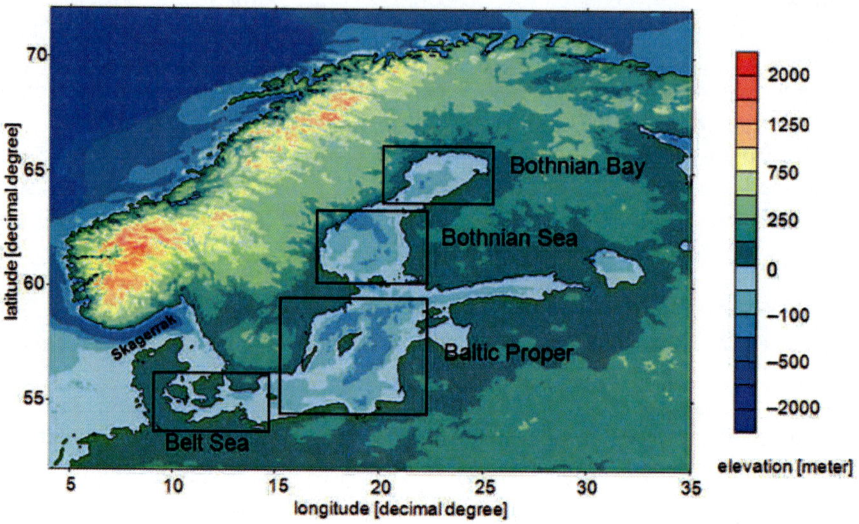

Fig. 5.1 Earth surface relief of the Baltic area

strong winds, whereas precipitation and river discharge reduce the salinity in the Gulf of Bothnia, preventing a halocline there.

In this study we concentrate on the sediments of the Eastern Gotland Basin at the Baltic Proper. In Fig. 5.3, the bottom relief of the Baltic Sea is displayed. It can clearly be seen that near-bottom currents (inflowing dense saline water) are ruled by the bottom relief. After having entered the Baltic Sea the dense water masses have to proceed to the Bornholm Basin and to pass the Stolpe Channel before they enter the Eastern Gotland Basin.

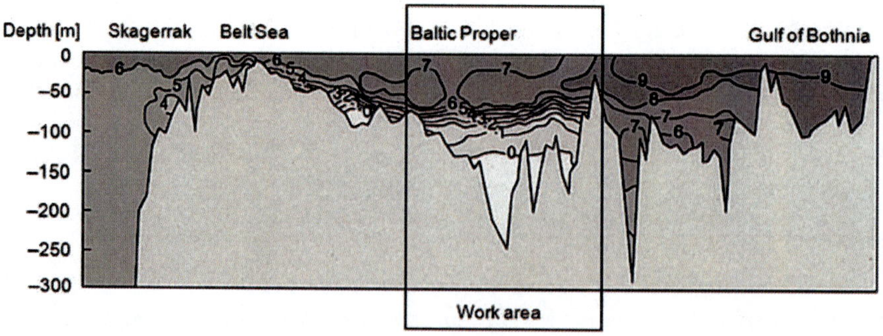

Fig. 5.2 Vertical oxygen concentration (ml/l) in the Baltic Sea from the Skagerrak to the Gulf of Bothnia, summer 1988 (modified from Sjöberg 1992)

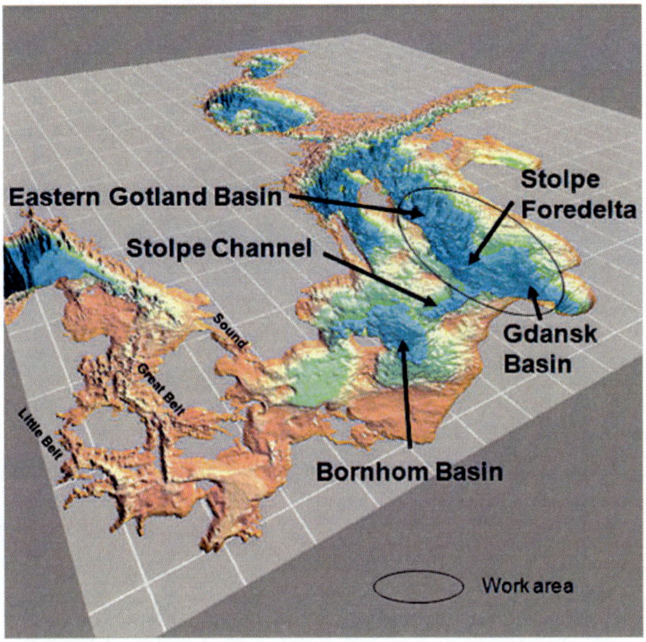

Fig. 5.3 Relief of the Baltic Sea basin (http://www.io-warnemuende.de/profile-of-the-baltic-sea.html) and work area

Suspended matter transported through the Stolpe Channel is being deposited and forms a sediment body of the "Stolpe Foredelta" where the channel merges with the Eastern Gotland Basin. Within the centre of the Eastern Gotland Basin pelagic deposition dominates the sediment accumulation.

The current system is described in Fig. 5.4. The illustration of Fig. 5.4 shows the current system as a result of numerical modelling and as paleoreconstruction after sediment proxies. Figure 5.4a illustrates the current field within the Baltic Proper at a water depth of 60 m. The arrows stand for mean current vectors from modelling results 1960–2005. The MOM3 code (Pacanowski and Griffies 2000) was used for modelling. The resolution of the grid is 3 nm. The source of the meteorological forcing is the era40 data file (Uppala et al. 2005). The counterclockwise direction of the currents is clear as well as the decreasing velocity when the water leaves the Stolpe Channel. It is noticeable that north of the "Stolpe Mouth" the currents describe a separate gyre within the southern part of the Eastern Gotland Basin. The "delta" sediments are accumulated within the centre of this north–south elongated gyre. Emelyanov (2006) reconstructed the late Holocene near-bottom currents after stratigraphic thickness analysis of post-Littorina sediments (Fig. 5.4b). The similarity of the current pattern in Fig. 5.4a, b is obvious and supports the assumption that the recent current system has been stable for a longer time period.

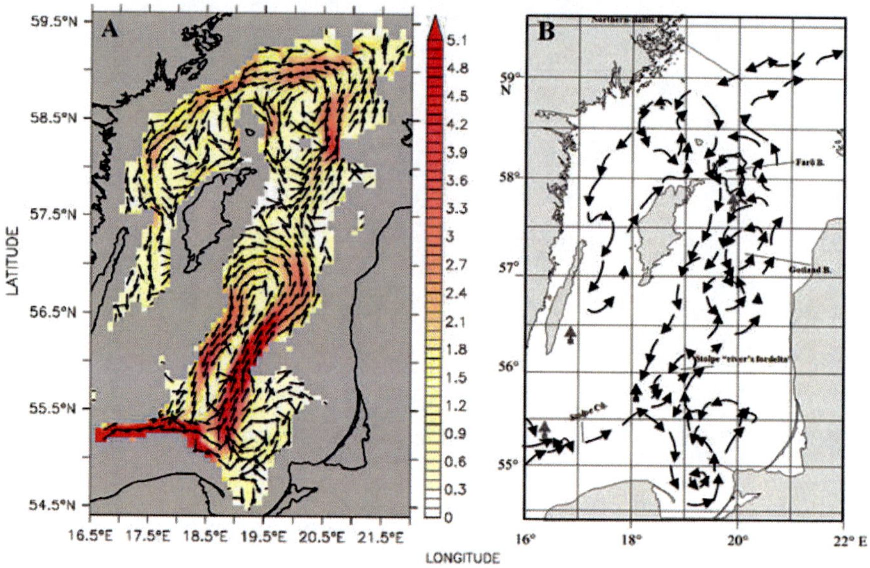

Fig. 5.4 **a** Current field within the Baltic Proper, water depth: 60 m, mean value from modelling results 1960–2005, model: MOM3 (Pacanowski and Griffies 2000), resolution: 3 nm, meteorological forcing: data file era40 (Uppala et al. 2005) courtesy, T. Seifert. **b** Late Holocene near-bottom currents (beneath the halocline). Paleoreconstruction by Emelyanov (2006) on the basis of Littorina mud thickness and proxies for resuspension and redistribution of sediments

5.3 Methodology

For the solution of the scientific task a special methodology elaborated mainly in basin analysis has been applied (Harff et al. 2001a). The target was to develop a spatial/temporal model of the basin fill under investigation based on measured data. These data derived from geophysical surveying and measuring of samples from sediment cores (facies variables) have to be connected spatially (interpolation) and allocated to the time of sediment formation. Coring sites were selected based on sediment profiling (sediment echosounder). The so-called master stations play a key role, representing the development of the basin through continuous sedimentary records. The variables can be measured for different cores taken at the master station, which later on are combined for a "composite" sediment sequence describing the master station. Continuously measured data sets ordered linearly along the sediment sequence within the cores are grouped according to the similarity of facies using multivariate classification methods. Contiguous samples showing a similar facies are put together to facies zones. This procedure is called "zonation" and defines core depth boundaries between lithostratigraphic units. These depth boundaries have to be converted to age data. Different dating methods for sediment samples can be used here. The age of the facies boundaries are determined by interpolation between the age data above and below the boundary. In a next step, the zonation of the sediment profile at the master station has to be transferred to contiguous sediment cores along seismic cross-sections using lithostratigraphic correlation methods for continuous logs of facies variables. We used a numerical method deploying the principle of multiple cross-correlation of sediment physical core logs (MSCL). The software CORRELATOR used (Olea and Sampson 2002) is an implementation of machine correlation that mimics the more conventional manual correlation of logs, which traditionally involves the simultaneous visual inspection of two logs per well, one of which is sensitive to the amount of shale. Given a stratigraphic interval A in well X, interval A is compared to intervals of the same length in well Y in order to find the interval in well Y that displays the maximum similarity both in terms of the amount of shale and in the pattern of fluctuations in the second log that are combined to produce a weighted correlation coefficient. In addition to the simplicity and efficiency of the approach, use of a weighted correlation coefficient has the advantage that the coefficient is an index for the quality of matches. Thus, the weighted coefficient can be used in combination with a threshold to eliminate correlations of low reliability. The program was originally developed for situations typical in the oil and gas industry. Yet the method has proved to be robust enough to satisfactorily work for the circumstances prevailing in marine geology. Having carried out the correlation for a grid of cross-sections the subsurface depths of stratigraphic zone boundaries can be spatially connected by numerical interpolation methods. In the result we receive digital elevation models of the subsurfaces of stratigraphic units. Different thicknesses of stratigraphic layers can now be interpreted in terms of paleo-dynamics in hydrography and sediment accumulation. Within the basinwide model detailed studies on the downcore variation of proxy variables can be carried out using methods of multivariate statistics. The (core-) depth to time transformation of the data is the main prerequisite for a time series analysis. We used the age data for the boundaries of physico-stratigraphic units as "anchor point" and interpolated between these points by applying the "piecewise cubic Hermite interpolating polynomial" method (PCHIP command in MatLab). This method finds values of an underlying interpolating function $P(x)$ at intermediate points providing smooth interpolation. This space to time transformation is supposed to produce smooth, monotonic functions honouring all of the tie-points. Time series analysis allows the extraction of information about these periodic components from time series data. We applied periodicity analysis based on spectral density estimates by means of fast Fourier transform (Bloomfield 2000). Results have been additionally enhanced with the help of "Hamming windowing and zeros padding technique". All the data records have been "de-noised" and detrended prior to periodicity analysis. For the signal-to-noise enhancement, we used a singular spectral analysis (SSA) method specially designed for noisy and not very long series (see Ghil et al. 2002), which originated from consideration of the theory of dynamical systems and multivariate statistics.

5.4 Data

5.4.1 Seismoacoustic Survey

High-resolution sub-bottom profiling was performed using the parametric sediment echosounder SES96 deployed during an expedition with R/V Poseidon in 2005 within the frame of the project GISEB (Harff 2005). It has a high system bandwidth and can therefore transmit short pulses without ringing (e.g. 1 period of 12 kHz). Short pulses, narrow beams and the absence of side lobes result in less volume and bottom surface reverberation compared to linear systems. This improves the signal-to-noise ratio and therefore the usable depth range (penetration depth). The primary transmitter frequency is about 100 kHz. During the profiling a secondary transmitter frequency was selected in the range of 6–12 kHz (preferably 8 kHz) depending on the water depth and the sediment type. All data are stored digitally on hard disk including navigational data. A motion reference unit was used to correct for ship's movement. A more detailed description of the SES96 sediment echosounder system is available at www.innomar.com.

Coring sites and coring device parameters (load, core barrel length, steering of the winch) were selected based on a first interpretation of the acoustic data. Profiling lines and stations are plotted in the cruise track plot of Fig. 5.5.

An echosounder record imaging the structure of the postglacial sediment sequences of the Gotland Basin is depicted in Fig. 5.6. The picture displays the colour-coded acoustic echo strength, with red colours for strong reflections and blue for weak echoes. The strength of the acoustic echoes (the reflectivity of a sediment sequence) depends on the vertical gradient of the acoustic impedance (the product of sound velocity times wet bulk density). A strong change in the vertical density profile will therefore cause a strong reflection in the echosounder record. The range of the density values of the Baltic Sea sediments extends from about 1,100 kg/m^3 (soft mud) up to 2,300 kg/m^3 (packed sand) whereas the sound velocity extends from about 1,400 m/s (mud, soft clay) up to about 1,900 m/s (sand). Therefore, most of the acoustic echoes reflect a change in density. In general, the echosounder records are interpreted using core data. Selected echo bands are identified and traced horizontally to map the thickness of Holocene and postglacial deposits as shown below.

5.4.2 Sampling and Sediment Data

Most of the Baltic sediment data have been acquired within the frame of international research projects. The first cores were taken in 1997 using R/V "Petr Kotsov" (Project BASYS, Winterhalter 2001a; Harff and Winterhalter 1996, 1997, Harff et al. 2001a), but most of them have been sampled by R/V "Poseidon" (Project GISEB, Harff 2005). Within the map of Fig. 5.6 a master station is marked (57.28°N, 20.11°E). This position was selected (Winterhalter 2001b) since the sediments here

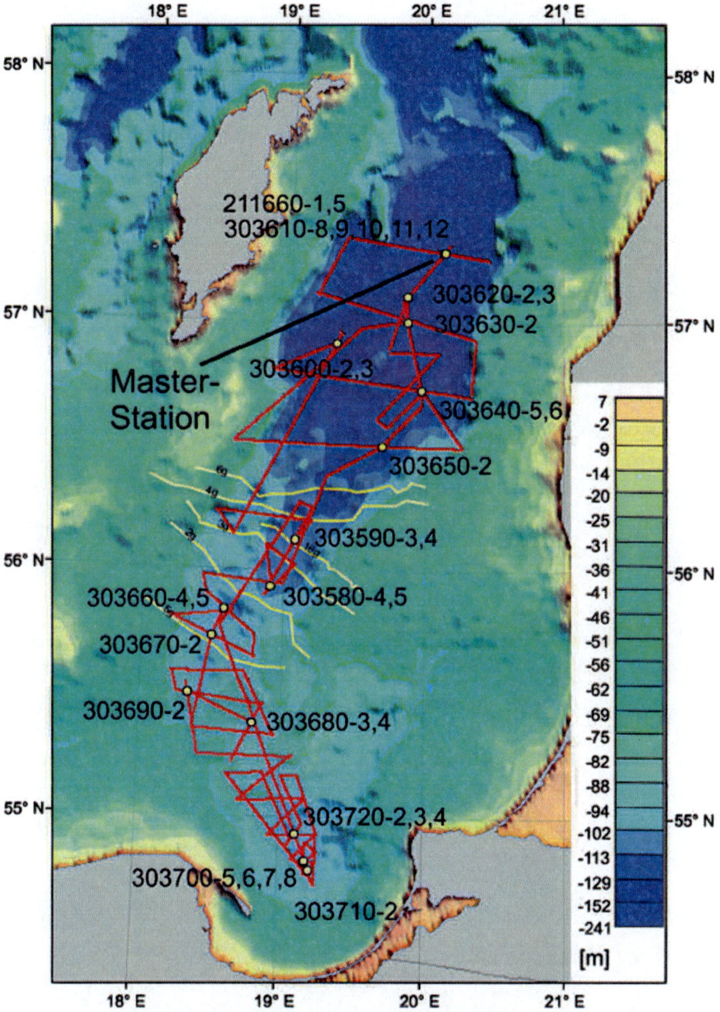

Fig. 5.5 Sediment echosounder SEL96 tracks (*red lines*), sampling stations (*yellow dots*) of cruise POS 323-1 (5–18 June 2005), and lithostratigraphic profiles after Emelyanov (2007) (*yellow lines*)

have been continuously accumulated during the Late Pleistocene and Holocene. Gravity core 211650-5 and piston core 211660-1 were sampled at that site in 1997 and core 303610-12 in 2005. Here we have used these three cores as references for the age model and stratigraphic analysis.

For coring we used gravity corers from 6 to 12 m length. The facies of the sediments have been analysed with respect to three main categories: sediment physical, geochemical, and microfossil (diatomological) analysis. Sediment physical variables provided the base for stratigraphic zonation and spatial correlation of sediment

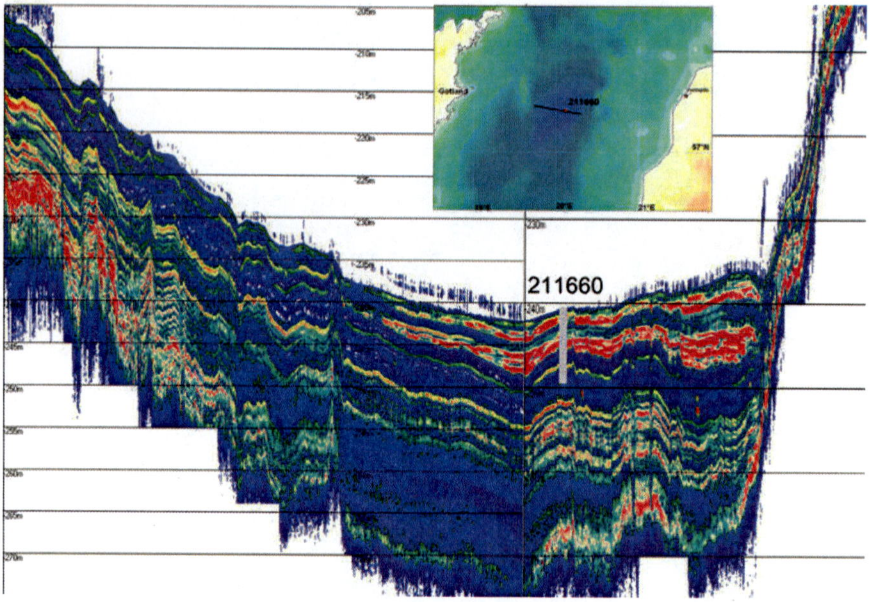

Fig. 5.6 E-W echosounder – profile crossing the "master station" within the Eastern Gotland Basin (sediment cores taken at the master station are marked)

cores. Geochemical and diatomological parameters have been used in particular as proxies for the paleo-environmental interpretation. Table 5.1 gives an overview about the variables measured including age models for each of the cores used in this study.

5.4.3 Physical Properties

Non-destructive logging (p-wave velocity, wet bulk density, magnetic susceptibility) of sediment cores was performed using a multi-sensor core logger (MSCL) from GEOTEK Ltd., UK.

For more detailed information about multi-sensor core logging, see Boyce (1973), Gunn and Best (1998), Schultheiss and Weaver (1992).

Sediment physical property data (see Figs. 5.7 and 5.8) have been used in this chapter for core zonation, correlation, and interpretation of echosounder records. As wet bulk density is sensitive to changes in the depositional regime, particularly an increased input of sandy (terrigenous) particles will directly cause an increase in density. A gradual decrease in density due to compaction occurs only in soft clay to silt sediments (not in sands) and is easily recognized. Magnetic susceptibility also reflects changes in the depositional regime (e.g. pelagic to terrigenous), but diagenetic formation of minerals like greigite will produce high values too. Comparing

Table 5.1 Variables measured including age model for sediment cores used in this study (GC, gravity corer; PC, piston corer)

Core-ID	Core doc	Age model	RGB scan/core photography	p-Wave velocity V_p	Density		Magnsusc	XRF	Diatoms	Device	Project
					MSCL	Sample					
211660-1	X	X	X			X				PC	BASYS
211660-5	X	(x)	X	X	X		X			GC	BASYS
303610-12	X	(x)	X		X		X	X	X	GC	GISEB
303580-5	X				X		X			GC	GISEB
303590-3	X				X		X			GC	GISEB
303620-3	X				X		X			GC	GISEB
303640-6	X				X		X			GC	GISEB
303650-2	X				X		X			GC	GISEB
303660-5	X				X		X			GC	GISEB
303670-2	X				X		X			GC	GISEB
303680-4	X				X		X			GC	GISEB
303690-2	X				X		X			GC	GISEB
303700-9	X				X		X			GC	GISEB
303710-2	X				X		X			GC	GISEB
303720-3	X				X		X			GC	GISEB

BASYS, Baltic Sea System Studies (Winterhalter 2001a); GISEB, GIS for Time/Space Modeling of Sediment Distribution as a Function of Changing Environment in the Baltic Sea (Harff 2005)

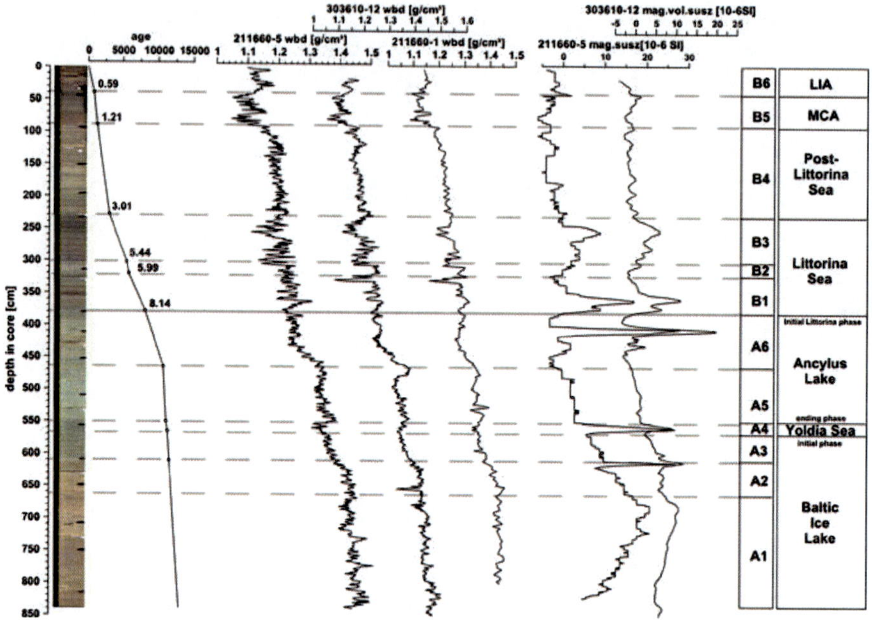

Fig. 5.7 "Composit" of cores 211660-1, 211660-5, and 303610-12 for the master station within the Eastern Gotland Basin. Colour scan of core 211660-5, updated age model after Harff et al. (2001a), for B zones the ages of the boundaries are given in cal. years BP, density and magnetic susceptibility displayed as functions of 211660-5 depth scale (for more detailed information, see text.). Physico-stratigraphic zones after Harff et al. (2001a), climate stages and Baltic Sea stages after Andrén et al. (Chap. 4, this book)

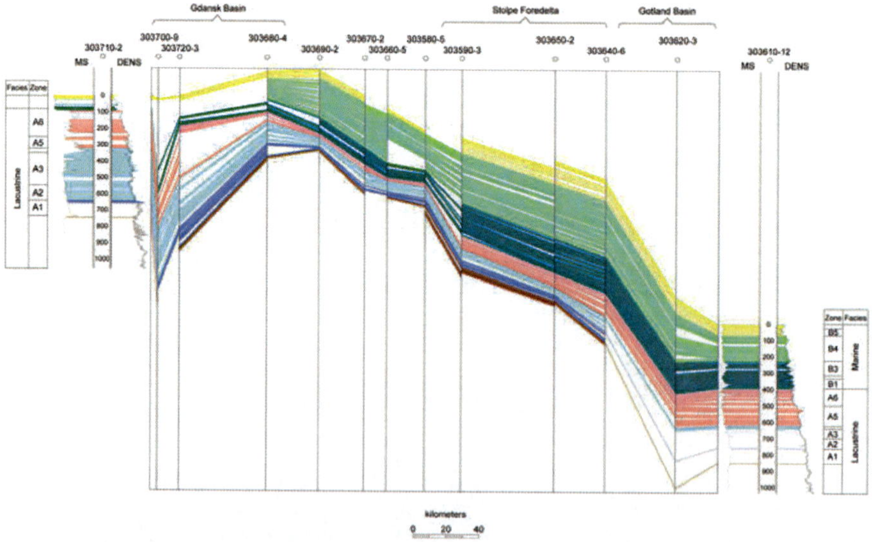

Fig. 5.8 Gdańsk to Gotland Basin lithostratigraphic correlation. Datum: sea level; depth origin: top of core; vertical exaggeration: 10,000 times; vertical water depth exaggeration: 1,000 times

close-by cores shows that these greigite spots are situated in specific layers and depth ranges and may also, with some caution, be used for zonation and correlation. p-Wave velocity V_p depends on the strength K (compression modulus) and the density (dwb) of the sediments ($V_p^2 = K/\text{dwb}$).

In soft homogeneous mud deposits, the sound velocity will drop below the sound velocity of water because the compression modulus remains about the same, but the density increases. In case of the deposition of laminated sediments, the thin layers have a higher strength but about the same density as the homogeneous mud resulting in a higher sound velocity in comparison to the homogeneous mud. Harff et al. (2001a) used this phenomenon to investigate the succession of laminated and homogeneous sediments using an acoustic index, which is the detrended and normalized p-wave velocity.

5.4.4 Geochemical Data

X-ray fluorescence (XRF) logging has been deployed to describe the down-core changes in chemical composition of the sediments of core 303610-12. For measurements an Avaatech XRF Core Scanner of the Royal Netherlands Institute for Sea Research (NIOZ) has been used. XRF analyses were carried out on the surface of split sediment cores. The surface of the split cores has been carefully flattened and covered finally with a thin (4 μm) Ultralene film, further diminishing surface roughness and preventing contamination of the measurement unit during core logging. While measuring the scanning system is flushed with helium to prevent partial or complete absorption of emitted radiation by air. The X-ray fluorescence signal which arrives at the detector originates from a sediment depth from about 50 μm for light elements up to 1 mm for heavy elements. The following components have been measured in a 0.5 cm step size: Al, Si, P, S, Cl, K, Ca, Ti, Cr, Mn, Fe, Co. The raw data were processed with WinAxil PC XRF analysis software. Data acquired are qualitative, given in numbers of counts per 30 s of measurement time. For method descriptions, see Richter et al. (2006).

5.4.5 Diatomological Data

Diatom analysis covered the sediment interval between 20 and 520 cm of the core 303610-12. Subsamples for the diatom analysis were taken at a sample space of 1–10 cm, depending on the lithology. In general the first 100 samples were collected in core interval of 20–330 cm at every 3 cm. The total number of samples analysed amounted to 132. Approximately 300 valves were counted in each sample. Based on the diatom counting results auto-ecological properties of particular taxa were determined and the grouping of diatoms in terms of habitat, salinity, and trophy was performed. The percentages of particular ecological groups were computed by means of the OMNIDIA ver. 3 software, which has a database (Omnis7) with information on more than 11,000 species. Diagrams showing the percentages of dominant taxa, but also for the ecological groups, were constructed. This was done with Tilia® and C2 software.

In the studied material, a total of 219 diatom taxa were identified. The quality of their preservation varied. The best preserved flora was found in laminated, fine-grained sediments. On the contrary the diatom record in the homogeneous sediments was distinctly poorer. The diatom valves were either affected by dissolution or mechanical fragmentation. In the homogeneous sediments the intervals of 20–37 and 126–172 cm were barren in diatoms. Hence, to be able to count the recommended 300 valves and be able to perform statistic analysis two slides were merged for the counting.

5.5 Results

5.5.1 Zonation of Basin Sediments

All three cores sampled at the master station have been used for the litho- and chronostratigraphic subdivision of the postglacial sediments in the Eastern Gotland Basin: 211660-1 (to develop the age model), 211660-5 (for the physico-stratigraphic zonation), and 303610-12 (as start profile for the basinwide lithostratigraphic cross-section). The cores have penetrated undisturbed sediments mirroring the development of the basin from the Late Pleistocene to the Holocene.

A lithostratigraphic zonation and correlation between cores taken at the master station was performed using core photography and downcore measured sediment physical properties. The principles of the method have been described by Harff et al. (1999). In the result we obtain a "composite" of the late Pleistocene to Holocene sedimentary sequence at the master station referred to the 211660-5 depth scale. Figure 5.7 shows curves of physical properties of the master station cores used for the correlation. For cores 211660-5 and 303610-12 wet bulk density and magnetic susceptibility were available for the correlation. The physico-facies of core 211660-1 (piston corer) has been described by density values measured from samples taken in a distance of about 2.5 cm.

The physico-stratigraphic zonation defined by Harff et al. (2001a) for core 211660-5 (using p-wave velocity, wet bulk density, and magnetic susceptibility) has been transferred to the whole set of cores sampled at the master station and is marked in Fig. 5.7 with an RGB colour scan of core 211660-5.

Within the lower parts of the cores the varved sediments of the Baltic Ice Lake (A1/A2 zone) are clearly visible. The short Yoldia Phase is marked by an initial phase (upper part of zone A3), a 3 cm mud layer enriched in organic matter (A4 zone), and an end phase (lower part of A5 zone). It is overlain by the homogeneous bioturbated fine-grained sediments of the Ancylus Lake with black Fe-sulphide spots. The lower part (A5 zone) is grey, whereas the upper part (A6 zone) is brownish in the scan due to oxygenated iron. Diagenetic mineral formation (greigite) causes typical anomalies in the magnetic susceptibility. The transition from the late

Pleistocene/early Holocene freshwater to brackish–marine environment can be seen at a core depth of about 380 cm where almost homogeneous sediments at the bottom are replaced to the top by laminated sequences. The reason for the general difference in the sedimentary facies is mainly caused by the halocline established by the inflow of saline water. During the late Pleistocene and early Holocene, the Baltic Basin was – except for the short Yoldia Sea Stage – disconnected from the world ocean. Within the lacustrine, partly isolated basins, the sedimentary facies were controlled primarily by variation in atmospheric temperature and precipitation. These parameters controlled changing water levels and contours of coastlines and changing supply of detrital sedimentary material from the drainage area. Lateral exchange of water masses between the basins did not play the important role assumed for the late Holocene when the permanent connection between the Baltic Sea and the Atlantic Ocean was opened.

For the B zones laminated sediments prevail. However, it should be stressed that zones B1, B3, and B5 are clearly laminated, while zones B2 and B4 (and B6) are more homogeneous. This is interpreted as a result of bioturbation due to good supply of oxygen to the bottom water. Zone B4 shows in its lower part still some lamination and can be regarded as a transition from laminated zone B3 to the more homogeneous structure of zone B4 (upper part). According to the dating, zone B5 represents the Medieval Climate Anomaly (MCA), while zone B6 denotes the Little Ice Age (LIA). The recent warm period is not displayed because the gravity coring system does not preserve the uppermost layers. The last 1,000 years is represented by multi-corer (MUC) samples that Leipe et al. (2008) describe from the Eastern Gotland Basin (site: 56°55′N, 19°20′E).

In Fig. 5.7 also the age model used for the master station is given as a curve. The age data of the zone boundaries have been projected from core 211660-1 to cores 211660-5 and 313610-12. The age model of core 211660-1 which combines data from paleomagnetic studies, AMS dating, and glacial varve analyses is explained by Harff et al. (2001a). Kotilainen et al. (2000) used inclination and declination of magnetic measurements of sediment cores for comparison to the secular variation recorded in varved lake sediments in Finland to date the sedimentary sequence over the past 3,000 years. Littorina Sea sediments were dated by Andrén et al. (2000) by AMS ^{14}C analysis. These dating results are still used here and by Andrén et al. (Chap. 4) in this book since no more reliable data for Littorina Sea sediments for the master station have been published during the last years. Dating of glacial varves came from measuring and correlating the Swedish time scale to the Greenland GRIP ice core $\delta^{18}O$ record (Andrén 1999, Andrén et al. 2000).

5.5.2 Spatial Correlation of Late Pleistocene to Holocene Sediments

In order to correlate the lithostratigraphic zonation from the basin centre along the basin axis to the SW, the Stolpe Foredelta, and the Gdańsk Basin (Fig. 5.3), the cores along a profile marked in Fig. 5.5 have been compared based on their MSCL data (density and magnetic susceptibility, Table 5.1) from core to core. The following sequence of core comparisons has been used along the profile:

303610-12 → 303620-3 → 303640-6 → 303650-2 → 303590-3 → 303580-5 → 303660-5 → 303670-2 → 303690-2 → 303680-4 → 303720-3 → 303700-9 → 303710-2.

In Fig. 5.8 the results of core-to-core correlation are displayed with regard to the lithostratigraphic correlation. The zones are colour coded whereby correlation with a strength >0.4 are graphically shown only. It is clearly visible that the freshwater sediments of the A zones (brown, blue and purple coloured) are well developed within the basins whereas on the ridge between Gotland and Gdańsk Basin these sequences are represented by thinner sediment successions. This fact can be explained by the higher proportion of detritus within the sediments and the closer distance of the basin centres to the terrestrial sediment sources. The brackish sequences of the B zones (green to yellow colours) display a different pattern. An overall thickness of 1 m within the Gdańsk Basin is to be compared with 4 m thickness within the Gotland Basin and 7 m on the down basin part of the ridge. This can be explained by the opening of the Öresund Strait about 8,000 cal. year BP (Björck 2008). As the Littorina transgression began the local coast-to-basin transport is replaced by a lateral basin-to-basin transport. Driven by west to east atmospheric energy transfer dense marine water enters the Baltic Basin and follows the counterclockwise transport path including basin-to-basin flow (paragraph 2). At the same time a halocline starts to develop inducing the typical estuarine (vertical) current system. Suspended matter including particles from bottom (and coastal) erosion and from biologic production in the uppermost part of the water column is transported with the water masses from the west to the east and is being deposited when transport energy slows down. This is the case in front of submarine channels and in the deeper basins explaining the sediment accumulation of the Stolpe Foredelta and of the Eastern Gotland Basin.

5.5.3 Thickness Analysis

In order to extend the first results in thickness analysis of the main Late Quaternary stratigraphic units within the Baltic Proper achieved by sediment core correlation, the study has been extended to a spatial 2D analysis. The main focus was directed to a comparison of thickness evolution of the early Holocene (pre-Littorina) sediments of the A zones with the brackish Littorina sediments of the B zones within acoustic cross-sections marked in Fig. 5.5. The base of the early Holocene (psammitic) Ancylus sediments (A zones) and the brackish muds (B zones) are clearly defined by reflectors within the seismoacoustic signals. For the identification of these two lithostratigraphic boundaries within the SES profiles the correlation results of sediment cores located on the profiles have been used. In Fig. 5.9 the procedure is explained by using the central basin profile Pr-0.

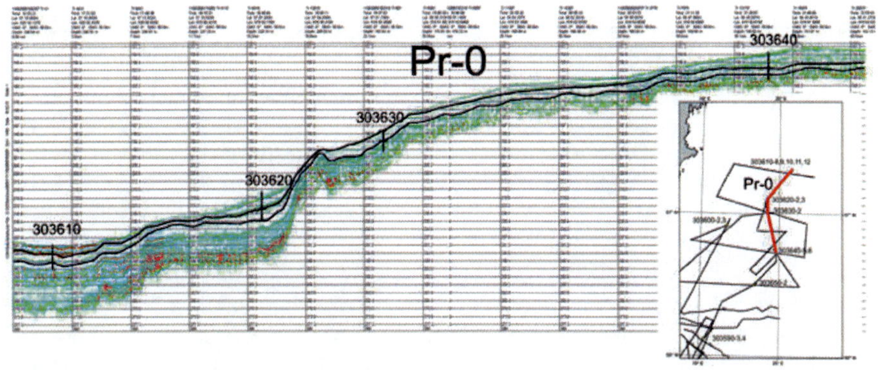

Fig. 5.9 NE–SW echosounder cross-sectional profile 0. *Dark lines* mark the stratigraphic base boundaries A and B zones. Coring locations and core depths are schematically displayed

If one compares the seismoacoustic profile in Fig. 5.9 with the lithostratigraphic correlation scheme of the cores displayed in Fig. 5.8 as well as within the seismic signal in the MSCL data, the bases of A and B zones can be recognized easily by the contrast in their sedimento-physical properties. The boundaries within each of the seismic profiles (bold red lines in Fig. 5.5) have been digitized and marked by thin dark lines as it is shown in Fig. 5.9. Digitized and geo-referenced stratigraphic boundaries were stored in a database for mapping the subsurfaces. However, the seismic survey of the POSEIDON expedition POS 323-1 did not sufficiently cover the area of the Stolpe Foredelta. As this basin structure plays a key role in understanding the depositional system of the central Baltic, additional lithostratigraphic cross-sections (yellow lines in Fig. 5.5) have been incorporated in the analysis. Emelyanov (2007) analysed sediment echosounder and sediment core data on NW–SE tracks crossing the basin axis perpendicularly. The digitized data from these profiles have also been integrated into the database so that finally an adequate data set was available for mapping of the subsurface of the base of the A3 zone (top of glacial sediments) and the base of the B zones (top of the Ancylus Lake sediments). The sea bottom surface is given by the bathymetric data of the Baltic Sea (Seifert et al. 2001). Having modelled the surface and subsurfaces the thicknesses of the A3 to A6 zone sediments (top of Baltic Ice Lake to top of Ancylus Lake) and B zones (Littorina to recent Baltic Sea sediments) can easily be calculated and mapped. In Fig. 5.10 both thickness maps can be compared.

For the A3 to A6 zone sediments the Eastern Gotland Basin and, in particular, the Gdańsk Basin form the depo-centres. The thickness maps support the results achieved by core-to-core correlation (Fig. 5.12). The transport and deposition are obviously dominated by terrestrial (fluvial) sediment sources. The (paleo-) Wisla River discharged its load to the Gdańsk Basin whereas the Eastern Gotland Basin sediments also descend from the (uplifting) Gotland Island in the west and from the mainland in the east. In contrast, the depositional pattern of the younger B zones is determined by particle load delivered by the currents entering the Gotland Basin

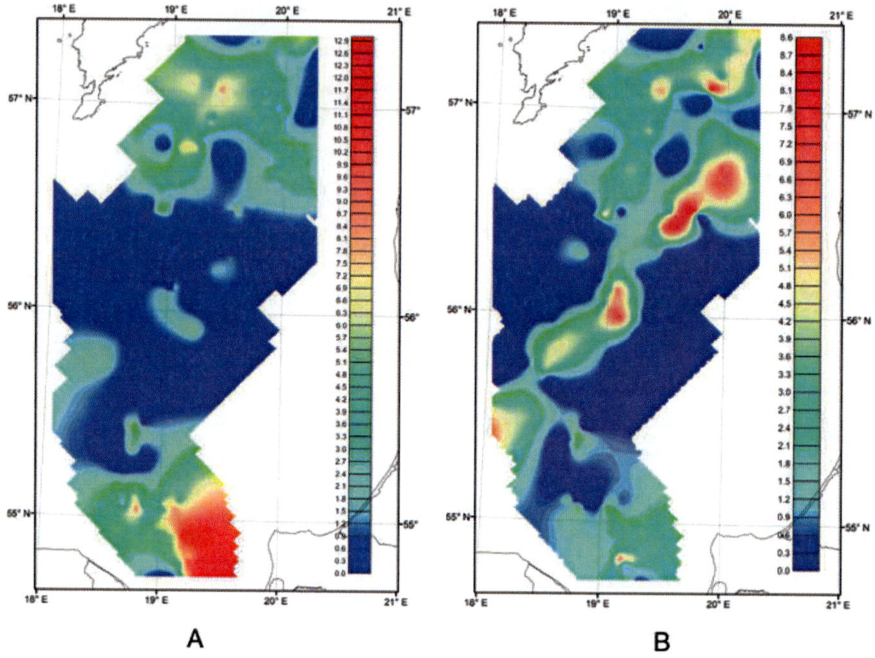

Fig. 5.10 Thickness (in m) of A zone sediments (Baltic Ice Lake to Ancylus Lake) and B zone sediments (Littorina to LIA)

through the Stolpe Channel. This is due to the opening of the entrance to the Baltic 8,000 cal. year (Harff et al. 2005), and with the Littorina transgressions the general basin-to-basin current system of the Baltic Sea (Fig. 5.4) controlled the particle dynamics. As a result the SW–NE trending sediment body of the Stolpe Foredelta accumulated and this structure can be identified in map B of Fig. 5.10.

5.5.4 Downhole Facies Variation at the Central Eastern Gotland Basin as Indicator for Holocene Environmental Change

In a previous work, Harff et al. (2001a) have mentioned that the change in the abundance of lamination in central Eastern Gotland Basin cores can be used to reconstruct the oxygen supply to the bottom water during deposition. Changes in lamination are reflected well in the acoustic MSCL p-wave velocity. An acoustic index as detrended (0, 1) standardized p-wave velocity turned out to show values close to 1.0 in laminated sediments, reflecting anoxic environment of deposition, whereas homogeneous sediments deposited under oxic conditions of bottom water show an acoustic index near 0 (Harff et al. 2001a). This implies that the acoustic index can be used as a qualitative proxy variable indicating the ventilation of near-bottom water during sediment deposition. Looking at the p-wave velocity curve

Table 5.2 Factor (principle components, PC) loadings, XRF data for master station, core 303610-12

	PC 1	PC 2	PC 3
Al	−0.949103	0.202158	0.054130
Si	−0.958174	0.145036	0.041052
P	−0.452261	−0.149140	0.413834
S	0.346463	−0.100103	−0.851123
Cl	0.677206	−0.593041	0.223976
K	−0.969052	0.182937	0.048005
Ca	0.018410	0.791265	−0.298696
Ti	−0.926965	0.129994	0.029933
Cr	−0.489651	−0.336852	0.132195
Mn	0.659964	0.608977	0.093181
Fe	−0.665403	−0.242609	−0.598784
Co	−0.411766	−0.618346	−0.256555
Expl. var	5.688367	2.047864	1.493389
Prp. totl	0.474031	0.170655	0.124449

from A6 to the B6 zone in core 211660-5 (in Harff et al. 2001a) this assumption can be confirmed as the homogeneous, bioturbated sediments of zones A6, B2, B4, and B6 show in general lower values than those at laminated zones B1, B3, and B5 (see Sect. 5.4.3). In order to specify the depositional environment for the different zones, from Ancylus Lake sediments to the recent Baltic Sea, geochemical parameters as well as diatom data have to be included in the analysis.

For the facies interpretation we have furthermore conducted a PCA (principle component analysis) for the data on concentration of geochemical elements. Table 5.2 shows the factor loadings. High negative loadings for Al, Si, K, Ti, (Fe) identify factor 1 as a proxy for detritical minerals derived from terrestrial sources. K has the highest loading of this factor and points to an illitic clay component (Gingele and Leipe 1997). Factor 2 is determined by high loadings for Mn and Ca which represents the early diagenetic formation of a Ca-Mn-carbonate (rhodochrosite) phase (see Neumann et al. 1997, Alvi and Winterhalter 2001; Sohlenius et al. 1996, 2001, Burke and Kemp 2002, Sternbeck et al. 2000). Factor 3 expresses the dominant position of sulphur (negative loadings) which is at least partly connected to a reduced iron sulphide phase, but additionally strongly bound to the organic-rich laminated mud sequences (organic sulphur) and can therefore regarded as proxy for the oxygen depletion in the paleo-bottom water. In contrast, in factor 3 P is known to be released from the sediment during anoxic conditions (Emeis et al. 2003; Conley et al. 2009).

We concentrate here on the principle components 1 and 3, which express syngenetically controlled processes.

In Fig. 5.11, the downcore concentration of K, Ti, and S is presented for core 303610-12 together with the physico-stratigraphic zonation of this core. Compared to zones A6, B2, B4, and B6 being characterized by relatively high values of K and Ti, zones B1, B3, and B5 show clearly lower concentrations of these elements. The reason is attributed to higher terrestrial discharge to the basin during the deposition of B2, B4, and B6 zone sediments compared to those of zones B1, B3, and B5.

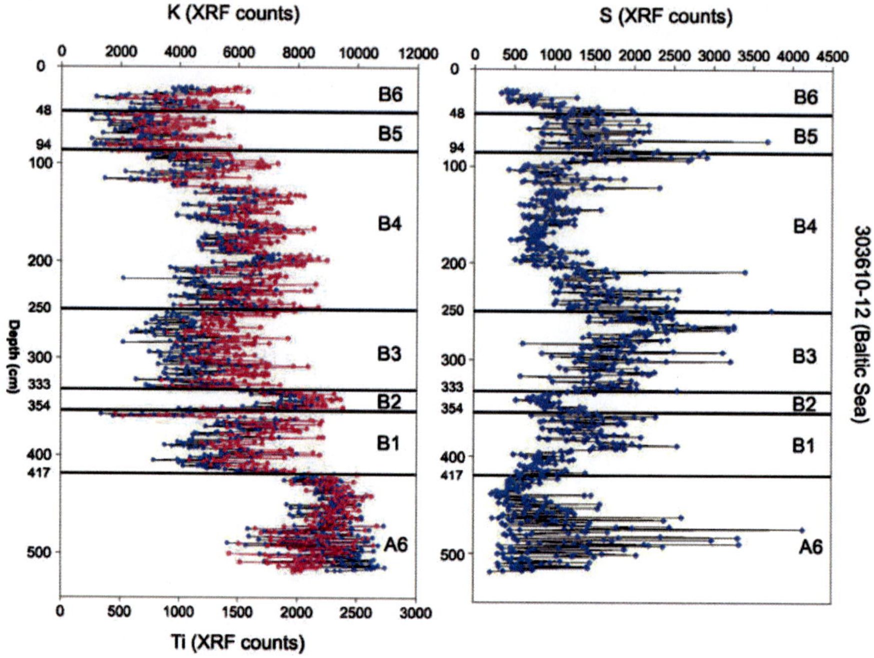

Fig. 5.11 Concentration of K (blue) and Ti (purple) (*left panel*) and S (*right panel*), expressed by XRF counts, and physico-stratigraphic zonation of sediments in core 313610-12 at the "master station" of the Eastern Gostland Basin

This higher terrestrial discharge can be explained either by higher precipitation and river runoff or stronger erosion of the southern coasts due to storm-driven wave and coastal current activity.

The lower concentrations of K and Ti in B1, B3, and B5 sediments stand for a relative decrease in terrestrial discharge together with pelagic deposition. Additionally, the K-concentrations can be interpreted as a function of aeolian dust deposition.

In Fig. 5.11, also the concentration of S as a function of core depth together with the physico-stratigraphic zonation is presented for core 313610-12. For our interpretation we have to take into account that the high S-concentrations in zone A6 are due to Fe-sulphides formed diagenetically within Ancylus Lake sediments. In contrast, the sulphur content in the B zones is regarded to be bound to organic sulphur complexes as well as to diagenetic iron sulphide phases (pyrite) being formed in anoxic environment (Sternbeck and Sohlenius 1997). Therefore, high sulphur concentration is interpreted as an indicator for anoxic environment. The sulphur concentrations range between an average of 500 XRF counts for homogeneous (oxic) sediments and 2000 XRF counts for laminated (anoxic) sediments. According to these values we interpret the top of the (Ancylus) zone A6, and zones B2, B4, and B6 deposited under oxic conditions, whereas zones B1, B3, and B5 originate from anoxic bottom water. It is, however, also visible from the S-concentration curve that the lower parts of zone B1 show still some ventilation of the bottom water (dysoxic), whereas a lower subzone of B4 is deposited under suboxic conditions. The change of laminated and homogeneous sediments in Gotland Basin sediments and its relation to changing oxygen supply to the bottom water has been discussed already by Ignatius et al. (1981) (see also Conley et al. 2002), but there is still a debate ongoing whether the ventilation of the bottom water is caused by inflow events of higher saline water or by vertical convection during fresher phases of the water body (see, e.g. Meier and Kauker 2003, Zillén et al. 2008). To answer this question we have used diatom analyses for the reconstruction of paleosalinity. Westman and Sohlenius (1999), Sohlenius et al. (2001), Emeis et al. (2003), and among others had already shown the potential of diatoms for paleosalinity studies in the Baltic. Figure 5.12 shows the result of a corresponding paleo-environmental study for core 303610-12 where dominant species are displayed.

Based on physico-stratigraphic zonation and changes in the distribution of the dominant species two diatom assemblage zones (zones A and B) were distinguished. In addition, in zone B, six subzones (B1–B6) were distinguished.

Zone A (520–417 cm)

In this sediment interval freshwater forms, the so-called large lake species predominate and they attain over 80% of the diatom assemblage. Among them *Aulacoseira islandica* (O. Műller) Simonsen, *Aulacoseira subarctica* (Műller)

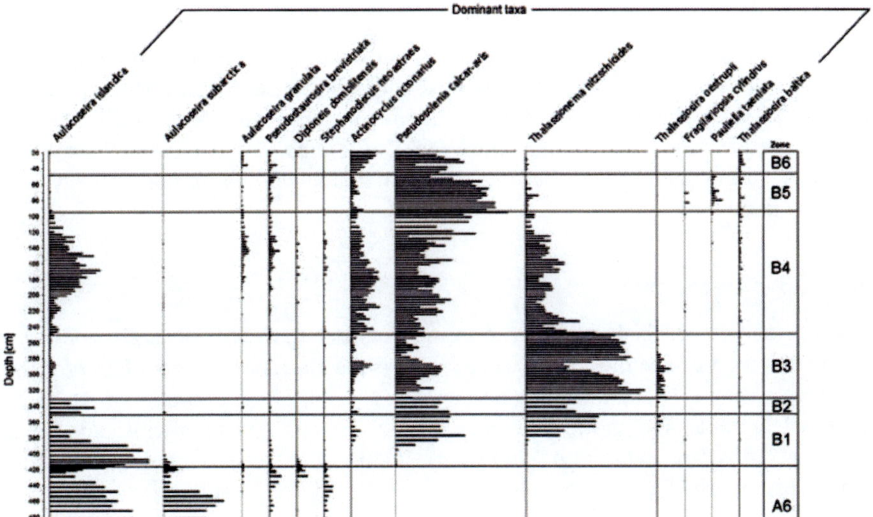

Fig. 5.12 Diatomological paleo-environmental indicators and physico-stratigraphic zonation in sediments of master station core 313610-12 (Eastern Gotland Basin)

Krammer, and *Stephanodiscus alpinus* Hustedt are dominant. The above-enumerated taxa are planktonic and the content of benthic ones is low. However, at the depth of 436 cm a drastic decrease in the content of planktonic forms is observed. They are replaced by benthic, typical freshwater forms showing in that part of the profile their maximum abundance, e.g. *Cymatopleura elliptica* (Brebisson) Smith, *Diploneis dombilitensis* (Ehrenberg) Cleve, *Gyrosigma attenuatum* (Kützing) Cleve.

At the boundary between zones A and B a distinct change in diatom preferences with respect to salinity can be observed.

Zone B1 (417–354 cm)

In this subzone a major decrease in benthic forms is recorded with planktonic forms becoming the most abundant ones. They reach up to 90% of relative abundance at the B1/B2 boundary. In the lower part of the subzone still freshwater and brackish water forms dominate. At the depth of ca. 380 cm a drastic increase of brackish–marine and marine–brackish forms and simultaneously a drastic decrease of freshwater species is recorded. Dominant above this depth a rapid increase in relative abundance in marine planktonic forms, e.g. *Pseudosolenia calcar-avis* (Schultze) Sundstrom, *Thalassionema nitzschioides* (Grunow) Grunow, is observed. There are also indicator species implying inflow of warm oceanic waters, e.g. *Actinocyclus octonarius* Ehrenberg and *Thalassiosira oestrupii* (Ostenfeld) Hasle.

Zone B2 (354–333 cm)

In this subzone a significant increase of freshwater, planktonic diatoms, e.g. *A. islandica*, is observed. It is accompanied with a decrease in brackish–marine and marine–brackish diatoms. The hitherto dominant marine–brackish and brackish–marine species, such as *P. calcar-avis* and *T. nitzschioides*, decrease. Freshwater diatom species which often occur together with diatoms living in more salty waters increase in this subzone.

Zone B3 (333–250 cm)

Marine, marine–brackish, planktonic forms reaching up to 65% strongly dominate the diatom record. The most abundant of them is the marine species *T. nitzschioides*, but a decrease in its abundance is observed at a depth of ca. 285 cm. At the same time a clear but also temporary increase of brackish–marine and brackish species is noted, e.g. *P. calcar-avis* and *A. octonarius*. In this unit a general increase in marine diatoms is recorded, e.g. *T. oestrupii*.

Zone B4 (250–94 cm)

Planktonic forms dominate in this zone, but their abundance decreases systematically up to 52% at the depth of ca. 130 cm, followed by an upward increase and amounts to 80% at the B4/B5 limit. An increase in the content of benthic species (in general up to 20–30%), and later on a drastic decrease to 5% (e.g. *Diploneis dombilitiensis*), is noteworthy.

The diatom record shows substantial differences within this subzone. In the lower part marine–brackish and brackish–marine forms predominate similar to the preceding unit. In the central part an increase in freshwater taxa from ca. 10% up to ca. 35% is observed, but then decrease to 2% at the B4/B5 limit. The most abundant of them is the planktonic species *A. islandica*. More brackish preferring species reveal a similar trend. Here the *P. calcar-avis* dominates; its content increases towards the top where it reaches its maximum abundance. Likewise the percentage of *T. nitzschioides* is high, but it shows a decrease towards the next subunit from ca. 40% to only 5% at a depth of 94 cm.

Zone B5 (94–48 cm)

In this subzone, brackish–marine planktonic diatoms are the most abundant species, but their content decreases towards the core top. A slight increase in benthic diatoms is observed in this subzone. This small increase continues also in zone B6. In addition, we can see an increase in brackish and brackish–fresh diatoms. *T. nitzschioides*, *A. octonarius* and *A. islandica* show a strong decline, whereas *P. calcar-avis* increases initially though its content finally decreases from c.a. 60 to 35%. An interesting feature is the appearing of sea-ice species, which reach up to 7%, and indicate inflow of cold, marine water, e.g. *Fragilariopsis cylindrus* (Grunow) Krieger and *Pauliella taeniata* (Grunow) Round and Basson.

Zone B6 (48–20 cm)

The lower part of this subzone is still dominated by brackish–marine planktonic species, although they show a strong upward decreasing trend. In general taxa of higher salinity requirements decrease. Regarding salinity in zone B6, brackish–fresh, brackish and brackish–marine species (*P. calcar-avis, A. octonarius*, *Thalassiosira baltica* (Grunow) Ostenfeld) are dominants, respectively, in the uppermost part. Notable is a steady increase of halophilus taxa, which according to salinity classification by Van der Werff and Hulls (1957–1974) prefer salinity between 0.18 and 0.9 psu, e.g. *Aulacoseira granulata* (Ehrenberg) Simonsen, *Pseudostaurosira brevistriata* (Grunow in Van Heurck) Williams and Round, *Staurosira construens* var. *binodis* Ehrenberg. Characteristic for this subzone is also very low percentages of the marine *T. nitzschioides*.

We cannot exclude that the diatom flora composition (which abundance is very low in this subzone) is affected by sediment disturbance caused by coring procedure or a gravity slide in zone B6 indicated by the sediment texture in the uppermost part of the core.

5.5.5 Periodicity (Frequency) Analysis

In order to investigate if the periodical facies changes have regional or even global signatures, we have carried out periodicity analysis of two proxy variables: p-wave velocity and the PC1 score of the geochemical XRF data (standing for the relation between pelagic and terrigenous deposition). Time series of climatic phenomena contain periodic components related to forcing at a wide range of time scales – from decades to millennia. The (core) depth to time transformation of the data is the main prerequisite for a time series analysis.

Figure 5.13 shows the variables' p-wave velocity of core 211660-5 and PC1-score of XRF data of core 303610-12 as time series after depth to time transformation. Additional columns summarize results from paleo-environmental reconstructions based on geochemical and diatom analyses. In Fig. 5.14 the spectral densities are displayed. As in particular the dating of Littorina Sea sediments are regarded uncertain in detail (Sect. 5.5.1) we do not interpret here the high-frequency periodicities, but concentrate instead on centennial scale periods where smaller uncertainties in the age model can be neglected.

In a previous study (Kotov and Harff 2006), we have investigated the periodicity in the grey scale values of colour scans of core 211660-5. The 900-year period appeared to be the most prominent peak of spectral densities. Additionally the 400- and 500-years periods were identified to be significant for the grey-scale time series of core 211660-5. These results are confirmed by the analysis carried out here. The

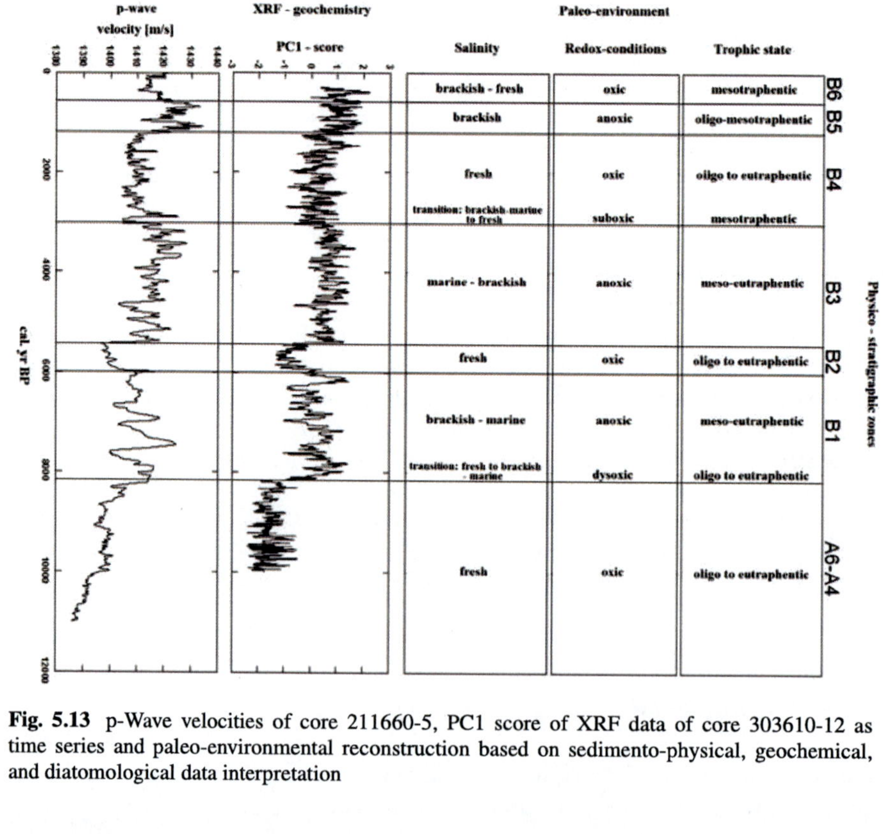

Fig. 5.13 p-Wave velocities of core 211660-5, PC1 score of XRF data of core 303610-12 as time series and paleo-environmental reconstruction based on sedimento-physical, geochemical, and diatomological data interpretation

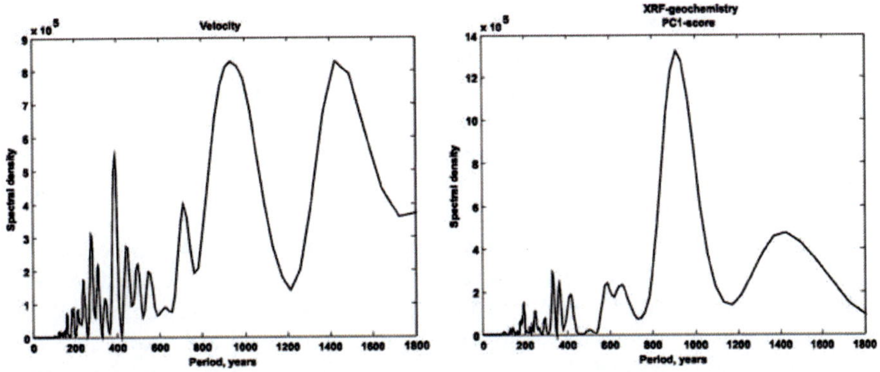

Fig. 5.14 Spectral density of p-wave velocity of core 211660-5, and PC1 score of XRF data of core 303610-12

900-year peak is dominant for both the PC1 score of geochemical data of core 303610-12 and the p-wave velocity of core 211660-5. Periodicities of higher frequencies are comparable, but do not completely coincide. The reason may be that the age model of core 211660-1 was transferred to the cores at the master using a lithostratigraphic correlation method which may have caused deviations in the high-frequency fluctuations. In addition to the 900-year period we would like to note the 1500-years cycle, which can be read from the spectral densities in Fig. 5.14. Even if this frequency shows a lower significance than the 900-years period we regard it notable as it is the first evidence of the "Bond Cycle" (Bond et al. 1997, 2001) in Holocene sediments of the Baltic Sea.

5.6 Discussion

We can conclude that the A zones have been deposited under well-ventilated conditions relatively poor in organic matter production and preservation. At 8.16 cal. years BP the lacustrine environment changed within the central Baltic basin rapidly to brackish conditions by inflowing saline water of the Littorina transgression leading to the deposition of laminated sediments. As a result, benthic fauna emigrated, making the accumulation of laminated sediments possible. This change in the environment is caused by a sea level rise to be correlated with the atmospheric warming phase after the significant cooling phase 8.8–8.2 k years ago (Sarnthein et al. 2003).

Within the Littorina sedimentary facies physico-stratigraphic zonation indicates shifts in the depositional environment on centennial time scales. According to the diatom analysis within the B zones brackish–marine phases (B1, B3, B5) are replaced periodically by phases of fresher water conditions (B2, B4, B6). Also the influx of terrigenous matter is intensified. This can be concluded not only from the geochemical data but also from the diatom record that shows an increase of acidophilus species indicating erosion of coastal peat. Due to this fact we interpret the terrigenous discharge by storms inducing coastal erosion and increasing coast-to-basin transport. In contrast, the brackish–marine phases are characterized mainly by pelagic deposition and basin-to-basin transport. The diatom record and the reconstruction of European paleotemperatures from pollen data (Davis et al. 2003) also support the interpretation of warmer sea water temperature during brackish phases B1, B3, and B5 whereas during the deposition of the more lacustrine phases of zones B2, B4, and B6 colder water masses prevailed. By dating, zone B5 can be allocated to the Medieval Climate Anomaly (MCA), whereas B6 mirrors the Little Ice Age (LIA). The uppermost sediments have been investigated by Leipe et al. (2008) who have shown that the MUC sediment succession reflects the change of climatic conditions from MCA through LIA to MWP. Within that core sediments from the MCA and the MWP are represented by dark laminated sediments interrupted by a layer of 30 cm homogeneous grey (bioturbated) sediments of the LIA. Paleosalinity proxies identify the laminated MCA and MWP sediments as brackish–marine, while the homogeneous LIA sediments have been deposited in a fresher water. This change in the sediment texture makes the MUC core analogue to the B zones deposited within the eastern Gotland Basin after the Littorina transgression. In order to identify the driving force of the changing depositional environment we compared the sedimentary facies with a reconstruction of the NAO oscillation mode reconstructed by a multi-proxy approach for the last millennium by Trouet et al. (2009). According to this study, the NAO mode was positive (maritime) for the MCA. It shifted to predominantly negative values for the time span from the beginning of the fifteenth to middle of the nineteenth century (LIA) before it returned to positive values for the modern warm period (with a negative excursion within the last third of the twentieth century). According to these results we have to assume that during the MCA, warm winters with westerly winds reduced ice coverage, dominated the meteorological and hydrographic regime, whereas during the LIA easterly winds with extended ice coverage during winter time prevailed. These differences have consequences on the supply of saline water to the central basins of the Baltic Sea (Baltic Proper). We have to assume that the baroclinic and barotropic inflows from the North Sea are the main reasons for "renewing" of the saline bottom water of the Baltic Sea basins (Matthäus et al. 2008). Both of them can reach the central Baltic. Strong barotropic inflows are more coupled to strong westerly winds (winter half). Time series analyses of major Baltic inflows from 1880 to today, which represents the modern warm period, prove the exceptional importance of strong barotropic inflows for central Baltic deepwater renewal and salinity (Matthäus 2006). The baroclinic inflows can occur during summer time and during calm periods. It means that under a general negative NAO situation (cold periods), at least barotropic inflows and therefore the supply of saline water to the Baltic Basin is reduced whereas at positive NAO and forced baroclinic inflow the salinity would increase. This assumption seems to yield for the last millennium according to the most recent publication of Trouet et al. (2009) in comparison to the investigation of the MUC from the Eastern Gotland Basin (Leipe et al. 2008). It does, however, not agree with the results of Zorita and Laine (2000), Meier and Kauker (2003), and Meier (2005, 2007) who investigated by statistical analysis and numerical process modelling hydrographic and meteorological processes of the Baltic area. These authors claim that under positive NAO, increasing inflow of freshwater due to intensified precipitation cause a decrease in the salinity of the surface and bottom water, the latter by increased vertical mixing. The results are based mainly on modelling of processes of the Baltic Sea and statistical analysis of hydrographic and meteorological data for the last century, which may not be relevant for longer time scales. Zorita and Laine (2000) mention that saline water entering the Baltic is distributed on a monthly up to an annual time scale. Meteorological processes vary even on shorter time scales than the hydrography of the Baltic Sea. This might be the reason why due to findings of Mariotti and Arkin (2007) a general and direct correlation between positive NAO and a high precipitation to the Baltic catchment area (freshwater inflow) is questionable. The latter authors found by an analysis of global meteorological and oceanographic databases that the North Sea and the Baltic precipitation is positively correlated to the NAO only for December to February. Even during these months zones of positive correlation do regionally not cover whole Scandinavia and the Baltic Sea basin. During spring and the fall months the correlation is not specific, and even negative during June, July, and August. Erikssen et al. (2007a) did not find any statistically significant trend in the annual river runoff to the Baltic Sea during the last half millennium. Erikssen et al. (2007b) claim regarding the analysis of hydrographic data of the Baltic Sea the "statistical methods by themselves are incomplete to identify physical mechanisms for the centennial variation". When the oxygenation is included into the analysis the processes even become more complex (Zillén et al. 2008). The oxygen consumption in the deep water is mainly caused by degradation of organic matter, annually produced in the euphotic zone and sinking down to the sea floor. The formation of long-term anoxic bottom water thus depends on the presence of a density (pycno-) cline and the "competition" between oxygen consumption and (lateral) oxygen supply. After more recent studies (Matthäus 2006, Matthäus et al. 2008) we know that the oxygen decline in, e.g. the Gotland deep after an inflow event (1993) is faster (a few months or a year only) than the sequence of (new) inflows (years to decade). The critical region for long-term anoxia is the central Baltic Sea because the northern Baltic has regular vertical convection and towards the western Baltic Sea, saline (oxygen) water inflows become more frequent. Thus the Gotland Basin is the typical region of formation of laminated sediments, representing long-term anoxia. For this area Pers and Rahm (2000) have clearly postulated that the deepwater inflow is the "main supply of oxygen except during periods with stagnant conditions in which case the diffusive supply from surface waters is dominant". This dominance is even intensified by downward convection which ventilates the water column in particular during the strong north to northeasterly wind when the NAO turns negative to the continental mode (Hagen and Feistel 2005). On the other side, sediment data investigated here record processes with a decadal resolution. The sampling and measuring spacing is between 0.5 cm (XRF scans) and 2 cm (MSCL data). Taking into account a sedimentation rate of 1 mm/year, the geological record mirrors variations of hydrographic processes in a resolution of 5–20 years. That means, in this study the geological data even of laminated sediments do not reflect high process variability on the monthly or seasonal scale. NB: laminated sediment texture of the brackish Baltic sediments does not represent varves. The data do reflect system shifts (Hagen and Feistel 2005) between different modes on the decadal to centennial time scale recording the invariant (average) component of the facies. These data do not reflect relatively high frequently changing oceanographic and meteorological conditions that have been analysed by the authors mentioned above.

In conclusion, we assume that for the last millennium, on average, periods on the centennial time scale of predominant positive NAO (maritime mode) are linked with saline bottom water in the Eastern Gotland Basin, oxygen deficiency, and the formation of laminated sediments, whereas predominant negative (continental) NAO is linked with fresher oxygenated bottom water and bioturbated sediments. Due to the similarity between the uppermost sediments representing the last millennium and the whole sequence of the brackish Holocene sediments at the master station in the Eastern Gotland Basin we extrapolate the model of the last millennium to the whole of series of B zones. Consequently, B1, B3, and B5 zones represent, in our interpretation, periods of a maritime NAO mode whereas B2, B4, and B6 stand for a continental NAO mode. This assumption holds as a rule, but exceptions may occur. Exceptionally, even during negative NAO mode strong westerlies may occur due to an expanded sea-ice cover in the Greenland Sea (Dawson et al. 2002). Such situations might be the reason for the diatom record pointing at brackish–marine conditions at the base of the (continental) B4 zone.

Consequently, we search for the driving force of the changing depositional environment and try to find some hints in the results of the periodicity analysis of sediment proxy data (Fig. 5.14). There are two periodicities indicated in both of the proxy variables investigated by time series analysis: the dominant 900-years period and the 1500-years period. The 900-years period identified also in the grey-scale time series at the master station of the Eastern Gotland Basin correlates well with a 900-years component of the oxygen isotope records from the Greenland site GISP2 (Kotov and Harff 2006). Schulz and Paul (2002) have noted the significant correlation of the Greenland oxygen isotope records with the 900-year signal component in summer insolation at 65°N in the time span 3.5–8 k years BP. Loutre et al. (1992) referred this cyclicity to an orbital (eccentricity-linked) period modulating incoming solar radiation. Sarnthein et al. (2003) found a very similar cycle (885 years) in sediments of the western Barents shelf. The authors reported about cyclic injections of coarser layers into the marine sediment succession over the whole Holocene which are interpreted as a result of storm-induced erosion along the northern coast of the Kola Peninsula with a periodicity due to solar forcing. This effect might also be seen in the basin sediments of the central Baltic investigated here. The second cycle, less dominant, but clearly visualized by the time series analysis, seems to reflect a global climate signal. Bond et al. (1997, 2001) called attention to this cyclicity as a Holocene climate phenomenon, which was known before for the Pleistocene ocean dynamics as Heinrich/Bond cycles with its Dansgaard Oeschger events (Rahmstorf, 2002). These cycles have now been found in many marine Holocene sediment sequences (for instance, Bianchi and McCave 1999, Andresen et al. 2005, Moros et al. 2009) indicating general periodical changes in ocean dynamics even after the deglaciation of the continents. The 1470 cycle in the oxygen isotope record of the GISP2 ice core was used by Stuiver et al. (1997) to point at the relation between atmospheric and salt circulation pattern in the North Atlantic. Mayewski et al. (1997, 2004) called attention on the variability of global storm intensity following the 1500-years cycle. In particular the latter one seems to be reflected by facies change in the Baltic Sea. The elevated K-concentration in zones A6, B2, B4, and B6 of core 303610-12 (Fig. 5.11) can tentatively be correlated to periods of aeolian erosion of the central Asian deserts during cold periods (Mayewski et al. 1997, 2004, O'Brien et al. 1995). Accordingly, we assume we may have a mixed regional to global climate signal in the periodicity of Baltic Sea basin sediments.

5.7 Summary

The Baltic Sea and its sediments serve as a textbook in climate and environmental history of the Baltic area and the North Atlantic realm. High sedimentation rates in the central parts of the Baltic basin qualify the sediments and their facies to reflect the dynamics of the atmospheric circulation of the North Atlantic, and also its modification due to the variation of Eurasian anticyclones on the inter-annual time scale. Methods of basin analysis have been applied to draw a picture of the development of the Eastern Gotland Basin in space and time from the Late Pleistocene to the modern warm period. Sediment echosounders have been used for the identification of coring stations, where gravity corers have been used for sampling up to 12 m sediments representing the 12,000 years of basin history. Sediment physical parameters measured with a multi-sensor core logger (MSCL) serve as reference variables for the physico-stratigraphic zonation and basinwide correlation sediment cores. In particular

(a) The physico-stratigraphic zones determined for a "master station" coincide with the main stages of the geological development of the Baltic Basin. The older sequences (A zones) consist mainly of freshwater sediments from the Baltic Ice Lake and the Ancylus Lake. Rapid sea level rise through the entrance of the Baltic Sea caused a sudden increase of ocean water inflow into the Baltic Basin (Littorina Sea with its transgression(s)) changing the environment of the Baltic Basin permanently to a brackish–marine one. This shift is marked in the sediment column by a change from homogeneous to laminated sediments of the B zones due to the establishment of a halocline and anoxic bottom water. The facies shift from zones A to B can be lithostratigraphically correlated basinwide as well as within the sediment cores as within the sediment echosounder profiles. By interpolation of these stratigraphic boundaries thickness maps of A and B zone sediments have been constructed. The different locations of the depositional centres for A and B zones show that the Littorina transgression caused a general shift of the hydrographic system of the central Baltic Basin. Whereas during the deposition of the lacustrine A zone sediments, a coast-to-basin system prevailed, the basin was dominated by a basin-to-basin transport after the gates to the North Sea opened during the Littorina transgression(s).

(b) The basin-to-basin transport from the Bornholm to the Gotland Basin resulted in the accumulation of the "Stolpe Foredelta" in front of the "mouth" of the Stolpe Channel within the southern Eastern Gotland Basin. The latter estuarine inflow dynamics driven by the atmospheric circulation varies obviously on centennial time scales. The resulting laminated sediments of zones B1, B3, and B5 alternate with more homogeneous sediments of zones B2, B4, and B6.

(c) According to diatom analysis the facies types stand for different paleosalinities. Whereas laminated sediments have been deposited under brackish–marine conditions, the more homogeneous sediments are mainly bound to a fresher water depositional environment.

(d) We relate the periodical facies shifts to changes of the NAO on centennial time scales. During phases of a predominantly maritime NAO mode, westerly winds drive more saline water to the Baltic Basin. The effect is a permanent halocline precluding vertical water mixing and oxygen supply to the bottom water. The poorly ventilated bottom water leads to the accumulation of laminated sediments not disturbed by bioturbation. During phases of a predominantly continental NAO the influence of westerlies to the Baltic Basin is reduced, the salinity drops, and a weak (or none) halocline allows the transport of oxygen from the surface to the bottom by vertical mixing. A benthic fauna, due to the bioturbation, results in homogeneous sediments.

(e) Time series analysis of sediment physical and chemical proxies of the depositional environment reveals remarkable periodicities of about 900 and 1500 years. Similar periods are reported from marine sediments from the Northern Atlantic and the Greenland ice cores. According to our hypothesis, these periodicities in Baltic Sea sediments stand for global climate signals.

Acknowledgements The study has been supported by the German Federal Ministry of Education and Research. The authors express their gratitude to the captain and the crew of the R/V "Poseidon" for the excellent co-operation during the expedition in June 2005.

We thank Dr. Torsten Seifert from the Leibniz-Institute for Baltic Sea Research Warnemünde, Germany, who provided results from numerical modelling of the current system in the central Baltic Sea.

We also thank Dorota Kaulbarsz, Polish Geological Institute Gdańsk, and Irina Taranenko, St. Petersburg State University, for her co-operation within the frame of this project.

References

Alheit J, Hagen E (1997) Long-term climate forcing of European herring and sardine populations. Fisheries Oceanography 6(2):130–139

Alvi K, Winterhalter B (2001) Authigenic mineralisation in the temporally anoxic Gotland Deep, the Baltic Sea. Baltica 14:74–83

Andrén E (1999) Holocene environmental changes recorded by diatom stratigraphy in the southern Baltic Sea. Meddelanden fran Stockholms Universitets Institution foer Geologi och Geokemi 302:22

Andrén E, Andrén T, Kunzendorf H (2000) Holocene history of the Baltic Sea as a background for assessing records of human impact in the sediment of the Gotland Basin. The Holocene 10(5):621–636

Andrén E, Andrén T, Kohly A (2001) Development of Anoxia during the Holocene fresh-brackish water transition in the Baltic Sea. Marine Geology 177:221–242

Andrén T, Lindeberg G, Andrén E (2002) Evidence of the final drainage of the Baltic Ice Lake and the brackish phase of the Yoldia Sea in glacial varves from the Baltic Sea. Boreas 31:226–238

Andresen CS, Bond G, Kuijpers A, Knutz PC, Bjoerck S (2005) Holocene climate variability at multidecadal time scales detected by sedimentological indicators in a shelf core NW off Iceland. Marine Geology 214:323–338

Bianchi GG, McCave N (1999) Holocene periodicity in North Atlantic climate and deep ocean flow south of Iceland. Nature 397:515–517

Björck S (1995) A review of the history of the Baltic Sea, 13.0–8.0 ka BP. Quaternary International 27:19–40

Björck S (2008) The late Quaternary development of the Baltic Sea basin. In: TBATeam (eds) Assessment of climate change for the Baltic Sea Basin. Springer, Berlin, Heidelberg, pp 398–407

Bloomfield P (2000) Fourier analysis of time series: an introduction. Wiley, New York, NY

Bond G, Showers W, Cheseby M, Lotti R, Almasi P, deMenocal P, Priore P, Cullen H, Hajdas I, Bonani G (1997) A pervasive millennial-scale cycle in North Atlantic Holocene and Glacial climates. Science 278:1257–1265

Bond G, Kromer B, Beer J, Muscheler R, Evans M, Showers W, Hoffmann S, Lotti-Bond R, Hajdas I, Bonani G (2001) Persistent solar influence on North Atlantic climate during the holocene. Science 294:2130–2136

Boyce RE (1973) Appendix I. Physical property methods. In: Edgar NT, Sounders JB, et al. (eds) Initial reports deep sea drilling project 15. US Government Printing Office, Washington, DC, pp 1115–1128

Burke JT, Kemp AES (2002) Microfabric analysis of Mn-carbonate laminae deposition and Mn-sulfide formation in the Gotalnd Deep, Baltic Sea. Geochimica and Cosmochimica Acta 66(9):1589–1600

Conley DJ, Humborg C, Rahm L, Savchuk OP, Wulff F (2002) Hypoxia in the Baltic Sea and basin-scale changes in phosphorous and biogeochemistry. Environmental Science and Technology 36:5315–5320

Conley DJ, Björck S, Bonsdorff E, Carstensen J, Destouni G, Gustafsson BG, Hietanen S, Kortekaas M, Kuosa H, Meier HEM, Müller-Karulis B, Nordberg K, Norkko A, Nürnberg G, Pitkänen H, Rabalais NN, Rosenberg R, Savchuk OP, Slomp CP, Voss M, Wulff F, Zillen L (2009) Hypoxia-related processes in the Baltic Sea-critical review. Environmental Science and Technology 43:3412–3420

Davis BAS, Brewer S, Stevenson AC, Guiot J (2003) The temperature of Europe during the Holocene reconstructed from pollen data. Quaternary Science Review 22:1701–1716

Dawson AG, Hickey K, Holt T, Elliott L, Dawson S, Foster IDL, Wadhams P, Jonsdottir I, Wilkinson J, McKenna J, Davis NR, Smith DE (2002) Complex North Atlantic Oscillation (NAO) Index signal of historic North Atlantic storm-track changes. The Holocene 12:363–369

Dippner JW, Voss M (2004) Climate reconstruction during the MWP in the Baltic Sea from biogeochemical proxies in a sediment record. Baltica 17(1):5–16

Emeis KC, Struck U, Leipe T, Pollehne F, Kunzendorf H, Christiansen C (2000) Changes in the C, N, P burial rates in some Baltic Sea sediments over the last 150 yearas – relevance to P regeneration rates and the phosphorous cycle. Marine Geology 167(1–2):43–59

Emeis KC, Dawson AG (2003) Holocene palaeoclimate records over Europe and the North Atlantic. The Holocene 13:305–309

Emeis KC, Struck U, Blanz T, Kohly A, Voß M (2003) Salinity changes in the central Baltic Sea (NW Europe) over the last 10000 years. The Holocene 13(3):411–421

Emelyanov EM (1994) Baltic Sea: geology, geochemistry, paleoceanography, pollution. Shirshov Institute Oceanology RAS, Atlantic Branch, Kaliningrad, pp 119

Emelyanov EM (2006) The thickness of the marine Holocene sediments of the Baltic Sea and their resuspension and redeposition. The Baltic Sea Geology 9 Mar. Geol. Conf. Riga University of Latvia, p 22

Emelyanov EM (2007) The geochemical and geoecological situation of the Gotland Basin in the Baltic Sea where chemical munitions were dumped. Geologija 60:10–26

Erikssen C, Hansson D, Omstedt A, Chen D (2007a) Reconstruction of river runoff to the Baltic Sea for the period 1500–1995. GEWEX News 17(3):3–4

Erikssen C, Omstedt A, Overland JE, Percival DB, Mofjeld AO (2007b) Characterizing the European Subarctic winter climate since 1500 using ice, temperature, and atmospheric circulation time series. Journal of Climate 20:5316–5334

Ghil M, Allen MR, Dettinger MD, Ide K, Kondrashov D, Mann ME, Robertson AW, Saunders A, Tian Y, Varadi F, Yiou P (2002) Advanced spectral methods for climatic time series. Reviews of Geophysics 40(1):3.1–3.41

Gingele FX, Leipe T (1997) Clay mineral assemblages in the western Baltic Sea: recent distribution and relation to sedimentary units. Marine Geology 140:97–115

Gunn DE, Best AI (1998) A new automated nondestructive system for high resolution multi-sensor core logging of open sediment cores. Geo-Marine Letters 18:70–77

Hagen E, Feistel R (2005) Climatic turning points and regime shifts in the Baltic Sea region: the Baltic winter index (WIBIX, 1659-2002). Boreal Environment Research 10:211–224

Hagen E (2006) Volcano dust and long-term modes of the Baltic Sea winter climate. BALTEX Newsletter 9:14–15

Hagen E, Feistel R (2008) Baltic climate change. In: Feistel R, Nausch G, Wasmund N (eds) State and evolution of the Baltic Sea 1952–2005. Wiley, Hoboken

Harff J, Winterhalter B (1996) Cruise report R/V PETR KOTTSOV Sept 2–Sept 12, 1996. Baltic Sea Research Institute, Warnemünde

Harff J, Winterhalter B (eds) (1997) Cruise report R/V PETR KOTTSOV July 22–Aug 1 1997. Baltic Sea Research Institute, Warnemünde

Harff J, Bohling GC, Endler R, Davis JC, Olea RA, Schwarzacher W (1999) Holocene sediments from the Baltic Sea basins as indicators for the paleoenvironment. In: Lippard SJ (ed) Proceedings of the 5th annual conference of the International association of mathematical geology, Trondheim 6–11th Aug 1999, vol 1, pp 195–200

Harff J, Bohling G, Davis J, Endler R, Kunzendorf H, Olea R, Schwarzacher W, Voss M (2001a) Physico-chemical stratigraphy of Gotland basin Holocene sediments, the Baltic Sea. Baltica 14:58–66

Harff J, Frischbutter A, Lampe R, Meyer M (2001b) Sea level change in the Baltic Sea – interrelation of climatic and geological processes. In: Gerhard LC, Harrison WE, Hanson BM (eds) Geological perspectives of global climate change. AAPG-Studies in Geology 47:231–250

Harff J (ed) (2005) Cruise report R/V POSEIDON, cruise POS 07/05/01 (POSEIDON 323). Baltic Sea Research Institute, Warnemünde

Harff J, Lampe R, Lemke W, Lübke H, Lüth F, Meyer M, Tauber F (2005) The Baltic Sea – a model ocean to study interrelations of geosphere, ecosphere and anthroposphere in the coastal zone. Journal of Coastal Research 21(3):441–446

Hurrell JW (1995) Decadal trends in the North Atlantic Oscillation: regional temperatures and precipitation. Science 269:676–679

Ignatius H, Axberg S, Nemistö L, Winterhalter B (1981) Quaternary geology of the Baltic Sea. In: Voipio A (ed) The Baltic Sea. Elsevier Oceanography Series 30:54–104

Justino F, Peltier W (2005) The glacial North Atlantic Oscillation. Geophysical Research Letters 32:L21803. doi:10.1029/2005GL023822

Kotilainen AT, Saarinen T, Winterhlter B (2000) High resolution paleomagnetic dating of sediments deposited in the Central Baltic Sea during the last 3000 years. Marine Geology 166:51–64

Kotov SR, Harff J (2006) A comparison of Greenland Ice and Baltic Sea Sediment Record: a contribution to climate change analysis. Mathematical Geology 38:6

Leipe T, Dippner JW, Hille S, Voss M, Christiansen C, Bartholdy J (2008) Environmental changes in the central Baltic Sea during the past 1000 years: inferences from sedimentary records, hydrography and climate. Oceanologia 50:23–41

Loutre MF, Berger A, Bretagnon P, Blanc P-L (1992) Astronomical frequencies for climate research at the decadal to century time scale. Climate Dynamics 7:181–194

Mariotti A, Arkin P (2007) The North Atlantic Oscillation and oceanic precipitation variability. Climate Dynamics 28:35–51

Matthäus W (2006) The history of investigation of salt water inflows into the Baltic Sea from the early beginning to recent results. Marine Science Reports 65. Baltic Sea Research Institute, Warnemünde

Matthäus W, Nehring D, Feistel R, Nausch G, Mohrholz V, Lass H-U (2008) The inflow of highly saline water into the Baltic Sea. In: Feistel et al. (eds) State and evolution of the Baltic Sea, 1952–2005. John Wiley & Sons, Inc., Chapter 10, pp 265–309

Mayewski PA, Meeker LD, Twickler MS, Whitlow S, Yang Q, Lyons WB, Prentice M (1997) Major features and forcing of high-latitude northern hemisphere atmospheric circulation using 110,000-year long glaciochemical series. Quaternary Research 102:26345–26366

Mayewski PA, Rohling EE, Stager E, Karlén K, Maasch K, Meeker LD, Meyerson E, Gasse F, van Kreveld S, Holmgren K, Lee-Thorp, J, Rosquist G, Rack F, Staubwasser M, Schneider R (2004) Holocene climate variability. Quaternary Research 62:243–255

Meier HEM, Kauker F (2003) Sensitivity of the Baltic Sea salinity to the freshwater supply. Climate Research 24:231–242

Meier HEM (2005) Modeling the age of Baltic Seawater masses: quantification and steady state sensitivity experiments. Journal of Geophysics Research 110:C02006. doi:10.1029/2004JC002607

Meier HEM (2007) Modeling the pathway and ages of inflow salt- and freshwater in the Baltic Sea. Estuarine, Coastal and Shelf Science 74:610–627

Moros M, De Decker P, Jansen E, Perner K, Telford RJ (2009) Holocene climate variability in the Southern Ocean recorded in a deep-sea sediment core off South Australia. Quaternary Science Review 28:1932–1941

Neumann T, Christansen C, Clasen S, Emeis K-C, Kunzendorf H (1997) Geochemical records of salt-water inflows into the deep basins of the Baltic Sea. Continental Shelf Research 17(1): 95–115

O'Brien SR, Mayewski PA, Meeker LD, Meese DA, Twickler MS, Whitlow S (1995) Complexity of Holocene climate as reconstructed from a Greenland ice core. Science 270: 1962–1964

Olea RA, Sampson RJ (2002) CORRELATOR 5.2 – computer program and user's manual. Open-File Report: 2002-51, Kansas Geological Survey, Lawrence, Kansas. Also electronically available as http://www.kgs.ku.edu/Publications/OFR/2002/OFR02_51/ManualOFR2002-51.pdf

Pacanowski RC, Griffies SM (2000) MOM 3.0 manual. Technical Report, Geophysical Fluid Dynamics Laboratory, Princeton, NJ, pp 680

Pers C, Rahm L (2000) Changes in apparent oxygen removal in the Baltic Proper deep water. Journal of Marine System 25:421–429

Rahmstorf S (2002) Ocean circulation and climate during the past 120,000 years. Nature 419: 207–213

Repecka M (2001) Physical properties of bottom sediments from three Baltic Sea basins. Baltica 14:24–29

Richter TO, van der Gaast S, Koster B, Vaars A, Gieles R, de Stigter HC, De Haas H, van Weering TCE (2006) The Avaatech XRF Core scanner: technical description and applications to NE Atlantic sediments. In Rothwell RG (ed) New techniques in sediment core analysis. Geological Society, London (Special Publication), vol 267, pp 39–50

Sarnthein M, van Kreveld S, Erlenkeuser H, Grootes PM, Kucera M, Pflaumann U, Schulz M (2003) Centennial-to-millennial-scale periodicities of Holocene climate and sediment injections off the western Barents shelf, 75°N. Boreas, Oslo, 32:447–461

Schultheiss PJ, Weaver PPE (1992) Multi-sensor core logging for science and industry. Proceedings of the Oceans 92. Mastering the Oceans through Technology 2:608–613
Schulz M, Paul A (2002) Holocene climate variability on centennial-to-millennial time scales: 1 Climate records from the North-Atlantic realm. In: Wefer G, Berger W (eds) Climate development and history of the North Atlantic realm. Springer, Berlin, pp 41–54
Seifert T, Tauber F, Kayser B (2001) A high resolution spherical grid topography of the Baltic Sea, 2nd edn. Baltic Sea Science Congress, Stockholm 25–29 November 2001. Abstract vol Poster #147
Sohlenius G, Stenbeck J, Andrén E, Westman P (1996) Holocene history of the Baltic Sea as recorded in a sediment core from the Gotland Deep. Marine Geology 134:183–201
Sohlenius G, Westman P (1998) Salinity and redox alterations in the northwestern Baltic Proper during the late Holocene. Boreas 27:101–114
Sohlenius G, Emeis K-C, Andrén E, Andrén T, Kohly A (2001) Development of anoxia during the Holocene fresh—brackish water transition in the Baltic Sea. Marine Geology 177:221–242
Sternbeck J, Sohlenius G (1997) Authigenic sulfide and carbonate mineral formation in Holocene sediments of the Baltic Sea. Chemical Geology 135:55–73
Sternbeck J, Sohlenius G, Hallberg RO (2000) Sedimentary trace elements as proxies to depositional changes induced by a Holocene fresh-brackish transition. Aquatic Geochemistry 6:325–345
Stuiver M, Braziunas TF, Grootes PM, Zielinski GA (1997) Is there evidence for solar forcing of climate in the GISP2 oxygen isotope record? Quaternary Research 48:259–266
Trouet V, Esper J, Graham NE, Baker A, Scource JD, Frank DC (2009) Persistent positive North Atlantic oscillation mode dominated the Medieval Climate Anomaly. Science 324:78–80
Uppala S, Kallberg P, Simmons A, Andrae U, da Costa Bechtold V, Fiorino M, Gibson J, Haseler J, Hernandez A, Kelly G, Li X, Onogi K, Saarinen S, Sokka N, Allan R, Andersson E, Arpe K, Balmaseda M, Beljaars A, van de Berg L, Bidlot J, Bormann N, Caires S, Chevallier F, Dethof A, Dragosavac M, Fisher M, Fuentes M, Hagemann S, Holm E, Hoskins B, Isaksen L, Janssen P, Jenne R, McNally A, Mahfouf J-F, Morcrette J-J, Rayner N, Saunders R, Simon P, Sterl A, Trenberth K, Untch A, Vasiljevic D, Viterbo P, Woollen J (2005) The ERA-40 reanalysis. Quarterly Journal of the Royal, Meteorological Society 131:2961–3012
Van der Werff A, Hulls H (1957–1974) Diatomeeën flora van Nederland. Otto Koeltz Science Publishers, Koenigstein, pp 1–142
Westman P, Sohlenius G (1999) Diatom stratigraphy in five offshore sediment cores from the northwestern Baltic Proper implying large scale circulation changes during the last 8500 years. Journal of Paleolimnology 22:53–69
Winterhalter B, Flóden T, Ignatius H, Axberg S, Niemistö L (1981) Geology of the Baltic Sea. In: Voipio A (ed) The Baltic Sea. Elsevier Oceanography Series 30, pp 1–121
Winterhalter B (ed) (2001a) The BASYS project and the paleoenvironment of the Baltic Sea. Baltica 14:143
Winterhalter B (2001b) On sediment patchiness at the BASYS coring site, Gotland Deep, the Baltic Sea. Baltica 14:18–23
Zillén L, Conley DJ, Andrén T, Andrén E, Björck S (2008) Past occurrences of hypoxia in the Baltic Sea and the role of climate variability, environmental change and human impact. Earth Science Reviews 91:77–92
Zorita E, Laine A (2000) Dependence of salinity and oxygen concentrations in the Baltic Sea on large-scale atmospheric circulation. Climate Research 14:25–41

Chapter 6
Geological Structure of the Quaternary Sedimentary Sequence in the Klaipėda Strait, Southeastern Baltic

Albertas Bitinas, Aldona Damušytė, and Anatoly Molodkov

Abstract The Klaipėda Strait is located between the Curonian Spit and the mainland coast of Lithuania. It links the Curonian Lagoon with the Baltic Sea. The Quaternary sequence is represented here by Pleistocene sediments formed during a few glaciations and interglacials. Its uppermost part is composed of Late glacial and Holocene sediments originating from different stages of the Baltic Sea development. One of the main problems of Quaternary geology in the vicinities of the Klaipėda Strait, as well as in the whole Lithuanian Coastal Area, is the reliable geochronology and stratigraphic correlation of sediments. To contribute to the solution of this problem, the infrared optically stimulated luminescence (IR-OSL) dating of the lacustrine inter-till sandy sediments was done during the engineering geological mapping of the Klaipėda Strait. The absolute majority of the IR-OSL ages obtained for the investigated inter-till sediments fall within the age range of marine isotope stages (MIS) 5d-5a. The subsequent more detailed examination of geological setting of Quaternary sequence has led to the assumption that the sampled inter-till sediments occur not in situ, i.e. they are found as blocks (rafts) in a thick till bed that have been formed by the ice advance during the Weichselian early pleniglacial maximum (MIS 4). This conclusion does not support the former standpoint that the till beds beneath the bottom of the Klaipėda Strait were formed during the Warthanian (Medininkai, MIS 6) glaciation.

Keywords Klaipėda Strait · Late Pleistocene · Till · Stratigraphy · IR-OSL dating · Glaciodislocations

A. Bitinas (✉)
Coastal Research and Planning Institute, Klaipėda University, LT-92294 Klaipėda, Lithuania;
Department of Geology and Mineralogy, Faculty of Natural Sciences, Vilnius University, LT-03101 Vilnius, Lithuania
e-mail: albertas.bitinas@gmail.com; albertas.bitinas@corpi.ku.lt

J. Harff et al. (eds.), *The Baltic Sea Basin*, Central and Eastern European Development Studies (CEEDES), DOI 10.1007/978-3-642-17220-5_6,

6.1 Introduction

The Klaipėda Strait links the Curonian Lagoon (Kuršių Marios) with the Baltic Sea, i.e. separates the Curonian Spit (Kuršių Nerija) from the continental part of Lithuania (Fig. 6.1). The only seaport of Lithuania is located in the Klaipėda Strait. The length of the strait from the port gates on the Baltic Sea coast to the Kiaulės Nugara isle in the Curonian Lagoon is 12 km. The strait is 1,500 m wide at its

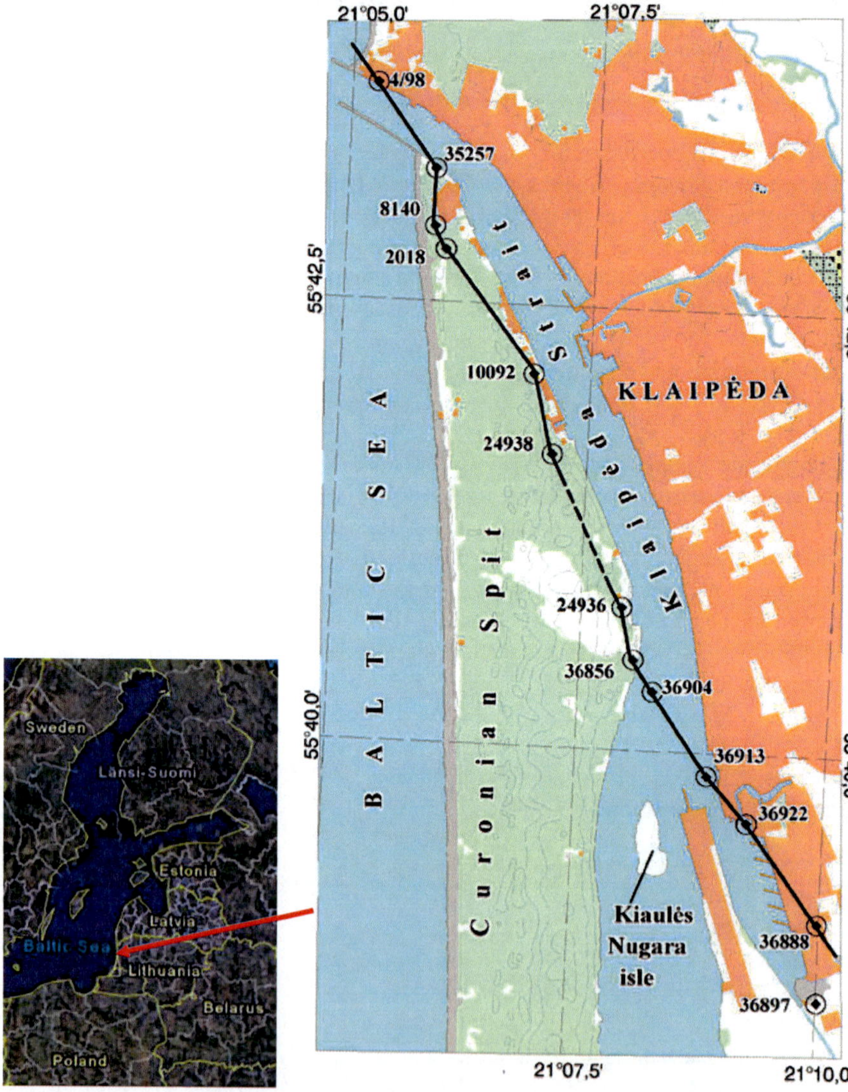

Fig. 6.1 Map of the study area, location of investigated boreholes and line of geological cross-section

widest point and 385 m at its narrowest. Due to permanent cleaning and dredging of the harbour basin area its depth varies from 8.0 to 14.5 m.

Geological setting of the Klaipeda Strait region is complicated. The lowermost part of the Quaternary sedimentary sequence was formed during the last few glacial–interglacial cycles and is represented by layers of glacial, glaciofluvial, glaciolacustrine, limnic and organogenic sediments, while the uppermost part was formed in the Late glacial and Holocene during a few stages of the Baltic Sea development (Fig. 6.2). The dredging of the strait opens the layers of fine-grained sand filled by groundwater. Some of these layers are under high hydrostatic pressure that causes sub-aquatic suffusion posing a threat to the jetties of the seaport. Thus, the complicated geological structure and hydrogeological conditions were the valid reason to start a detailed (at a scale of 1:5 000) engineering geological mapping of the Klaipėda Strait area. The vast majority of geological information presented in this chapter was collected during this mapping.

The Klaipėda Strait and surroundings, investigated in detail, can be considered as an important key area for the Lithuanian Coastal Area and whole Western Lithuania. During the different stages of the Baltic Sea development – the Baltic Ice Lake, Yoldia Sea, Ancylus Lake, Littorina and Post-Littorina Seas – the paleogeographic situation in the Klaipėda Strait environs was very different and changeable, but this

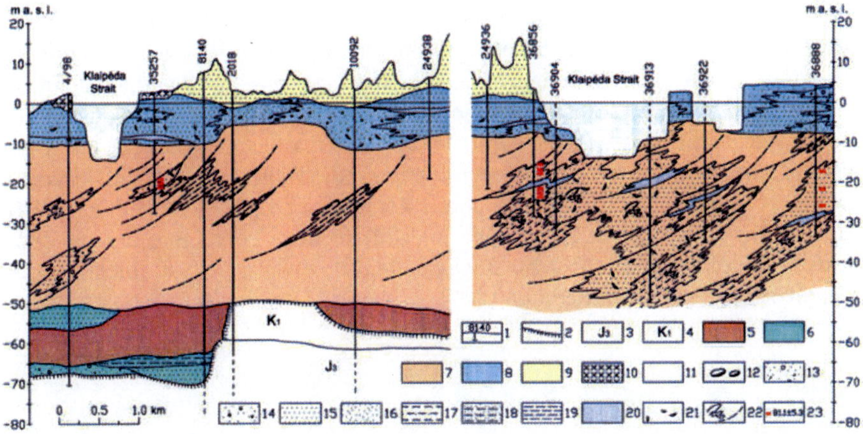

Fig. 6.2 Geological cross-section along the Klaipėda Strait: *1* – borehole and its number; *2* – surface of pre-Quaternary sediments; *3* – upper Jurassic sediments; *4* – lower Cretaceous sediments; *5* – middle Pleistocene glacigenic sediments; *6* – middle Pleistocene glaciofluvial and glaciolacustrine sediments; *7* – upper Pleistocene glacigenic sediments with glaciotectonized blocks of inter-till limnic sediments; *8* – late glacial and Holocene marine and lagoonal sediments; *9* – Holocene aeolian sediments; *10* – anthropogenic sediments. Lithology of sediments: *11* – till; *12* – boulders; *13* – sand with gravel; *14* – various-grained sand; *15* – fine-grained sand; *16* – very fine-grained sand; *17* – silty sand; *18* – sandy silt; *19* – clay; *20* – gyttja, peat; *21* – fine dispersal remnants of organic matter; *22* – glaciotectonic features (folds, thrust faults); *23* – sampling point for infrared optically stimulated luminescence (IR-OSL): number indicates the luminescence age of sediment (in ka)

issue has been investigated only superficially and is still waiting for solution. A reliable geochronology of Pleistocene deposits and their stratigraphic correlation is another so far unsolved problem: a precise number of glacial advances and their stratigraphic rank have been the objects of intensive scientific discussions until the present time. The petrographic composition of the gravel part of glacigenic sediments (tills) which has traditionally been applied for stratigraphic subdivision and correlation of Pleistocene tills in Lithuania has proved to be poorly effective (Gaigalas et al. 1987, 1997). Other lithostratigraphic methods and criterions, such as geochemical composition of fine-grained part of tills (less than 1 mm) or variation of well-rounded hornblende grains in tills (fraction Ø 0.25–0.1 mm), are more effective, but, unfortunately, also do not give a clear-cut answer (Majore et al. 1997, Bitinas et al. 1999); the fabric measurements of tills are available only in a few cliff sections on the Baltic Sea coast (Bitinas 1997). The only most positive results were obtained by using thermoluminescence (TL) dating of inter-till sediments: the Pamarys Interstadial sediments which were formed at the end of the Medininkai (Varthanian, MIS 6) glaciation (i.e. around 140–160 kyears BP) were identified in the big part of Lithuanian Coastal Area (Satkūnas and Bitinas 2002). The Pamarys stratigraphic unit mentioned separates sediments of the middle and upper Pleistocene, but these sediments have not been used for solution of stratigraphic problems because they are not prevalent in the area of the Klaipėda Strait. Notwithstanding this factor, the methods of absolute geochronology were used in determining the stratigraphy and geological structure of Pleistocene sediments in the Klaipeda Strait area. This chapter presents the results of IR-OSL dating of the lacustrine inter-till sandy sediments. Besides, investigations aimed at finding out the possibilities of till age estimation by the IR-OSL method were also carried out. For this purpose, two till layers from boreholes dislocated along the Klaipeda Strait and three till layers from the Olando Kepurė outcrop (the Baltic Sea cliff), dislocated a few kilometres to the north from the strait, were examined (Molodkov et al. 2010). Additionally, a number of Pleistocene inter-till sections containing organic sediments were examined paleobotanically by pollen and diatom analysis. Identification of mollusc species was carried out.

6.2 Geological Setting

The thickness of the Quaternary cover in the Klaipėda Strait and surroundings varies from 60 to 90 m. The upper Jurassic and lower Cretaceous sediments are outcropping beneath the Quaternary sedimentary sequence. The Pleistocene sedimentary sequence is generally composed of alternating till and inter-till sediments. According to the results of previous geological investigations – state geological mapping at a scale of 1:50,000 – the sediments of four different glaciations have been detected in the sequence of Quaternary sediments (unpublished data, report in the archive of Lithuanian Geological Survey). Till layers beneath the bottom of the Klaipėda Strait were generally attributed to the middle Pleistocene Warthanian (Medininkai, MIS 6) glaciation, and in some cases they were attributed to the first glacial advance of late Weichselian (late Nemunas, MIS 2) glaciation. The uppermost part of glacigenic sediments along the Baltic Sea coast (including the vicinities of the Klaipėda Strait as well) is covered by sediments of the Baltic Ice Lake, Ancylus Lake, Littorina and Post-Littorina Seas and also by recent aeolian sediments (Gelumbauskaitė and Šečkus 2005, Kabailienė et al. 2009).

The boreholes in the Klaipėda Strait and surroundings drilled during the engineering geological mapping generally uncovered only the upper part of the Quaternary sequence to an altitude of 30, in some cases 50 m below sea level (Fig. 6.3). Alternating till and inter-till sediments were established in this part of the Quaternary. According to visual description of borehole cores, two types of till layers were distinguished in the geological sections: brown-grey or grey-brown till and dark grey till (at intervals with a greenish tinge). The inter-till sediments are represented by laminated silt, sandy or clayey silt and fine-grained sand with inter-layers of organogenic sediments – dark grey or black gyttja and dark brown peat. Traces of glaciotectonic disturbances (micro-folds, thrust faults, micro-rafts, i.e. glaciodislocations) were observed in the cores of inter-till sediments (Figs. 6.4 and 6.5). The structure of sediments was possible to establish only in the compact laminated (sand, silt, clay) sediments, because the cores of incompact powdery-like sandy sediments were withdrawn disordered due to drilling technology. This technology does not enable to collect the samples in plastic tubes, so the textures of the sand samples removed from the core barrel are destroyed. The upper part of the borehole sections (to the depth of 2–8 m below sea level) is composed of sandy sediments of the Baltic Ice Lake, organogenic sediments (like gyttja, clayey gyttja, etc.) formed in the lagoons of the Ancylus Lake and the Littorina Sea, as well as layers of marine sand with molluscs formed in the Littorina and Post-Littorina Seas. Recent aeolian sediments are widely prevalent on the western coast of the Klaipėda Strait – the Curonian Spit. In some places, 2–3 m thick layers of anthropogenic sediments occur in the uppermost parts of the borehole sections (Fig. 6.3).

6.3 Methods

6.3.1 Sampling

Fine-grained inter-till sand, in some intervals with minor inclusions of tiny particles of organic matter (limnic sediments), was sampled for IR-OSL analysis in four borehole sections (Fig. 6.3). It was very important to establish the absolute age of organic (gyttja and peat) sediments. Therefore, three samples beneath and three samples above them were taken for IR-OSL dating in borehole 36856. Sampling of sand layer only above the organic sediments was available in borehole 36888. In the other two boreholes (35257 and 36897) where the samples were taken from sandy sediments, the borehole sections did not contain inter-layers of organic sediments. All

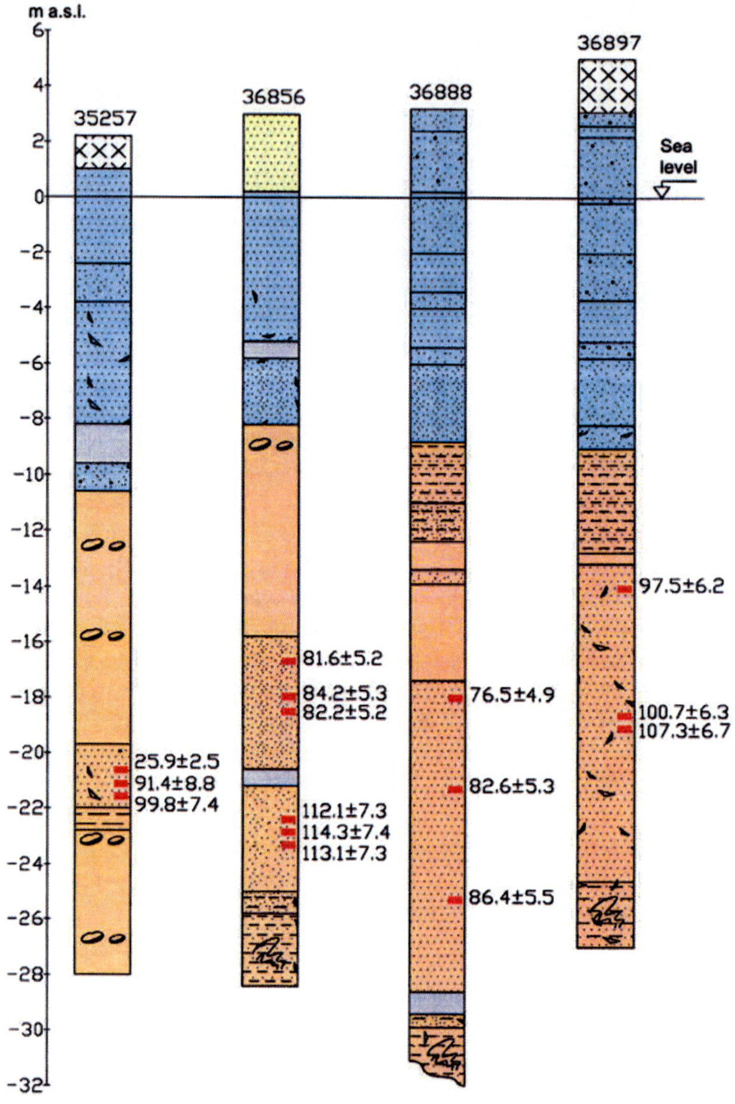

Fig. 6.3 Geological sections of key boreholes from the Klaipėda Strait surroundings, showing the location of the IR-OSL sampling points and luminescence ages of inter-till sediments. For conventional signs, see Fig. 6.2

Fig. 6.4 Glaciodislocated inter-till sediments represented by deformed micro-layers of fine sand, silt and clay in the core of borehole 36888 (depth 33.5 m)

the inter-till sandy sediments sampled for IR-OSL dating were very similar in terms of geological setting and lithological composition. Drilling technology, despite the fact that it does not allow to remove the undisturbed core samples of sand, is quite suitable for correct sampling for IR-OSL dating.

6.3.2 IR-OSL Measurements

All samples were prepared for the luminescence analysis according to standard laboratory procedures (see, e.g. Molodkov and Bitinas 2006). Briefly, alkali feldspar grains of 100–150 μm size were extracted from the sediment under subdued filtered light in the laboratory by a procedure including wet sieving, heavy liquid floatation (collecting 2.54–2.58 g/cm^3 fraction) and treatment by 10% HF for 15 min and finally etching by 20–40% HCl. The IR-OSL measurements were carried out with an Ingrid-Type SLM-1 reader using 860 nm stimulation by short laser pulses. Detection was in the 380–430 nm wavelength range using a combination of 3 mm SZS-22

Fig. 6.5 Glaciodislocated inter-till organogenic sediments in the core of boreholes 36917 and 36922

(blue-green), 3 mm PS-11 (purple) and 2 mm FS-1 (violet) colour glass filters manufactured by the LZOS, JSC (Lytkarino Optical Glass Factory), Russian Federation. For laboratory irradiation a calibrated ^{60}Co source delivering 6.5×10^{-2} Gy/s of gamma radiation was used. After irradiation all samples were kept for about 1 month at room temperature to allow the decay of post-irradiational phosphorescence and to eliminate some anomalous fading-like effects (Jaek et al. 2007).

The paleodose D_e was determined by extrapolating the dose–response curves to zero IR-OSL intensities using the multiple-aliquot additive dose (MAAD) technique (up to 66 aliquots, 15 mg/aliquot, 11 dose points). Dose rate data are based on a laboratory NaI (Tl) gamma spectrometry (for details see, e.g. Molodkov and Bitinas 2006) taking into account the in situ water content and the contribution from cosmic rays. The internal beta dose from the decay of potassium and rubidium within K-feldspar grains was obtained from the concentration estimates reported by Huntley and Baril (1997) and Huntley and Hancock (2001).

IR-OSL dating was performed in the Research Laboratory for Quaternary Geochronology, Institute of Geology, Tallinn University of Technology.

6.3.3 Other Investigations

In the Klaipėda Strait and surroundings, inter-till organogenic sediments were found in 10 boreholes. Paleobotanical analysis was carried out in the Department of Quaternary Researches of the Institute of Geology and Geography, Vilnius, Lithuania. The pollen content of seven borehole sections was identified by O. Kondratienė (including boreholes 36856, 36922 and 36888 shown in the geological cross-section, Fig. 6.2). V. Šeirienė analysed the diatoms in three borehole sections (not included in the presented geological cross-section).

In seven borehole sections, remnants of mollusc shells were found inside the layers of organic sediments. Identification of mollusc species was carried out in the Lithuanian Geological Survey by A. Damušytė.

The adaptation of other methods of absolute geochronology for determining the absolute age of organogenic inter-till sediments was unsuccessful: the sediments were too old for the radiocarbon (^{14}C) method, whereas the uranium–thorium (U–Th) method was unsuitable due to the very low content of organic matter in the sediments.

6.4 Results

The results of IR-OSL analysis in the four dated borehole sections fall into a relatively narrow time span: from 76.5 ± 4.9 to 114.1 ± 7.3 ka (Table 6.1, Fig. 6.3). The average ages in the five inter-till layers are as follows: 95.6 ± 8.1 ka in borehole 35257, 82.7 ± 5.2 and 113.2 ± 7.3 ka in borehole 36856, 81.8 ± 5.2 ka in borehole 36888 and 101.8 ± 6.4 ka in borehole 36897. Each result presented here is an average of three dating obtained on samples taken from three different sedimentary levels. The single young data of 25.9 ± 2.5 ka in borehole 35257 is probably an anomaly due to mistakes in sampling or labelling or due to the influence of some uncontrollable factor. Therefore, this date can be regarded as an outlier and omitted from consideration.

The results of the investigations aimed at finding the features of glacigenic tills which allow to temporally constrain Pleistocene tills in the Klaipėda Strait region are discussed in our companion article (Molodkov et al. 2010). The data of the pollen analysis of organogenic inter-till sediments show that the sedimentation of the examined deposits took place under interglacial conditions (Kondratienė and Damušytė 2009).

The results of diatom analysis of inter-till organogenic sediments indicate that these sediments accumulated in a freshwater basin: for example, in borehole 36917 (depth 14.2–15.0 m) freshwater planktonic species like *Aulacoseira granulata* and *Aulacoseira ambigua* prevail.

Molluscs were very poorly preserved. They were represented only by shell fragments. As a result, in all investigated borehole sections it was possible to identify only two species – *Pisidium* sp. and *Bithynia tentaculata* (Linnaeus 1758). Both species are of freshwater origin.

Table 6.1 IR-OSL results and radioactivity data for inter-till sand samples from boreholes along the Klaipėda Strait

No.	Laboratory code	Borehole no.	Sample no.	U (ppm)	Th (ppm)	K (%)	D_e (Gy)	IR-OSL age (ka)
1.	RLQG 1667-065	35257	2	0.31	0.23	0.87	44.7	25.9 ± 2.5
2.	RLQG 1669-065	35257	3	0.23	0.33	0.55	123.7	91.4 ± 8.8
3.	RLQG 1670-065	35257	4	0.08	0.63	0.65	139.1	99.8 ± 7.4
4.	RLQG 1786-028	36856	1	0.03	1.50	0.74	126.0	81.6 ± 5.2
5.	RLQG 1787-038	36856	2	0.08	0.51	0.58	114.3	84.2 ± 5.3
6.	RLQG 1788-038	36856	3	0.01	1.08	0.69	118.5	82.2 ± 5.2
7.	RLQG 1789-038	36856	4	0.49	2.12	1.15	231.0	112.1 ± 7.3
8.	RLQG 1790-038	36856	5	0.31	1.21	1.14	229.5	114.3 ± 7.4
9.	RLQG 1791-038	36856	6	0.33	2.24	0.96	214.5	113.1 ± 7.3
10.	RLQG 1792-038	36888	1	0.00	0.48	0.66	109.5	76.5 ± 4.9
11.	RLQG 1793-038	36888	6	0.15	1.42	0.71	129.0	82.6 ± 5.3
12.	RLQG 1794-038	36888	10	0.01	0.87	0.72	128.3	86.4 ± 5.5
13.	RLQG 1784-028	36897	1	0.30	1.72	0.85	169.1	97.5 ± 6.2
14.	RLQG 1785-028	36897	4	0.37	1.13	0.68	154.4	100.7 ± 6.3
15.	RLQG 1783-028	36897	6	0.24	1.23	0.67	161.4	107.3 ± 6.7

Notes: U, Th and K content in sediments are determined from laboratory gamma-ray spectrometry; water content corrections, calculated cosmic ray contribution and internal feldspar dose rate were taken into account on calculation of the IR-OSL ages.

6.5 Discussion

One of the main problems of Quaternary geology in the vicinities of the Klaipėda Strait and in the whole Lithuanian Coastal Area is the reliable stratigraphic subdivision and correlation of sediments. The problem is that there are no reliable criteria for stratigraphic correlation, especially for glacial sediments (tills). The colour of tills is very changeable and cannot serve as a correlative. According to the experience of large-scale geological mapping of Quaternary sediments in the Lithuanian Coastal Area, other indicators, such as the petrographic composition of gravel part tills or other lithological components, are also very changeable and not suitable for the above-mentioned purposes. Thus, indirect methods are to be used to solve the problems of stratigraphic correlation of tills.

The results of IR-OSL studies show that the inter-till sediments investigated in the Klaipėda Strait were formed during the ice-free interval MIS 5d-5a. The sampled inter-till sediments are occurring not in situ but as blocks (rafts) in the till bed (Fig. 6.2). This opinion is confirmed by an abundance of micro-glaciodislocations observed in borehole cores (Figs. 6.4 and 6.5). Based on the geotechnical properties of sediments, some additional conclusions about till age could also be drawn. It was established that the geotechnical properties of the lowermost complex of tills both in the Klaipėda Strait area and in the whole Klaipėda City region at altitudes close to zero or below sea level significantly differ from those of the relief-forming tills situated at higher altitudes (Gadeikis 1998). There are some differences in the density of tills (1.96–2.20 g/cm^3 for the younger and 2.21–2.24 g/cm^3 for the older ones, respectively), but the biggest distinction is the module of deformation, which varies from 16 to 74 MPa for the beds of relief-forming tills and reaches up to 100–110 MPa for older till beds. A big difference is observed in the values of cone resistance, which are 1.1–5.0 and 5.0–14.0 MPa, respectively. According to the presented geotechnical properties, the above-mentioned separate group of tills was in different conditions of consolidation – the older one was additionally influenced by compression from the glaciers and long-lasting lithification processes, i.e. this till was formed significantly earlier than the relief-forming till beds that belong to the late Weichselian (late Nemunas). This difference is very obvious in the above-mentioned Olando Kepurė section (Molodkov et al. 2010).

Hence, we may conclude that the till containing these incorporated inter-till sediments could be formed only during the Weichselian (Nemunas) glaciation. Some other indications corroborating this hypothesis are also reported (ibid.).

The limnic sediments – sand alternating with silty-clayey or organogenic sediments – are widespread in the Klaipėda Strait area where they have been established in tens of boreholes. Thus, it is possible to presume that during the MIS 5d-5a time span a quite big freshwater sedimentary basin (or basins) existed within our study area – very likely in the depression of the Baltic Sea; lately it served as a source of terrigenic material for till formation during the Weichselian glacial advances.

According to the interpretation of results of pollen analysis, the pattern of the vegetation development including the immigration of particular tree species is different from those typical for Holsteinian (Butėnai) and Eemian (Merkinė) interglacials, but is in good agreement with the biostratigraphical records of the Drenthe-Warthe (Snaigupėlė, Lubavian, Schöningen) interglacial, late middle Pleistocene that suggest the similar age of the investigated inter-till sediments (Kondratienė and Damušytė 2009). However, such interpretation is in disagreement with our IR-OSL data.

Taking into account all the above-mentioned factual data, it is possible to maintain that the till bed beneath the bottom of the Klaipėda Strait was most probably formed by a glacier advance during MIS 4, i.e. during the Weichselian early pleniglacial. This till bed can be correlated with the lowermost till complex in the

Olando Kepurė outcrop that also was most probably formed by a glacier advance during MIS 4 (Molodkov et al. 2010). According to fabric measurements in the Olando Kepurė outcrop, this till was formed by a glacier advancing from the north (Bitinas 1997). All these data are in good agreement with a reconstruction made by Svendsen et al. (2004) according to which part of the southwestern margin of the Eurasian ice sheet could have been situated in the Lithuanian Coastal Area or in the whole Western Lithuania during the Weichselian early pleniglacial maximum (MIS 4). The till bed formed during the early pleniglacial has been distinguished in the neighbouring Western Latvia: in stratigraphic schemes it has been identified as Talsi Stadial (Zelčs and Markots 2004). In the more southern region – territory of Poland – the till bed of the same age has been identified as Świecie Stadial (Lindner and Marks 1995, Ber 2006). According to the assumptions of some former researchers, the glacial advance could reach Lithuania during the early Weichselian – the corresponding till bed was distinguished as Varduva Stage in the stratigraphic scheme of Lithuania (Vonsavičius 1984). Later this opinion was not confirmed by factual data and the mentioned stratigraphic unit was rejected from the stratigraphic schemes (Gaigalas 2001, Satkūnas et al. 2007). The results of geochronological investigations presented in this chapter suggest that the Quaternary stratigraphic scheme of Lithuania should be supplemented by a new stratigraphic unit (for instance, it could be named as Melnragė Stadial) valid for Western Lithuania.

Thus, the evidence reported in this study does not support an opinion that the till layer beneath the bottom of the Klaipėda Strait and those at the same altitudes in the surroundings formed during the Warthanian (Medininkai) glaciation (MIS 6). The till layers in the northern part of the Klaipėda Strait, lying between the above-mentioned middle Weichselian till and pre-Quaternary sediments (Fig. 6.2, boreholes 4/98, 8140, 10092), most probably belong to the middle Pleistocene.

6.6 Conclusions

The results obtained in this work show that the absolute majority of the IR-OSL ages of investigated inter-till sediments beneath the bottom of the Klaipėda Strait fall within the age range of MIS 5d-5a, i.e. these sediments were formed during the early Weichselian. The sampled inter-till sediments are occurring not in situ, but as blocks (rafts) within the till bed formed during the Weichselian (Nemunas) glaciation. According to a reconstruction by Svendsen et al. (2004), the latter most probably can be associated with the ice movement during MIS 4 – part of the southwestern margin of the Eurasian ice sheet could have been situated in the Lithuanian Coastal Area and, probably, in the whole Western Lithuania during the Weichselian early pleniglacial maximum (MIS 4). This conclusion does not support the standpoint that the till beds beneath the bottom of the Klaipėda Strait were formed during the Warthanian (Medininkai, MIS 6) glaciation.

Acknowledgements We are grateful to colleagues Tatyana Balakhnichova and Marina Osipova for their contribution to IR-OSL dating laboratory work reported here, to Helle Kukk for checking the English text and to Eglė Šinkūnė and Donė Gribytė for help in preparing the illustrations. This research was supported by grant no. 6112 from the Estonian Science Foundation, by Estonian State Target Funding Project No. 0320080s07, by grant no. LEK-10005 from the Research Council of Lithuania by Klaipėda State Seaport and Lithuanian Geological Survey.

References

Ber A (2006) Pleistocene interglacials and glaciations of northeastern Poland compared to neighbouring areas. Quaternary International 149:12–23

Bitinas A (1997) Quaternary deposits on the outcrop Olando Kepurė. The 5th Marine geological conference "The Baltic", Abstracts, Excursion Quide, Vilnius, Lithuania, pp 109–110

Bitinas A, Repečka M, Kalnina L (1999) Correlation of tills from the South-Eastern Baltic Sea bottom and near shore boreholes. Baltic Special Publication 12:5–10

Gadeikis S (1998) Engineering geological conditions of Klaipėda City. Doctors thesis, Vilnius University, Lithuania, 35p

Gaigalas A, Melešytė M, Gulbinskas S (1987) The Pleistocene tills in the area of Nida structure. Geological structure of Quaternary deposits in the bottom of the Baltic Sea and distribution of useful minerals, Abstracts of the seminar, Palanga, Vilnius, April 11–12, pp 11–13 (in Russian)

Gaigalas A, Melešytė M, Gulbinskas S (1997) Petrographic composition of Pleistocene tills of the Baltic Sea coastal zone. Proceedings of the 2nd conference of Lithuanian oceanologists "Lithuania – Marine State", Abstracts, Klaipėda, May 28–30, p 3. Klaipėda. (In Lithuanian)

Gaigalas A (2001) Stratigraphy and geochronology of the Upper (Late) Pleistocene. In: Baltrūnas V (ed) Stone age in the South Lithuania. Geologijos Institutas, Vilnius, pp 7–24

Gelumbauskaitė LŽ, Šečkus J (2005) Late Quaternary shore formations of the Baltic basins in the Lithuanian sector. Geologija 52:34–45

Huntley DJ, Baril MR (1997) The K content of the K-feldspars being measured in optical dating or in thermoluminescence dating. Ancient TL 15:11–13

Huntley DJ, Hancock RGV (2001) The Rb contents of the K-feldspar grains being measured in optical dating. Ancient TL 19:43–46

Jaek I, Molodkov A, Vasilchenko V (2007) On the possible reasons of anomalous fading in alkaline feldspars used for luminescence dating of Quaternary deposits. Estonian Journal of Earth Sciences 56(3):167–178

Kabailienė M, Vaikutienė G, Damušytė A, Rudnickaitė E (2009) Post-Glacial stratigraphy and paleoenvironment of the northern part of the Curonian Spit, Western Lithuania. Quaternary International 207:69–79

Kondratienė O, Damušytė A (2009) Pollen biostratigraphy and environmental pattern of Snaigupėlė Interglacial, Late middle Pleistocene, western Lithuania. Quaternary International 207:4–13

Lindner L, Marks L (1995) Correlation of glacial episodes of the Wisla (Vistulian) glaciation in the Polish Lowland and mountain regions, and in Scandinavia. Bulletin of the Polish Academy of Sciences, Earth Sciences 43(1):5–15

Majore J, Rinke R, Savvaitov A, Veinbergs I (1997) Lithostratigraphical identification of tills in the south-eastern part of the Baltic sea by the method of the rounded hornblende grains. Baltica 10:9–12

Molodkov A, Bitinas A (2006) Sedimentary record and luminescence chronology of the Lateglacial and Holocene aeolian sediments in Lithuania. Boreas 35(2):244–254

Molodkov A, Bitinas A, Damušytė A (2010) IR-OSL studies of till and inter-till deposits from the Lithuanian Maritime region. Quaternary Geochronology 5:263–268

Satkūnas J, Bitinas A (2002) State-of-art of Quaternary stratigraphy of Lithuania. Proceedings of the 5th Baltic Stratigraphic conference "Basin stratigraphy – modern methods and problems", Extended abstracts, Vilnius, Lithuania, September 22–27, 2002, pp 179–181. Geological Survey of Lithuania, Vilnius

Satkūnas J, Grigienė A, Bitinas A (2007) Stratigraphical division of the Lithuanian Quaternary: the present state. Geologijos akiračiai 1:38–46. (In Lithuanian)
Svendsen JI, Alexanderson H, Astakhov VI et al (2004) The Late Quaternary ice sheet history of Northern Eurasia. Quaternary Science Reviews 23:1229–1271
Vonsavičius VP (1984) The structure of Quaternary deposits in the Lithuania and problems of their stratigraphic division. In: Kondratienė OP, Mikalauskas AP (eds) Palaeogeography and stratigraphy of Quaternary of the Baltic and adjacent areas, Vilnius, pp 88–96 (in Russian)
Zelčs V, Markots A (2004) Deglaciation history of Latvia. In: Ehlers J, Gibbard PL (eds) Quaternary glaciations – extent and chronology. Elsevier BV, Oxford, pp 225–243

Part IV
Coastline Changes

Chapter 7
Coastlines of the Baltic Sea – Zones of Competition Between Geological Processes and a Changing Climate: Examples from the Southern Baltic

Jan Harff and Michael Meyer

Abstract Relative sea level change is, besides the geological buildup and hydrographic parameters, the main controlling factor in shaping the coastlines on the centennial timescale and beyond. Vertical displacement of the earth's crust and eustasy serve as main components driving the relative sea level (RSL) change during the Quaternary. Whereas the eustatic change mirrors mainly climatic factors, the vertical displacement of the earth's crust has to be regarded in former glaciated areas as a result of glacio-isostatic adjustment superimposed by the regional tectonic regime or land subsidence due to local factors. A simple model is applied to reconstruct the palaeogeographic development of a coastal area and to generate future projections as coastline scenarios. For the hindcast relative sea level curves have to be compared with recent digital elevation models. For future projections data of vertical crustal displacement received from gauge measurements and eustatic changes based on climate scenarios have to be superimposed. The model has been applied to the Baltic Basin, considered as a natural laboratory for coastal research as it is extending from the uplifting Fennoscandian Shield to the subsiding southern Baltic lowlands. Subsidence, climatically driven sea level rise, and meteorologically induced coastal flooding provoke permanent coastal retreat at the southern sinking coasts. Predictions of coastal hazards are made with the model by using neotectonical data and long-term sea level change data superimposed with extreme sea level data measured during the storm surge in November 1872.

Keywords Sea level · Climate · Eustacy · Isostacy · Coastline history · Hazard · Future projection

---

J. Harff (✉)
Leibniz Institute for Baltic Sea Research Warnemünde, D-18119 Rostock, Germany; presently at Institute of Marine and Coastal Sciences, University of Szczecin, PL-70-383 Szczecin, Poland
e-mail: jan.harff@io-warnemuende.de

J. Harff et al. (eds.), *The Baltic Sea Basin*, Central and Eastern European Development Studies (CEEDES), DOI 10.1007/978-3-642-17220-5_7,

7.1 Introduction

The problem of sea level change is one of the most important topics of scientific research programmes and intergovernmental discussions (Metz et al. 2007). The anthropogenic driving forces and future development of global sea level have been described by Cubasch (2001). In addition, scenarios of secular sea level rise have to be superimposed with the effect of short-term events as storm surges. While long-term (secular) sea level scenarios are derived from climate and neotectonic modelling, events have to be described by empirical data. For an effective coastal defence and catastrophe management, local authorities need reliable information of future development along coastal zones. For this, geoscientists have to take into consideration not only global sea level changes but also regional vertical crustal movement, coastal morphogenesis, and regional/seasonal characteristics in climatically driven water level regularities. New results have been published during the last two years. For instance, a prediction of the deformation of the earth's crust caused by loading and unloading of inland glaciers was given by Peltier (2007). Tarasov and Peltier (2002) describe the interrelation of subsidence and sediment formation for the Lagoon of Venice. But, there is still a need for interdisciplinary studies of the interrelation of crust deformation, climatically driven sea level variations, and catastrophic events. The chapter presented here contributes to the understanding of the complex interrelation between geo-system and climate along changing coastlines. We deal with the cause and effect relation between climate change, vertical crustal movement, and the change of the coastlines. We approach the reconstruction of palaeogeographic scenarios as well as future coastline scenarios coupled with IPCC sea level projections and empirical data of vertical crustal movements and gauge measurement of hazardous events.

The need of tools for investigating coastal change processes requires the development of models that display cause and effect relation in a changing coastal environment. Despite that need it has to be stated that modelling results of the complex interrelation of processes of the earth's crust, sea level change, climate, and socio-economic development on timescales of millennia are scarce by now. With the research project SINCOS (Sinking Coasts – Geosphere, Ecosphere, and Anthroposphere of the Holocene Southern Baltic Sea) funded by the German Research Foundation (DFG), a pace forward has been done in filling this gap (Harff and Lüth, 2007). The basis for an interdisciplinary approach in coastline change modelling is a data management system that allows the integration of data from quite different scientific sources describing coastal systems from different points of view. This database system serves as the main prerequisite for an analysis of an interrelation of variables measured (or received by modelling) from different disciplines.

Modelling has been carried out in two directions: hindcasting and projective scenarios on a time span between 5700 years BP and 2100 years AD. While palaeomodelling depends on the construction of relative sea level curves, the concept of projective modelling involves climate scenario data. These data are provided by the German Climate Research Centre (Voß et al. 1997) and Intergovernmental Panel on Climate Change (IPCC) reports (Metz et al. 2007). To display the temporal and spatial dependencies of variables mirroring the complex structure, a coastal 4D GIS is used for this study.

7.2 Area of Investigation

Figure 7.1 shows the Baltic Sea as a semi-enclosed marginal sea surrounded by the Scandinavian Caledonides and the Fennoscandian Shield in the north, the Russian Plate in the southeast, and the Northeast-German Depression in the south and southwest. The Baltic area including the sea basin was shaped by the Quaternary glaciations: glaciers have abraded the Baltic Sea Basin (water depth 55 m on average) forming several separate sub-basins and shallower sills. Within the Baltic Basin and along its southern coastlines Weichselian glacial deposits form the main sources for the Late Pleistocene and Holocene sediment formation. The Baltic Sea is connected with the North Sea through the Belt and the Sound which serve as a "bottleneck" for the water exchange with the world ocean.

The type of coasts around the Baltic Sea depends on the geological structures and the geotectonic setting. Fjord-like coasts and sea bottom coasts (Gulf of Bothnia) as well as archipelagos (northern Gulf of Finland, East Sweden) prevail at the Fennoscandian Shield built up by Proterozoic crystalline bedrock. At the southern Gulf of Finland and the Estonian coast, cliffs can be found, built from Palaeozoic sediments, whereas in the southern Baltic Sea, moraine cliffs and sandy Holocene spits and lowland coasts are dominating.

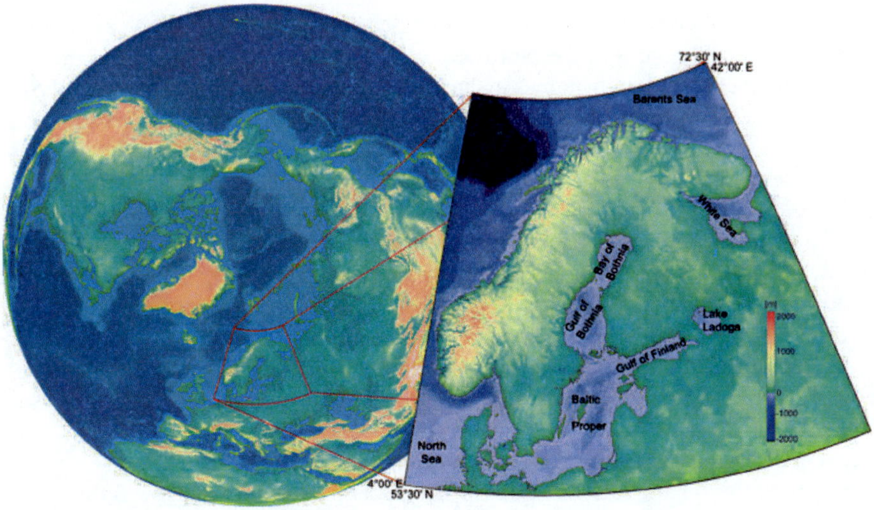

Fig. 7.1 Relief map of the rigid earth (Digital Elevation Model – DEM_0) for the Baltic Sea area. Original data are provided by NGDC (2001)

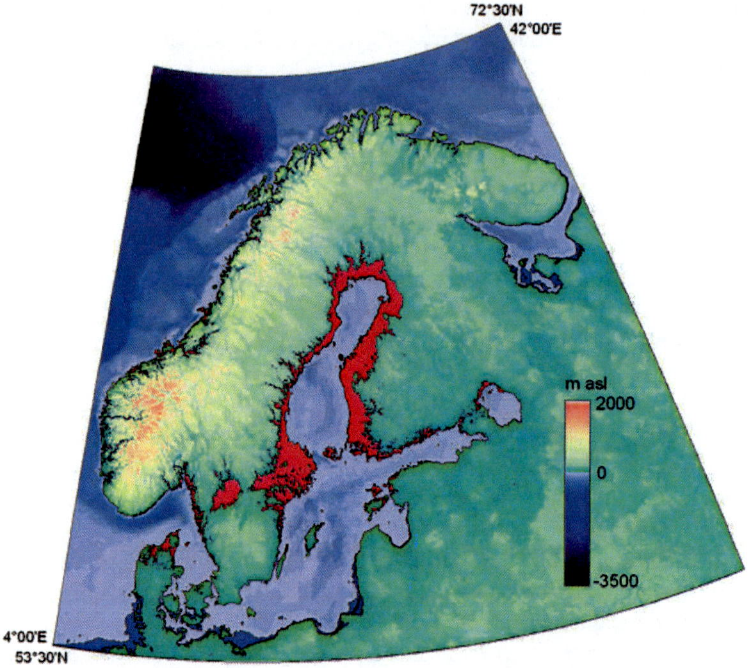

Fig. 7.2 The Baltic Sea and the change of coastlines since the onset of the Littorina Transgression about 7700 years BP (modified from Harff et al. 2007). *Red colours mark* areas of regression and *blue colours* areas of transgression

For studies of coastline change the Baltic Sea serves as an excellent natural laboratory as isostatic uplift in the North has caused continuous regression of the sea during the last 8000 years, whereas in the South climatically controlled sea level rise superimposed with subsidence of the earth's crust is responsible for a transgression between the Belt Sea and the Curonian spit in the Southeast. Figure 7.2 shows areas of Holocene transgression and regression based on a map published by Harff et al. (2007).

The main environmental change within the areas of investigation was due to the inflow of marine water via the Danish straits about 8000 years BP changing the freshwater environment into a brackish-marine one. This salt water inflow is called "Littorina Transgression" named by the fossil beach snail *Littorina littorea.*

Along the subsiding coasts the permanent transgression has affected also processes of morphogenesis that can be studied in an exceptional manner here. Therefore, for a subregional study the southern Baltic Sea coast has been investigated in detail within the frame of the research project SINCOS, "Sinking Coasts – Geosphere, Ecosphere, and Anthroposphere of the Holocene Southern Baltic Sea" (Harff and Lüth 2009). In this study light is also shed on the Wismar Bight at the southwestern Baltic coast where detailed geological and archaeological studies have revealed not only the natural history of that area but also the change of the socio-economic environment in reaction to the coastal retreat. In this chapter an extreme sea level scenario for 2100 AD as a future projection is given for the Wismar Bight.

7.3 Regional Transgression/Regression Model

Within the Baltic Sea area, the interaction of crustal subsidence and uplift (glacio-isostatic adjustment) and climatically driven eustatic sea level changes can be studied in an exceptional manner.

For any time point $t \in T$ the elevation of an area can be expressed by a digital elevation model DEM_t or geographic surface terrain model (Harff et al. 2005) covering as a grid an area of investigation $\mathbf{R}$. The DEM_0 is the "recent" digital elevation model ($t = 0$) for the area under investigation.

$$\mathrm{DEM}_t = \mathrm{DEM}_0 - \begin{cases} \mathrm{RSL}_t, & \text{if } t < 0 \\ \mathrm{EC}_t + \mathrm{GIA}_t, & \text{if } t \geq 0 \end{cases}. \quad (1)$$

We can explore the surface terrain model DEM_t in two different ways: for the geological past ($t < 0$) and for future projections. For $t < 0$ time is measured in conventional radiocarbon years. $t = 0$ stands for the reference year 1950 AD. For the time ($t > 0$) we apply the annual (calendar) scale. $\mathrm{rsl}_t(\mathbf{r}) \in \mathrm{RSL}_t$ marks a relative sea level curve at a location $\mathbf{r} \in \mathbf{R}$. RSL_t has to be determined by spatial interpolation of data from shoreline displacement curves (relative sea level data, rsl) to a grid covering the area of investigation.

The relative sea level change RSL consists of two components: RSL = EC+GIA. Here, EC marks the eustatic component and GIA (glacial isostatic adjustment) stands for the vertical deformation of the earth's crust. EC is controlled mainly by the change of the palaeoatmospheric temperature which affects the volume of the oceanic water body not only by thermal expansion but also by melt water inflow from the decaying continental ice shields. GIA expresses the vertical movement of the earth's crust due to loading and unloading caused by accumulation and melting of inland ice masses. For the Fennoscandian Shield this process is described regarding the last glaciations by Lambeck et al. (1998a, b), Amatov et al. (Chap. 3) in this book, and more generally by Peltier (2007). Also the gravitational influence on the sea level change caused by compensational mass flow below uplifting crust should be mentioned (Ekman 2009). As a function of time $t \in T$, ec_t is regarded constant for the whole area of investigation ($\forall \mathbf{r} \in \mathbf{R}$). The isostatic component $\mathrm{gia}_t(\mathbf{r}) \in \mathrm{GIA}_t$ of a relative sea level curve $\mathrm{rsl}_t(\mathbf{r}) \in \mathrm{RSL}_t$ at a location $\mathbf{r}$ is expressed for each time step t by the difference between the value of the relative sea level curve and the corresponding eustatic value.

$$\mathrm{gia}_t(\mathbf{r}) = \mathrm{rsl}_t(\mathbf{r}) - \mathrm{ec}_t, \; \mathbf{r} \in \mathbf{R}, \; t \in T \quad (2)$$

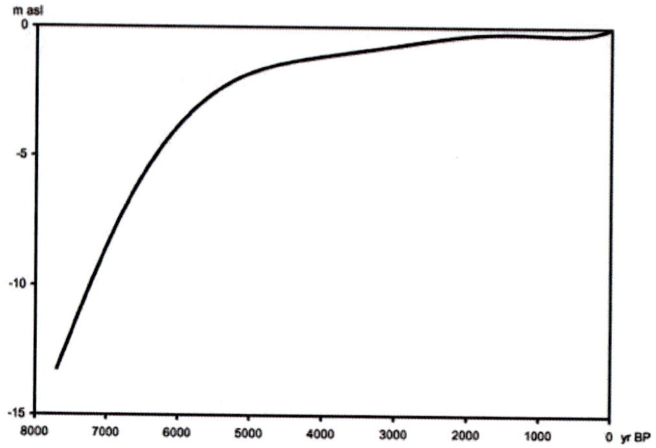

Fig. 7.3 Relative sea level change curve for the Darss Peninsula, western Baltic Sea (data published by Lampe et al. 2007), expressing at a neotectonically stable position mainly the climatically driven sea level change. The original data have been fitted by a polynomial trend function of 6th degree

The eustatic curve ec is identical with a relative sea level curve rsl determined at a tectonical stable coastal site (Harff et al. 2001). Such a position is situated at the root of the Darss-Zingst Peninsula (SW of Rügen Island, marked by an arrow within Fig. 7.7, explanation below). The corresponding rsl curve (Fig. 7.3) displays the eustatic change for the Baltic Sea since the Littorina Transgression onset. Deploying Eq. (2) and using the rsl curves and the eustatic curve as input data it becomes possible to calculate the glacio-isostatic adjustment (gia) curve for each of the sites the rsl curves are allocated to. Figure 7.4 shows a selection of curves along the whole Baltic coast. For each of the selected sites the local rsl curve, the regional (blue) ec curve (after Lampe et al. 2007), and the (red) gia curve according to the calculation after Eq. (2) are shown. The shape of these gia curves reveals the character of glacio-isostatic behaviour. Sites 6, 7, and 8 in the northern part of the basin show a continuous uplift signal. Also sites 4, 5, and 9 show a predominantly uplift signal, but remarkably weaker (see also Berglund et al. 2005 Miettinen et al. 2007), regarded as an (uplift) transition type (Harff et al. 2001). At the southern Baltic coast, sites 1 and 3 are characterized by subsidence which can be explained by its position south of the hinge line (Fig. 7.5) at the subsiding belt. The gia curve of site 2 located at Rügen Island shows a shape similar to the transition type. We interpret this fact by the position north of the hinge line at the uplifting part of the crust (Fig. 7.7).

7.4 Sea Level Change and Palaeogeographic Scenarios

Long-term sea level changes are expressed for special sampling sites near coast areas by relative sea level (rsl) curves. For the regional palaeogeographic scenarios,

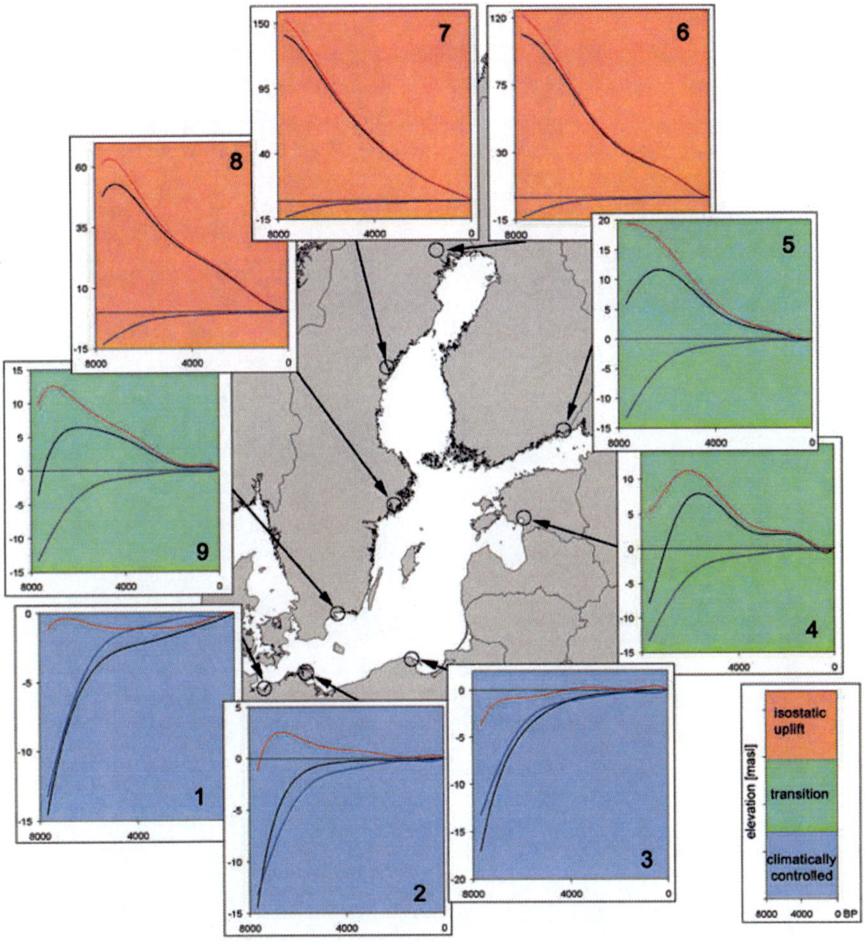

Fig. 7.4 Relative sea level change curves (*black*) compared to the eustatic curve (*blue*) and the isostatic component (*red*) at nine locations in the Baltic Sea coast area. Locations *1–3* are dominated by climatically controlled sea level rise, whereas *6–8* are mainly determined by isostatic uplift exceeding the sea level rise clearly. Sites *4*, *5*, and *9* are allocated to a transition type. Curves 1 and 2: Lampe et al. (2007), curve 3: Uścinowicz (2006), curve 4: Veski et al. (in press), curve 5: Miettinen (2004), curve 6: Linden et al. (2006), curve 7: Berglund (2004), curve 8: Karlsson and Risberg (2005), curve 9: Berglund (1964)

published relative sea level curves were used as model input. Rosentau et al. (2007) have described how the data from these curves can be digitized and interpolated in any time resolution providing data grids covering the area of investigation. We refer here to the selected set of relative sea level curves given in Fig. 7.4 (black curves). Each rsl curve covers the time span between 8000 years BP ($t = 8000$, the Littorina Transgression onset) and recent time ($t = 0$). The different shapes of curves 5–9

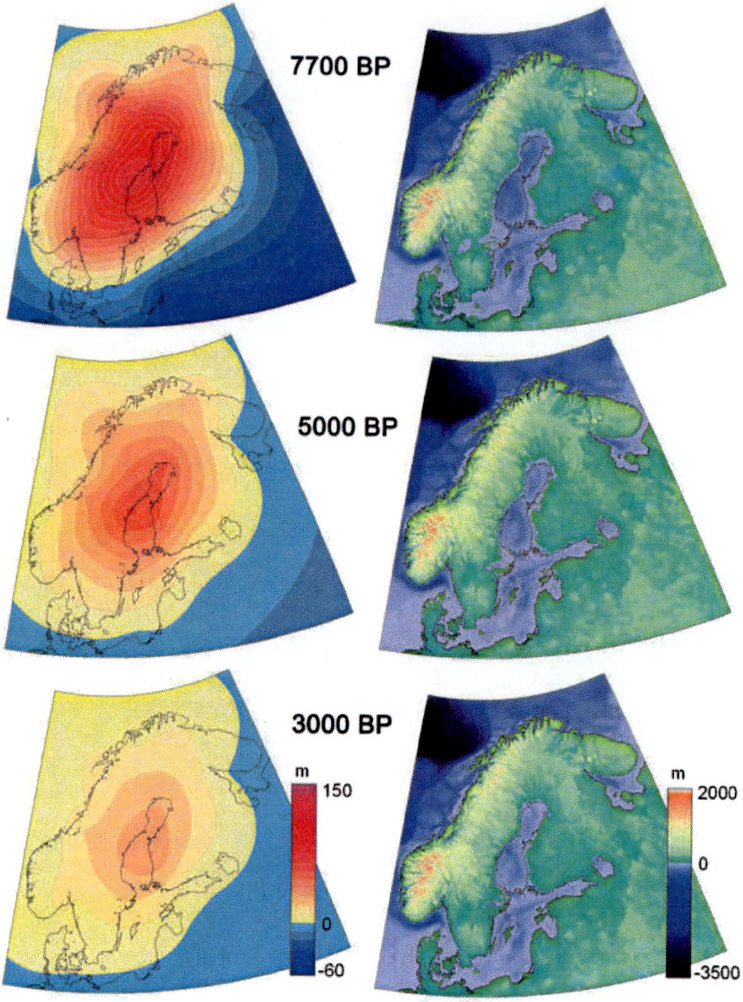

Fig. 7.5 *Left panel*: Differences of earth elevations (RSL_{7700}, RSL_{5000}, RSL_{3000}) to present time elevation of the Baltic Sea area. *Right panel*: Palaeogeographic scenarios (palaeo-digital elevation scenarios DEM_{7700}, DEM_{5000}, DEM_{3000}) used to generate Fig. 7.2. The comparison of DEM_{7700}, DEM_{5000}, and DEM_{3000} shows clearly the synchronous regression of the sea at the northern rim of the Baltic Basin and transgression at its southern coast. The process of transgression has been systematically analysed within the frame of a research project SINCOS (Sinking Coasts) between 2003 and 2008 (Harff and Lüth 2009)

(representing the uplifting Fennoscandian Shield) and curves 1–4 (standing for the glacio-isostatically subsiding belt) are evident.

For the investigated time span three time points have been selected: 7700 years BP at the early stage of the brackish Baltic Sea history, 5000 years BP when the postglacial sea level rise was decelerated, and 3000 years BP standing for the warm period of the Roman Climate Optimum (RCO). For each of these time slots the sea level values have been picked from the curves cited by Rosentau et al. (2007) and interpolated to grids RSL_t, $t \in \{7700, 5000, 3000\}$. These grids are displayed as isoline maps at the left panel in Fig. 7.5 and show clearly the deceleration in relative sea level change since the Littorina Transgression onset. Within the maps the zero-isoline (hinge line) marks the transition between falling sea level in the centre of the Fennoscandian Shield and rising sea level in its circumjacent belt. According to Eq. (1) palaeogeographic scenarios have been generated for the three time slots of 7700, 5000, and 3000 years BP. These scenarios are given at the right panel in Fig. 7.5.

7.5 Vertical Displacement of the Earth's Crust

Rosentau et al. (2007) have published an isobase map as a compilation of tide gauge measurements (rsl curves) in the area of the Baltic Sea, combined with data from Ekman (1996) for the central Baltic and new sea level data from the Kattegat to the Gulf of Gdańsk provided by R. Dietrich and A. Richter from the Technical University Dresden for the western Baltic Sea area. So, the resulting map of Rosentau et al. (2007) is an update of the one published by Ekman (1996) for the area south of 57.5°N. As shown in Eq. (1) the relative sea level change consists of the glacio-isostatic component and the eustatic (climatically driven) one. The eustatic component can be regarded constant for the western Baltic Sea during the last century (Hupfer et al. 2003). Therefore, one can separate quantitatively the isostatic field by subtracting a constant from the field displayed by Rosentau et al. (2007). Hupfer et al. (2003) and Ekman (2009) in his complete treatment of the eustatic sea level rise in the Baltic Sea during the last centuries suggested 1.0 mm/year eustatic sea level rise for the western Baltic Sea during the twentieth century. Based on this assumption we subtracted this value from the data mapped by Rosentau et al. (2007) and received a map of vertical crustal movement (Fig. 7.6). The uplifting Fennoscandian Shield caused by unloading of Scandinavia due to the melted Weichselian ice sheet is clearly marked by its centre at the Gulf of Bothnia. The hinge line between the rising Fennoscandian Shield and its subsiding belt follows at the southern Baltic the coastline.

In order to get a map in a higher resolution for the southern area we have corrected the gauge data in the southern Baltic by subtracting the eustatic component and re-interpolated these data to a grid covering the western Baltic Sea. The corresponding contour map is displayed in Fig. 7.7. It is eye catching that the hinge line which strikes almost coast-parallel WSW in the eastern part of the area bends NW south of Rügen Island and forms a step south of the Danish Islands. This pattern is mirroring the generally NW striking tectonic elements close to the SW border of the Eastern European Platform (Krauß 1994).

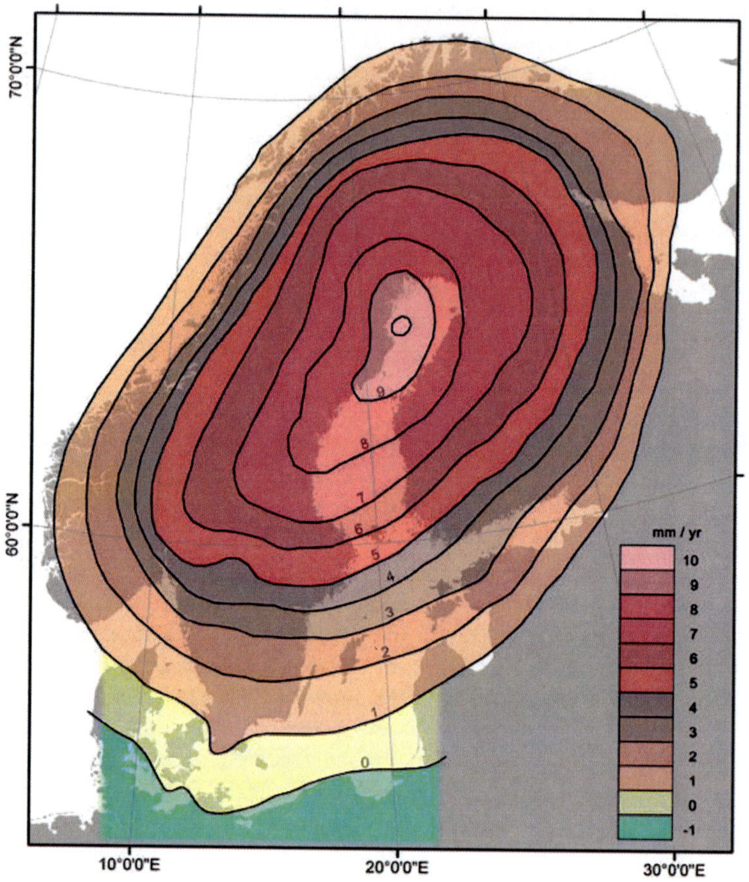

Fig. 7.6 Vertical crustal displacement (mm/year) for the central Baltic and for the western Baltic Sea for the twentieth century

7.6 Extreme Sea Level Scenarios (Future Projections)

Scenarios of future extreme sea level events are needed not only for planning of sustainable coastal development but also for catastrophe management. Generalizing Eq. (1) we introduce a field MAX into the formula

$$\mathrm{DEM}_t = \mathrm{DEM}_0 - \mathrm{EC}_t + \mathrm{GIA}_t + \mathrm{MAX}, \quad \forall t > 0 \tag{3}$$

MAX stands for the highest sea level measured within the area of interest. This approach is similar to the recommendation to estimate the defence water level for coastal protection constructions (Oumeraci and Schüttrumpf 2002). For the western Baltic Sea we model a scenario for $t = 150$, which means a possible sea level event

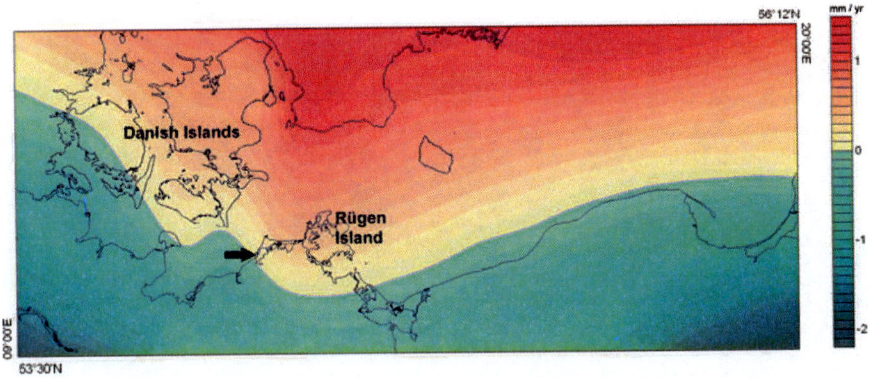

Fig. 7.7 Vertical movement of the earth's crust in the western Baltic Sea for the twentieth century. The *arrow* shows the Darss Peninsula site of the relative sea level curve published by Lampe et al. (2007)

in 2100 AD. For the eustatic change component EC_{150}, data given by Voß et al. (1997) for the North Sea have been used. Meyer et al. (Chap. 14, this book) show a sea level curve as future scenario for the North Sea based on IPCC scenario A for CO_2 emission. Due to the permanent connection between North and Baltic Sea since the Littorina Transgression we assume a 1:1 transfer function of the secular sea level development from the North Sea to the Baltic Sea and apply the value $EC_{150} = 20$ cm as constant parameter to our model. This is a conservative value compared to more recent (still debated) estimations (Church and White 2006, Metz et al. 2007) of global sea level rise for the twenty-first century. But, as Ekman (2009) and Hupfer et al. (2003) give a value of 1 mm/year sea level rise for the twentieth century, we regard a doubled value for the twenty-first century reasonable. As an estimate for the field of extreme sea level, a reconstruction of the storm surge from 4 to 14 November 1872 – the highest sea level field ever measured in the western Baltic – has been applied, according to the methods for the estimation of the defence water level (Oumeraci and Schüttrumpf 2002). Rosenhagen and Bork (2009) have re-modelled the wind field and the water level at the western Baltic based on air pressure data measured during the storm period (using the circulation model of the Bundesamt für Seeschiffahrt und Hydrographie Hamburg, Version v4, Model of the North- and Baltic Sea with integrated coastmodel).

From the reconstructed sea level data the maximum values have been picked and a map of the maximum water level for the time between 4 and 14 November 1872 has been generated (Fig. 7.8).

Superimposing the data by deploying Eq. (3) we receive the regional sea level scenario for the southern Baltic coast given in Fig. 7.9. It is clearly visible that large coastal areas would be endangered to be flooded in the case of a storm surge at the end of this century. Taking into account that cities like Wismar, Rostock, Stralsund, Greifswald, and Szczecin would be directly affected by such an extreme water level, it is beyond doubt that strict efforts for the protection of the coast are necessary.

Fig. 7.8 Reconstruction of the sea level within the western Baltic Sea during the storm event and flood of November 1872. The map presents the maximum values for the time between 4 and 14 November 1872 of the hindcast given by Rosenhagen and Bork (2009)

For planning of coastal protection activities local models in a higher resolution are needed. In Fig. 7.10, a local extreme water level scenario is given for the Wismar Bight. In the centre of the Bay the navigational channel directing to Wismar Harbour is clearly visible. Areas endangered to be flooded during an extreme storm surge are marked red. It is obvious that coastal meadows, in particular fragile peninsulas and bars, are at risk of flooding and erosion. The core of Wustrow Peninsula detaching the "Salzhaff" lagoon from the Baltic Sea would be separated in case of a storm surge from the mainland and the sandy bar in the NW of the map in Fig. 7.10 would be washed over and exposed to the physical

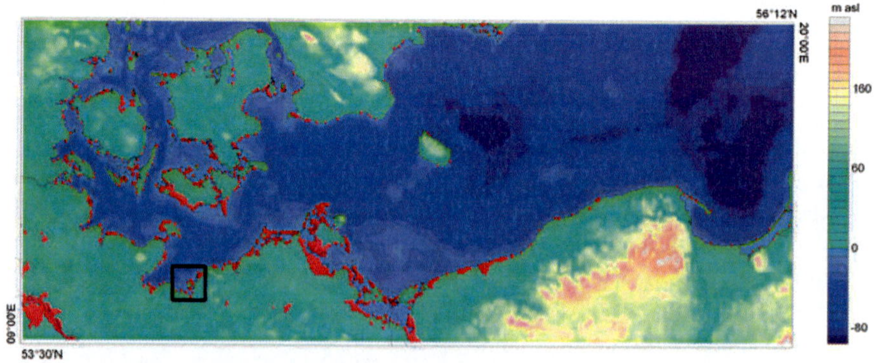

Fig. 7.9 Extreme sea level scenario for 2100 AD at the western Baltic Sea, combining secular trends in neotectonic displacements (vertical crustal movement), climatically controlled sea level rise based on IPCC scenario, and gauge reconstruction for the coastal flood in November 1872. *Red colour* marks areas of potential coastal hazards. The Wismar Bight used for the local scenario in Fig. 7.10 is marked at the southwestern coast. The scenario is generated under the theoretical assumption that no coastal protection activities will take place

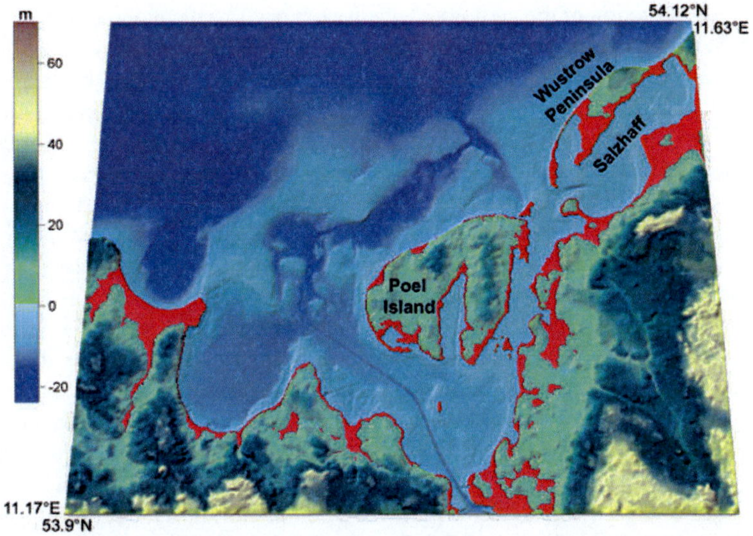

Fig. 7.10 Extreme sea level scenario for 2100 AD for the Wismar Bight, western Baltic Sea, combining secular trends in neotectonic displacements (vertical crustal movement), climatically controlled sea level rise based on IPCC scenario, and gauge reconstruction for the coastal flood in November 1872. *Red colour* marks areas of potential coastal hazards. The scenario is generated under the theoretical assumption that no coastal protection activities will take place

stress of eroding waves and currents. In order to save the environment of the lagoon and settlements along its coast, this area deserves special effort of protection as beach re-nourishment, erection of dykes, and the installation of groyne fields.

7.7 Conclusion

The Baltic Sea Basin serves as a natural laboratory for the investigation of regional coastline change. For the Holocene, transgression and regression of the sea can be studied at the same time here. The northern Baltic has been uplifted by more than 100 m over the last thousands of years. On the contrary, in the southern Baltic the sea level rise and isostatic subsidence cause a permanent transgression of the sea there. In addition to the continuously rising sea level, storm surges result in catastrophic events of coastal erosion. We have developed a transgression/regression model that allows the

- reconstruction of the palaeogeographic development of the Baltic area since the Littorina transgression onset (8000 years BP),
- elaboration of future scenarios of coastline change on the decadal to century scale with special focus on hazard events as storm floods.

For the reconstruction of the geological history of coastline development relative sea level curves combining the signal of eustatic sea level change and vertical crustal displacement have to be determined for sites surrounding the basin. The palaeogeographic scenarios are generated by spatial interpolation of synchronous rsl values. As future projection, scenarios of extreme (catastrophic) sea level events become crucial in sustainable management of the coastal zone. For the projection of maximum sea level events secular trends as vertical crustal movements and eustatic sea level change have to be superimposed with empirical extreme historical sea level data. Here, the separation of the eustatic and tectonic component in relative sea level change data plays an important role. We propose to use sea level change data from neotectonically stable areas for an estimation of the eustatic change. As an example, future scenarios for a time span of 100 years have been elaborated for the southern Baltic Sea. Predictions for vertical displacement of the earth's crust are derived from gauge measurements along the coastline. The projection of the eustatic rise was provided by climate model runs based on an IPCC scenario of CO_2 emission.

The combination of these data sets with gauge measurements of the extreme flood in November 1872 provides a predictive digital elevation model for the coasts along the western Baltic Sea. As "defence level" the data can be used for long-term planning of coastal protection constructions as dykes. The models developed can be deployed for the generation of coastal scenarios outside the Baltic Sea. As a prerequisite for an application in coastal zone management the procedure has to be completed by modelling of sediment transport and deposition on timescales from decades to millennia. An elaboration of appropriate methods requires the faithful cooperation between geologists, physical oceanographers, and coastal engineers.

Acknowledgement The research has been conducted within the frame of the project SINCOS (www.sincos.org) funded by the German Research Foundation. The authors express thanks for the support.

References

Berglund BE (1964) The post-glacial shore displacement in eastern Blekinge, southeastern Sweden. Sveriges Geologiska Undersoekning 47:1–599

Berglund BE, Sandgren P, Barnekow L, Hannon G, Jiang H, Skog G, Yu SY (2005) Early Holocene history of the Baltic Sea, as reflected in coastal sediments in Blekinge, southeastern Sweden. Quaternary International 130:111–193

Berglund M (2004) Holocene shore displacement and chronology in Ångermanland, eastern Sweden, the Scandinavian glacio-isostatic uplift center. Boreas 33:48–60

Church JA, White NJ (2006) A 20th century acceleration in global sea-level rise. Geophysical Research Letters 33:L01602, 4. doi:10.1029/2005GL024826

Cubasch U (2001) Climate projections, including regional projections and sea level. http://www.ipcc.ch/present/COP65/cubasch.ppt

Ekman M (1996) A consistent map of the postglacial uplift of Fennoscandia. Terra Nova 8:158–165

Ekman M (2009) The changing level of the Baltic Sea during 300 years: a clue to understanding the Earth. Summer Institute for Historical Geophysics, Åland Islands, 155 pp

Harff J, Frischbutter A, Lampe R, Meyer M (2001) Sea level change in the Baltic Sea: interrelation of climatic and geological processes. In: Gerhard J, Harrison WE, Hanson BM (eds) Geological perspectives of climate change. AAPG Studies in Geology 47:231–250

Harff J, Lampe R, Lemke W, Lübke H, Lüth F, Meyer M, Tauber F (2005) The Baltic Sea – a model ocean to study interrelations of geosphere, ecosphere, and anthroposphere in the coastal zone. Journal of Coastal Research 21(3):441–446

Harff J, Lemke W, Lampe R, Lüth F, Lübke H, Meyer M, Tauber F, Schmölcke U (2007) The Baltic Sea coast – a model of interrelations among geosphere, climate, and anthroposphere. In: Harff J, Hay WW, Tetzlaff DM (eds) Coastline changes: interrelation of climate and geological processes. The Geological Society of America, Special Paper 426:133–142

Harff J, Lüth F (2009) The SINCOS project – geosphere ecosphere and anthroposphere of the Holocene Southern Baltic Sea. Baltica 22(2):133–134

Harff J, Lüth F (eds) (2007) Sinking coasts – geosphere ecosphere and anthroposphere of the Holocene Southern Baltic Sea. Rep Roman Germanic Commission

Hupfer P, Harff J, Sterr H, Stigge HJ (2003) Der Wasserstand an der Transgressionsküste der südwestlichen Ostsee. Entwicklung – Sturmfluten – Klimawandel, Sonderband Küste, 311 pp

Karlsson S, Risberg J (2005) Växthistoria och strandförskjutning i området kring Fjäturen och Gullsjön, södra Uppland. In: Johansson A, Lindgren C (eds) En introduktion till det arkeologiska projektet Norrortsleden. Birger Gustafsson, Stockholm, pp 71–126 (in Swedish)

Krauß M (1994) The tectonic structure below the southern Baltic Sea and its evolution. In: Krauß M, Bankwitz P, Harff J (eds) Rügen-Bornholm, International conference, Binz-Prora, Bornholm, 5th–10th Oct 1993. Zeitschrift für Geologische Wissenschaften 22:19–32

Lambeck K, Smither C, Johnston P (1998a) Sea-level change, glacial rebound and mantle viscosity for northern Europe. Geophysical Journal International 134:102–114

Lambeck K, Smither C, Ekman M (1998b) Tests of glacial rebound models for Fennoscandinavia based on instrumented sea- and lake-level records. Geophysical Journal International 135: 385–387

Lampe R, Meyer H, Janke W, Ziekur R, Janke W, Endtmann E (2007) Holocene evolution of an irregularly sinking coast: the interplay of eustasy, crustal movement and sediment supply. In: Harff J, Lüth F (eds) Sinking coasts – geosphere ecosphere and anthroposphere of the Holocene Southern Baltic Sea. Bericht der RGK 88:15–46

Linden M, Möller P, Björck S, Sandgren P (2006) Holocene shore displacement and deglaciation chronology in Norrbotten, Sweden. Boreas 35:1–22

Metz B, Davidson O, Bosch P, Dave R, Meyer L (eds) (2007) Contribution of Working Group III to the Fourth assessment report of the Intergovernmental panel on climate change, Cambridge

Miettinen A (2004) Holocene sea-level changes and glacio-isostasy in the Gulf of Finland, Baltic Sea. Quaternary International 120:91–104

Miettinen A, Savelieva L, Subetto DA, Dzhinoridze R, Arslanov K, Hyvārinen H (2007) Palaeoenvironment of the Karelian Isthmus, the easternmost part of the Gulf of Finland, during the Litorina Sea stage of the Baltic Sea history. Boreas 34(4):441–458

NGDC – National Geophysical Data Center (2001) 2-minute gridded global relief data (ETOPO2). CD-ROM

Oumeraci H, Schüttrumpf H (2002) Äußere Belastung als Grundlage für Planung und Bemessung von Küstenschutzwerken. Die Küste 65:1–302

Peltier WR (2007) Postglacial coastal evolution: ice-ocean-solid earth interactions in a period of rapid climate change. In: Harff J, Hay WW, Tetzlaff DM (eds) Coastline changes: interrelation of climate and geological processes. The Geological Society of America, Special Paper 426: 5–28

Rosenhagen G, Bork I (2009) The extreme storm surge at the German coasts of the Baltic Sea in November 1872 – reanalysis of the wind fields for coastal purposes. In: Witkowski A, Harff J, Isemer H-J (eds) Conference proceedings of the Climate change – the environmental and socio-economic response in the southern Baltic region, Szczecin, 25–28 May 2009. International BALTEX Secretariat Publication No. 42:125–126

Rosentau A, Meyer M, Harff J, Dietrich R, Richter A (2007) Relative sea level change in the Baltic Sea since the Litorina Transgression. Zeitschrift für Geologische Wissenschaften 35(1/2):3–16

Tarasov L, Peltier WR (2002) Greenland glacial history and local geodynamic consequences. Geophysical Journal International 150(1):198–229

Uścinowicz S (2006) A relative sea-level curve for the Polish Southern Baltic Sea. Quaternary International 145–146:86–105

Veski S, Poska A, Talviste P, Hang T, Rosentau A, Hiie S, Heinsalu A, Teiter K (in press). Investigations for reconstructing the landscape. In: David E, Kriiska A, Lõugas L (eds) The early Holocene in the Eastern Baltic with special emphasis on the Mesolithic Pulli site (Pärnu region, Estonia). Muinasaja Teadus, Tallinn

Voß R, Mikolajewicz U, Cubasch U (1997) Langfristige Klimaänderungen durch den Anstieg der CO_2-Konzentration in einem gekoppelten Atmosphäre-Ozean-Modell. Annalen der Meteorologie 34:3–4

Chapter 8
Palaeogeographic Model for the SW Estonian Coastal Zone of the Baltic Sea

Alar Rosentau, Siim Veski, Aivar Kriiska, Raivo Aunap, Jüri Vassiljev, Leili Saarse, Tiit Hang, Atko Heinsalu, and Tõnis Oja

Abstract The authors combined geological, geodetic and archaeological shore displacement evidence to create a temporal and spatial water-level change model for the SW Estonian coast of the Baltic Sea since 13,300 cal. years BP. The Baltic Sea shoreline database for Estonian territory was used for the modelling work and contained about 1,200 sites from the Baltic Ice Lake, Ancylus Lake and Littorina Sea stages. This database was combined with a shore displacement curve from the Pärnu area (in SW Estonia) and with geodetic relative sea-level data for the last century. The curve was reconstructed on the basis of palaeocoastline elevations and radiocarbon-dated peat and soil sequences and ecofacts from archaeological sites recording three regressive phases of the past Baltic Sea, interrupted by Ancylus Lake and Littorina Sea transgressions with magnitudes of 12 and 10 m, respectively. A water-level change model was applied together with a digital terrain model in order to reconstruct coastline change in the region and to examine the relationships between coastline change and displacement of the Stone Age human settlements that moved in connection with transgressions and regressions on the shifting coastline of the Baltic Sea.

Keywords Shore displacement · Coastline reconstruction · Stone Age settlements · Estonia

8.1 Introduction

The use of digital terrain models (DTM) and GIS-based spatial calculations has opened up new perspectives for the reconstruction of palaeo-water bodies in formerly glaciated areas. Such palaeoreconstructions are based on spatial calculations in which glacioisostatically deformed water-level surfaces are subtracted from the

A. Rosentau (✉)
Department of Geology, University of Tartu, 51014 Tartu, Estonia; Institute of History and Archaeology, University of Tartu, Tartu, Estonia
e-mail: alar.rosentau@ut.ee

J. Harff et al. (eds.), *The Baltic Sea Basin*, Central and Eastern European Development Studies (CEEDES), DOI 10.1007/978-3-642-17220-5_8,

DTM (cf. Leverington et al. 2002). There are two main techniques available for water-level surface interpolation. The first uses the geostatistical correlation of coastal landform elevations of the same age (Saarse et al. 2003, Rosentau et al. 2007, Jakobsson et al. 2007), whereas the second technique utilizes interpolated shore displacement curve data (Harff et al. 2005, Påsse and Andersson 2005, Rosentau et al. 2007). The advantage of geostatistical correlation is the generally good spatial coverage of the surface with proxy data, and the major shortcoming is the small number of available time slices. The problem mainly appears in subsidence and near-zero uplift areas where older coastal landforms are destroyed or buried under younger transgressive sediments. The interpolated shore displacement technique allows more detailed time resolution and thus a better interpolation, but does not commonly have as large a spatial data set.

This study examines the possibilities of combining these two techniques in order to create a spatial and temporal water-level change model of the SW Estonian coast of the Baltic Sea (Fig. 8.1). For the modelling exercise, the interpolated Baltic Sea

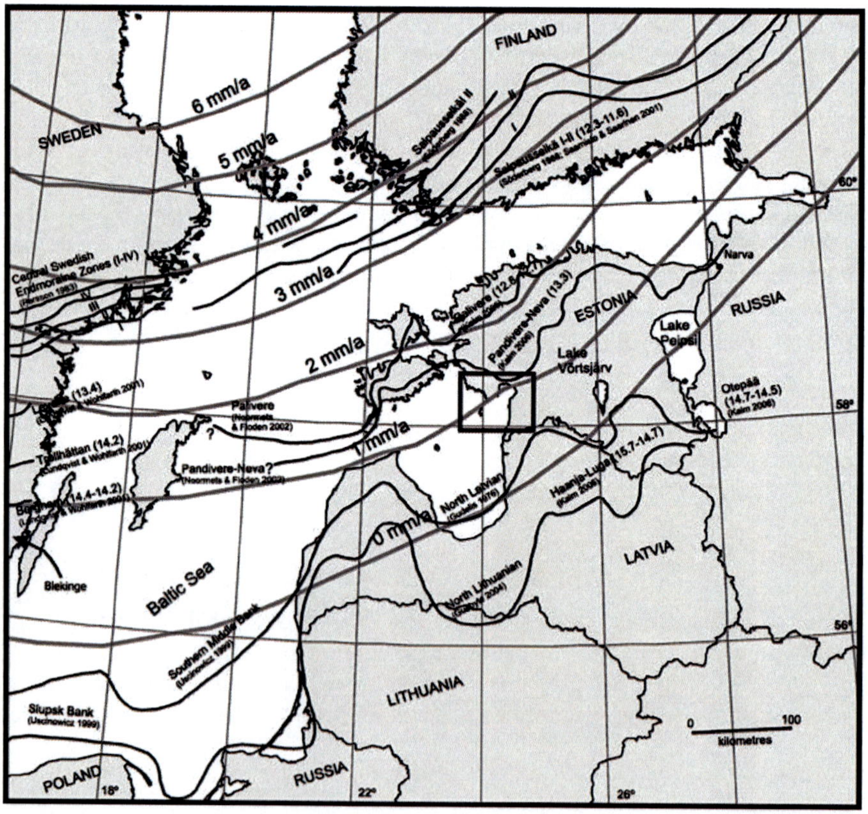

Fig. 8.1 Overview map with apparent land uplift isobases (mm/a; Ekman 1996) and main late glacial ice marginal positions with ages (cal. kyears BP) according to Kalm (2006), Lundqvist and Wohlfarth (2001) and Saarnisto and Saarinen (2001). The study area is marked with *square*

water-level surfaces will be combined with shore displacement curve data from the Pärnu region in SW Estonia. Previous palaeo-environmental and shore displacement data are summarized in this chapter in order to reconstruct the curve (Raukas et al. 1999, Heinsalu et al. 1999, Veski et al. 2005, Kriiska and Lõugas 2009). The water-level change model will be applied together with DTM to reconstruct the coastline change in SW Estonia and to examine the relationships between coastline change and the displacement of early human settlements in the area.

8.2 Study Area

The study area was chosen to meet certain requirements: first of all slow postglacial isostatic rebound with present-day apparent (relative to the mean sea level) uplift rates of around 1 mm/year (Fig. 8.1). The region is relatively flat, rising to ca. 30 m above present-day sea level. As a result, even small increases in sea level can easily lead to the flooding of substantial areas. A complex deglaciation history of the Baltic Sea area, with up-dammed lakes and early phases of postglacial seas, has periodically caused SW Estonia to be submerged by the waters of the Baltic Sea basin and to emerge in other periods as terrestrial land. Thus, deposits of water-laid sediments formed during the transgression of the Ancylus Lake or the Littorina Sea have led to repeated soil burials and to peat and/or gyttja formations, often associated with the cultural layers of Stone Age settlement sites. Our study area in SW Estonia is rich in sites from different prehistoric periods. Coastal habitation is characteristic of the Stone Age. The Pulli, Sindi-Lodja I and II and Jõekalda settlement sites in the lower reaches of the Pärnu River and the Malda, Lemmetsa I and II settlement sites in the lower reaches of the Audru River are important in this context (Fig. 8.2; Kriiska 2001, Kriiska et al. 2002, 2003, Kriiska and Saluäär 2000, Kriiska and Lõugas 2009).

8.3 Modelling of Water-Level Change and Palaeocoastlines

8.3.1 Reconstruction of Water-Level Surfaces

The interpolated surfaces of water levels were derived using the late glacial (Saarse et al. 2007) and Holocene Baltic Sea shoreline databases (Saarse et al. 2003). In this study we used six interpolated surfaces of water levels for different Baltic stages: the Baltic Ice Lake (stages A_1, BI, BIII) around 13,300, 12,300–12,100 and 11,700 cal. years BP (Saarse et al. 2007); Ancylus Lake transgression maximum around 10,200 cal. years BP (Saarse et al. 2003), Littorina Sea transgression maximum around 7,300 cal. years BP (Veski et al. 2005) and the modern Baltic Sea over the period of last 100 years. The interpolated water-level surface for the modern Baltic Sea is based on sea-level measurements complemented with geodetic data (Ekman 1996).

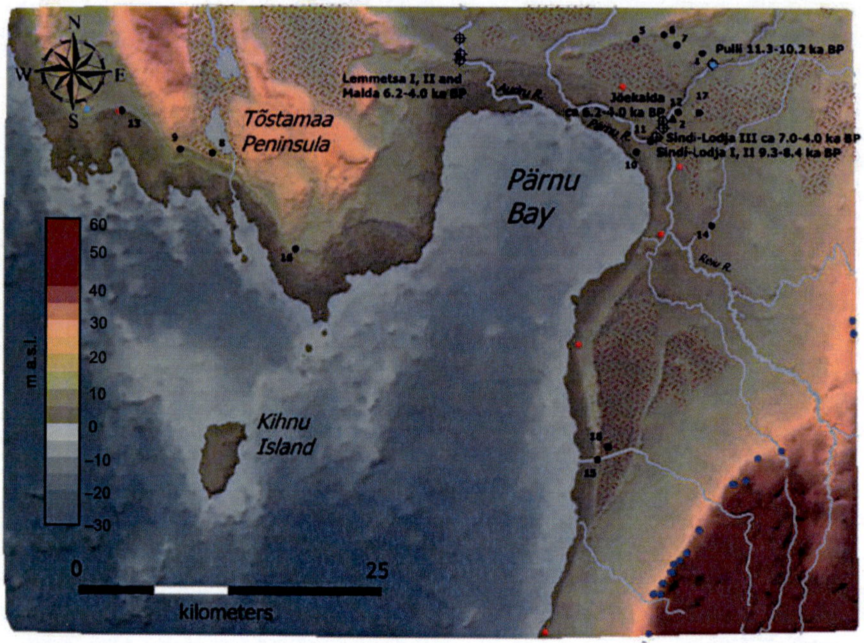

Fig. 8.2 Digital terrain model of the study area in SW Estonia and the location of the investigated geological and archaeological sites. Sites with buried organic matter and dated peat sequences are marked with *black dots*. The locations of the coastal landforms of the Baltic Ice Lake (*blue dots*), Ancylus Lake (*light blue dots*) and Littorina Sea (*red dots*) are also shown on the map. Peat bogs are marked by *brown hatching* and the reference site for the water-level curve at Paikuse by a *triangle*

At present the late glacial and Holocene shoreline databases cover more than 1,200 sites in Estonia, although statistical analyses show that roughly half of this data does not match water-level reconstruction requirements due to inaccurate coordinates, elevations or the erroneous correlation of different shore marks. Therefore the reliability of shoreline displacement data was verified using different methods. First, sites with altitudes that did not match neighbouring sites were eliminated. Second, point kriging interpolation with linear trend was used to create interpolated surfaces of water level, with a grid size of 5 × 5 km. Kriging is useful because it interpolates accurate surfaces from irregularly spaced data and shows the outliers in the data set. Residuals (the difference between the actual site altitude and the interpolated surface) were calculated and used to check data reliability, so that sites with residuals more than ±1 m were discarded. Then the final interpolated water-level surfaces were calculated using for BIL stages A_1 – 52, BI – 111, BIII – 164 sites; for Ancylus Lake 110 sites; and for Littorina Sea 176 sites. Timing of the surfaces was derived from the ages of the ice marginal positions and varvochronology for the late glacial (Rosentau et al. 2009, Saarse et al. 2007) and radiocarbon dating for the Holocene (Saarse et al. 2007).

A map with the isobases of the recent postglacial rebound of Fennoscandia and Baltic compiled by Ekman (1996) was used to reconstruct the relative sea-level surface for the 100-year period (1892–1991). Apparent uplift rates on Ekman's map were calculated from the sea-level and lake-level records combined with repeated high-precision levelling results, and the uncertainty of these rates was estimated to be ± 0.5 mm/a and less (Ekman 1996). The uplift rates of Ekman's map were recently compared to the velocities of the permanent GPS stations, and overall agreement (consistency) was found at the 0.5 mm/a level (Lidberg et al. 2009).

8.3.2 Water-Level Change Curve for the Pärnu Area

A set of 18 sites within an area of 3,500 km^2 displaying 66 radiocarbon dates from different stages of the Baltic Sea at different levels (Table 8.1) was used to reconstruct the water-level curve for the area (Veski 1998, Heinsalu et al. 1999, Veski et al. 2005, Saarse et al., 2003, 2006). Before reconstructing the curve, the correction for the spatial spread of the sites was applied using interpolated surfaces of water levels with different shoreline tilting gradients. All sites were transposed to the Paikuse location (Fig. 8.2). The elevations of the pre-Ancylus Lake and Ancylus Lake sites (sites 1–27 in Table 8.1) were corrected in respect to the Ancylus Lake surface and the pre-Littorina and Littorina Sea sites (sites 28–61 in Table 8.1) in respect to the Littorina Sea surface. For correction of the Littorina Sea regression sites (sites 16–18 in Table 8.1), the Littorina Sea surface was combined with the Baltic Sea surface at 100 years ago (Ekman 1996) assuming a linear decay in shoreline tilting gradient and the differences in elevation were calculated depending on the age of each site (for details see Sect. 3.3).

The data can be divided into six groups that delimit the various stages of the Baltic Sea in the past (Fig. 8.3). Baltic Ice Lake coastal landforms at different levels form the first group, representing the time span from the deglaciation of the area to the Billingen drainage (Figs. 8.2 and 8.4). The second group represents organic matter from the lowstand of the Baltic Sea during the Yoldia Sea and Ancylus Lake stages buried under the transgressive Ancylus Lake waters (Table 8.1), and the third group embraces the coastal landforms from the culmination of the Ancylus Lake transgression (Figs. 8.2 and 8.3). The fourth group represents buried organic matter of the period between the transgressions of the Ancylus Lake and the Littorina Sea at altitudes above 0 m a.s.l. A subgroup of this set is the cluster of dated organic matter from Uku and Reiu (Fig. 8.2) at altitudes distinctly below 0 m a.s.l. (Table 8.1), which is discussed separately due to suspected redeposition. The coastal landforms from the culmination of the Littorina Sea make up the fifth group, and the few sites that define the water level after the Littorina Sea transgression form the last group (Figs. 8.2 and 8.3). Thus the described groups record three regressive phases interrupted by two transgressive phases (Ancylus Lake and Littorina Sea transgressions) in the Baltic Sea water-level change history in the Pärnu area (Fig. 8.3).

Table 8.1 Radiocarbon datings and altitudes of organic sediments in Pärnu area used to reconstruct the shore displacement curve in Fig. 8.3. Location of sites is given in Fig. 8.2

No	Site	Radiocarbon age	Lab. code	Elevation (m a.s.l.)		Material	References	Calibrated age BP (max–min)		Correct. to spatial spread (m)	Corrected elevation (m a.s.l.)
	Pre-Ancylus Lake and Ancylus Lake buried sediments										
1	Sindi-Lodja II	9,170±200	Ta-2784	4.4	4.4	▲ Peat	Veski et al. (2005)	10,750	10,200	0.1	4.5
2	Paikuse	9,575±90	TA-2547	5.1	5.2	▲ Peat	Veski (1998)	11,150	10,800	0.0	5.1
		9,350±75	Ua-11691	5.2	5.3	▲ Wood	Veski (1998)	10,740	10,540	0.0	5.2
		9,340±130	Ua-12446	5.0	5.1	▲ Seeds	Veski (1998)	10,770	10,420	0.0	5.0
3	Pulli	9,095±90	Ua-13352	9.0	9.0	■ *Elk bone*	Poska and Veski (1999)	10,450	10,230	–0.4	8.6
		9,385±105	Ua-13351	8.9	8.9	■ *Charcoal*	Poska and Veski (1999)	10,810	10,470	–0.4	8.5
		9,145±115	Ua-13353	9.3	9.3	■ *Seeds*	Poska and Veski (1999)	10,550	10,260	–0.4	8.9
		9,575±115	TA-176	9.0	9.0	▲ *Cult. layer*	Kessel and Punning (1969b)	11,160	10,790	–0.4	8.6
		9,300±75	TA-175	9.3	9.3	▲ Peat	Kessel and Punning (1969b)	10,700	10,430	–0.4	8.9
		9,350±60	TA-949	9.0	9.0	▲ *Charcoal*	Jaanits and Jaanits (1978)	10,720	10,540	–0.4	8.6
		9,600±120	TA-245	9.0	9.0	▲ *Cult. Layer*	Punning et al. (1971)	11,190	10,820	–0.4	8.6
		9,285±120	TA-284	9.3	9.3	▲ *Charcoal*	Ilves et al. (1974)	10,700	10,330	–0.4	8.9
		9,620±120	Hel-2206A	9.0	9.0	▲ *Soil, INS*	Haila and Raukas (1992)	11,230	10,830	–0.4	8.6
		9,290±120	Hel-2206B	9.0	9.0	▲ *Soil, SOL*	Haila and Raukas (1992)	10,700	10,340	–0.4	8.6
4	Urge	9,125±85	Tln-1691	11.0	11.0	▲ Peat	Raukas et al. (1999)	10,460	10,260	–0.7	10.3
5	Lõpe	9,215±70	Tln-1631	11.2	11.2	▲ Peat	Raukas et al. (1999)	10,490	10,320	–1.5	9.7
		9,260±70	Tln-1632	11.2	11.2	▲ Wood	Raukas et al. (1999)	10,600	10,340	–1.5	9.7
6	Pressi	9,135±70	Tln-1991	11.5	11.5	▲ Peat	Raukas et al. (1999)	10,450	10,270	–1.6	9.9
7	Kõdu	8,480±90	Tln-66	11.7	11.7	▲ Peat	Kessel and Punning (1974)	9,600	9,460	–1.6	10.1
		9,340±45	Tln-1993	11.7	11.7	▲ Peat	Raukas et al. (1999)	10,700	10,540	–1.6	10.1
8	Ermistu	9,595±130	Ua-13034	12.5	12.5	▲ Peat	Veski (1998)	11,230	10,580	–5.6	6.9
		9,515±120	Tln-1378	12.5	12.6	▲ Peat	Veski (1998)	11,180	10,520	–5.6	6.9
		9,745±85	Tln-1137	12.4	12.5	▲ Peat	Veski (1998)	11,320	11,060	–5.6	6.8
		9,345±90	Ua-13035	12.5	12.5	▲ Peat	Veski (1998)	10,780	10,250	–5.6	6.9

Table 8.1 (continued)

No	Site	Radiocarbon age	Lab. code	Elevation (m a.s.l.)		Material	References	Calibrated age BP (max–min)		Correct. to spatial spread (m)	Corrected elevation (m a.s.l.)
		9,635±100	Tln-1380	12.3	12.4	▲ Peat	Veski (1998)	11,220	10,710	–5.6	6.7
		9,850±165	Ua-13036	12.3	12.3	▲ Peat	Veski (1998)	11,830	10,760	–5.6	6.7
9	Kastna	8,780±50	Tln-1824	16.4	16.4	▲ Fen Peat	Veski (1998)	9,900	9,600	–5.7	10.7
	Pre-Littorina Sea and Littorina Sea buried sediments										
10	Uku	7,910±90	Tln-1187	–4.2	–4.0	▲ Bulk peat	Veski et al. (in press)	9,030	8,650	0.3	–3.9
		8,030±180	Ta-2828	–2.1	–1.7	▲ Peat	Veski et al. (in press)	9,300	8,650	0.3	–1.8
		8,270±120	Ta-2829	–2.4	–2.1	▲ Peat	Veski et al. (in press)	9,520	9,080	0.3	–2.1
		8,080±150	Ta-2830	–2.5	–2.4	▲ Peat	Veski et al. (in press)	9,350	8,750	0.3	–2.2
		8,420±150	Ta-2831	–3.0	–2.8	▲ Peat	Veski et al. (in press)	9,600	9,190	0.3	–2.7
		7,580±120	Ta-2832	–3.2	–3.0	▲ Peat	Veski et al. (in press)	8,510	8,240	0.3	–2.9
		7,250±150	Ta-2833	–3.4	–3.2	▲ Peat	Veski et al. (in press)	8,330	7,920	0.2	–3.2
		7,610±180	Ta-2834	–4.6	–4.3	▲ Peat	Veski et al. (in press)	8,650	8,240	0.2	–4.4
		7,440±150	Ta-2835	–4.9	–4.6	▲ Peat	Veski et al. (in press)	8,440	8,100	0.2	–4.7
11	Reiu	7,175±100	Ta-2836	–1.2	–0.9	▲ Peat	Veski et al. (in press)	8,210	7,910	0.2	–1.0
		7,570±150	Ta-2837	–2.4	–2.2	▲ Peat	Veski et al. (in press)	8,590	8,230	0.2	–2.2
		7,743±150	Ta-2838	–2.8	–2.6	▲ Peat	Veski et al. (in press)	8,850	8,400	0.2	–2.6
		7,560±150	Ta-2839	–3.0	–2.8	▲ Wood	Veski et al. (in press)	8,590	8,230	0.2	–2.8
		7,730±150	Ta-2840	–3.2	–3.0	▲ Peat	Veski et al. (in press)	8,810	8,410	0.2	–3.0
		7,800±150	Ta-2841	–3.4	–3.2	▲ Wood	Veski et al. (in press)	8,700	8,450	0.2	–3.2
		8,320±150	Ta-2842	–3.6	–3.4	▲ Peat	Veski et al. (in press)	9,540	9,140	0.2	–3.4
		8,570±150	Ta-2843	–3.8	–3.6	▲ Peat	Veski et al. (in press)	9,950	9,350	0.2	–3.6
1	Sindi-Lodja I, II	7,300±150	Ta-2785	4.7	4.7	▲ Wood	Veski et al. (2005)	8,360	8,020	0.1	4.8
		7,630±120	Ta-2783	3.3	3.3	▲ Peat	Veski et al. (2005)	8,630	8,380	0.1	3.4
		7,780±120	Ta-2737	3.3	3.3	▲ *Wood*	Veski et al. (2005)	8,770	8,460	0.1	3.4

Table 8.1 (continued)

No	Site	Radiocarbon age	Lab. code	Elevation (m a.s.l.)		Material	References	Calibrated age BP (max–min)		Correct. to spatial spread (m)	Corrected elevation (m a.s.l.)
		7,870±80	Ta-2774	3.3	3.3	▲ Wood	Veski et al. (2005)	9,000	8,600	0.1	3.4
		7,980±100	Ta-2736	3.3	3.3	▲ Wood	Veski et al. (2005)	9,050	8,710	0.1	3.4
		8,035±80	Ta-2788	3.3	3.3	▲ *Wood*	Veski et al. (2005)	9,080	8,810	0.1	3.4
		8,035±80	Ta-2769	4.6	4.6	▲ Wood	Kriiska et al. (2002)	9,080	8,810	0.1	4.7
		8,070±70	Ua-17013	4.6	4.6	■ *Charcoal*	Kriiska (2001)	9,180	8,820	0.1	4.7
		8,190±80	Ta-2789	3.3	3.3	▲ Wood	Veski et al. (2005)	9,310	9,070	0.1	3.4
		8,210±80	Ta-2786	3.2	3.3	▲ Wood	Veski et al. (2005)	9,330	9,080	0.1	3.3
		8,250±150	Ta-2787	3.2	3.2	▲ Peat	Kriiska et al. (2002)	9,460	9,080	0.1	3.3
12	Sindi	6,710±110	Ta-55	7.0	7.0	▲ Wood	Kessel and Punning (1969a)	7,720	7,530	0.0	7.0
		7,215±90	Ta-133	7.0	7.0	▲ Peat	Punning et al. (1977)	8,210	8,000	0.0	7.0
2	Paikuse	7,535±80	Tln-2603	7.0	7.0	▲ Peat	Veski et al. (2005)	8,470	8,260	0.0	7.0
		7,120±120	Ua-12447	6.8	6.8	▲ Seeds	Veski (1998)	8,100	7,840	0.0	6.8
		7,030±120	Ta-2548	6.6	6.7	▲ Peat	Veski (1998)	8,020	7,780	0.0	6.6
13	Kolga	7,555±44	Tln-1822	6.7	6.8	▲Fen Peat	Veski (1998)	8,400	8,300	−3.0	3.7
14	Vaskrääma	7,580±170	TA-140	6.5	6.5	▲ Peat	Kessel and Punning (1969a)	8,800	8,000	1.3	7.8
15	Rannametsa	8,080±110	Hel-2207A	6.5	6.5	▲ Peat	Hyvärinen et al. (1992)	9,300	8,600	2.3	8.8
	Littorina Sea regression organic sediments										
16	Seliste	5,950±60	TA-183	8.5	8.5	▲ Gyttja	Kessel and Punning (1969a)	6,900	6,600	−1.4	7.1
17	Kõrsa	5,790±80	TA-1986	12.0	12.0	▲ Peat	Orru (1992)	6,670	6,490	0.3	12.3
18	Tolkuse	2,320±100	TA-1990	1.5	1.5	▲ Peat	Orru (1992)	2,480	2,150	0.6	2.1

▲ Conventional ^{14}C dates on charcoal/wood/bulk sediment/peat. ■ AMS dates on terrestrial microfossils. *Italicized text:* Dates on Stone Age settlement sites.

Ta – ^{14}C Laboratory, Tartu University, Estonia; Tln – ^{14}C Laboratory, Institute of Geology at Tallinn Technical University, Estonia; Hel – Radiocarbon Dating Laboratory, Helsinki University, Finland; Ua – Ångström Laboratory, Uppsala University, Sweden. Radiocarbon ages are calibrated according to the IntCal04 curve (Reimer et al. 2004) within 1 sigma deviation.

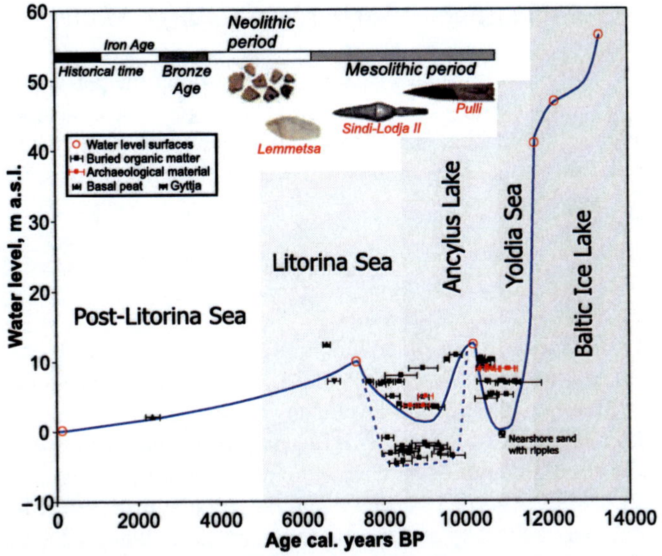

Fig. 8.3 Water-level curve for the Pärnu area. Water-level elevations of all sites were corrected by spatial spread and referenced to the Paikuse location given in Fig. 8.2. Baltic Sea stages are according to Andren et al. (2000). Radiocarbon dates of organic sediments are given in Table 8.1 and water-level surface ages and elevations in Table 8.2. *Dashed line* represents the hypothetical low water level, discussed in detail in the text, according to Uku and Reiu sites

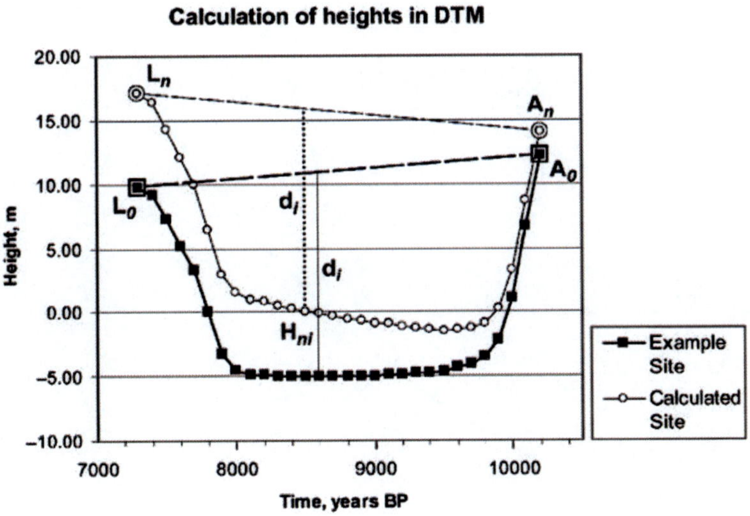

Fig. 8.4 Principle scheme for calculation of water-level change for any new grid cell

8.3.3 Temporal and Spatial Water-Level Change Model

Temporal interpolation of interstage surfaces for a certain time period was provided by linear calculation according to the water-level change curve developed using the data from the Paikuse site (Fig. 8.3). Through prior simplifications, we were able to compute the elevation H_{ni} of every grid cell n for a certain time period i (Fig. 8.4) using the following equation:

$$H_{ni} = A_n + \frac{L_n - A_n}{T} T_i + d_i,$$

where A and L are the section's older and younger reference surfaces, respectively, T is the length of time between stages A and L, T_i is the time from initial stage A, and d_i is the difference in the water-level change curve of the sample site from the linear trend line. We had two assumptions in using the simple linear model: first, the study area was small enough to be characterized by homogeneous dynamics, and second, the six reference surfaces inserted into the calculation describe the temporal behaviour of the water level by sufficiently frequent stages that gradient differences in a section do not produce deviations that exceed uncertainties from elevation and dating (Fig. 8.5).

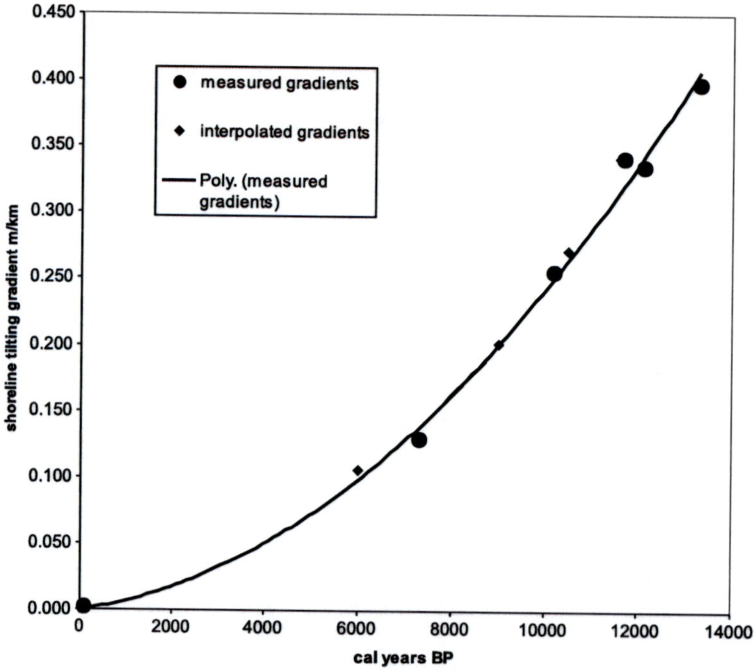

Fig. 8.5 Water-level surface tilting gradients for different times and polynomial trend line showing the decay of land uplift over time. Mean tilting gradients of water-level surfaces and the directions of fastest uplift are given in Table 8.2

8.3.4 Reconstruction of Palaeocoastlines

The reconstruction of palaeocoastlines and bathymetry were based on GIS analysis, from which interpolated surfaces of water levels were subtracted from the modern DTM (Fig. 8.6). The modern DTM with a grid size of 20×20 m was generated using the linear solution of the Natural Neighbour interpolation using different sources of elevation data. Elevation data for the mainland were derived from the Estonian Basic map on a scale of 1:10,000 (western part), the Soviet military topographic map on a scale of 1:25,000 (eastern part) and the Baltic seabed from the bathymetric maps on a scale of 1:50,000 (Estonian Maritime Administration 2001a–c, 2002a, b). All maps were transformed into L-EST national reference system. The vertical datum for the elevation data and DTM modelling was national height system BK77 based on Kronstadt zero level.

DTM-based palaeoreconstructions have some limitations due to the impact of deposition subsequent to the time being modelled. Therefore the thicknesses of Holocene peat (Orru 1995) and gyttja (Veski 1998) deposits were removed from the DTM before the palaeocoastline reconstruction.

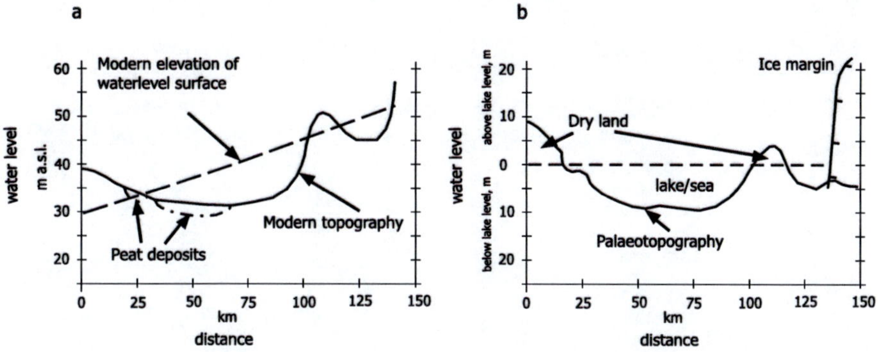

Fig. 8.6 General cross-sections showing the principles of palaeoreconstructions. Topography related to the isostatically deformed (uplifted) sea/lake water-level surface today (**a**) and during sea/lake formation (**b**)

8.4 Modelling Results

The distribution of the Baltic Ice Lake water-level surface isobases and shorelines in the Pärnu area is presented in Fig. 8.7a–i for nine time slices since the deglaciation of the area. The created spatial and temporal model made it possible to reconstruct the palaeo-water levels and coastlines for the times for which coastal landforms data are lacking, for instance the lowstands of the Ancylus Lake and Littorina Sea (Fig. 8.7d, f, g), and to relate the palaeocoastlines with Stone Age settlement sites in SW Estonia (Fig. 8.7d, f, i).

The main characteristics of interpolated water-level surfaces are summarized in Table 8.2. Calculated mean tilting gradients decrease exponentially over time as a

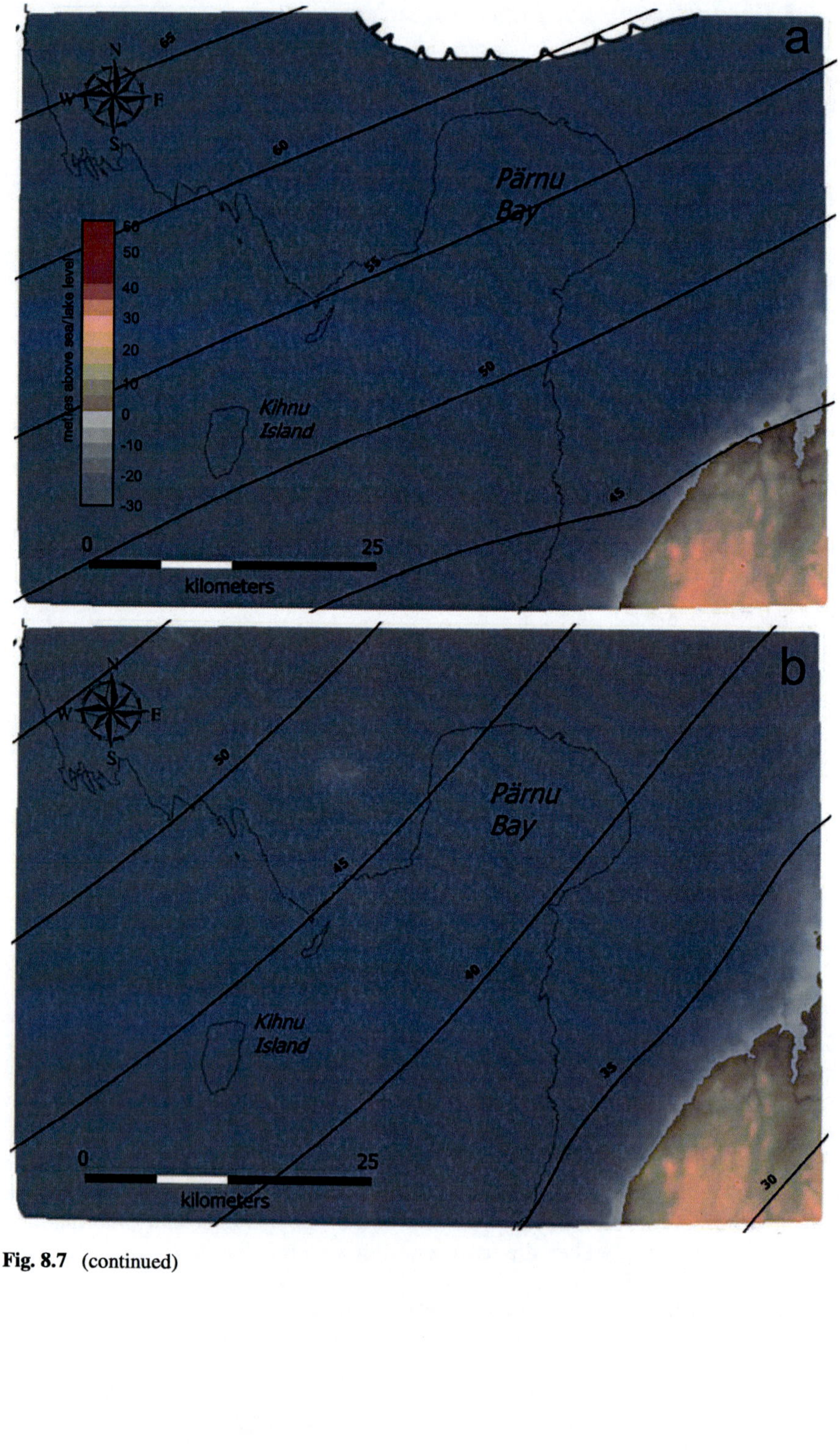

Fig. 8.7 (continued)

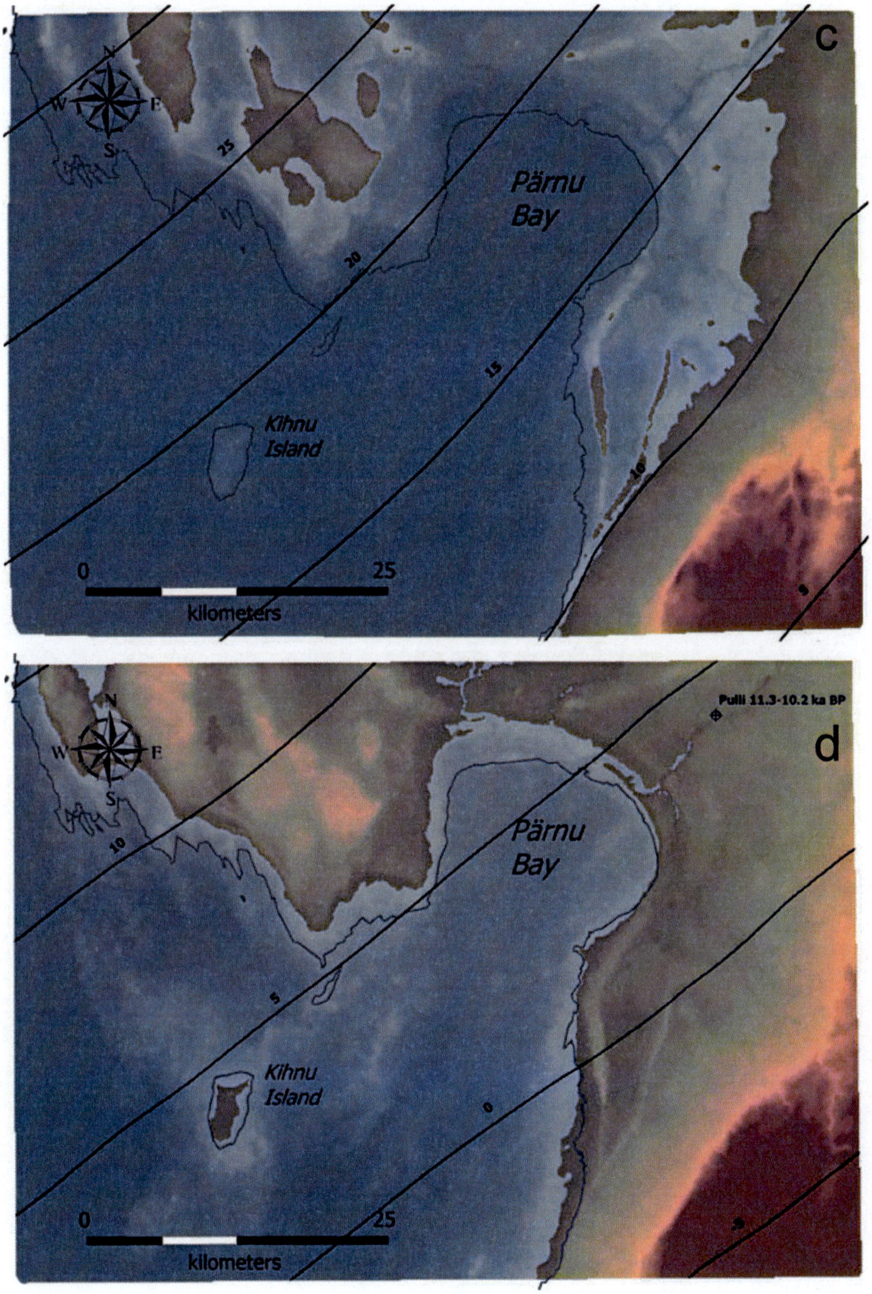

Fig. 8.7 (continued)

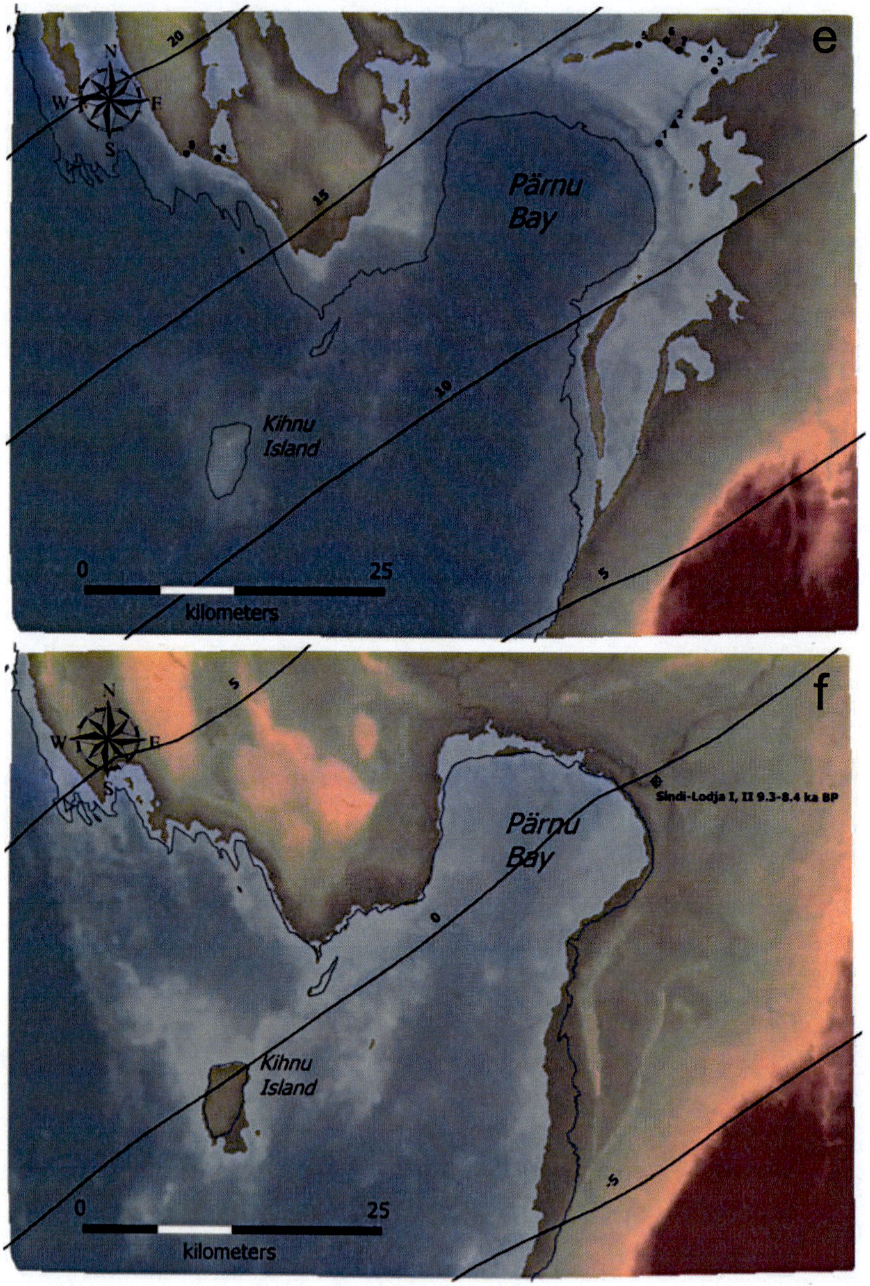

Fig. 8.7 (continued)

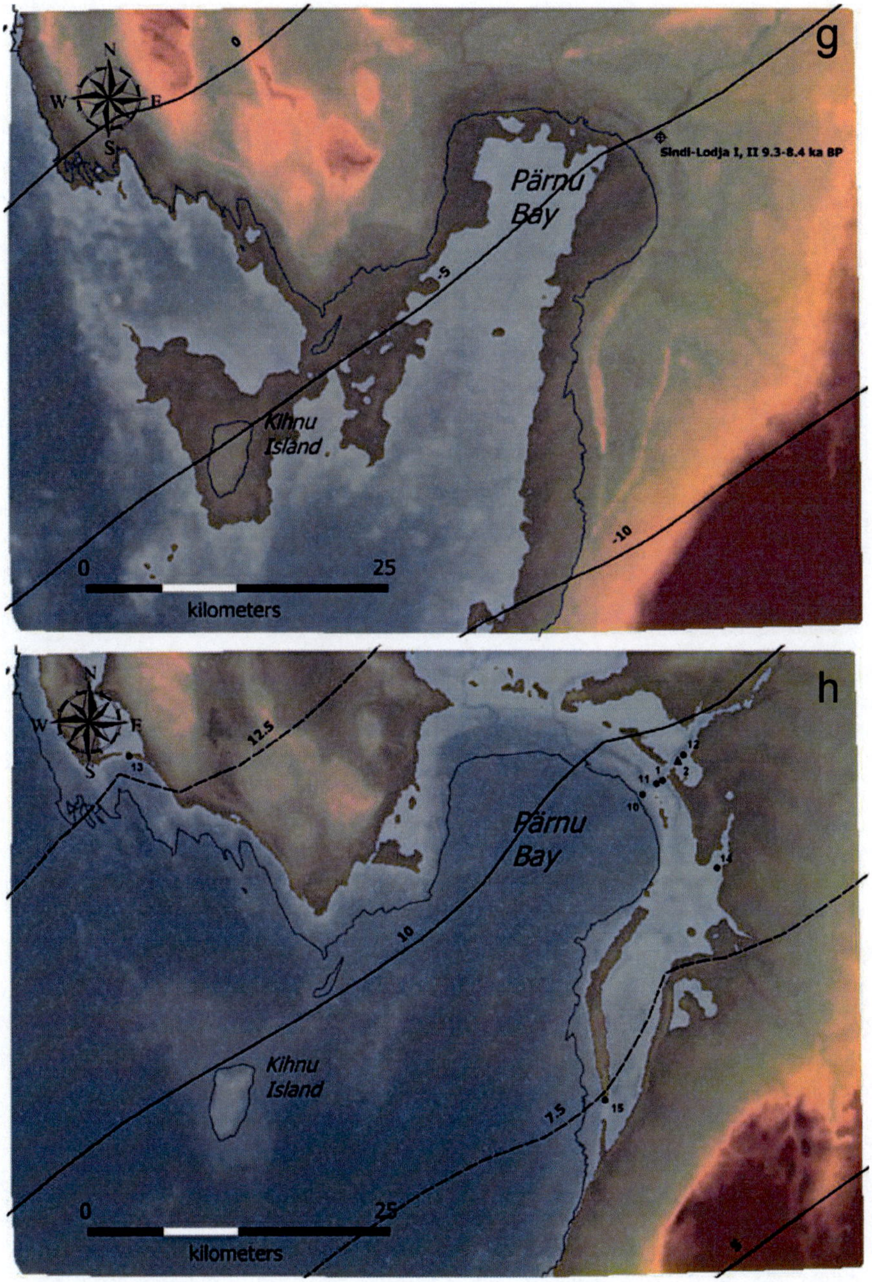

Fig. 8.7 (continued)

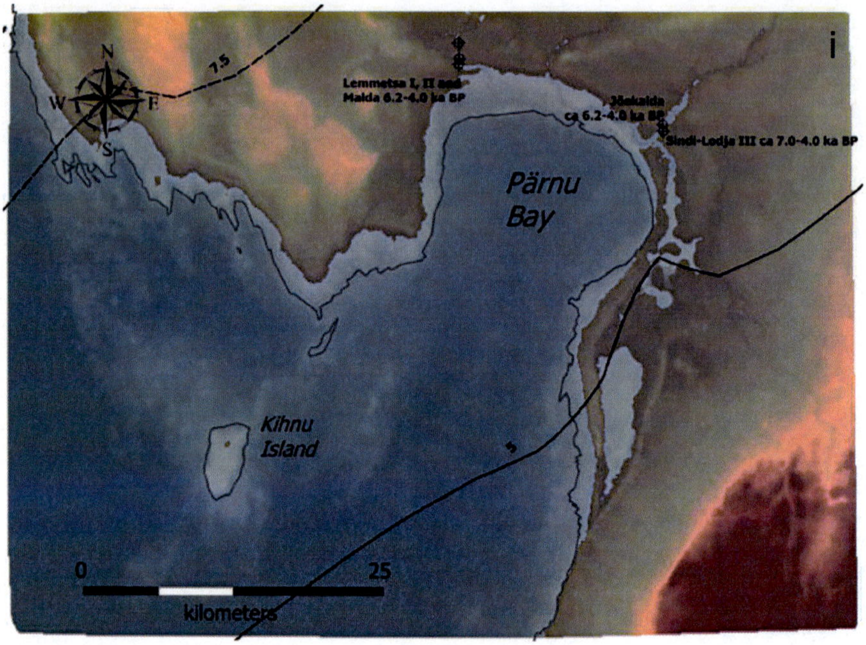

Fig. 8.7 Palaeogeographic reconstruction of the Baltic Sea palaeocoastlines and water depths with indication of water-level isobases (m a.s.l.) during its different stages: (**a**) the Baltic Ice Lake during the deglaciation of the Pärnu area and formation of the end-moraines of the Pandivere-Neva ice marginal zone at about 13,300 cal. years BP (Kalm 2006), (**b**) the Baltic Ice Lake prior to the Billingen drainage at about 11,700 cal. years BP, (**c**) the Baltic Ice Lake after the Billingen drainage at about 11,600 cal. years BP, (**d**) Ancylus Lake at the beginning of the transgression and during the Pulli settlements at about 10,500 cal. years BP, (**e**) Ancylus Lake during its maximum in the Pärnu area at about 10,200 cal. years BP, (**f**) the Littorina Sea before the transgression and during the Sindi-Lodja I and II settlements at about 9,000 cal. years BP, (**g**) alternative low water-level (–5 m a.s.l. at Paikuse) scenario for the Littorina Sea before the transgression at about 9,000 cal. years BP, (**h**) the Littorina Sea during its maximum in the Pärnu area at about 7,300 cal. years BP, (**i**) the Littorina Sea after the transgression and during the Lemmetsa, Malda, Jõekalda and Sindi-Lodja III settlements at about 6,000 cal. years BP

result of the slowdown in uplift (Fig. 8.5). The only section with which we encountered minor difficulties to match actual shoreline tilting gradient to linear regression was the long period from the Littorina Sea culmination to the present (Fig. 8.5). Because of the applied linear regression, it seems that our model slightly overestimates the shoreline tilting gradient for 6,000 cal. years BP. However, due to the relatively small study area, this deviation is smaller than uncertainties from elevation and dating, and we can use this approximation to interpolate the water-level surface for this time slice. Baltic Ice Lake and Littorina Sea tilting gradients differ more than threefold (Table 8.2), which is also reflected in palaeocoastline positions, if one compares the SE and NW parts of the maps (Fig. 8.7d, i). The results also show that the direction of fastest uplift was migrated slightly westward during the Baltic Ice Lake and then back north during the Holocene, ranging between 336 and 314°.

Table 8.2 Main characteristics of the interpolated water-level surfaces

Water-level surface	Age, cal. years BP	Water level, m (max–min)		Water level at Paikuse (m a.s.l.)	Mean tilting gradient (m/km)	Mean tilting direction (°)	References
Baltic Ice Lake (stage A1)	13,300	40.3	68.3	56.1	0.398	336	Saarse et al. (2007)
Baltic Ice Lake (stage B1)	12,100	34.1	62.0	46.7	0.335	324	Saarse et al. (2007)
Baltic Ice Lake (stage B3)	11,700	28.5	57.8	40.8	0.342	314	Saarse et al. (2007)
Baltic Ice Lake (drainage)	11,600	3.5	32.8	15.8	0.342	314	This study
Ancylus Lake	10,500	-7.1	15.5	3.7	0.272	323	This study
Ancylus Lake culmination	10,200	1.7	22.7	12.2	0.256	325	Saarse et al. (2003)
Pre-Littorina Sea transgression	9,000	-8.4	8.2	0.0	0.202	325	This study
Littorina Sea culmination	7,300	4.2	14.6	9.7	0.129	325	Saarse et al. (2003)
Post-Littorina Sea	6,000	2.0	10.5	6.5	0.106	325	This study
Recent Baltic Sea	100	0.04	0.14	0.1	0.002	330	Ekman (1996)

Although the peat and gyttja deposits of the Holocene age were removed from the DTM, other postglacial deposits and landforms influence the palaeoshoreline positions and water depth. This influence relates mainly to the marine and eolian deposits. For example, the impact of the Ancylus Lake and Littorina Sea sediments "withdraws" pre-Ancylus Lake and pre-Littorina Sea palaeocoastlines to lower position as expected (Fig. 8.7d, f, g). Such an impact is highest in Pärnu River valley, where the thickness of these deposits is up to 6 m, whereas outside of the valley it is typically less than 2 m (Veski et al. 2005). The impact of the superimposed coastal dunes on the palaeocoastline position is visible on the modelled Ancylus Lake (Fig. 8.7e) and Littorina Sea (Fig. 8.7h) coastlines SE of Pärnu Bay. Unfortunately, our geological information on the age and spatial distribution of marine and eolian sediments is insufficient to subtract them from the DTM.

8.5 Development of the Baltic Sea Coastline and Stone Age Human Occupations in SW Estonia

During the deglaciation of SW Estonia, the Baltic Ice Lake formed between the retreating Scandinavian Ice Sheet and emerged land in the southeast at about 13,300 cal. years BP (Fig. 8.7a). The Baltic Ice Lake water was deep enough for the formation of annually laminated varved clays over a vast area in Pärnu Bay and the present-day mainland area (Fig. 8.7a, b). The correlation of ice-proximal coastal landforms with varve – chronologically dated ice – marginal zones makes it possible to reconstruct the shore displacement of the Baltic Ice Lake. The Billingen drainage event lowered the water level by approximately 25 m (Fig. 8.3, from 42 to 17 m a.s.l. in the area) to the ocean level terminating the varved clay accumulation. Due to the drainage event, the landscape of SW Estonia changed dramatically. New land emerged from the waters in the east, and an archipelago formed in the Tõstamaa area (Fig. 8.7b, c). The water level of the Yoldia stage, following the "Billingen" event, was in equilibrium with the ocean and was quite stable. Therefore new land emerged from the Yoldia Sea owing to the land uplift and seemingly regressive shore displacement. The moderate land uplift in SW Estonia exceeded the water-level rise in the Baltic Sea basin; as a result, the shoreline displacement near the Pärnu area was regressive during the whole Baltic Ice Lake and Yoldia Sea stages (Fig. 8.3). It is difficult to estimate the minimum level of the Yoldia Sea shoreline in the Pärnu area, but it was certainly below 3 m a.s.l. (Fig. 8.3). Indications of near-shore or shallow water ripples and microlayers of sand and resedimented organic matter at around 0 m a.s.l. at Sindi-Lodja II may point to the retreat of the Yoldia Sea shoreline to that level (Veski et al. 2005). Thus the drainage of the Baltic Ice Lake contributed to the regression with 25 m and the subsequent fall of another ca. 16 m during the Yoldia Sea (Fig. 8.3).The total regression since the beginning of the Baltic Ice Lake to the Yoldia Sea lowstand was about 55 m (Fig. 8.3).

Environmental conditions, including sea-level changes, have undoubtedly influenced the human settlement pattern in the region. The Pulli settlement site is the oldest known human occupation in Estonia and has been dated to between 11,300 and 10,200 cal. years BP (Kriiska and Lõugas 2009). Recent AMS dates of ecofacts from the cultural layer suggest that the Pulli settlement site was most probably inhabited slightly later, during the Ancylus Lake transgression period, at about 10,800–10,200 cal. years BP (Table 8.1; Fig. 8.3). If one considers the AMS mean age of the cultural layer (10,500 cal. years BP), the Pulli people settled at about 10 km from the coast, on the lower reaches of the ancient Pärnu River (Fig. 8.7d). However, over the next 200–300 years the coastline was displaced quickly towards the mainland due to the rapid transgression that took place at that time (Fig. 8.3). The water-level change model shows that the transgressive waters of Ancylus Lake passed the Pulli site at about 10,300–10,200 cal. years BP, just before the culmination of the transgression. Terrestrial conditions were interrupted in the Pulli and other buried organic matter sites when the rising level of Ancylus Lake submerged the area (Fig. 8.7e). Our palaeogeographic model shows that most buried organic matter sites (Seliste, Kastna, Lõpe, Kõdu, Pulli, Urge and Pressi in Table 8.1) were located directly in the coastal zone (±1.5 m), probably in the storm surge zone, of the transgressive Ancylus Lake, which might be explained by the good preservation conditions in this zone due to the rapid burial (Fig. 8.7e). It is difficult to estimate the total amplitude of the transgression, but considering the elevations of pre-Ancylus Lake near-shore sand facies in Sindi-Lodja II and the highest coastal landforms in the area, it is at least 12 m (Fig. 8.3). However, the comparison of the presented transgression amplitude with corresponding data from Blekinge in SE Sweden (Ancylus Lake transgression from –15 to 5 m a.s.l.; Berglund et al. 2005) also leaves space for the lower pre-Ancylus Lake level (Fig. 8.1).

Following the rapid regression of Ancylus Lake due to lake drainage into the Kattegat (Björck 1995, Bennike et al. 2004) the land was exposed and allowed the formation of peat deposits in the area. The organic sedimentation between the transgressions of Ancylus Lake and the Littorina Sea occurred at minimum altitudes to about –5 m a.s.l. (Uku and Reiu sites). Water level dropped at least 12 m in the Pärnu area during the regression, as shown by the elevation of the lowermost pre-Littorina Sea organic layers at Paikuse and Sindi-Lodja (Fig. 8.3). The fall in water level during the regression in isostatically similar areas in Narva and Blekinge (Fig. 8.1) was about 11–9 m (from 12–10 to 1 m a.s.l.; Lepland et al. 1996) and 5.5 m (from 5 to –0.5 m a.s.l.; Berglund et al. 2005), respectively. This shows that a hypothetical fall in water level to –5 m a.s.l (Fig. 8.3) is rather unlikely in the Pärnu area, and the question of the origin of the Uku and Reiu peat layers below present sea level remains open. Relocation along the palaeo-Pärnu River valley is suspected to have transported the Uku and Reiu organic layers to a deeper location than that supported by the model. Further investigations are needed to clarify the origin of these peat layers and to discuss their relation with the history of Baltic Sea basin water-level change.

The next footprints of ancient human activity originating from the Sindi-Lodja I and II settlement sites have been dated to 9,300–8,400 cal. years BP (Kriiska and Lõugas 2009). A single AMS date of charcoal from the cultural layer suggests that Sindi-Lodja settlement sites were most probably inhabited during the pre-Littorina

Sea transgression lowstand at about 9,200–8,800 cal. years BP (Table 8.1; Fig. 8.3). Our reconstruction shows that at about 9,000 cal. years BP these dwelling sites were situated about 0.5–4.5 m above and about 2 km from the coastline on the left bank of the ancient Pärnu River (Fig. 8.7f). Dwelling sites were located closer to the seashore than in the case of the Pulli settlement, probably due to the seal diet, which was not the case for the people of Pulli, whose main means of subsistence were elk and beaver hunting and pike-perch fishing (Veski et al. 2005). Judging from the animal bones, one may assume that the sites were at least inhabited in spring – the best time for taking ringed seal (*Phoca hispida*) and pike-perch (*Sander lucioperca*) – although the choice of location in the river mouth (Fig. 8.7f) and general Late Mesolithic contexts might even justify the assumption of year-round base camps (Kriiska and Lõugas 2009). Terrestrial conditions were interrupted in the Sindi-Lodja and in other buried organic matter sites (Table 8.1), when the rising level of the Littorina Sea submerged the area (Fig. 8.7g). Similar to Ancylus Lake, several Littorina Sea buried organic matter sites (Kolga, Vaskrääma, Rannametsa in Table 8.1) were also located in the reconstructed coastal zone (Fig. 8.7g). Our model of water-level change suggests that the Littorina Sea inundated settlement sites at about 8,500–8,400 cal. years BP just before the culmination of the transgression (Fig. 8.3).

Water-level rise during the Littorina Sea transgression was slower compared with the Ancylus Lake transgression, as reflected by inundated peat layers from different altitudes (Fig. 8.3). The Littorina Sea transgression culminated in the Pärnu area at about 7,300 cal. years BP. Sediment stratigraphies show only one pre-Littorina buried organic layer for the Pärnu area (Veski et al. 2005) and do not assert the multi-transgressive pattern of the Littorina Sea, which is reported from Blekinge (Berglund et al. 2005) and the Karelian Isthmus in NW Russia (Miettinen et al. 2007). These low-magnitude (around 1 m) short-term oscillations did not result in extensive peat formation in the Pärnu area, which could be evidence for a multi-transgressive Littorina Sea.

The relatively rapid global sea rise slowed down and isostatic uplift began to dominate in the Pärnu area after 7,300 cal. years BP, causing regressive shore displacement and peatland formation between the highest Littorina Sea and present-day coastlines. The beginning of peat formation in Kõrsa and Tolkuse bogs (Fig. 8.3; Table 8.1) combined with shoreline tilting data (Fig. 8.5) suggests that the fall in water level was most rapid immediately after the transgression and gradually slowed down during the late Holocene. The relative fall in sea level (taking place at an average rate of 1 mm/year) together with regressive shore displacement still continues in the area, as shown by the sea-level data for the last century (Vallner et al. 1988; Ekman 1996).

The late Mesolithic and Neolithic settlement sites at Sindi-Lodja III and Neolithic sites Jõekalda, Lemmetsa I and II and Malda all formed in conditions of a regressive coastline (Kriiska and Lõugas 2009; Fig. 8.7i). Sindi-Lodja III (dated typologically between 7,000 and 4,000 cal. years BP) and Jõekalda (dated typologically between 6,200 and 4,000 cal. years BP) settlement sites were located about 2–3 m above the Littorina Sea at the mouth of the ancient Pärnu River (Kriiska and

Lõugas 2009; Fig. 8.7i). The Lemmetsa II and Malda (dated typologically between 6,200 and 4,000 cal. years BP) and Lemmetsa I (dated typologically between 5,600 and 4,000 cal. years BP) settlement sites were situated about 2–3 m above the Littorina Sea, at the estuary-like mouth of the ancient Audru River (Kriiska and Lõugas 2009; Fig. 8.7i). Numerous finds of ringed seal bones demonstrate that all sites have been inhabited at least during the early spring, when the seals breed on the ice, or in late summer/autumn, when they make feeding tours in bays and rivers. Our reconstruction of palaeoshoreline and topography also shows natural conditions that are well suited to year-round base camps behind the protective Littorina coastal landforms at the mouths of the ancient Pärnu and Audru rivers (Fig. 8.7i). Cultural layers rich in finds, the diversity of the artefacts and the large size of dwelling sites support this suggestion (Kriiska and Lõugas 2009).

8.6 Conclusions

The most important conclusions to emerge from the project reported here could be listed as follows:

- Temporal and spatial water-level change model for the SW Estonian coastal zone of the Baltic Sea was compiled by combining the interpolated water-level surfaces for the different Baltic stages with a reconstructed shore displacement curve.
- We presented a displacement curve for the Pärnu area (SW Estonia), which records three regressive phases of the past Baltic Sea interrupted by Ancylus Lake and Littorina Sea transgressions with magnitudes of 12 and 10 m, respectively.
- Due to uncertainties in stratigraphy and chronology the two sites in the Pärnu area with buried organic beds displaying possible pre-Littorina Sea transgression water level below present-day sea level were not considered in the current shore displacement reconstructions.
- Palaeogeographic situations for different Baltic Sea stages were reconstructed by subtracting the water-level change model from the modern digital terrain model in order to understand preferences in the selection of settlement sites of Stone Age man at the shifting coastline of the Baltic Sea in SW Estonia.
- Reconstructions show that most buried organic matter sites lay at or slightly above the highest coastlines of the modelled Ancylus Lake and Littorina Sea, probably as a result of the good preservation conditions due to rapid burial. This may make it possible to discover new sites of buried organic matter.
- Uncertainties in palaeogeographic reconstructions described in this chapter are related to subsequent deposition and erosion since the time that was modelled. Holocene peat and gyttja were removed from the digital terrain model, although postglacial marine, eolian and fluvial deposits influence palaeoreconstructions.

Acknowledgements The authors express their thanks to Mrs. Annika Veske and Mrs. Evelin Lumi for help in digitalizing the elevation and sediment thickness data and to Alexander Harding for checking the language. We also thank Dr. Antoon Kuijpers and an anonymous reviewer for their comments and suggestions to improve the manuscript. This multidisciplinary study was primarily supported by Estonian Science Foundation Grant "Development of the Baltic Sea Coastline Through Time: Palaeoreconstructions and Predictions for Future". The research was also financed by Estonian target-funding projects SF0180150s08, SF0180048s08 and SF0332710s06, Estonian Science Foundation Grants no 7375, 6736 and 7029 and by the European Union through the Center of Excellence in Cultural Theory.

References

Andren E, Andren T, Kunzendorf H (2000) Holocene history of the Baltic Sea as a background for assesing records of human impact in the sediments of the Gotland Basin. Holocene 10:687–702

Bennike O, Jensen JB, Lemke W, Kuijpers A, Lamholt S (2004) Late- and postglacial history of the Great Belt, Denmark. Boreas 33:18–33

Berglund BE, Sandgren P, Barnekow L, Hannon G, Jing H, Skog G, Yu S-Y (2005) Early Holocene history of the Baltic Sea, as reflected in coastal sediments in Blekinge, southeastern Sweden. Quaternary Inaternational 130:111–139

Björck S (1995) A review of the history of the Baltic Sea, 13.0–8.0 ka BP. Quaternary International 27:19–40

Ekman M (1996) A consistent map of the postglacial uplift of Fennoscandia. Terra Nova 8:158–165

Estonian Maritime Administration (2001a) Depth chart 1:50 000 (59°): Gulf of Riga. Sõrve peninsula to Allirahu Islet

Estonian Maritime Administration (2001b) Depth chart 1:50 000 (59°): Gulf of Riga. Ruhnu Is land to Kihnu Shoal

Estonian Maritime Administration (2001c) Depth chart 1:50 000 (59°): Gulf of Riga. Kõiguste-Kübassaare-Sõmeri

Estonian Maritime Administration (2002a) Depth chart 1:50 000 (59°): Gulf of Riga. Kihnu Shoal to Heinaste (Ainaži)

Estonian Maritime Administration (2002b) Depth chart 1:50 000 (59°): Gulf of Riga. Pärnu-Kihnu-Sõmeri

Gudelis V (1976) History of the Baltic Sea. In: Gudelis V, Emelyanov E (eds) Geology of the Baltic Sea. Mokslas, Vilnius, pp 95–116 (in Russian with English summary)

Guobyte R (2004) A brief outline of the Quaternary of Lithuania and the history of its investigation. In: Ehlers J, Gibbard PL (eds) Quaternary glaciations—extent and chronology, Part I: Europe. Elsevier, Amsterdam, pp 245–250

Harff J, Lampe R, Lemke W, Lübke H, Lüth F, Meyer M, Tauber F (2005) The Baltic Sea – a model ocean to study interrelations of geosphere, ecosphere, and anthroposphere in the coastal zone. Journal of Coastal Research 21(3):441–446

Heinsalu A, Veski S, Moora T (1999) Bio- and chronostratigraphy of the Early Holocene site of double-storied buried organic matter at Paikuse, Southwestern Estonia. Proceedings Estonian Academy of Sciences, Geology 48(1):48–66

Haila H, Raukas A (1992) Ancylus lake. In: Raukas A, Hyvärinen H (eds) Geology of the Gulf of Finland. Estonian Academy of Sciences Institute of Geology, Academy of Finland, Helsinki University, Tallinn, pp 283–296 (in Russian)

Hyvärinen H, Raukas A, Kessel H (1992) Mastogloia and Litorina Seas. In: Raukas A, Hyvärinen H (eds) Geology of the Gulf of Finland. Estonian Academy of Sciences Institute of Geology, Academy of Finland, Helsinki University, Tallinn, pp 296–303 (in Russian)

Ilves E, Liiva A, Punning J-M (1974) Radiocarbon dating in the quaternary geology and archaeology of Estonia. Academy of Sciences Estonian SSR, Tallinn (in Russian with English summary)

Jaanits L, Jaanits K (1978) Ausgrabungen der frühmesolithischen Siedlung von Pulli. Eesti NSV Teaduste Akadeemia Toimetised, Ühiskonnateadused 27:56–63

Jakobsson M, Björck S, Alm G, Andrén T, Lindeberg G, Svensson N-O (2007) Reconstructing the Younger Dryas ice dammed lake in the Baltic Basin: Bathymetry, area and volume. Global and Planetary Change 57:355–370

Kalm V (2006) Pleistocene chronostratigraphy in Estonia, southeastern sector of the Scandinavian glaciation. Quarternary Science Reviews 8:960–975

Kessel H, Punning J-M (1969a) Über das Absolute Alter der holozänen transgressionen der Ostsee in Estland. ENSV Teaduste Akadeemia Toimetised, Keemia, Geoloogia 18:140–153 (in Russian with German summary)

Kessel H, Punning J-M (1969b) Über dieVerbreitung und Stratigraphie der sedimente des Joldiameeres in Estland. ENSV Teaduste Akadeemia Toimetised, Keemia, Geoloogia 18:154–163 (in Russian with English summary)

Kessel H, Punning J-M (1974) About the age of the Ancylus stage in Estonia (radiometric datings). ENSV TA Toimetised, Keemia, Geoloogia 23:59–64 (in Russian)

Kriiska A, Saluäär U (2000) Lemmetsa ja Malda neoliitilised asulakohad Audru jõe alamjooksul. Artiklite kogumik, 2. Pärnumaa ajalugu. Vihik 3. Pärnu, 8–38 (in Estonian)

Kriiska A (2001) Archaeological field work on Stone Age settlement site of SW Estonia. Archaeological fieldwork in Estonia 2000, Tallinn, pp 19–33

Kriiska A, Saluäär U, Lõugas L, Johanson K, Hanni H (2002) Archaeological excavations in Sindi-Lodja. Archaeological fieldwork in Estonia 2001, Tallinn, pp 27–40

Kriiska A, Johanson K, Saluäär U, Lõugas L (2003) The results of research of Estonian Stone Age. Archaeological fieldwork in Estonia 2002, Tallinn, pp 25–41

Kriiska A, Lõugas L (2009) Stone Age settlement sites on an environmentally sensitive coastal area along the lower reaches of the River Pärnu (south-western Estonia), as indicators of changing settlement patterns, technologies and economies. In: McCartan SB, Schulting R, Warren G, Woodman P (eds) Mesolithic horizons: papers presented at the 7th international conference on the Mesolithic in Europe, Belfast 2005. Oxbow Books, Oxford, pp 167–175

Lepland A, Hang T, Kihno K, Sakson M, Sandgren P (1996) Holocene sea-level changes and environmental history in the Narva Area. PACT 51:205–216

Leverington DW, Teller JT, Mann JD (2002) A GIS method for reconstruction of late Quaternary landscapes from isobase data and modern topography. Computers & Geosciences 28:631–639

Lidberg M, Johansson JM, Scherneck H-G, Milne GA, Davis JL (2009) New results based on reprocessing of 13 years continuous GPS observations of the Fennoscandia GIA process from BIFROST. Observing our Changing Earth, IAG Symposium Series, vol 133. Springer, Berlin, pp 557–568

Lundqvist J, Wohlfarth B (2001) Timing and east-west correlation of south Swedish ice marginal lines during the Late Weichselian. Quaternary Science Reviews 20:1127–1148

Miettinen A, Savelieva L, Subetto DA, Dzhinoridze R, Arslanov K, Hyvärinen H (2007) Palaenvironment of the Karelian Isthmus, the easternmost part of the Gulf of Finland, during the Litorina Sea stage of the Baltic Sea history. Boreas 36:441–458

Noormets R, Floden T (2002) Glacial deposits and ice-sheet dynamics in the north-central Baltic Sea during the last glaciation. Boreas 31:362–377

Orru M (1992) Estonian peat resources. RE Eesti Geoloogiakeskus, Tallinn (in Estonian)

Orru M (1995) Handbook. Estonian mires. Geological Survey of Estonia, Tallinn (in Estonian with English summary)

Påsse T, Andersson L (2005) Shore-level displacement in Fennoscandia calculated from empirical data. GFF 127(4):233–296

Persson C (1983) Glacial deposits and the Central Swedish end moraine zone in eastern Sweden. In: Ehlers J (ed) Glacial deposits in north-west Europe. Balkema, Rotterdam, pp 131–139

Poska A, Veski S (1999) Man and environment at 9500 BP. A palynological study of an Early-Mesolithic settlement site in south-west Estonia. Acta Palaeobotanica, Supplement 2: 603–607

Punning J-M, Ilves E, Liiva A, Rinne T (1971) Tartu radiocarbon dates V. Radiocarbon 13:78–83
Punning J-M, Rajamäe R, Ehrenpreis M, Sarv L (1977) Tallinn radiocarbon dates IV. Radiocarbon 19:111–117
Raukas A, Moora T, Karukäpp R (1999) The development of the Baltic Sea and Stone Age settlement in the Pärnu area of Southwestern Estonia. PACT 57:15–34
Reimer PJ, Baillie MGL, Bard E, Bayliss A, Beck JW, Bertrand CJH, Blackwell PG, Buck CE, Burr GS, Cutler KB, Damon PE, Edwards RL, Fairbanks RG, Friedrich M, Guilderson TP, Hogg AG, Hughen KA, Kromer B, McCormac G, Manning S, Bronk Ramsey C, Reimer RW, Remmele S, Southon JR, Stuiver M, Talamo S, Taylor FW, van der Plicht J, Weyhenmeyer CE (2004) IntCal04 Terrestrial radiocarbon age calibration, 0–26 Cal kyr BP. Radiocarbon 46:1029–1058
Rosentau A, Vassiljev J, Saarse L, Miidel A (2007) Palaeogeographic reconstruction of proglacial lakes in Estonia. Boreas 36:211–221
RosentauA, Vassiljev J, Hang T, Saarse L, Kalm V (2009) Development of the Baltic Ice Lake in the eastern Baltic. Quaternary International 206:16–23
Saarnisto M, Saarinen T (2001) Deglaciation chronology of the Scandinavian Ice Sheet from the Lake Onega Basin to the Salpausselkä End Moraines. Global and Planetary Change 31:387–405
Saarse L, Vassiljev J, Miidel A (2003) Simulation of the Baltic Sea Shorelines in Estonia and Neighbouring Areas. Journal of Coastal Research 19:261–268
Saarse L, Vassiljev J, Miidel A, Niinemets E (2006) Holocene buried organic sediments in Estonia. Proceedings Estonian Academy of Sciences, Geology 55(4):296–320
Saarse L, Vassiljev J, Miidel A, Niinemets E (2007) Buried organic sediments in Estonia related to the Ancylus Lake and Litorina Sea. In: Johansson P, Sarala P (eds) Applied Quaternary research in the central part of glacial terrain. Geological Survey of Finland, Special Paper 46:87–92
Söderberg P (1988) Notes on continuation of the Salpausselkä ice marginal zone in the northern Baltic Proper. Geological Survey of Finland Special Paper 6:69–72
Vallner L, Sildvee H, Torim A (1988) Recent crustal movements in Estonia. Journal of Geodynamics 9:215–233
Veski S (1998) Vegetation history, human impact and palaeogeography of West Estonia. Pollen analytical studies of lake and bog sediments. STRIAE 38:1–119
Veski S, Poska A, Talviste P, Hang T, Rosentau A, Hiie S, Heinsalu A, Teiter K (in press) Investigations for reconstructing the landscape. In: David E, Kriiska A, Lõugas L (eds) The Early Holocene in the Eastern Baltic with special emphases on the Mesolithic Pulli site (Pärnu region, Estonia). Muinasaja Teadus
Veski S, Heinsalu A, Klassen V, Kriiska A, Lõugas L, Poska A, Saluäär U (2005) Early Holocene coastal settlement and palaeoenvironment on the shore of the Baltic Sea at Pärnu, southwestern Estonia. Quaternary International 130:75–85
Uścinowicz S (1999) Southern Baltic area during the last deglaciation. Geological Quarterly 43:137–148

Chapter 9
Palaeoreconstruction of the Baltic Ice Lake in the Eastern Baltic

Jüri Vassiljev, Leili Saarse, and Alar Rosentau

Abstract A GIS-based palaeogeographic reconstruction of the development of the Baltic Ice Lake (BIL) in the eastern Baltic during the deglaciation of the Scandinavian Ice Sheet is presented. A Late Glacial shoreline database containing sites from Finland, NW Russia, Estonia, Latvia and modern digital terrain models was used for palaeoreconstructions. The study shows that at about 13,300 cal. years BP the BIL extended to the ice-free areas of Latvia, Estonia and NW Russia, represented by the highest shoreline in this region. Reconstructions demonstrate that BIL initially had the same water level as the Glacial Lakes Peipsi and Võrtsjärv because these water bodies were connected via strait systems in central and northeast Estonia. These strait systems were gradually closed at about 12,700–11,700 cal. years BP due to isostatic uplift, prior to the final drainage of the BIL. Glacial Lake Võrtsjärv isolated from the BIL at about 12,400–12,000 cal. years BP. Exact timing of Glacial Lake Peipsi isolation is not clear, but according to the altitude of the threshold in northeast Estonia and shore displacement data, it was completed at about 12,400–11,700 cal. years BP.

Keywords Baltic Ice Lake · Water level reconstruction · Palaeogeography

9.1 Introduction

The last termination of the Scandinavian Ice Sheet (SIS) produced a huge volume of meltwater that led to the formation of large proglacial lakes, like a Baltic Ice Lake (BIL). BIL was first recognized by Munthe (1910), who showed that in the Baltic Sea basin, an ice lake dammed up during the late glacial, when the ice margin was located at the central Swedish moraines. Later, when the ice retreated from the central Swedish moraines, the water level dropped to the Yoldia Sea or

J. Vassiljev (✉)
Institute of Geology, Tallinn University of Technology, 19086 Tallinn, Estonia
e-mail: vassilje@gi.ee

J. Harff et al. (eds.), *The Baltic Sea Basin*, Central and Eastern European Development Studies (CEEDES), DOI 10.1007/978-3-642-17220-5_9,

ocean level. Ramsay (1917) found similar results from the Salpausselkkä area in Finland, showing that the BIL had the highest shoreline. However, Ramsay (1928, 1929) later studied shorelines in Estonia and Ingermanland (NW Russia) and found that uppermost shores there are older than BIL shores in Finland and proposed that they represent the shores of the local ice lakes. Since then, late-glacial shorelines in Estonia and NW Russia were divided between local ice lakes and BIL (Markov 1931, Pärna 1960, Kessel and Raukas 1979). It was assumed that the local ice lake shorelines developed during Alleröd and BIL shorelines during Younger Dryas. In contrast, in Latvia BIL started to develop in Alleröd (Grinbergs 1957, Veinbergs 1979), at the same time when in Estonia and NW Russia local ice lakes existed.

Estonian local ice lakes were studied in detail by Pärna (1960), who found that the largest proglacial lakes developed during the Pandivere/Neva stage (13,300 corrected varve years BP; Saarnisto and Saarinen 2001, Hang 2003, Kalm 2006;

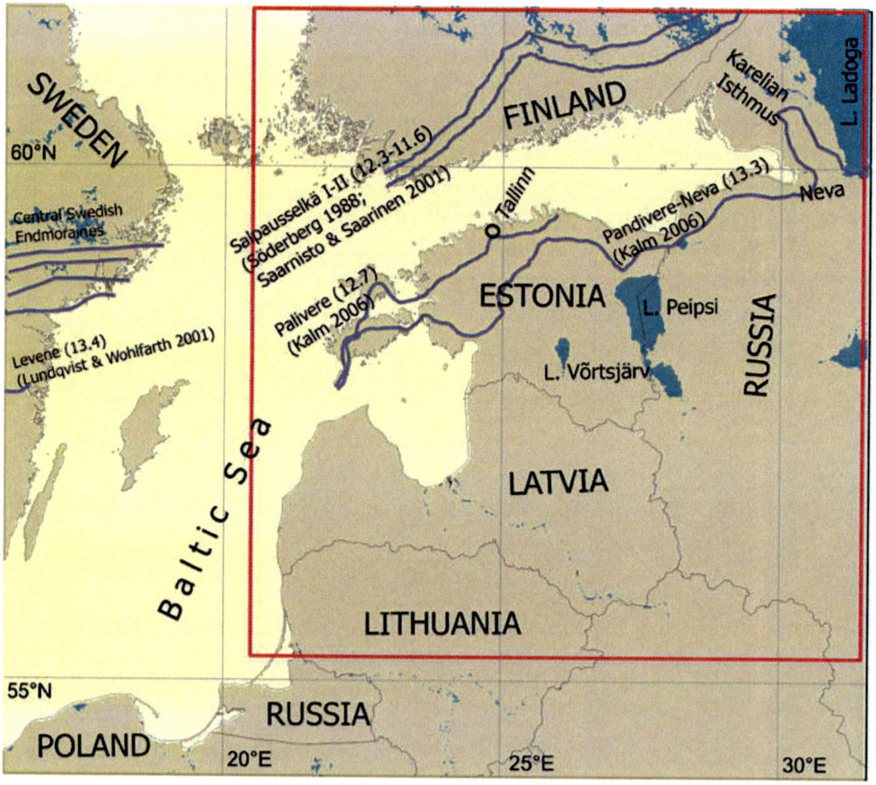

Fig. 9.1 Overview map of the study area. *Blue lines* indicate ice-marginal positions, discussed in the text, with ages (cal. kyears BP) according to Kalm (2006), Lundqvist and Wohlfarth (2001) and Saarnisto and Saarinen (2001). *Red box* indicates the area shown in Figs. 9.2, 9.3, 9.4, 9.5 and 9.6

Fig. 9.1). Pärna (1960) also suggested that water level lowered ca. 30 m when the ice retreated from the Pandivere/Neva to the Palivere ice-marginal zone (12,700 corrected varve years BP; Kalm 2006; Fig. 9.1). However, this low water level is marked only by few glaciofluvial flat plains near Tallinn (Pärna 1960; Fig. 9.1). Similar low water level, the so-called g-delta level, also existed in Finland (Sauramo 1958) as a result of a connection between the Baltic Sea and White Sea. Later studies (Kvasov and Raukas, 1970), however, showed that such a connection was unlikely and there was no physical reason for a low water level before the BIL stage BI (Donner 1995).

The aim of the current chapter is to correlate the Baltic Ice Lake coastal formations in eastern Baltic using a shoreline database and GIS analyses. The palaeogeographical reconstructions were used to study drainage routes of proglacial lakes and BIL, especially how and when Lake Peipsi and Võrtsjärv isolated from the BIL.

9.2 Methods

Reconstruction of the BIL shorelines and bathymetry in eastern Baltic area were based on GIS analysis, by which interpolated surfaces of water levels were removed from the modern digital terrain model (DTM; Rosentau et al. 2004).

The interpolated surfaces of water levels were derived using the late-glacial shoreline database (Vassiljev et al. 2005, Saarse et al. 2007). The late-glacial shoreline database covers more than 1,200 sites from eastern Baltic, including shore displacement data for Estonia (Vassiljev et al. 2005, Saarse et al. 2007), Latvia (Grinbergs 1957, Veinbergs 1979), NW Russia (Markov 1931, Shmaenok et al. 1962) and southern Finland (Donner 1978). However, statistical analyses showed that roughly half of this data does not match water level reconstruction requirements due to inaccurate elevations or erroneous correlation of different shore marks. The reliability of shoreline data was verified by different methods. First, sites with altitudes not matching with neighbouring sites were eliminated. Then, point kriging interpolation with linear trend was used to interpolate water level surfaces. Kriging is advantageous because it interpolates accurate surfaces from irregularly spaced data and it is easy to identify outliers in the data set. Residuals, the difference between the actual site altitude and the interpolated surface, were calculated and used to check the shoreline data reliability so that sites with residuals more than ± 1 m were discarded. Finally, interpolated water level surfaces were calculated using for A_1 – 52, A_2 – 77, BI – 111, BII – 88 and BIII – 164 sites.

BIL stage A_1 correlates with Pandivere/Neva ice-marginal zone (Fig. 9.1) dated 13,300 corrected varve years BP (Saarnisto and Saarinen 2001, Hang 2003, Kalm 2006). BIL stage A_2 correlates with Palivere ice-marginal zone (Fig. 9.1) dated 12,700 corrected varve years BP (Hang 2003, Kalm 2006). The age of stage A_2 in earlier studies was older than Palivere stade; however, recent studies (Rosentau et al. 2007) indicate that it formed during the Palivere stade.

BIL stages BI, BII and BIII are contemporaneous with the Salpausselkä I (SsI) and Salpausselkä II (SsII) endmoraines (Donner 1969, 1978, Glückert 1995), dated, respectively, 12,250–12,030 and 11,770–11,590 corrected varve years BP (Saarnisto and Saarinen 2001, Rinterknecht et al. 2004). BIL stage BI correlates with Ss I, stage BII corresponds to the time when the ice retreated from Ss I to Ss II and stage BIII corresponds to Ss II (Donner 1978). Accordingly, the age of stage BI is about 12,200, BII 12,000 and BIII 11,600 corrected varve years BP. Stage BIII corresponds to the BIL water level just before the Billingen drainage, which occurred 11,560 corrected varve years BP (Andrén et al. 2002).

Modern DTM with a grid size of 200×200 m was generated using the linear solution of the Natural Neighbour interpolation using different sources of elevation data. Estonian elevation data were derived from Digital Base Map of Estonia on a scale of 1:50,000 (Estonian Land Board 1996) and complemented with more detailed elevation data at critical threshold areas using Soviet military topographic maps on a scale of 1:10,000 and 1:25,000. Elevation data for Latvia and NW Russia were derived from Shuttle Radar Topography Mission (International Centre for Tropical Agriculture 2004) and for southern Finland from GTOPO30 data (US Geological Survey 1996). Baltic Sea topography data were derived from Seifert et al. (2001).

DTM-based palaeoreconstructions have some limitations pointed out by Leverington et al. (2002) due to the impact of deposition subsequent to the time being modelled. The deposition influence on our modelling relates in critical threshold areas mostly to the Holocene peat deposits, which were removed from DTM according to the digital soil maps in scale 1:10,000 and different peat investigations (Orru 1995). After the removal of the Holocene peat and the interpolated surfaces of water level from DTM, the shorelines and bathymetry were reconstructed for five different stages.

9.3 Results

9.3.1 BIL 13,300 cal. years BP (A_1)

Reconstruction shows that during Pandivere–Neva stade in front of the ice margin the BIL extended to ice-free areas of Estonia, Latvia, Lithuania and NW Russia (Fig. 9.2). Several ice-contact slopes, flat-topped glaciolacustrine and glaciofluvial landforms mark the ice-proximal coast of the lake up to 90 m above present sea level in northern Estonia. Accumulative and abrasional coastal landforms developed in more ice-distal areas. Water level near SW Lithuanian coast and further south was below present sea level.

The Glacial Lake Peipsi was connected via strait system through the Lake Võrtsjärv basin with BIL in the Baltic Sea basin (Fig. 9.2). The water level in Glacial Lake Peipsi was similar to BIL water level. A narrow strait could also exist in northern Estonia and NW Russia directly in front of the Pandivere–Neva ice margin.

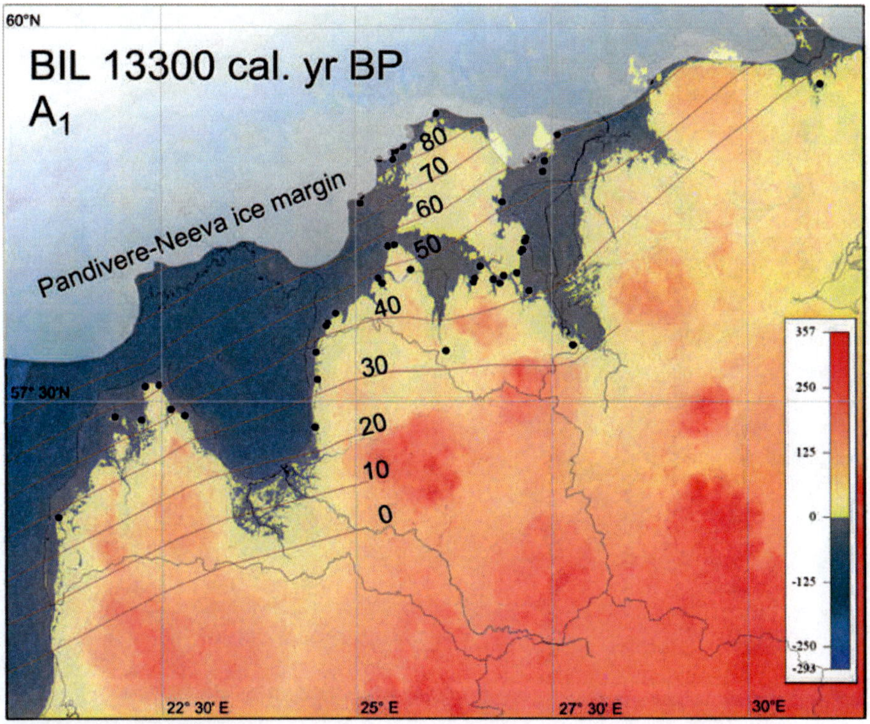

Fig. 9.2 BIL 13,300 cal. years BP, stage A_1. Isobases show the reconstructed water level above present sea level in metres. *Black dots* show the location of used shoreline sites

9.3.2 BIL 12,700 cal. years BP (A_2)

During the Palivere stade, BIL covered ice-free areas of Estonia, Latvia, Lithuania and NW Russia (Fig. 9.3). Compared to stage A_1 the water level was about 10–15 m lower in northern Estonia and only a few metres lower in southern Estonia and Latvia. Water level near SW Lithuanian coast and further south was below present sea level. The main connection between BIL in Baltic Sea basin and Glacial Lake Peipsi was located in northern Estonia. The strait in southern Estonia via Lake Võrtsjärv still existed, but it was considerably narrowed. Reconstruction indicates that in NW Russia the BIL extended to Lake Ladoga (Fig. 9.3).

9.3.3 BIL 12,200 cal. years BP (BI)

During the standstill of ice margin at Salpausselkä I BIL stage BI, waters covered southern Finland, western Estonia and coastal areas of Latvia and NW Russia (Fig. 9.4). Water level was up to 140–150 m a.s.l. in Finland, but near SW Lithuanian coast and further south it was below present sea level. During the ice retreat from

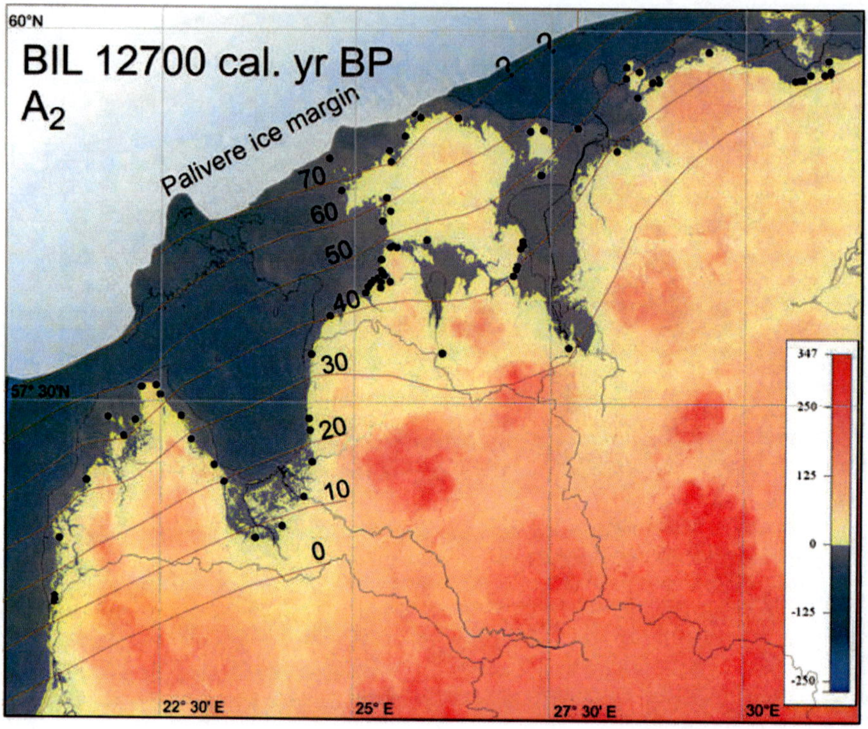

Fig. 9.3 BIL 12,700 cal. years BP, stage A_2. Isobases show the reconstructed water level above present sea level in metres. *Black dots* show the location of used shoreline sites

the Palivere ice-marginal zone to the Salpausselkä I endmoraines, considerable rearrangement of the proglacial lake drainage system occurred in Estonia: connection of Lakes Võrtsjärv and Peipsi with the BIL was terminated via straits and these lakes started to develop as isolated water bodies (Fig. 9.4). The water bodies in Lakes Peipsi and Võrtsjärv were most likely slightly larger than shown in Fig. 9.4, because their thresholds were about 0.5–2 m higher than BI water level in Baltic Sea basin. The exact location of Lake Võrtsjärv threshold is difficult to determine because of the relief flatness: its altitude is about 42 m a.s.l. The BIL water level at threshold was about 40.5±1 m a.s.l. Lake Peipsi threshold was located at the NE end of the lake. Its altitude was at minimum 28 m a.s.l. as limestone surface, which however at that time was covered by sandy deposits, so that the maximum altitude of the threshold was about 34 m a.s.l. The BIL water level at threshold was about 31±1 m a.s.l., so that Lake Peipsi could have very shallow connection with BIL depending on the exact threshold altitude. In NW Russia, BIL extended to the Lake Ladoga basin over Karelian Isthmus and Neva valley. However, the strait in Neva valley was considerably narrowed compared to the previous stage A_2.

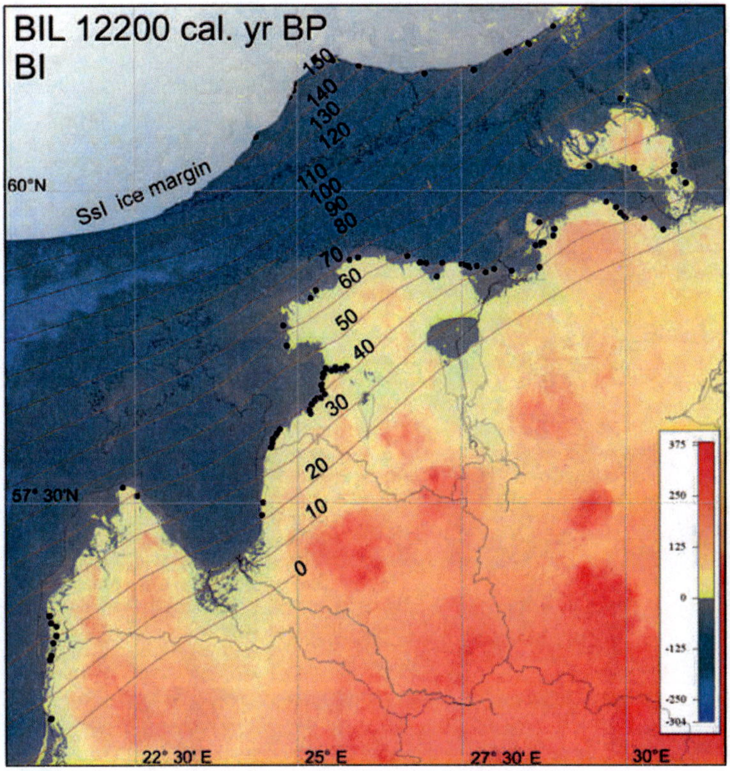

Fig. 9.4 BIL 12,200 cal. years BP, stage BI. Isobases show the reconstructed water level above present sea level in metres. *Black dots* show the location of used shoreline sites

9.3.4 BIL 12,000 cal. years BP (BII)

After the retreat of the ice margin from the Ss I to the Ss II, BIL stage BII developed (Fig. 9.5). Its coastal formations are located lower than previous BI ones: in Finland about 10 m, in northern Estonia about 5 m and in southern Estonia and Latvia about 1–2 m. However, in NE Estonia in the northern part of Lake Peipsi, BII isobases are up to 0.5 m higher than BI ones. The reason for that is not yet clear. The BIL water level at Võrtsjärv threshold was 40±1m a.s.l. and at Peipsi threshold 31.5±1 m a.s.l. Lakes Peipsi and Võrtsjärv were closed basins with considerably lower water levels than today, especially Lake Võrtsjärv, which was most likely dry land (Fig. 9.5). Lake Peipsi had the water only in the northern part of the lake and it could still have very shallow connection with the BIL depending on the threshold altitude. Lake Ladoga in NW Russia was connected with BIL over Karelian Isthmus and Neva valley.

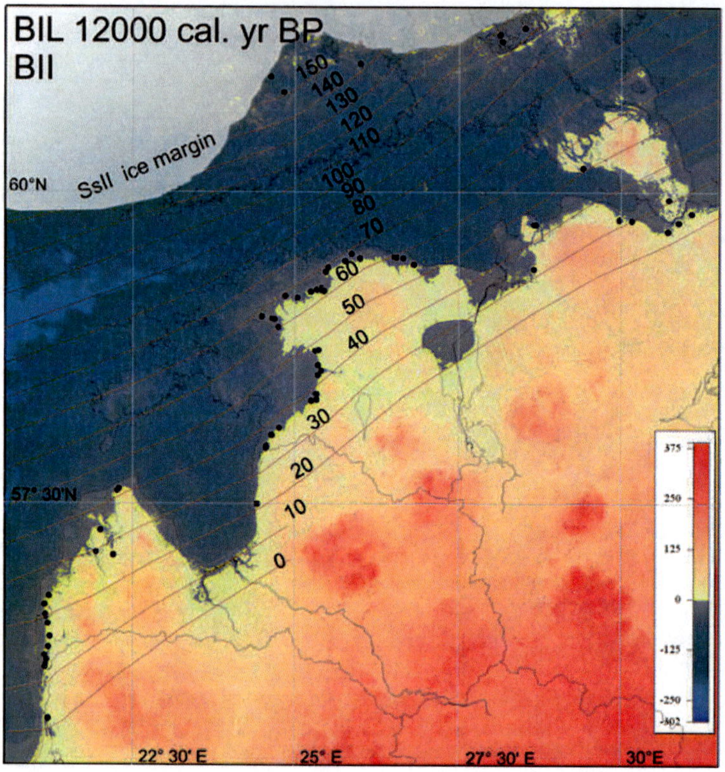

Fig. 9.5 BIL 12,000 cal. years BP, stage BII. Isobases show the reconstructed water level above present sea level in metres. *Black dots* show the location of used shoreline sites

9.3.5 BIL 11,600 cal. years (BP/BIII)

During the ice margin standstill at the Salpausselkä II about 11,770–11,590 cal. years BP (Saarnisto and Saarinen 2001), a well-developed BIL coastline BIII was formed. BIII waters covered southern Finland, western and northern Estonia and coastal areas of Latvia and NW Russia (Fig. 9.6). The water level was up to 150 m a.s.l. in Finland, but below present sea level south from Latvia. The BIL water level at Lake Peipsi threshold was 26±1 m a.s.l., which is at least 1–3 m below the minimum threshold altitude (limestone surface). The water body in the Lake Peipsi basin was slightly larger than shown in Fig. 9.6 because its threshold was higher than BIII water level in the Baltic Sea basin. The southern part of Lake Peipsi was dry. Lake Võrtsjärv was most likely dry or had very shallow water body only in the northern part. Lake Ladoga in NW Russia was connected with BIL; however, the strait in Neva valley was further narrowed and a main connection existed over Karelian Isthmus (Fig. 9.6).

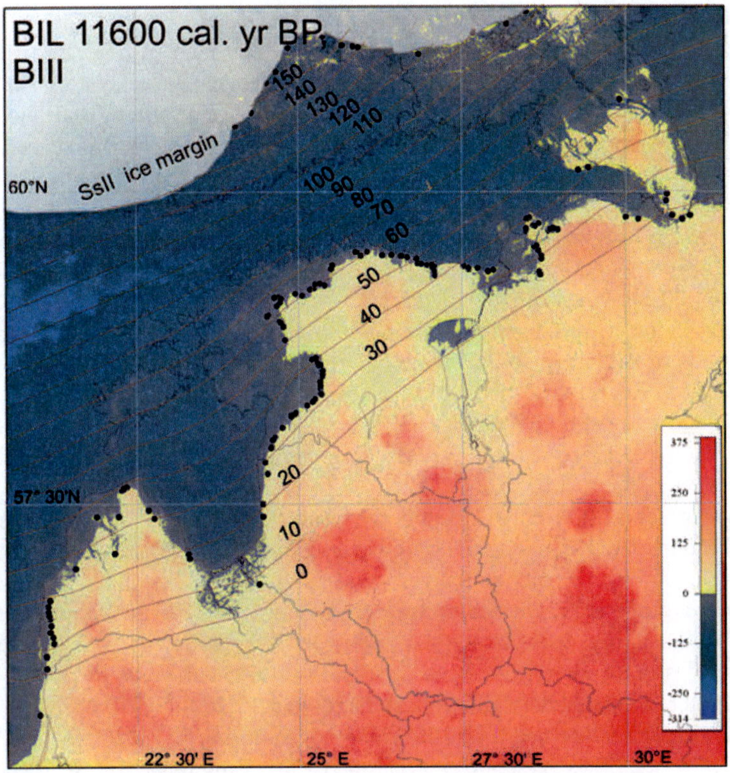

Fig. 9.6 BIL 11,600 cal. years BP, stage BIII. Isobases show the reconstructed water level above present sea level in metres. *Black dots* show the location of used shoreline sites

9.4 Discussion

Modelling results demonstrate BIL at five different levels showing that land uplift induced regressive shore displacement between 13,300 and 11,600 corrected varve years BP. Shoreline data of Latvia, Estonia, NW Russia and Finland are generally in good agreement. Latvian and NW Russian shoreline proxies helped to improve water level reconstruction for Lakes Võrtsjärv and Peipsi. The isobases of stages BI, BII and BIII show a generally regular pattern of uplift all over the study area. The isobases of the stages A_1 and A_2 (Figs. 9.2 and 9.3) show a regular pattern of the uplift in NW; however, in the Lake Peipsi basin, isobases curve remarkably towards SE, being up to 8 m higher than expected from the regional pattern. The main reason of curving is anomalous lowering of the tilting gradient that was mentioned by correlation of over-deepened river mouths (Miidel et al. 1995) and the late-glacial river terraces (Hang et al. 1964, Hang et al. 1995). This phenomenon can reflect the forebulge effect during glaciation and its later collapse (Rosentau et al. 2007).

Reconstructions of BIL stages A_1, A_2 and BIII are comparable with those given earlier by Björck (1995); however, there are differences concerning revised (calibrated) ages given in an unpublished Baltic history summary (http://www.geol.lu.se/personal/seb). The deglaciation chronology on these maps was related to conventional ^{14}C years BP (Björck et al. 1988, Lundqvist 1986) so that the ages of the BIL stages were originally 12,000, 11,000 and 10,300 ^{14}C years BP (Björck 1995). The revised ages suggested by Björck are 14,000, 13,000 and 11,700–11,600 cal. years BP, respectively. However, the first age is not supported by new datings. According to Björck (1995), during BIL stage 12,000 uncal./14,000 cal. years BP, the ice margin was located in Sweden at Levene and in Estonia at Palivere line (Fig. 9.1), although the water level in Estonia in that map corresponds to BIL stage A_1, developed during Pandivere–Neva stade (Kessel and Raukas 1982, Rosentau et al. 2007). Levene moraine formed around 13,400 cal. years BP (Lundqvist and Wohlfarth 2001), which is approximately the same age as Pandivere–Neva ice margin: 13,300 corrected varve years BP (Saarnisto and Saarinen 2001, Hang 2003, Kalm 2006).

BIL stage at 11,000 uncal./13,000 cal. years BP has approximately the same age as reconstructed BIL stage A_2; however, ice-marginal positions are different: ice margin in Estonia was at Palivere during that time, but in Björck's (1995) map ice margin was in Sweden at Younger Dryas (YD, Skövde) moraines and in Finland at Ss I (Fig. 9.1). Lundqvist and Wohlfarth (2001) have shown that the YD moraines in Sweden formed around 12,475–11,615 cal. years BP, which corresponds to the Salpausselkä age (Ss I–Ss II) around 12,250–11,590 corrected varve years BP (Saarnisto and Saarinen 2001). Moreover, during Ss I (12,250–12,030 corrected varve years BP (Saarnisto and Saarinen 2001, Rinterknecht et al. 2004)) BIL stage BI developed (Donner 1978). Thus, Björck's (1995) reconstruction around 13,000 cal. years BP represents the BIL at the supposedly first drainage, but the ice margin position seems to be questionable. In previous studies, the age of BIL stage A_2 was unclear: it was considered to be in between Pandivere and Palivere stades (Kessel and Raukas 1982). Rosentau et al. (2007) compared the water levels of BIL stage A_2 in Estonia with shore displacement curve data from eastern Småland (Svensson 1991) and Blekinge (Björck 1995) in Sweden and concluded that formation of the coastal landforms of BIL stage A_2 occurred before the first drainage of the BIL (Björck 1995) at about 12,800 cal. years BP, which is equivalent to the age of the Palivere ice-marginal zone (12,700 corrected varve years BP; Kalm 2006).

Earlier studies in Estonia (cf. Kessel and Raukas 1982) have shown that during Palivere stade, proglacial lake with ca. 30 m lower water level than A_2 existed. However, there is no clear evidence of the low water level in eastern Baltic between BIL stages A_2 and BI as has been suggested on the basis of low-lying glaciofluvial plateau-like plains in front of the Palivere (Kessel and Raukas 1982) and Salpausselkä (Sauramo 1958, Donner, 1982) ice margins (Fig. 9.1). Fyfe (1990) showed that these plains in Finland are overlapping fans formed under the water. Most likely, the above-mentioned plateau-like marginal formations in Estonia have also been deposited below the water surface. The wide distribution of subaquatic waterlain glacial diamictons formed during the Palivere stade at the depth of

50–60 m below the Baltic Ice Lake water table (Kalm and Kadastik 2001) suggests a higher water level during deposition of the ice-marginal plains.

Our results match well with the reconstructions published by Jakobsson et al. (2007) for the last stage of the BIL (BIII). However, in the southern part of the Latvian coast, our data show about 4–6 m lower water level as compared to Jakobsson et al. (2007). Further in the south, their inferred shoreline elevations (Jakobsson et al. 2007) indicate higher water levels than the observed shore displacement data from the Lithuanian (Gelumbauskaite and Seckus 2005) and the Polish coasts (Uścinowicz 2006). Our reconstructions indicate that there are problems to model water levels in the southern part of the Baltic Sea, where the water level was below present sea level, because of the lack of data.

Our reconstructions show that the so-called Estonian local ice lakes A_1 and A_2 were connected with BIL. Kvasov and Raukas (1970) proposed that the beginning of the BIL in eastern Baltic was earlier than Ss I (BIL stage BI). They found that when the proglacial lakes east and west of the Pandivere Upland joined up about 13,000 cal. years BP, proglacial lake A_2 formed as a first BIL stage. It was assumed that Glacial Lake Peipsi and proglacial lakes in NW Russia had a higher water level before that event. Our reconstructions indicate that water levels were similar and Glacial Lake Peipsi and BIL were already connected during stage A_1 about 13,300 cal. years BP (Vassiljev et al. 2005). Moreover, reconstructions suggest that A_2 proglacial lake emerged later, about 12,700 cal. years BP during the Palivere stade (Rosentau et al. 2007).

Reconstruction indicates that Lakes Võrtsjärv and Peipsi isolated from the BIL clearly before Billingen drainage. Earlier it has remained unclear whether the isolation was caused by the Billingen drainage or by uplift. Lake Võrtsjärv isolated from BIL about 12,400–12,000 cal. years BP and it is also supported by the low water level data from the southern part of the lake: buried peat layer in the bottom of the recent Lake Võrtsjärv (Moora et al. 2002) and the water level reconstructions (Pirrus et al. 1993, Hang et al. 1995). Isolation of Lake Peipsi from the BIL was about 12,400–11,700 cal. years BP and it is in good agreement with low water level indicators from the southern part of the lake (Sarv and Ilves 1975, Poska and Saarse 2006, Hang et al. 2008). Our modelling results show that Lake Peipsi isolated from BIL before the Billingen drainage at about 12,400–11,700 cal. years BP. However, the Billingen drainage probably enhanced the threshold erosion and the water level lowering in Lake Peipsi continued according to the amount of erosion estimation which needs further investigation.

9.5 Conclusions

BIL in the eastern Baltic was reconstructed at five levels between 13,300 and 11,600 cal. years BP.

Around 13,300 cal. years BP, the BIL extended to the ice-free areas of Latvia, Estonia and NW Russia, represented by the well-developed shoreline in this region.

BIL had 13,300 cal. years BP, the same water level as in Lakes Peipsi and Võrtsjärv, because these water bodies were connected via strait systems in central Estonia.

According to shoreline and bathymetry reconstructions, the land uplift induced gradual isolation of Lakes Peipsi and Võrtsjärv from the BIL, which occurred before the Billingen drainage.

Glacial Lake Võrtsjärv isolated from BIL at about 12,400–12,000 cal. years BP, most likely before the BI stage.

Glacial Lake Peipsi isolated from BIL in between 12,400 and 11,700 cal. years BP, thus before BIII stage.

Shore displacement of the BIL in eastern Baltic was regressive due to the land uplift.

Acknowledgements This study was supported by Estonian target funding projects SF0332710s06 and SF0180048s08 and Estonian Science Foundation Grants 6736, 6992, 7029 and 7294.

References

Andrén T, Lindeberg G, Andrén E (2002) Evidence of the final drainage of the Baltic Ice Lake and the brackish phase of the Yoldia Sea in glacial varves from the Baltic Sea. Boreas 31:226–238

Björck S, Berglund BE, Digerfeldt G (1988) New aspects on the deglaciation chronology of South Sweden. Fossils and Strata 14:1–93

Björck S (1995) A review of the history of the Baltic Sea, 13.0–8.0 ka BP. Quaternary International 27:19–40

Donner J (1969) Land/sea level changes in southern Finland during the formation of the Salpausselkä endmoraines. Bulletin of the Geological Society of Finland 41:135–150

Donner J (1978) The dating of the levels of the Baltic Ice lake and the Salpausselkä moraines in South Finland. Societas Scientarum Fennica, Commentationes Physico-Mathematicae 48:11–38

Donner J (1982) Fluctuations in water level of the Baltic Ice Lake. Annales Academiae Scientiarum Fennicae, AIII 134:13–28

Donner J (1995) The quaternary history of Scandinavia. Cambridge University Press, Cambridge, 200pp

Estonian Land Board (1996) Digital base map of Estonia on a scale of 1:50000 on 112 sheets

Fyfe GJ (1990) The effect of water depth on ice-proximal glaciolacustrine sedimentation: Salpausselkä I, southern Finland. Boreas 19:147–164

Gelumbauskaite LP, Seckus J (2005) Late Quaternary shore formations of the Baltic basins in the Lithuanian sector. Geologija 52:34–45

Glückert G (1995) The Salpausselkä End Moraine in southwestern Finland. In: Elhers J, Kosarski S, Gibbard P (eds) Glacial deposits in North-East Europe. A. A. Balkema, Rotterdam, pp 51–56

Grinbergs EF (1957) Late- and postglacial history of the coast of the Latvian SSR. Academy of Sciences of the Latvian SSR, Institute of Geology and Mineral Resources, Riga, 127 pp (in Russian)

Hang E, Liblik T, Linkrus E (1964) On the relations between Estonian valley terraces and lake and sea levels in the Late-Glacial and Holocene periods. Transactions of the Tartu State University 156. Publications on Geography IV:29–42

Hang T, Miidel A, Pirrus R (1995) Late Weichselian and Holocene water-level changes of Lake Peipsi, eastern Estonia. PACT 50:121–131

Hang T (2003) A local clay-varve chronology and proglacial sedimentary environment in glacial Lake Peipsi, eastern Estonia, Boreas 32:416–426

Hang T, Kalm V, Kihno K, Milkevicius M (2008) Pollen, diatom and plant macrofossil assemblages indicate a low water level phase of Lake Peipsi at the beginning of the Holocene. Hydrobiologia 599(1):13–21

International Centre for Tropical Agriculture (2004) Hole-filled seamless SRTM data V1. http://gisweb.ciat.cgiar.org/sig/90m_data_tropics.htm; http://srtm.usgs.gov/

Jakobsson M, Björck S, Alm G, Andrén T, Lindeberg G, Svensson NO (2007) Reconstructing the Younger Dryas ice dammed lake in the Baltic basin: bathymetry, area and volume. Global and Planetary Change 57:355–370

Kalm V, Kadastik E (2001) Waterline glacial diamicton along the Palivere ice-marginal zone on the West Estonian Archipelago, Eastern Baltic Sea. Proceedings of the Estonian Academy of Sciences, Geology 50:114–127

Kalm V (2006) Pleistocene chronostratigraphy in Estonia, southeastern sector of the Scandinavian glaciation. Quarternary Science Reviews 8:960–975

Kessel H, Raukas A (1979) The Quaternary history of the Baltic. Estonia. In: Gudelis V, Königsson L-K (eds) The Quaternary history of the Baltic. Acta Universitatis Upsaliensis, Uppsala, pp 127–146

Kessel H, Raukas A (1982) On geological development of the Baltic Sea in Late-Glacial time on the basis of the east Baltic evidence. Peribaltic II:131–143 (in Russian)

Kvasov DD, Raukas A (1970) Postglacial history of the Gulf of Finland. Proceedings Geographical Society of Soviet Union 102:432–438 (in Russian)

Leverington DW, Teller JT, Mann JD (2002) A GIS method for reconstruction of late Quaternary landscapes from isobase data and modern topography. Computers & Geosciences 28:631–639

Lundqvist G (1986) Late Weichselian glaciation and deglaciation in Scandinavia. In: Sibrava V, Bowen DQ, Richmond GM (eds) Quaternary glaciation in the northern hemisphere. Quaternary Science Reviews 5:269–292

Lundqvist J, Wohlfarth B (2001) Timing and east-west correlation of south Swedish ice marginal lines during the Late Weichselian. Quaternary Science Reviews 20:1127–1148

Markov KK (1931) The development of the relief in the north-western part of the Leningrad district. Trudy Glavnogo geologo-razvedochnogo upravlenija VSNH SSSR 117, 256pp (in Russian)

Miidel A, Hang T, Pirrus R, Liiva A (1995) On the development of the southern part of Lake Peipsi in the Holocene. Proceedings of the Estonian Academy of Sciences, Geology 44:33–44

Moora T, Raukas A, Tavast E (2002) Geological history of Lake Võrtsjärv. Proceedings of the Estonian Academy of Sciences, Geology 51:157–179

Munthe H (1910) Studies in the Late-Quaternary history of Southern Sweden. Geologiska Föreningen i Stockholm Förhandlingar 32:1197–1292

Orru M (1995) Handbook. Estonian mires. Geological Survey of Estonia, Tallinn, 240 pp (in Estonian with English summary)

Pärna K (1960) Zur Geologie des Baltischen Eisstausees sowie der lokalen grossen Eisstauseen auf dem Territorium der Estnischen SSR. Eesti NSV Teaduste Akadeemia Geoloogia Instituudi uurimused V:269–278 (in Russian with German summary)

Pirrus R, Hang T, Liiva A (1993) On the geological development of the Väike-Emajõgi valley and the southern part of Lake Võrtsjärv. Proceedings of the Estonian Academy of Sciences, Geology 42:28–37 (in Russian with English summary)

Poska A, Saarse L (2006) New evidence of possible crop introduction to north-eastern Europe during the Stone Age. Cerealia pollen finds in connection with the Akali Neolithic settlement, East Estonia. Vegetation History and Archaeobotany 15:169–179

Ramsay W (1917) De sök. marine gränserna i södra Finland. Geologiska Föreningen i Stockholm Förhandlingar 42:243–262

Ramsay W (1928) Eisgestaute Seen und Rezession des Inlandeises in Südkarelian und im Nevatal. Fennia, 50, p 21

Ramsay W (1929) Niveauverschiebungen, Eisgestaute Seen und rezession des Inlandeises in Estland. Fennia, 52, p 48

Rinterknecht VR, Clark PU, Raisbeck GM, Yiou F, Brook EJ, Tschudi S, Lunkka JP (2004) Cosmogenic 10Be dating of the Salpausselkä I Moraine in southwestern Finland. Quaternary Science Reviews 23:2283–2289

Rosentau A, Hang T, Miidel A (2004) Simulation of the shorelines of glacial Lake Peipsi in Eastern Estonia during the Late Weichselian. Geological Quarterly 48(4):299–307

Rosentau A, Vassiljev J, Saarse L, Miidel A (2007) Palaeogeographic reconstruction of proglacial lakes in Estonia. Boreas 36:211–221

Saarnisto M, Saarinen T (2001) Deglaciation chronology of the Scandinavian ice sheet from the east of Lake Onega basin to the Salpausselkä end moraines. Global and Planetary Change 31:387–405

Saarse L, Vassiljev J, Rosentau A, Miidel A (2007) Reconstructed late glacial shore displacement in Estonia. Baltica 20(½):35–45

Sarv A, Ilves E (1975) Über das Alter der holozänen Ablagerungen im Mündungsgebiet des Flusses Emajõgi (Saviku). Proceedings of the Estonian Academy of Sciences, Geology 24:64–69 (in Russian with German summary)

Sauramo M (1958) Die Geschichte der Ostsee. Annales Academiae Scientiarum. Fennicae, Ser. A III, Geologica-Geographica 51:1–522

Seifert TT, Tauber F, Kayser B (2001) A high resolution spherical grid topography of the Baltic Sea, 2nd edn. Baltic Sea Science Congress 2001: past, present and future – a joint venture. Stockholm Marine Research Centre, Stockholm University, p 298

Shmaenok AI, Sammet EJ, Belenetskaya GA, Verbova IM, Korneeva TL, Ronshin NI, Feigelson MM (1962) Geology of the Narva, Luga and Sisty River lower courses. Complex geological mapping, scale 1:200 000. Manuscript in Northwest Geological Survey of Russia, St. Petersburg (in Russian)

Svensson NO (1991) Late Weichselian and Early Holocene shore displacement in the central Baltic Sea. Quaternary International 9:7–26

US Geological Survey (1996) GTOPO30 Global 30 arc elevation data. EROS http://edc.usgs.gov/products/elevation/gtopo30/gtopo30.html

Uścinowicz S (2006) A relative sea-level curve for the Polish Southern Baltic Sea. Quaternary International 145/146:85–105

Vassiljev J, Saarse L, Miidel A (2005) Simulation of the proglacial lake shore displacement in Estonia. Geological Quarterly 49(3):253–262

Veinbergs I (1979) The Quaternary history of the Baltic. Latvia. In: Gudelis V, Königsson LK (eds) The Quaternary history of the Baltic. Acta Universitatis Upsaliensis, Uppsala, pp 147–157

Chapter 10
Submerged Holocene Wave-Cut Cliffs in the South-eastern Part of the Baltic Sea: Reinterpretation Based on Recent Bathymetrical Data

Vadim Sivkov, Dimitry Dorokhov, and Marina Ulyanova

Abstract As the existing data on the location, number, and age of submerged Holocene wave-cut cliffs (submerged coastlines) in the Gulf of Gdańsk (SE Baltic Sea) are rather conflicting, earlier data were reanalysed and compared with recent information. Digital bathymetric and slope angle maps were developed from the modern 1:25,000, 1:50,000, and 1:100,000 nautical charts. The maximum slope lines were assumed to correspond to wave-cut cliff axes. A total of five axial lines of post-glacial wave-cut cliffs were identified: two dated to the Yoldia Sea (58–45 and 52–40 m), one assigned to the Ancylus Lake (38 m), and two dated to the Littorina Sea (29 and 21 m).

Keywords Holocene · Submerged wave-cut cliffs · Baltic sea · Gulf of Gdańsk · Axial lines · GIS

10.1 Introduction

Primarily as a result of interplay between glacio-isostatic movements and eustatic sea-level changes, evolution of the southern Baltic coast in the Late Pleistocene and Holocene was closely related to the presence of thresholds off Sweden and Denmark and to changes in the relative sea level. To some extent, the coast's evolution was also associated with the geological setting, sediment erosion and accumulation, and climatic oscillations.

Despite ample research on the history of the Baltic Sea (Björck 1995, Eronen 1988, Gudelis 1979, Gudelis and Königsson 1979, Harff et al. 2001, Harff and Meyer 2008, Ignatius et al. 1981, Kvasov 1975, Mörner 1980, Uścinowicz 2003), many questions related to the Late Quaternary events in the area still remain unanswered. The unresolved questions include some key palaeogeographic issues,

V. Sivkov (✉)
Atlantic Branch, P.P. Shirshov Institute of Oceanology, Russian Academy of Sciences (ABIORAS), Kaliningrad, Russia
e-mail: sivkov@kaliningrad.ru

J. Harff et al. (eds.), *The Baltic Sea Basin*, Central and Eastern European Development Studies (CEEDES), DOI 10.1007/978-3-642-17220-5_10,

including the location, number, and age of Holocene wave-cut cliffs (submerged coastlines).

The south-eastern (SE) Baltic is an excellent area to study fossil coastline formation, since it was not affected by tectonic isostasy-associated movements during the Holocene. The zero isobase at various evolutionary stages in the Baltic history has been assumed to be located in the southern Baltic (e.g. Eronen 1988, Ignatius et al. 1981, Uścinowicz 2003).

As opposed to the adjacent slopes, the SE slope of the Gdańsk Basin shows well-preserved traces of the fossil coast. However, we concur with Uścinowicz (2003) in contending that there is an absence of sufficient bathymetric information collected specifically for the purpose of adequate identification of fossil coasts. The erosion-accumulative platforms were assumed to have evolved below the cliff base when the sea level stabilized. Such interpretations were usually inferred from bathymetric charts developed by interpolating depth data from navigational charts. However, erosion surfaces of wave-cut platforms may be covered only by a thin layer of lag deposits or by later sediments, occasionally a few metres thick. In such cases, bathymetry-based spatial correlations of wave-cut platforms are inaccurate. Furthermore, some distinct changes in the bottom profile curve, previously interpreted as typical relics of cliff shores or simply as submerged wave-cut cliffs, are structure-dependent erosion formations. To render the matter still more complex, difficulties are encountered when dating wave-cut platforms, identifying corresponding forms in the modern cliff morphology, and determining their relationship with the mean sea level at the time of their formation.

Moreover, the existing knowledge concerning fossil coasts on the submerged slopes of the SE Baltic is strewn with conflicting information, as all reconstructions were based on fragmentary data collected in the mid-twentieth century during cruises involving very rough geographical positioning and primitive echo-sounders. At that time, before the computer era, the analogue echograms obtained were processed manually, which contributed significantly to the subjectivity of data interpretation.

Thus, the first step in future studies aimed at palaeogeographic reconstructions in the Russian sector of the SE Baltic should involve a survey and update of the existing data on the fossil coast levels, based mainly on depth measurements.

This study was aimed at gaining an insight into the ancient coast levels in the SE Baltic, based on modern verifiable bathymetric information and the use of geoinformation techniques.

The chronology used in this paper is based on the conventional carbon year (^{14}C) BP approach.

10.2 Study Area

The structural geology of the south-eastern Baltic Sea is characterized by a syncline slope of a complex block character. The top of the sedimentary bedrock consists of Upper Cretaceous marls, sandstones, and cemented siltstone. The highest block,

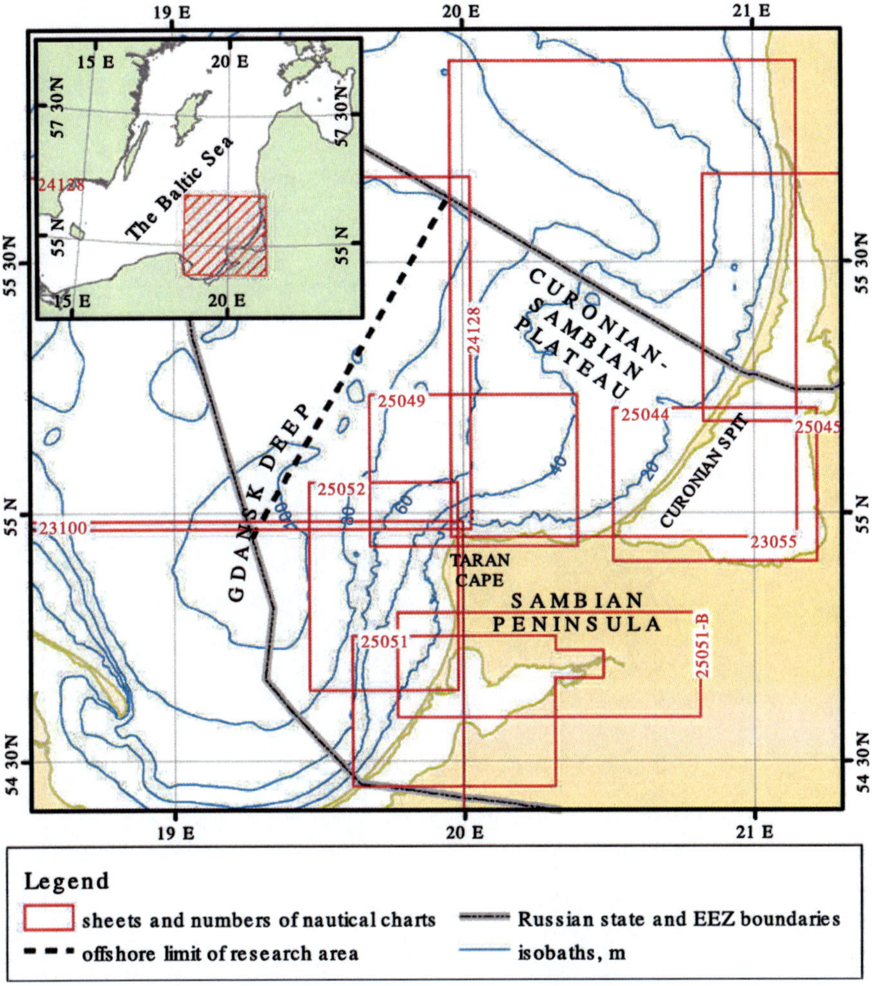

Fig. 10.1 A scheme of bathymetric data acquisition for the area of study (location and number of navigation charts used are shown)

which now forms the Sambian Peninsula in the southern part of the area (Fig. 10.1), features Palaeogene–Neogene, sandy–clayey sediments preserved from glacial washout. North of the Sambian Peninsula lies the extensive Curonian–Sambian Plateau, overlain by a thin cover of Quaternary (moraine) formations.

The coastal area constitutes an inclined surface of erosion–accumulation plain developed in the moraine and, in some places, in the bedrock; it extends offshore to the depths of 30–35 m. The Gdańsk Deep (a cup-shaped depression with depths of up to 115 m) is an accumulation plain covered by unconsolidated late- and post-glacial silts and clays (up to 15–20 m thick). Its slope forms an inclined surface, at some places showing low mounds inherited from the buried moraine topography, and extends over the depth range of 35–80 m.

The Baltic bottom topography features the first- and second-order morphostructural elements, i.e. geological structures formed by endogenous processes. The first-order morphostructure covers the entire central and southern part of the Baltic and is represented by the NW margin of the Russian plate (the Baltic syneclise).

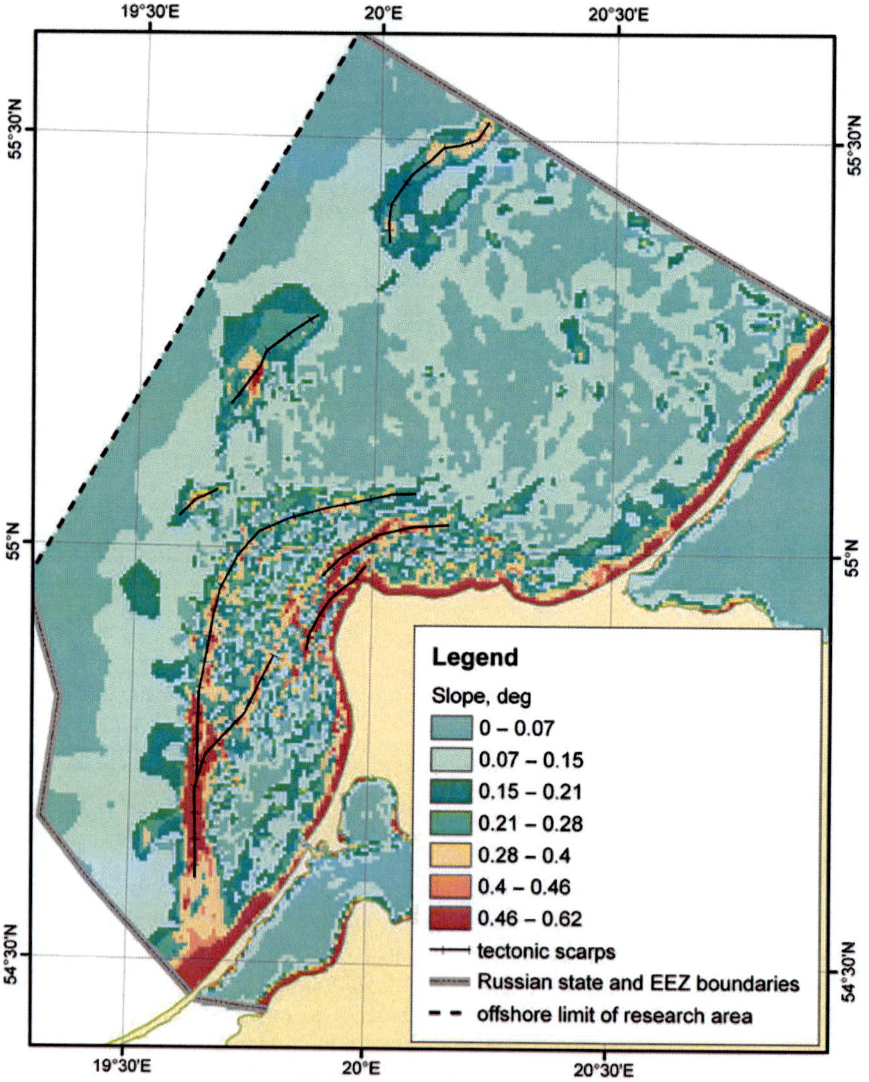

Fig. 10.2 Digital model of the sea bottom slopes. Expected tectonic scarp locations are shown after Gelumbauskaite et al. (1991) (modified)

The topographic cross section of the southern part of the Baltic shows the absence of step- or ridge-like formations. The first-order morphostructure mentioned contains complex second-order structures, represented in the area of study by the Curonian–Sambian Plateau, separated from the central Baltic depression in the west by linear morphostructures (tectonic scarps) which correspond to fault zones (Fig. 10.2). At places, these cliffs coincide with fossil coastlines. Marine accumulation formations are common at the bottom offshore, while erosion–accumulation structures occur in the coastal area. Erosion–accumulation surfaces are usually represented by seaward-inclined plains, the relief of which shows recent and ancient wave-cut cliffs and accumulation formations. These fossil coastlines correspond to different stages of the Baltic history. The Curonian–Sambian Plateau is characterized by having a complex late- and post-glacial erosion–accumulation-terraced topography, with relics of glacial-accumulation undulations. Ancient wave-cut cliffs are extremely common. The SE part of the Gulf of Gdańsk shows the presence of an extensive moraine plain banded by limno-glacial surfaces (Gelumbauskaite et al. 1991).

10.3 Previous Studies

The first attempt to determine the possible location of the former coastlines in the area of study was made by Gudelis in the early 1950s, who extrapolated the shore record of the Lithuanian coast (Gudelis et al. 1977). Subsequently, based on echo-sounding, vibrocoring, and dredging data collected in 1965–1978 by research vessels operated by the Atlantic Branch of P. P. Shirshov Institute of Oceanology (USSR Academy of Sciences) and VNIIMorGeo, spectrograms of fossil coastlines were developed (Gudelis et al. 1977, Gelumbauskaite 1982) and wave-cut terraced surfaces were mapped (Blazhchishin et al. 1982).

The spectrogram described by Gudelis et al. (1977) consisted of six submerged levels dated, from the bottom to the top, to the Yoldia Sea (Y) (on the Kaliningrad coast; 58–63 m); the first Ancylus transgression (Anc1) (36–42 m); the first Littorina transgression (Lit1) (27–33 m); the second Littorina transgression (Lit2) (15–20 m); the second Ancylus transgression (Anc2) (4–10 m); and the third Littorina transgression (Lit3) (2–7 m). As earlier levels were not preserved in the topography, they were drawn as the best estimates. No signs of ancient coastlines could be detected at depths exceeding 70 m, therefore that depth is considered to be the lowest post-glacial Baltic level. The ancient coast slopes were determined by extrapolating the coastal slopes of Lithuania and Latvia, i.e. north of the area of study. The vibrocores, dated by means of biostratigraphic proxies (diatom assemblages and fossil macrofauna remains), made it possible to outline transgression–regression contacts of differently aged sediment complexes originating at different stages of the Baltic history. However, Gudelis et al. (1977) maintained that the geomorphologic and biostratigraphic data they had were not adequate to be used for determining, tracing, and dating the various complexes of submerged fossil coast formations.

The spectrogram obtained by Gelumbauskaite (1982) revealed four levels: Y (52–62 m), Anc2 (40 m), Lit1 (28 m), and Lit2 (18 m). These levels were defined by mapping the axial lines of wave-cut platforms, platform edges, and rear seams, and were marked on transverse profiles of the underwater slope. These so-called distinctive lines are situated at similar depths and reflect the location of coastal formations of a similar age. Unlike in the spectrogram by Gudelis et al. (1977), all the levels are horizontal, except Y. This exception, presented in the deeper part of the lowest submerged ancient coastline, runs southward from 52 to 62 m, and – according to Gelumbauskaite (1982) – was caused by late- and post-glacial glacio-isostatic crust movements which virtually stopped at the Ancylus (Boreal) stage. Deviations from the mean levels identified (i.e. deformations of the distinctive lines) were interpreted as a result of relative vertical crust movements during the Holocene. The resultant schematic map shows the vertical movements to have proceeded at a rate of +1.4 to –0.8 mm/year, which is within the generally accepted, estimated range of tectonic movements in the area during the Holocene. At the same time, Gelumbauskaite (1982) listed some factors that affected the hypsometric level of the fossil coast formations and rendered their identification and mapping difficult. Those factors include various origins of the submerged erosion–exaration cliffs and the adjacent erosion–accumulation surfaces, coastal orography, and the lithological composition of bottom sediments (rocks), which influenced the intensity of lithodynamic processes. Besides, the wave-cut platform edges and their rear seams are usually poorly visible on the cross section of the submerged slope; moreover, it is not always possible to identify wave-cut platform levels on echograms and bathymetric maps.

Blazhchishin et al. (1982) defined the Holocene coast levels by analysing wave-cut platform surface depths. They identified a few underwater levels and one which lies above modern sea level. Unfortunately, their bathymetry and age designations often differ from the information in the text and in the schematic maps, casting doubts on their reliability. The map published by Blazhchishin et al. (1982) features five ancient coast lines: Y (55–62 m), Anc1 (35–42 m), Lit1 (27–32 m), Lit2 (16–20 m), and Lit3 (3–0 m).

According to Blazhchishin et al. (1982), it would be difficult to date the fossil coast formations in the sediment cores with a higher precision because those coasts were frequently shifting during the Holocene, with the transgression-subjected land edge being cut and transformed by the oncoming sea. Two submerged levels were also defined NW of Cape Taran at depths of 70–75 and 90–105 m: the first is visible as an accumulation plain and the other as a well-defined wave-cut cliff, 10 m high, buried by the sediment cover. Blazhchishin et al. (1982) suggest that the two levels are associated with sea-level fluctuations during the Pleistocene, since the lowest Holocene level of the southern Baltic is to be found at 60–65 m (Kvasov 1975).

Kharin (1987) reported detecting further four wave-cut cliffs north of Cape Taran, at the depths of 90–96, 16–24, 12–20, and 4–6 m; in addition, he noted the presence of faintly expressed steps (at 68–69, 62–63, 57–58, 42–45, 37–39, 30–34, 26–27, and 16–18 m depths) in the northern part of the area of study as well as wave-cut ancient cliff relics (at 58, 47, 41, 33, 26, and 19 m depths) in the western part of the Gulf of Gdańsk (Rosa 1970). In Kharin's (1987) opinion, discovery of other ancient coast levels was not improbable.

A schematic map showing the location of fossil coast levels in the eastern part of the Gdańsk Basin was also provided by Boldyrev (1992). The map is sketchy and lacks greatly in detail; the author provided no information as to the database and the mapping technique applied.

Emelyanov and Romanova (2002) reconstructed the fossil coastlines of the Gdańsk Basin by comparing eustatic levels and amplitudes of vertical tectonic movements. Their summary schematics of tectonic uplift or submergence were developed from 45 hypsometric curves of ancient coastlines found in different parts of the Baltic and the eustatic curves for tectonically stable regions of the World's ocean. Superimposition of these schematics over the generalized hypsometric map of the Baltic basin allowed Emelyanov and Romanova (2002) to develop a map of the Baltic bottom topography for any time period relative to the present-day sea level. They used Punning's (1982) eustatic curve of the Baltic to calculate differences between the present and past levels. Emelyanov and Romanova (2002) developed three palaeogeographic schemes: for the Baltic Ice Lake (BIL), the Yoldia Sea, and the Ancylus Lake (10.5–7.8 ka BP). The fossil coastline reconstructed for the BIL stage (10.5 ka BP) rises from the depth of 70 m west of the Sambian Peninsula to 40–30 m on the Curonian–Sambian Plateau and up to the modern sea level near the northern part of the Curonian Spit. During the Yoldia Sea (10–9.5 ka BP), the sea level rose (the relative southern Baltic level) and the coastlines occurred at the depths of 70–40 m west of the Sambian Peninsula and 30–25 m on the Curonian–Sambian Plateau; the Yoldia Sea coast almost coincided again with the present shoreline near the northern part of the Curonian Spit. Emelyanov and Romanova (2002) placed several locations of the Ancylus Lake coast (9–7.8 ka BP) at the present-day depths, also rising gradually, from 40 to 0 m, northwards. Clearly, the Emelyanov and Romanova's (2002) reconstruction differs considerably from the opinions expressed by other authors (Gudelis 1982, Uścinowicz 2003) that the SE Baltic did not experience isostasy-related neotectonic movements during the Holocene.

In his models of sea-level changes and glacio-isostatic rebound in the southern Baltic, Uścinowicz (2003) showed the coastline location west of the Vistula Spit and the Sambian Peninsula. At the early BIL phase (13 ka BP), the coastline was located at the present depths of 35–40 m and shifted up to 20–40 m during the final BIL phase (10.3 ka BP); during the early Yoldia Sea (10 ka BP), the maximum extent of the Ancylus Lake (9.2 ka BP), the early Littorina Sea (7.5 ka BP), and midway through the Littorina Sea (6 ka BP) and at the post-Littorina stage, the coastline occupied the present depths of 60, 20–25, 15–20, and about 0 m and above, respectively.

Thus, a brief review of previous studies on the SE Baltic fossil coastlines above evidences the absence of a common and generally accepted interpretation and demonstrates the necessity of clarifying the situation.

10.4 Materials and Methods

Initial bathymetrical data were processed and the maps were developed using the ArcGIS 9.2 software. The digital bathymetric and slope maps generated are based on the 500-m cell size raster model of data organization (GRID). A raster bathymetric GIS layer, serving as the baseline layer, was generated by reading depths of a total of 6,390 points from nautical charts of scales 1:25,000, 1:50,000, and 1:100,000 (Fig. 10.1). The points were irregularly spaced at 100–2,500 m distances from one another.

The depth data used can be fully relied upon, as the depths were measured, corrected, and mapped strictly in accordance with the procedures adopted by the Hydrographic Survey of the Russian Federation's Ministry of Defence, the only institution licensed to draw and distribute nautical charts of the Russian territorial waters. Under the procedures in question, the depths measured by an echo-sounder are corrected for the following: instrument error; deviation of the actual sound velocity from the reading; echo-sounder vibrator submergence; between-vibrator distance; the vessel's drag in shallow water; seafloor slope; and the offset with respect to the sea surface level. As a result, the standard deviation-based uncertainty factor of the depth reading does not exceed 0.1, 0.3, 0.5, and 1 m for the depth intervals of 0–10, 10–30, 30–50, and 50–100 m, respectively, and the uncertainty for depths larger than 100 m does not exceed 1% of the depth read-out. Important depth points (local minima and maxima, inflection sites) are marked on the nautical charts for a better representation of the seafloor bathymetry. The procedures described are strictly adhered to at all stages of bathymetric surveys and map development.

The initial data set was large enough for us to use the nearest neighbour interpolation technique in the bathymetric model development. The main advantage of this technique lies in the absence of high distortions in the input depth values.

The seafloor slopes (bathymetric surface gradients) were mapped using the ArcGIS 9.2 slope function which calculates the maximum depth gradient between adjacent cells. The output raster slope GIS layer was calculated using degrees as units. The raster bathymetric surface gradient model obtained made it possible to identify areas with both relatively high and relatively low gradients. The low gradients correspond to fossil wave-cut platform surfaces, whereas the high gradients correspond to the maximum inclination of ancient coastal cliffs. Axial lines corresponding to areas with the maximum slopes were drawn on the map as linear vector GIS layers.

The submerged wave-cut cliffs were located with the aid of echograms produced by a single-beam Simrad EA 400 SP echo-sounder (38 and 200 kHz) used during several cruises of RV PROFESSOR SHTOKMAN in 2007–2008. Hydrographic data collected during the cruises with a CTD were used to correct for the sound velocity error. Thus corrected, the echo-sounder readings were processed by the ArcGIS 9.2 software to produce seafloor topography profiles. All the GIS layers and the final maps were developed in the WGS_1984_UTM_Zone_34 N projected coordinate system.

10.5 Results

The seafloor slope map allows to visualize the step-like structure of the underwater slope of the Gdańsk Deep, with the structure being dependent on the fossil coast levels (Fig. 10.2). The maximum slope lines correspond to the wave-cut cliff axes and generally coincide with mid-lines of the cliffs. The exceptions include only

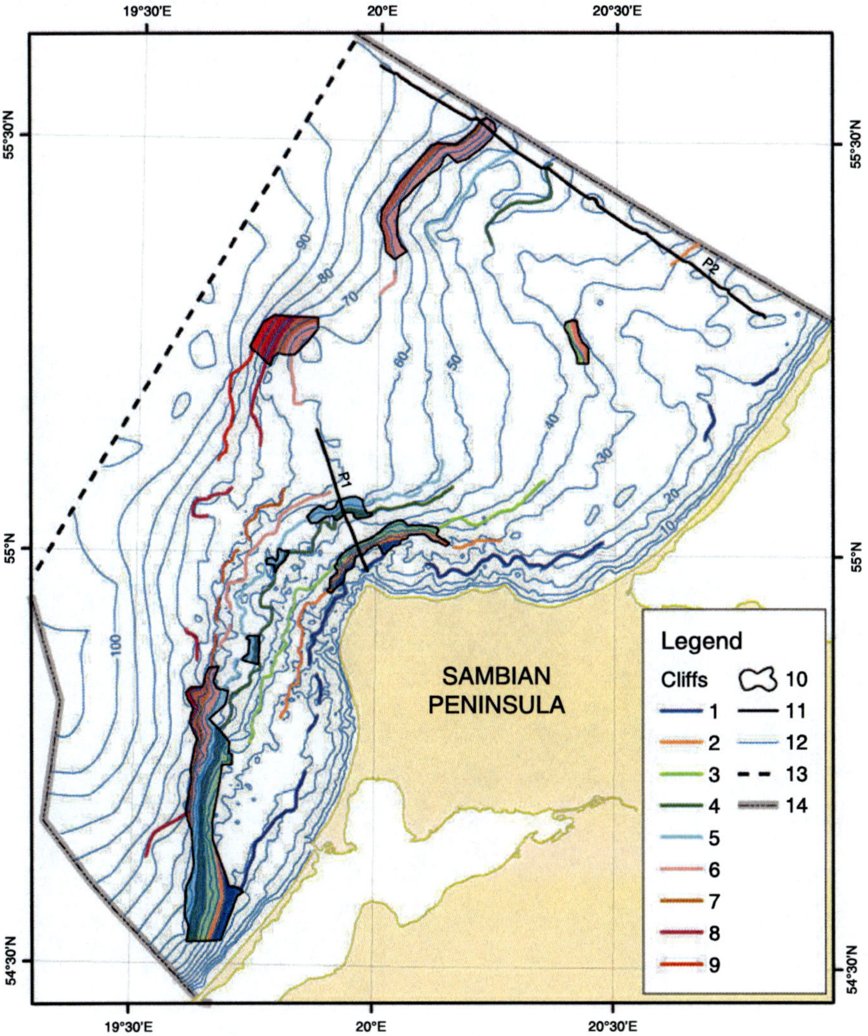

Fig. 10.3 Axial lines of submerged cliffs (average depths (m): *1*, 21; *2*, 29; *3*, 38; *4*, 48; *5*, 53; *6*, 62; *7*, 68; *8*, 76; *9*, 88; *10*, limits of polygenetic submerged cliffs with *colours* corresponding to individual cliffs; *11*, location of certain typical echo-sounder profiles (P1 and P2); *12*, offshore boundary of the study area; *13*, isobaths (m); *14*, Russian state and EEZ borders

the steep wave-cut cliffs in the area of Cape Taran, where the lines were moved upwards from the overlying wave-cut platform edges. Thus, the fossil coast levels as presented in this work are located between the levels defined by the wave-cut platform surface depths (Blazhchishin et al. 1982).

We identified nine axial lines of wave-cut cliffs with average depths of 21, 29, 38, 48, 53, 62, 68, 76, and 88 m (Fig. 10.3). Shallow levels of 15 m and less were not charted. The resolution of our method is too low to be applied to the high-energy coastal zone, where the relic topography is masked by modern lithodynamics. The highest (coastal) level (3–0 m) was therefore beyond the area of study.

Due to the scarcity of initial bathymetric data in certain areas of the bottom, not all coast levels could be laid out. Insufficient spatial resolution forced us to merge the coastlines which are then indicated as polygenic. An example of the polygenic wave-cut cliff and platform is shown in Fig. 10.4, with the cliff occurring at 25–40 m and the adjacent wave-cut platform constituting a seafloor segment between 3,000 and 7,000 m on the horizontal scale. The structures were initially tectonic in origin, with their subsequent evolution being affected by wave-driven erosion during

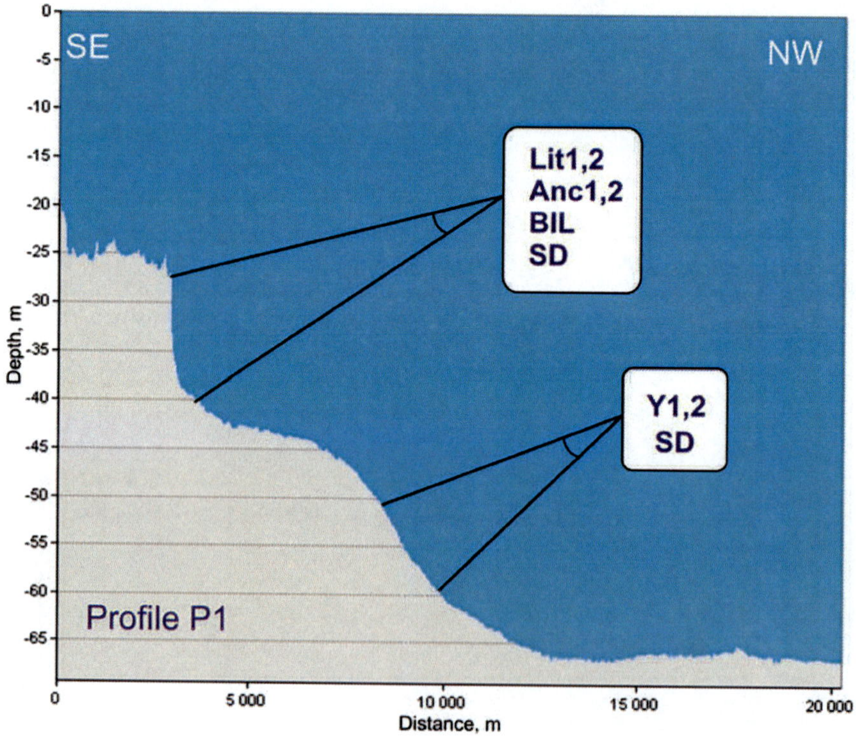

Fig. 10.4 An example of polygenic cliffs north-west of Sambian Peninsula (echo-sounder profile 1 in Fig. 10.3). Insets indicate abbreviations of the Baltic evolution stages; SD, structure-dependent cliffs

the Baltic Ice Lake, Ancylus Lake, and Littorina Sea stages. The linear structures described are overlain by post-glacial coast formations (the second-order morphostructure).

10.6 Discussion

When interpreting the results, we relied on the current knowledge of the southern Baltic history as described by Uścinowicz (2003). Geological evidence, first and foremost the 314 radiocarbon datings of terrestrial and marine sediments, backed up by numerous pollen, diatom, micro- and macro-faunal analyses as well as by the analysis of seismic profiles and sediment cores allowed Uścinowicz (2003) to reconstruct the history of relative sea-level changes as well as vertical crustal movements and changes in the shoreline location during the Late Pleistocene and Holocene in the southern Baltic. Every relative sea-level curve showed the joint effect of eustatic change and vertical crustal movement. The two processes controlled the appearance and disappearance of thresholds and played a prominent role in the relative coast-level history. The contributions of eustasy and isostasy were determined by comparing each relative sea-level curve with the eustatic curve.

The coast evolved at three stages: the Late Pleistocene–Early Holocene, a stage of several rapid and extensive shoreline displacements; the Middle Holocene, initially characterized by a rapid shoreline migration and the coast becoming stabilized between 5 and 6 ka BP; and the Late Holocene, with a narrow range of shoreline displacement and domination of coast levelling processes, resulting in the present location of the shoreline.

By comparing the southern Baltic relative sea-level curves and the eustatic ocean-level curves, Uścinowicz (2003) was able to reconstruct the glacio-isostatic rebound. The restrained rebound phase began c. 17.5 ka BP and lasted until c. 14.0 ka BP. Over that time, the sea level in the southern Baltic rose about 20 m. The basic post-glacial rise proceeded from c. 14.0 to c. 11.0 ka BP, with the southern Baltic area being raised by about 85 m. The rise proceeded at a maximum rate of about 45 mm/year around 12.2–12.4 ka BP. The residual rise began c. 11.0 ka BP and terminated c. 9.0–9.2 ka BP, the crust rising about 15 m. The southern Baltic rate of rising c. 10.0 ka BP slowed down to c. 5.5 mm/year and the uplift stopped c. 9.2–9.0 ka BP. The rapid termination of post-glacial uplift within c. 11.0–9.0 ka BP was probably caused by the restraining effect of hydro- and sediment isostasy. Around c. 4.0 ka BP, the Earth's crust regained its equilibrium.

However, the actual history of Holocene sea-level changes, particularly in the Atlantic, Subboreal, and Subatlantic, was undoubtedly more complex than that emerging from a smoothed-out relative sea-level curve. In the Late Holocene in particular, against the background of a slight and slow rise of the mean sea level, regional eustatic fluctuations associated with climatic changes could have substantially affected coastline evolution (Behre 2007). Neotectonic movements were another factor not associated with glacio-isostasy that produced regional differences in relative sea-level changes, and hence affected the shoreline evolution at that time (Dodonov et al. 1976, Sliaupa 2002).

The present understanding of sea-level changes, as discussed above, allows the fossil coast levels revealed to be assigned to the periods indicated below (see Fig. 10.3).

Fragments of the four deepest submerged wave-cut cliffs (62, 68, 76, and 88 m) are probably correlated with post-glacial fluctuations of the water level in the near-glacier lakes, or were caused by glacial exaration, or else their origin is structure dependent. A special study is required to resolve this question. At some sites, separation of wave-cut cliff locations was not possible. These wave-cut cliffs are most probably best described as polygenic.

The range of the first BIL regression (11.2–11.0 ka BP, Allerød–Younger Dryas) was estimated from the River Vistula progradational deltaic structures as well as from barrier structures on the western slopes of the Gdańsk Basin (Uścinowicz 2003). The two set of structures are found down to the depth of about 65 m, which speaks of the sea level being located lower when the structures were being developed. Subsequently, the BIL southern coast experienced a rapid transgression with the maximum occurring c. 10.3 ka BP.

Consequently, the breaks visible at the depths of 62 and 68 m may be inferred to have been related to the maximum BIL regression. However, the absence of any northward rise of the lines suggests that their origin is independent of the late- and post-glacial sea-level fluctuations and isostatic effect. Thus, they are probably structure-dependent breaks.

In our opinion, the next submerged coastline (at the average depth of 53 m), which occurs at the depth range of 58–55 m near the Sambian Peninsula and at 50–45 m on the Curonian–Sambian Plateau, corresponds to the beginning of the Yoldia–Ancylus transgression (Y1). Its depths correspond to the location of the lowest relative sea level (Gudelis 1979, Uścinowicz 2003) reached after the rapid BIL regression and at the beginning of the Yoldia stage. The isostatic depth difference matches Gudelis's (1979, 1982) estimates of the maximum at 10–12 m during the late- and post-glacial periods.

The next fossil coastline (at the average depth of 48 m), discovered near the Sambian Peninsula at the depth of 52–50 m and at 45–40 m on the Curonian Rise, corresponds to the Yoldia Sea stage (Y2) as well.

The 38-m-level coastline identified on the submerged slope of the Sambian Peninsula corresponds to the first Ancylus shoreline (Anc1), with the 29-m level corresponding to the initial stage of the Littorina transgression (Lit1). The wave-cut cliffs at those depths had formed earlier during the BIL transgression (BIL1, Allerød) and Anc2. The middle levels of 38 and 29 m probably merge on the Curonian–Sambian Plateau because of the difference between isostatic rise rates.

Finally, the 21-m middle level corresponds to the initial stage of the Littorina transgression.

Unfortunately, our reinterpretation has not eliminated the major discrepancies between wave-cut cliff locations put forth by different authors. This can be illustrated by comparing locations of the Y wave-cut cliff: while Gudelis et al.

(1977) placed it at 58–63 m, Gelumbauskaite (1982), Blazhchishin et al. (1982), and Uścinowicz (2003) placed it at 52–62, 55–62, and 27–52 m, respectively, with our interpretation being indicative of 40–58 m.

The discrepancies arose due to various reasons: one may invoke different origin and accuracy of the data (Gelumbauskaite 1982) and/or different, incomparable distinguishing lines (axial lines of the wave-cut cliff or platform; Blazhchishin et al. 1982). In addition, the authors quoted used different methods: while Gudelis et al. (1977) used extrapolation and biostratigraphy, the relative sea levels shown in Uścinowicz's (2003) figs. 6, 8, and 35 were determined by interpolating ^{14}C datings at reference sites distributed along the entire southern Baltic coast, from the Vistula Spit to the Pomeranian Bay (it has to be mentioned that this method, while providing only approximate shore location, may involve errors of up to 10 m). In addition, Uścinowicz's (2003) palaeoreconstructions account for a possibility of local deviations in the fossil sea level from the regional averaged curve (see figs. 42–47 in Uścinowicz 2003).

The resulting spatial layout of submerged coastline locations is certainly tentative and its details are far from accurate (Fig. 10.5), as illustrated by the echo-sounder profile 2 crossing the Curonian–Sambian Plateau. The locally increased seafloor inclination is interpreted as a signal of ancient wave-cut cliffs. However, corroboration of this interpretation requires additional surveys.

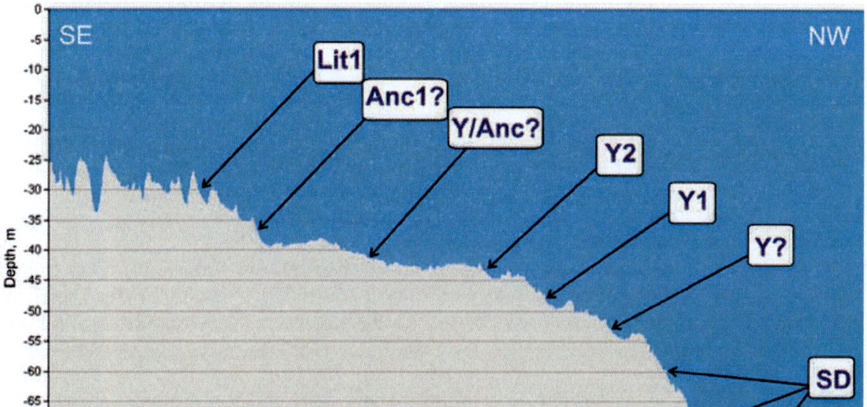

Fig. 10.5 Submerged cliffs on the Curonian–Sambian Plateau (echo-sounder profile 2 in Fig. 10.3). Insets indicate abbreviations of the Baltic evolution stages; SD, structure-dependent cliffs (*question marks* indicates submerged cliffs missing in the slope model)

10.7 Conclusions

The present state of knowledge on the fossil coastline levels in the Kaliningrad region of the southern Baltic is far from complete and deviates largely from what is known in other parts of the Baltic. No multibeam survey has ever been carried out in the area. This study's attempt at collating and assessing the existing information, using the available cartographic data and GIS techniques, showed the studies carried out in the 1960s and 1970s to have brought about conflicting results. However, drawing upon the recent advances in understanding the development of the Baltic as a whole and that of the sea's southern coast as put forth by Uścinowicz (2003), and using digital cartographic techniques for processing recent bathymetric data, we identified five submerged post-glacial shores the development of which was associated with the Yoldia-Ancylus and Littorina transgressions. When those coasts were being formed, the Earth's crust was subjected to glacio-isostatic rise, for which reason the coastlines Y1 (58–45 m) and Y2 (52–40 m) are elevated to the north. The younger Ancylus (38 m) and Littorina (29 and 21 m) shores are quasi-horizontal because the crust had virtually ceased to rise when they were forming.

Thus, to understand the climate-driven formation of the southern Baltic coast, vertical crustal movements, regarded as local scale neotectonic processes, have to be taken into account.

The results obtained so far are just the first step towards resolving the remaining open questions and addressing the misunderstandings which have accumulated over the years with respect to the problem of submerged Baltic coasts. The study is partly financed by RFBR 11-05-01093-a.

References

Behre K-E (2007) A new Holocene sea-level curve for the southern North Sea. Boreas 36:82–102

Björck S (1995) A review of the history of the Baltic Sea, 13.0–8.0 ka BP. Quaternary International 27:19–40

Blazhchishin AI, Boldyrev VL, Efimov AN, Timofeev IA (1982) Ancient coastal levels and formations in the south-eastern Baltic Sea. Baltica 7:57–64 (in Russian)

Boldyrev VL (1992) Forming, development and modern dynamics of Kaliningrad coast of the Baltic Sea. In: Orviku K (ed) Main regularities and tendencies of the Baltic Sea shoreline migration during the past 100 years. Estonian Academy of Science, Institute of Geology, Tallinn (in Russian)

Dodonov AE, Namestnikov YuG, Yakusheva AF (1976) The latest neotectonics of the south-eastern part of the Baltic syneclise. Moscow State University Publisher, Moscow (in Russian)

Emelyanov EM, Romanova EA (2002) Paleogeography of the Gdansk Basin in post-glacial period and bottom sediments. In: Emelyanov EM (ed) Geology of the Gdansk Basin, Baltic Sea. Yantarny Skaz, Kaliningrad

Eronen M (1988) A scrutiny of the Late Quaternary history of the Baltic Sea. In: Winterhalter B (ed) The Baltic Sea. Geological Survey of Finland, Special Papers, 11–18

Gelumbauskaite Zh (1982) Methods and results study of ancient coastal levels deformations of the SE Baltic Sea. Baltica 7:95–104

Gelumbauskaite ZhA, Litvin VM, Malkov BI, Moskalenko PE, Jushkevichs VV (1991) Geomorphology. In: Grigelis AA (ed) Geology and geomorphology of the Baltic Sea. Explanatory note of the geological maps, scale 1:500000. Nedra, Leningrad

Gudelis VK (1979) Lithuania. In: Gudelis VK, Konigsson LK (eds) The Quaternary history of the Baltic. Acta Univ Ups Symp Univ Ups Ann Quing Cel, 1. Almqvist & Wiksell International, Stockholm, Sweden

Gudelis VK, Lukoshevichus LS, Klejmenova GI, Vishnevskaya EM (1977) Geomorphology and late-after-glacial bottom sediments of the south-eastern Baltic. Baltica 6:245–256 (in Russian)

Gudelis VK, Königsson L-K (1979) The Quaternary history of the Baltic. Acta Univ Ups Symp Univ Ups Ann Quing Cel, 1. Almqvist & Wiksell International, Stockholm, Sweden

Gudelis VK (1982) The newest and recent movements of the earth crust at the south-eastern coast of the Baltic Sea. Baltica 7:179–186 (in Russian)

Harff J, Frischbutter A, Lampe R, Meyer M (2001) Sea level change in the Baltic Sea: interpretation of climatic and geological processes. In: Gerhard LC, Harrison WE, Hanson BM (eds) Geological perspectives of the global climate change. AAPG – Studies in Geology 47:231–250

Harff J, Meyer M (2008) Changing sea level at sinking coasts – competition between climate change and geological processes. Polish Geological Institute Paper 23:39–44

Ignatius H, Axberg S, Niemisto L, Winterhalter B (1981) Quaternary geology of the Baltic Sea. In: Voipio A (ed) The Baltic Sea. Elsevier Oceanogr Series 3. Elsevier, Amsterdam, London

Kharin GS (1987) Ancient coastlines and cliffs at the bottom of the Gdansk Gulf and central Baltic. In: Emelyanov EM, Vypyh K (eds) Sedimentary processes in the Gdansk Basin (the Baltic Sea). PP Shirshov Institute of Oceanology AS USSR (Atlantic Branch), Moscow (in Russian)

Kvasov DD (1975) Late-Quaternary history of large lakes and inner seas of the Eastern Europe. Nauka, Leningrad (in Russian)

Mörner NA (1980) Late Quaternary sea-level changes in north-western Europe: a synthesis. Geologiska Föreningens i Stockholm Förhandlingar 100(4):381–400

Punning Ya MK (1982) Eustatic oscillations of the Baltic Sea level during Holocene. In: Kaplin PA, Klige RK, Chepalyga AL (eds) Water level oscillations of the seas and oceans during 15 000 years. Nauka, Moskva, pp 134–143 (in Russian)

Rosa B (1970) Einige Probleme der Geomorphologie, Paläogeographie und Neotektonik des südbaltischen Küstenraumes. Baltica 4:197–210

Sliaupa S (2002) Influence of the last glaciation to stress field and tectonic activity of faults of the Baltic region. Geologija 39:12–24

Uścinowicz S (2003) Relative sea level changes, glacio-isostatic rebound and shoreline displacement in the southern Baltic. Polish Geological Institute Special Paper 10:1–79

Chapter 11
Drowned Forests in the Gulf of Gdańsk (Southern Baltic) as an Indicator of the Holocene Shoreline Changes

Szymon Uścinowicz, Grażyna Miotk-Szpiganowicz, Marek Krąpiec, Małgorzata Witak, Jan Harff, Harald Lübke, and Franz Tauber

Abstract This chapter presents a newly discovered locality of tree stumps occurring in situ at the bottom of the Gulf of Gdańsk. It focuses in particular on the age of the stumps and characterization of the palaeoenvironment, i.e. the nature of the plant communities in which the trees grew and also on their position in relation to the palaeo sea level. Tree stumps occurring in situ on the sea floor along with peat deposits are the most reliable indicators of sea level changes. The site is located about 6–7 km NE of the entrance to the Gdańsk harbour, in water depth of 16–17 m. The thickness of marine sand at the site is from a few to a dozen centimetres. Below the sand, gyttja with peat intercalations and wood fragments occur. Sixteen fragments of alder trunks and one oak trunk's fragment were extracted. Radiocarbon ages of the tree trunk fragments are 7,920 ± 50 BP, 7,940 ± 40 BP, 7,960 ± 40 BP and 8,000 ± 50 BP. The age of gyttja, according to pollen analyses, is of early Atlantic period. The characteristic forest composition of that time was the broad deciduous forest with oak (*Quercus*), elm (*Ulmus*) and lime (*Tilia*). The climate was characterized by good thermal and moisture conditions, which is confirmed by the presence of pollen grains of mistletoe (*Viscum*) and ivy (*Hedera*). The obtained data about the time of accumulation of the investigated sediments indicate that the sea level at that time was about 19–20 m lower than at present.

Keywords Shoreline displacement · Drowned forest · ^{14}C dating · Pollen and diatom analyses · Middle Holocene · Baltic Sea · Gulf of Gdańsk

11.1 Introduction

Stumps of trees occurring in situ on the sea bed are known from many places in the world. The first scientific reports about "sunken forests" appeared in the nineteenth century in Great Britain, where many of them were discovered in coastal areas (e.g.

S. Uścinowicz (✉)
Polish Geological Institute, National Research Institute, Gdańsk, Poland
e-mail: szymon.uscinowicz@pgi.gov.pl

J. Harff et al. (eds.), *The Baltic Sea Basin*, Central and Eastern European Development Studies (CEEDES), DOI 10.1007/978-3-642-17220-5_11,

James 1847, Fisher 1862). In the Baltic Sea region, tree stumps rooted in the sea bed were until recently known only from the coastal waters of Denmark and Germany (e.g. Christensen 1995, Lampe 2005, Lampe et al. 2005, Curry 2006, Tauber 2007). In the southwestern part of the Baltic Sea, subboreal tree stumps occur not deeper than 1 m below present sea level (Lampe 2005). Much more common are underwater sites of tree stumps from the Atlantic period drowned in German Baltic waters at depths between 2 and 14 m below present sea level (Lampe et al. 2005). The tree stumps of pine (*Pinus*) in situ have also been identified offshore in Lithuanian waters (Damusyte et al. 2004, Damusyte 2006). Conventional radiocarbon ages (^{14}C) for these trees are 9,160 ± 60 and 6,930 ± 130 years BP and the water depths 27.0 and 14.5 m, respectively.

In Poland, numerous localities of tree stumps on beaches between Rowy and Łeba have been known for many years (e.g. Tobolski et al. 1981, Krąpiec and Florek 2005). The ages of the stumps examined there (oak, ash, alder and pine) ranged from 4,610 to 210 years BP. The first tree stumps found in situ in the Polish coastal zone of the Baltic Sea were reported from Puck Lagoon. The wood of a stump excavated from a depth of about 3 m was dated to 9,370 ± 90 BP (Gd–7938) (unpublished data). The peat deposits at the bottom of Puck Lagoon are of a similar age, having been formed in the Preboreal and Boreal periods. Puck Lagoon itself is much younger, existing since the end of the Atlantic period (e.g. Kramarska et al. 1995, Uścinowicz and Miotk-Szpiganowicz 2003). During field work carried out by the Polish Geological Institute in 2006 in the Vistula Lagoon (Polish: *Zalew Wiślany*; German: *Frisches Haff*; Russian: *Kaliningradskiy Zaliv*), tree stumps rooted in peat have been recognized around the site with coordinates 54°24.03′N and 19°42.61′E (some 5 km NE of Frombork). The alder stumps rooted in subboreal peat at a depth of 2 m were dated to 4,770 ± 35 BP (Poz-15115) and 3,295 ± 35 BP (Poz-1516) (Łęczyński et al. 2007).

11.2 Area, Scope and Methods of Study

The Gulf of Gdańsk is the southernmost part of the Gdańsk Basin (southern Baltic Sea). The external sea boundary of the Gulf of Gdańsk is conventionally taken to be a straight line connecting the promontories of Rozewie on the Polish coast and Taran on the Sambian Peninsula (Russian exclave Kaliningrad). In the extreme western part of the Gulf of Gdańsk lies the Puck Bay, protected from the more open waters by about 32-km-long Hel Peninsula. The southeastern part of the Gulf of Gdańsk is fringed by about 55-km-long Vistula Spit, which forms the Vistula Lagoon. The area of the Gulf of Gdańsk is about 5,000 km^2. The maximum water depth is 107 m in the northern part of the Gulf, in the Gdańsk Deep.

The sea bed relief of the Gulf of Gdańsk is very diverse. In depths of 0–10 m, the near-shore slope is characterized by a system of bars. Outside the slope at a distance of 20–25 km from the shore and to depths of 30–40 m, the sea bed relief is flat or slightly hummocky with elevations of 0.5–5.3 m, locally up to 8 m with slope inclines from 0.5° to 1.5°, locally to 2.5°. The shallows are separated from the deeper parts of the gulf by slopes 30–35 m deep and inclined 6–12° in some places. The sea bed in the deep water part of the Gulf of Gdańsk is completely flat. Hummocky glacial relief is cropping out only very locally. In the southern part of the Gulf of Gdańsk, late Glacial and early and middle Holocene deltaic deposits occur at depths up to 60 m (Ejtminowicz 1982, Uścinowicz and Zachowicz 1993). The drowned part of the Vistula Delta, with an area of about 700 km^2, was formed during the early and middle Holocene when the water level was lower than at present.

The site of tree stumps and fallen trunks occurring in situ at the bottom of the Gulf of Gdańsk was discovered by the Polish Central Maritime Museum in 2006. In September 2007, the site was investigated by the German expedition on board of r/v "Professor Albrecht Penck" within the frame of the "SINCOS" project (e.g. Harff et al. 2005), in cooperation with Polish Geological Institute – National Research Institute.

The site is located about 6–7 km NE of the entrance to the Gdańsk harbour, in water depths of 16–17 m on the submerged part of the Vistula Delta (Fig. 11.1).

The first step of the investigation was a sidescan-sonar survey of the area with a size about 1.3 × 1.8 km. A dual-frequency digital sidescan sonar with a towed

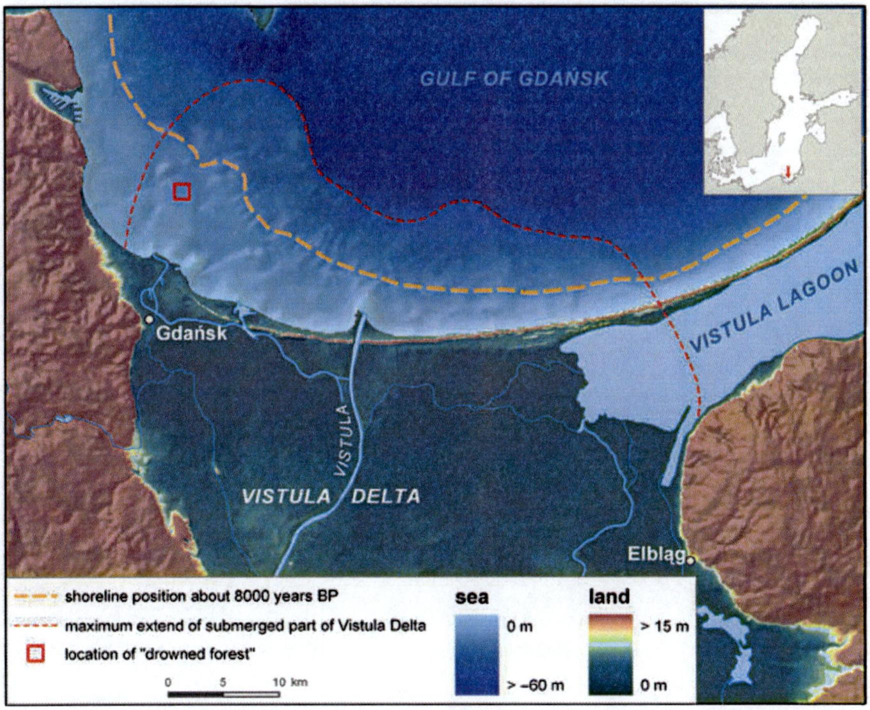

Fig. 11.1 Location of the area and reconstructed shoreline position at the beginning of the Atlantic period and maximum extent of the Vistula Delta

sidescan fish EG&G DF-1000 was used, emitting acoutic pulses with the frequencies 100 and 384 kHz. The slant range was 75 m on both sides. The highest resolution of sidescan images at the seabed amounts to about 20 × 30 cm per pixel, depending on the ship velocity. The ship antenna position was measured with differential GPS, but due to the changing length of the cable to the towfish and the movement of the fish relative to the ship, the accuracy of absolute position was estimated to be about 20 m. After completing the survey, the recorded data were processed and inspected visually. A sidescan mosaic of the whole area (Fig. 11.2) and detailed sidescan images of selected places (such as in Fig. 11.3) were created. Sites with promising textures of the seabed (Fig. 11.3) were then inspected by a remotely operated vehicle with video camera, and finally, scuba divers investigated the selected sites on the seabed (Fig. 11.4).

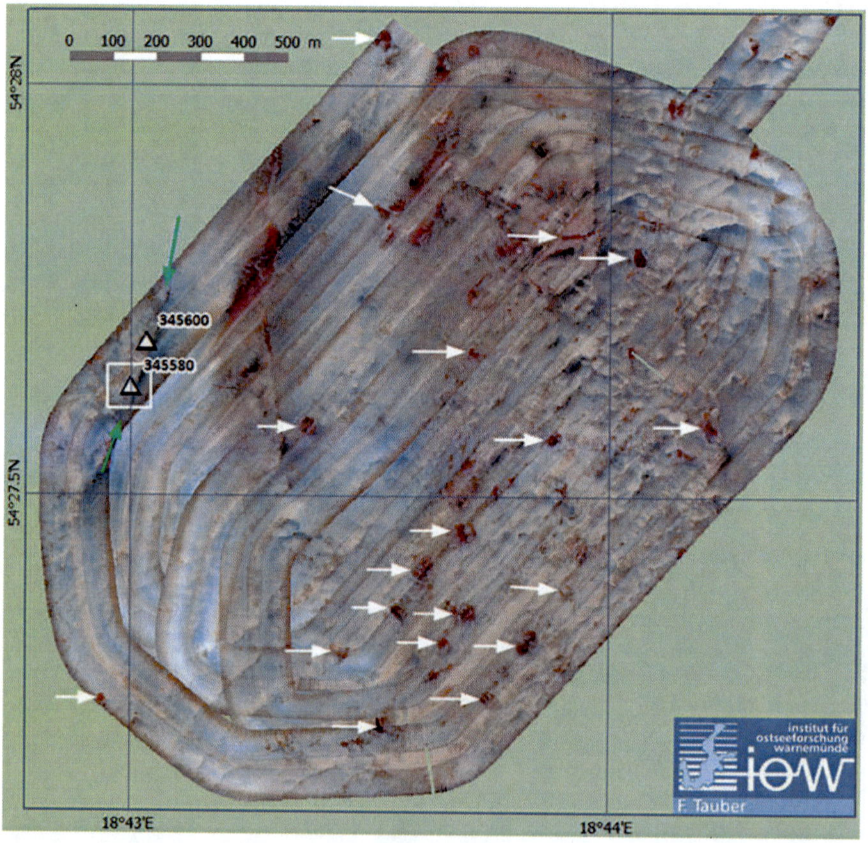

Fig. 11.2 Sidescan-sonar mosaic of the area of investigation in the Gulf of Gdańsk. *White arrows* point to dumped material (*brown patches*). The area with fossil tree trunks extends between the two *green arrows*. The *white square* is shown with higher resolution in Fig. 11.3. *Bluish patches* are coarse sand ripple fields and *bright grey patches* consist of fine sand. *Triangles* with *numbers* are locations explored by scuba divers

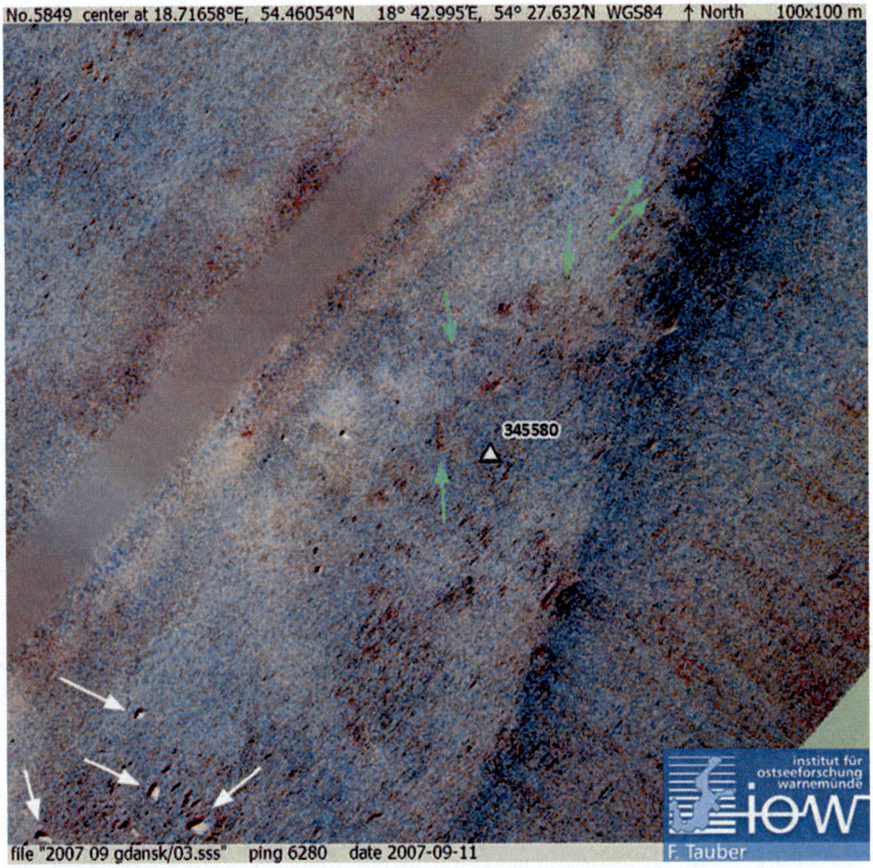

Fig. 11.3 A sidescan-sonar image of the site 345580 showing 100 × 100 m of seabed. *Green arrows* point to some of the tree trunks covered almost completely with a thin sand layer. *White arrows* point to dumped debris

Scientists from the Polish Central Maritime Museum extracted two tree stumps, and during the German expedition, 17 fragments of trunks were extracted. The German divers have also taken two short (30–35 cm) sediment cores.

Dendrological analyses were carried out on 17 wood samples. Four samples of wood were radiocarbon-dated by AMS at the Poznan Radiocarbon Laboratory. Fifteen samples of gyttja and peat from two cores were palynologically analysed, and 10 samples from two cores were analysed for diatom assembly composition, according to standard methods (Faegri and Iversen 1975, Berglund 1985, Battarbee 1986).

Sonar records were registered and analysed by Franz Tauber. Underwater work was carried out by Harald Lübke and a German team of scuba divers. The palynological analyses were performed by Grażyna Miotk-Szpiganowicz, the dendrological analysis by Marek Krąpiec and diatomological analyses by Małgorzata Witak.

Fig. 11.4 Scuba diver sampling a tree trunk

11.3 Results

The water depth in the area of "drowned forest" is between 16 and 17 m, and the sea bed is almost completely flat (Figs. 11.2 and 11.3), covered by fine and medium sand with current-wave ripple marks. During the study, the distance between the crests was about 10–20 cm and their height 1–2 cm. The thickness of sand varied from a few to a dozen centimetres. Below the sand, gyttja with many plant remains, wood fragments and intercalations of peat occur. Over an area of a few dozen hectares, many fallen tree trunks and stumps rooted in situ in gyttja were observed and sampled (Fig. 11.4).

According to dendrological data of 17 samples of wood, mainly alder (*Alnus* sp.) trunks were found, but only one sample of oak (*Quercus* sp.) trunk. The wood samples represent mainly relatively young trees. The oldest one was 54 years old. The ages of the fallen trees are as follows: three trunks – 50–54 years, five trunks – 40–50 years, four trunks – 30–40 years, five trunks – <30 years old.

The investigated tree trunks did not die at the same time. According to dendrograms of three trunks (Fig. 11.5), two of them died at the same time, whereas one tree fell several years earlier.

Four samples of wood among the three samples of alder and one sample of oak wood were dated by ^{14}C method (Table 11.1).

Radiocarbon dating indicates that the trees did grow between 8,000 and 7,920 BP (conventional radiocarbon years), i.e. in the early Atlantic period. According to

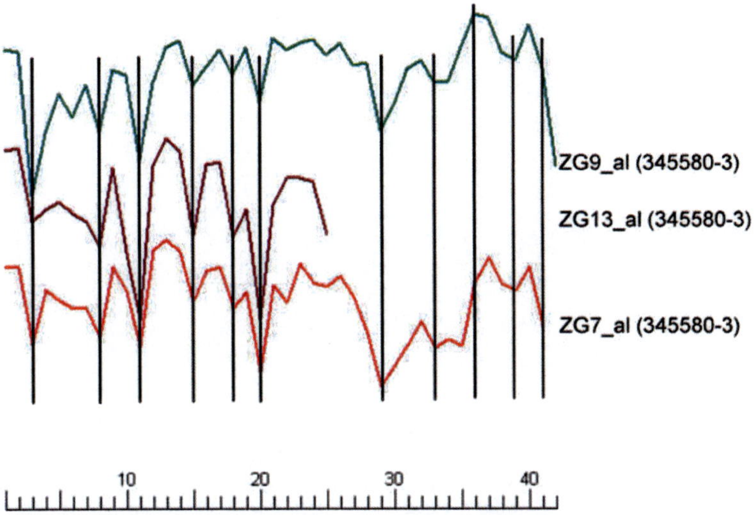

Fig. 11.5 Dendrogram of trunk samples (*vertical lines*: annual growth rings)

calendar years, this corresponds to a time span between 9,020 and 8,600 BP with 95.4% probability.

The pollen analyses of gyttja and peat sediments allow to distinguish five local pollen assemblage zones (LPAZ) (Figs. 11.6 and 11.7). These are *Pinus–Betula*, *Alnus* (core 345580-11) or *Pinus* (core 345600), *Ulmus–Salix*, *Pinus–Quercus* and *Alnus–Ulmus* L PAZ.

These two diagrams show that the development of vegetation at the two sites was generally similar, which means that the deposition of the investigated sediments took place at the same time. The presence of oak (*Quercus*), elm (*Ulmus*) and lime (*Tilia*) in the forest composition proves the existence of demanding, broad deciduous forest, typical for the Atlantic period. This type of forest association needs habitats with good thermal and moisture conditions, as well as the presence of non-leached soils. The existence of the warm climate is confirmed by the occurrence of pollen grains of mistletoe (*Viscum*) and ivy (*Hedera*) (Fig. 11.6). The most visible difference between these two investigated sites is connected with the dominance of alder (*Alnus*) forest association (Fig. 11.6) or pine (*Pinus*) forest (Fig. 11.7). This shows the existence of different hydrological conditions of local habitats. The association with alder (*Alnus*) did occupy wet, frequently flooded habitats rich in nutrients, while pine (*Pinus*) forest grows mainly on the dry, sandy areas.

According to pollen analyses, the gyttja and peat were deposited during the early Atlantic period. The palynological records are therefore in good agreement with the radiocarbon ages of tree trunks from this site.

The diatomological analyses show a deficiency of microflora in most of the samples. In two intervals (21 and 25 cm) of core 345600, the presence of only single

Table 11.1 Radiocarbon age of tree trunks (atmospheric data from Reimer et al. 2004)

Sample no.	Coordinates (Geoid: WGS 84) φ	λ	Tree species	^{14}C conventional year BP	Cal. year BP; 68.2% probability	Cal. year BP; 95.4% probability	Lab. code
345580-2	54°27.631′	18°43.000′	Alder	7,920±50	8,970–8,630	8,980–8,600	Poz-22484
ZG-d1	54°27.668′	18°43.043′	Oak	7,940±40	8,980–8,640	8,990–8,630	Poz-22483
345600-6	54°27.686′	18°43.033′	Alder	7,960±40	8,980–8,720	9,000–8,640	Poz-22485
ZG-0/1	54°27.886′	18°43.022′	Alder	8,000±50	9,000–8,770	9,020–8,650	Poz-15105

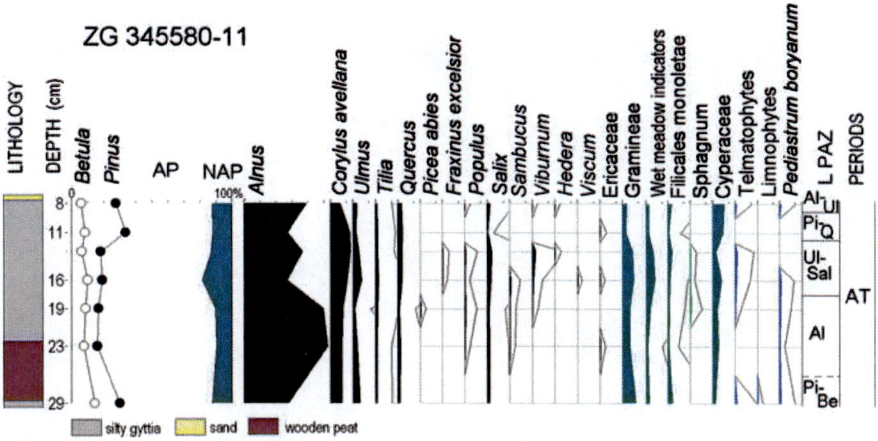

Fig. 11.6 Simplified pollen diagram of core 345580-11

diatom valves was observed. Their very low frequency and the poor state of preservation indicate the intensification of chemical dissolution processes and mechanical fragmentation. The diatom community in the sample at 11–16 cm in core 345580 was also poorly preserved, but quite abundant and represented by diverse taxons (Fig. 11.8).

Two ecological groups among the predominant benthic freshwater diatoms can be distinguished:

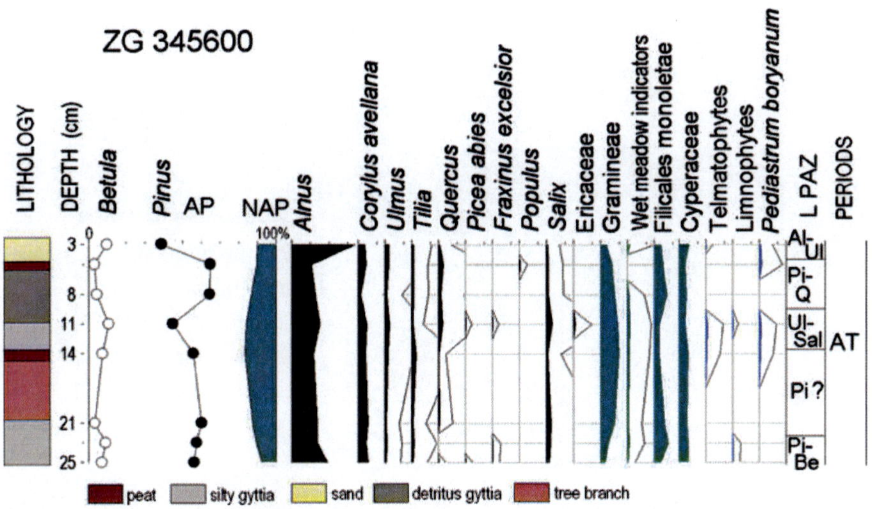

Fig. 11.7 Simplified pollen diagram of core 345600

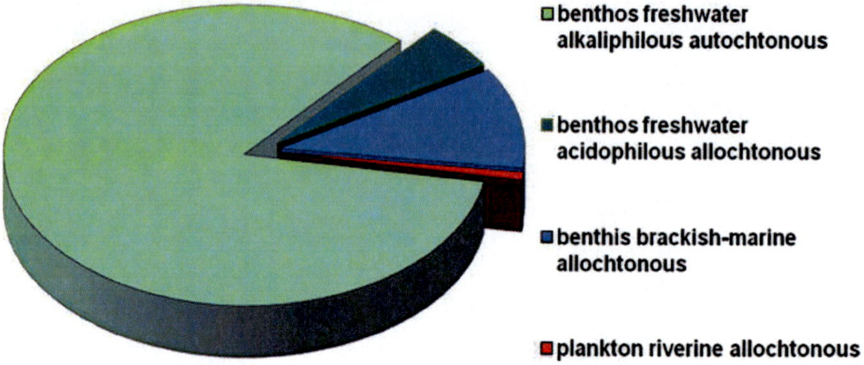

Fig. 11.8 The distribution of main ecological groups in core ZG 345580-11 (11–16 cm)

1. Allochthonous diatoms typical for acid dystrophic waters, represented by *Eunotia bilunaris*, *Tabellaria flocculosa*, and *Pinnularia* spp. – the most common species.
2. Autochthonous diatoms inhabiting waters rich in dissolved nutrients and organic matter, with low oxygen content. This group is represented by euryhaline forms *Amphora copulata*, *Cocconeis neodiminuta*, *C. placentula* var. *lineata*, *C. placentula* var. *euglypta*, *Cavinula scutelloides*, *Cymbella cistula*, *Diatoma vulgaris*, *Fragilaria martyi*, *Gomphonema acuminatum*, *G. angustatum*, *G. truncatum*, *Navicula oblonga*, *Pseudostaurosira brevistriata*, *Rhopalodia gibba*, *Staurosira construens* and *Ulnaria ulna.* Except for the latter species, these taxa have high potential of preservation.

Moreover, benthic allochthonous forms preferring more saline water were also recorded (Fig. 11.9).

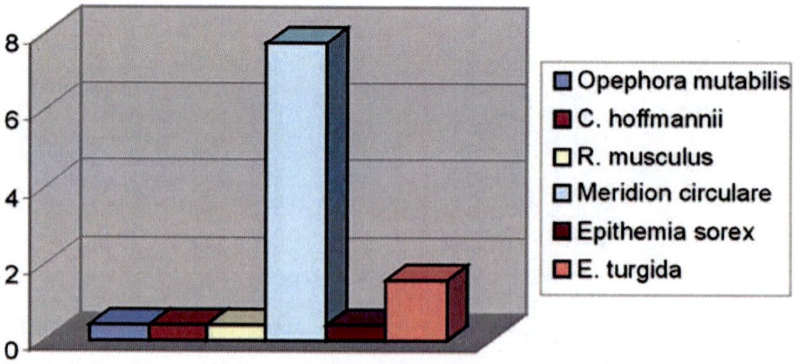

Fig. 11.9 The frequency of main diatoms belonging to a group of brackish-marine benthos in core ZG 345580-11 (11–16 cm)

Among oligohalobous, halophilous *Epithemia turgida*, *E. sorex* and *Meridion circulare* were noted. In addition, *Cocconeis hoffmannii* belonging to euhalobous as well as mesohalobous taxa *Opephora mutabilis* and *Rhopalodia musculus* were observed.

Planktonic diatoms were rarely represented by single valves of riverine species *Aulacoseira granulata*.

11.4 Discussion

Results of palynological studies as well as the lithological description of the sediments show that at the beginning of the Atlantic period most of the Vistula Delta was a swampy area (Figs. 11.6 and 11.7). Habitats with poor drainage prevailed and caused the domination of alder (*Alnus*) associations with hazel (*Corylus*), elm (*Ulmus*), willow (*Salix*), ash (*Fraxinus*), poplar (*Populus*) and locally also elderberry (*Sambucus*) and viburnum (*Viburnum*). Alder forests are usually present in flooded, fertilized areas where the water level remains high for longer time spans. The presence of diatoms typical of shallow, eutrophic and poorly oxygenated water (Fig. 11.8) supports the existence of an environment mentioned above. The admixture of riverine plankton and acidophilous benthos species could either result from riverine floods or be redeposited together with fluvial sediments and peat bogs. Drier habitats in the close vicinity of the delta were probably occupied by deciduous mixed forests with oak (*Quercus*), lime (*Tilia*), elm (*Ulmus*) and hazel (*Corylus*). The pine-oak forest, most probably, grew on the driest areas, on the dunes or the barrier which separated the low-lying delta from the sea. A similar composition of the forest association is also known from the dunes of Łeba Barrier on the middle Polish coast (Tobolski 1997). The barrier could be periodically overwashed during storm surges as indicated by the admixture of benthic allochthonous forms preferring more saline waters, i.e. oligohalobous, halophilous and mesohalobous, in diatom composition (Fig. 11.9).

Dendrochronological data (Fig. 11.5) indicate that the tree growth was not terminated by one catastrophic event, but rather during several flood episodes.

According to the relative sea level curve for the southern Baltic, the water level in the early Atlantic (8,000–7,900 BP) was about 20–19 m below present (Uścinowicz 2003, 2006). That means that the investigated area was lying about 2–4 m above water level in the Gulf of Gdańsk at that time. The termination of the vegetation was probably caused by few river floods (spring high water stand and ice flow), rather than by marine transgression. It could have happened during 200–300 years before the sea did enter into the area. We may infer that the gulf's waters passed the –16 m level no earlier than about 7,700–7,600 BP. This view is based on all the evidences explained above (radiocarbon dates, results of pollen, diatom and dendrological analyses).

Similar drowned swampy areas of Atlantic age have also been found close to the edges of the Gulf of Gdańsk (Miotk-Szpiganowicz 1997, Uścinowicz and Miotk-Szpiganowicz 2003, Łęczyński et al. 2007, Uścinowicz et al. 2007) as well as in nowadays shallow water areas of the German part of the Baltic Sea (Lampe et al. 2005). Nevertheless, the site in the Gulf of Gdańsk, discussed in this chapter, is located deeper than the above-mentioned sites from Polish and German coastal waters.

11.5 Summary

The investigated site of the Gulf of Gdańsk furnishes significant information about the palaeogeography and water level in this part of the Baltic Sea during the very early Atlantic period.

In the Gulf of Gdańsk at depths of 16–17 m, remains of alder forest dated to 8,000–7,920 years BP (9,020–8,600 calendar years) occur indicating that the water level in the southern Baltic Sea at that time was lower, at least 17 m below the present one.

Palynological and diatomological studies show that the investigated area was swampy, often flooded by river and sometimes by storm surges as well.

According to dendrological investigations, the trees did not die at the same time, so the termination of vegetation was probably caused by few river floods rather than by marine transgression.

Results of palaeoecological investigation confirm earlier known seismic and geological evidences that a large area of the Vistula Delta was submerged by the Baltic Sea transgression during the Atlantic period.

References

Battarbee RW (1986) Diatom analysis. In: Berglund BE (ed) Handbook of Holocene palaeoecology and palaeohydrology. Wiley, New York, NY, pp 527–570

Berglund BE (1985) Pollen analysis. In: Berglund BE (ed) Palaeohydrological changes in the temperate zone in the last 15,000 years. Subproject B, 2:133–167

Christensen C (1995) The litorina transgressions in Denmark. In: Fischer A (ed) Man and sea in Mesolithic. Oxbow Books, Oxford, pp 15–22

Curry A (2006) A stone age world beneath the Baltic Sea. Science 314:1533–1535

Damusyte A, Bitinas A, Damusyte A, Kiseliene D, Mapeika J, Petrodius R, Pulkus V, (2004) The tree stumps in the south eastern Baltic as indicators of Holocene water level fluctuations. Proceedings of the 32nd International Geological Congress, Florence, Abstracts (part 2), p 1167

Damusyte A (2006) Evolution of the Lithuanian coastal zone (south-eastern Baltic) during the Late Glacial and Holocene. In: Camoin G, Droxler A, Fulthorpe C, Miller K (eds) Sea level changes: records, processes and modeling – SEALAIX'06, Abstract book. Publication ASF, Paris, 55:31–32

Ejtminowicz Z (1982) Submarine delta of the Wisła river in the Bay of Gdańsk (some results of continuous seismic profiling). Baltica 7:65–74

Faegri K, Iversen J (1975) Podrecznik analizy pyłkowrej. Warszawa.

Fisher O (1862) On the Bracklesham Beds of the Isle of Wight Basin. Quarterly Journal of the Geological Society of London 18:65–94

Harff J, Lampe R, Lemke W, Lübke H, Lüth F, Meyer M, Tauber F (2005) The Baltic Sea – a model ocean to study interrelations of geosphere, ecosphere, and anthroposphere in the Coastal Zone. Journal of Coastal Research 21(3):441–446

James H (1847) On a section exposed by the excavation at the new steam basin in Portsmouth Dockyard. Quarterly Journal of the Geological Society of London 27:249–251

Kramarska R, Uścinowicz Sz, Zachowicz J (1995) Origin and evolution of the Puck Lagoon. Journal of Coastal Research, Special Issue 22:187–191

Krąpiec M, Florek W (2005) Subfossil tree stumps and trunks on the beaches in Rowy area [Eng. summ.]. Geologia i geomorfologia pobrzeża Południowego Bałtyku 6. Pomorska Akademia Pedagogiczna w Slupsku, pp 145–154

Lampe R (2005) Late Glacial and Holocene water level variations along the NE German Baltic Sea coast: review and new results. Quaternary International 133–134:121–136

Lampe R, Endtmann E, Janke W, Meyer M, Luebke H, Harff J, Lemke W (2005) A new relative sea-level curve for the Wismar bay, N-Germany Baltic coast. Meyniana 57:5–35

Łęczyński L, Miotk-Szpiganowicz G, Zachowicz J, Uścinowicz Sz, Krąpiec M (2007) Tree stumps from the bottom of the Vistula Lagoon as indicators of water level changes in the Southern Baltic during the Late Holocene. Oceanologia 49(2):245–257

Miotk-Szpiganowicz G (1997) Results of palynological investigations in the Rzucewo area. In: Król D (ed) The built environment of coast areas during the stone age. The Baltic Sea-Coast Landscape Seminar, Session 1, Gdańsk, pp 153–162

Reimer PJ, Baillie MGL, Bard E, Bayliss A, Beck JW, Bertrand C, Blackwell PG, Buck CE, Burr G, Cutler KB, Damon PE, Edwards RL, Fairbanks RG, Friedrich M, Guilderson TP, Hughen KA, Kromer B, McCormac FG, Manning S, Bronk Ramsey C, Reimer RW, Remmele S, Southon JR, Stuiver M, Talamo S, Taylor FW, van der Plicht J, Weyhenmeyer E (2004) IntCal04 terrestrial radiocarbon age calibration, 0–26 cal kyr BP. Radiocarbon 46:1029–1058

Tauber F (2007) Seafloor exploration with sidescan sonar for geo-archaeological investigations. Berichte der RGK 88:67–79

Tobolski K (1997) Fazy holoceńskich transgresji morskich. In: Tobolski K, Mocek A, Dzięciołowski W (eds) Gleby Słowińskiego Parku Narodowego w świetle historii roślinności i podłoża. Homini, Bydgoszcz – Poznań, pp 41–44

Tobolski K, Pazdur MF, Pazdur A, Awsiuk R, Bluszcz A, Walanus A (1981) datowania metodą ^{14}C subfosylnych drewien występujących na mierzejach Niziny Gardzieńsko-Łebskiej. Badania Fizjograficzne nad Polską Zachodnią 33A:133–148

Uścinowicz Sz, Zachowicz J (1993) Geological Map of the Baltic Sea Bottom, 1:200 000, sheet Gdańsk. Panstwowy Instytut Geologiczny, Warszawa

Uścinowicz Sz (2003) Relative sea level changes, glacio-isostatic rebound and shoreline displacement in the Southern Baltic. Polish Geological Institute Special Papers 10:1–79

Uścinowicz Sz, Miotk-Szpiganowicz G (2003) Holocene shoreline migration in the Puck Lagoon (Southern Baltic Sea) based on the Rzucewo Headland case study. Landform Analysis 4:81–95

Uścinowicz Sz (2006) A relative sea-level curve for the Polish Southern Baltic Sea. Quaternary International 145/146:86–105

Uścinowicz Sz, Zachowicz J, Miotk-Szpiganowicz G, Witkowski A (2007) Southern Baltic sea-level oscillations: new radiocarbon, pollen and diatom proof of the Puck Lagoon. In: Harff J, Hay WW, Tetzlaff DM (eds) Coastline changes: interrelation of climate and geological processes. Geological Society of America Special Paper 426:143–158

Chapter 12
Holocene Evolution of the Southern Baltic Sea Coast and Interplay of Sea-Level Variation, Isostasy, Accommodation and Sediment Supply

Reinhard Lampe, Michael Naumann, Hinrich Meyer, Wolfgang Janke, and Regine Ziekur

Abstract Coastal barriers and spits develop when the accumulation space available in the coastal sea for sediment deposition decreases and partly fills up. The accommodation space increases when sea level rises and decreases when sediment accumulates. In addition to the coastal relief prior to the sea-level rise, which determines the potential accommodation, the evolution depends on the volume and rate of sediment supply. The example from the north-eastern German Baltic coast shows how the course of Holocene sea-level rise (Littorina transgression) varied due to glacio-isostatic uplift of different coastal sections and thus the growth of accommodation space. Further, the role of the sediments which built up the shoreface and the coastal landforms is discussed. We also examine the influence of the main inclination of pre-transgressional relief on the development, aggradation and progradation of beach ridges, spits and barriers. The determination of the volume of the present barriers allows rough estimations regarding the volume of sediment supplied from eroding cliffs. In a final synopsis, the interplay of all factors is discussed, explaining the distribution, volume and stability of the barriers along the German Baltic coast.

Keywords Baltic Sea · Germany · Coastal evolution · Sea-level development · Isostatic adjustment · Structures and volumes of coastal barriers

12.1 Introduction

Late Quaternary sea-level history from north-western Europe reflects the influence of various eustatic, isostatic, tectonic and, to a minor extent, other factors like sediment compaction, halokinetics and hydrographic variations. The many

R. Lampe (✉)
Institut für Geographie und Geologie, Ernst-Moritz-Arndt-Universität Greifswald, D-17487 Greifswald, Germany
e-mail: lampe@uni-greifswald.de

J. Harff et al. (eds.), *The Baltic Sea Basin*, Central and Eastern European Development Studies (CEEDES), DOI 10.1007/978-3-642-17220-5_12,

combinations of these factors in a rather limited area have made north-west Europe an intensively studied natural sea-level laboratory (Mörner 1980).

This study focuses on the southern Baltic coast as an ideal object to study the interplay of these main natural driving forces of coastal development, the influences of which are controlled by the sediment supply from both cliffs and sea bottom and the space available for potential sediment accumulation, so-called accommodation space (Posamentier and Allen 1999). Due to the insignificant tidal variations as a source of uncertainty in sea-level determinations, the eustatic sea-level variations in this area can be identified more precisely than elsewhere. On a millennium timescale, neotectonic crustal movements are believed to be insignificant because over the last 34 My they have varied from 200 m subsidence in the west to 120 m uplift in the north (Ludwig 2001), which is on average equivalent to –0.006 and 0.004 mm/year. However, glacio-isostatic movements have to be considered because the study area is located in the transition area between the Fennoscandian uplift and the central European zone where the effects of a decaying glacial forebulge have to be assumed (Fjeldskaar 1994, Garetsky et al. 2001, Nocquet et al. 2005).

At first, three new relative sea-level curves for the north-east German Baltic coast will be presented to show the sea-level variation and the tendency and stability of crustal behaviour. Results from intense onshore and offshore investigations (drilling, geophysical surveys) will be described to show how the relief prior to the transgression was formed. These data will be used to calculate the sediment volume of the barriers, which will be related to the sea-level history. Finally, a preliminary model of coastal evolution along the southern Baltic will be established which considers eustatic and isostatic sea-level variation, sediment supply and accommodation.

12.2 Geographic Setting

Mecklenburg-Vorpommern's 354-km-long outer (Baltic Sea) coast (Fig. 12.1) consists of cliff sections composed of Pleistocene outwash and till, interspersed with low uplands, barriers, spits and accreting forelands composed of Holocene sand and, to a very minor extent, gravel. The sea coast provides shelter to a longer shoreline within the inner bays or lagoons (boddens). The low-lying coastal segments owe their existence to sediment supplied alongshore from eroding bluffs, which are less mobile and are believed to act as headlands (hinge points) that help stabilize adjacent shores. Approximately 70% of the German Baltic Sea coast erodes at an average rate of 0.34 m/year (Ministerium für Bau, Landesentwicklung und Umwelt 1994).

The inner shelf consists primarily of Pleistocene outwash, till and glacial lake sediment (fine sand and silt). The latter forms large flat sediment bodies offshore of Usedom (Pomeranian Bight), Zingst (Falster–Rügen plane) and Rostocker Heide. Extensions of these sediment bodies can also be found landwards of the present coastline below the barriers. The inclination of the glacial lake sediment surface is predominantly less than 0.1° and the depths reach from –8 m below the barriers to

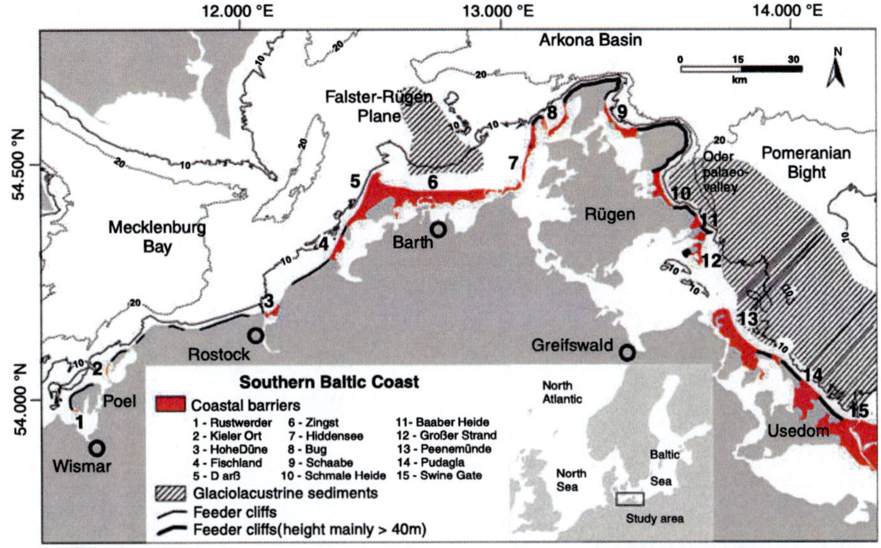

Fig. 12.1 Geographic setting along the southern Baltic Sea coast. Pleistocene substrate is shown in *grey*, Holocene coastal barriers are shown in *red*. Feeder cliffs are divided into two categories regarding altitude (*bold line* indicates cliff height mainly > 40 m). Note the course of the –10- and –20-m isobaths, which characterize the different offshore relief east and west of the Fischland barrier

–15 to –18 m at the edges to the proper basins in the Baltic, where marine mud accumulates. In some places, drowned river valleys such as the Oder palaeo-valley can be traced, incised during the Late Glacial and Early Holocene, when the water table in the Baltic basin was lowered to about 40 m below mean sea level (msl), i.e. to –40 m. During this period, large areas of the present sea bottom were characterized by a landscape of wetlands, shallow lakes and even forests.

When the eustatic sea-level rise had risen to the altitude of the thresholds of the Great Belt system in Denmark, the Baltic basin became connected to the North Sea. This first intrusion of saltwater into the Baltic basin took place at around 9,800–9,200 year cal BP when marine waters could enter through the Great Belt (Winn et al. 1998, Jensen et al. 1997, 2005, Bennike et al. 2004, Björck 2008). The subsequent sea-level rise is called the Littorina transgression in the Baltic Sea during which the landscapes of today's coast drowned. During the early transgression phase, the rise was rapid, more than 10 mm/year, but slowed later on. Earlier investigations have shown that on Rügen the sea level reached a position of –5 m by c. 8,000 year cal BP and a level between –1 and –0.5 m at c. 6,500 year cal BP (Kliewe and Janke 1982). This period, during which the rate of sea-level rise largely decreased, is believed to be the time when the main coastal sediment wedge accumulated between the Pleistocene headlands, thereby isolating lagoons from the Baltic. During the subsequent some thousand years, the sea level varied

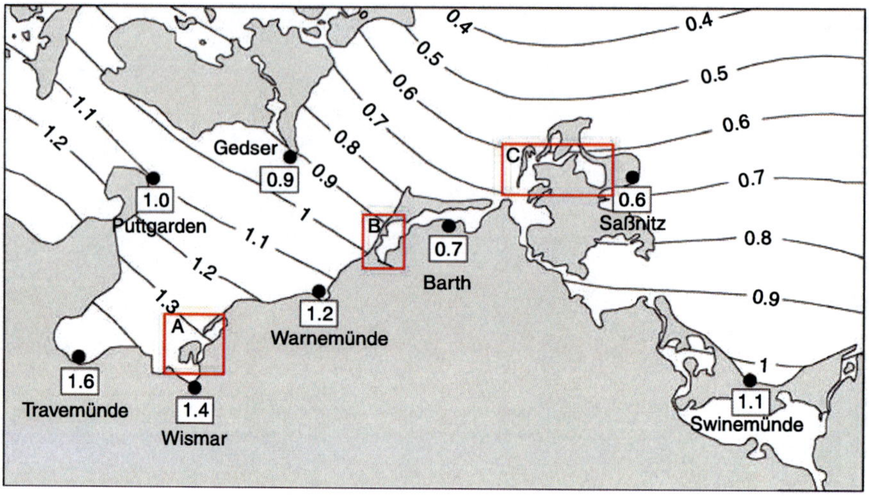

Fig. 12.2 The Wismar Bay (*A*), Fischland (*B*) and north Rügen/Hiddensee (*C*) study areas and current relative sea-level rise (mm/year) at different gauge stations in the southern Baltic Sea (Dietrich and Liebsch 2000)

only slightly, and shoreline evolution was characterized mainly by progradation and dune belt development.

The recent relative sea-level change was investigated using repeated precision levelling and long-term mareograph records (Montag 1967, Bankwitz 1971, Liebsch 1997, Dietrich and Liebsch 2000). The change in pattern, constrained from the latter, is shown in Fig. 12.2. It indicates a shoreline tilt with a relatively slower sea-level rise on Rügen than is at Wismar. The eustatic rise during the past 100 years is estimated to be 1–1.2 mm/year (Dietrich and Liebsch 2000, Stigge 2003). It corresponds to the relative rise between the Fischland and the coastal section west of Warnemünde and means that a slight but increasing crustal uplift occurs from there towards Rügen and a subsidence towards Wismar and Travemünde. This was already concluded by Kolp (1982) and Ekman (1996), who marked the –1 mm/year isobase as the isoline where the glacio-isostatic emergence fades out.

12.3 Data Acquisition

Sea-level curves deduced from regionally distributed data might be flawed by differential crustal motions. To avoid this source of error, Kíden et al. (2002) recommend sampling areas not larger than 15–30 km in diameter to guarantee that differences in the crustal movement within the area are small and negligible. For this investigation, data from the three study areas, Wismar Bay (Lampe et al. 2005), Fischland and North Rügen/Hiddensee (Lampe et al. 2007), were used which are located along the gradient of the recent relative sea-level change (Fig. 12.2) and are less than 25 km in diameter. The mareograph records from Wismar, Barth and Saßnitz which represent the relative sea-level rise in these study areas, show a secular rise of 1.4, 1.0 and 0.6 mm/year, respectively (Dietrich and Liebsch 2000).

In these areas, peat samples from both basal peat layers (sensu Lange and Menke 1967) and coastal mire profiles (Lampe et al. 2007) were taken, retrieved during onshore and ship-based offshore surveys. To evaluate the sea-level control, the samples were checked for the intensity of marine influences using pollen, diatoms and floral macro remains (Endtmann 2005, Lampe 2004, Lampe et al. 2005, Mandelkow et al. 2005). Floral macro remains from these samples or bulk subsamples were ^{14}C-AMS dated. The data set was extended by datings from both underwater in situ finds of tree stumps and archaeological finds (bones and woods). In a few cases, previous ^{14}C dates from basal peat layers were considered, which were conventionally analysed. All age data were calibrated to calendar years before present (year cal BP) using CalPal software (Danzeglocke et al. 2007). The 2σ confidence interval in the calendar age ranges was used in the construction of the sea-level curves. To estimate the altitude error, all sample depths were related to recent mean sea level. Considering the many errors possible when relating the position of the samples to the former sea level, its altitude can be determined with an accuracy of –0.1 to –0.5 m for precisely levelled sampling sites and +0.2 to –0.8 m for all other sites. These age–depth ranges were used for the construction of the relative sea-level curves and their error envelopes (Lampe et al. 2007).

To determine the pre-transgressional relief, the distribution of the coastal sediments, their thickness and facies were surveyed extensively by means of motor hammer-driven drilling equipment. Ground penetration radar (GPR) surveys were carried out for layer tracing between the auger holes and for recognizing internal structures (van Heteren et al. 1999; Jol et al. 2003; Neal 2004) except in artificially drained areas where introduced saltwater prevented useful measurements (Lampe et al. 2004). For ship-based offshore investigations, a 4-m vibrocorer was used. Sediment echosounding (SES) surveys were made off Zingst, Rügen and Usedom, using an INNOMAR SES-96 set. Signal recording was restricted to 10 or 15 m below the sediment surface.

To estimate the volumes of the barriers, all information from geological maps, drilling results and GPR surveys regarding the depth of the transgression contact were gathered and checked against each other. Based on these data the pre-transgressional land surface was modelled using an ordinary kriging algorithm from standard geostatistical software (Keckler 1997). For the present land surface a digital elevation model from the State Survey Office was used and for consistency reasons was recalculated to the same resolution as the surfaces modelled. On average the modelled surface deviates less than 10% from measured depths. The difference between the pre-transgressional land surface and the recent land surface represents the volume of the sediments accumulated under marine–brackish or aeolian conditions. This volume was recalculated to match a volume assumed to have been eroded from neighbouring Pleistocene feeder cliffs.

12.4 Investigation Results

12.4.1 Sea-Level Development

In all study areas the sea-level development shows the same tendency (Fig. 12.3). In all of them, the rapid rise ended at about 7,500 year cal BP as also found by other investigators (Fig. 12.4). The subsequent development was characterized by a very slow rise or even stagnation with minor variations. Neither desiccation horizons in coastal mire profiles nor intercalated peat horizons in the sandy barriers have been detected. This finding supports the statement that sea-level variations with amplitudes wider than the estimated error band of sea-level determination did not occur. Short-term fluctuations of more than 1 m as described by Behre (2003) or Yu (2003) and Yu et al. (2007) from neighbouring areas (Fig. 12.4) were not found and cannot be confirmed. During the last c. 1,000 years the rise became more important again (Late Subatlantic transgression; Lampe and Janke 2004) but was interrupted during the Little Ice Age. During this period a prominent black pitchy soil layer was formed in the coastal mires, possibly caused by peat desiccation and degradation. Comparable layers were described from many sites along the German and Danish North Sea coast (Freund and Streif 1999, Gehrels et al. 2006) and indicate that Little Ice Age sea-level changes were a widespread phenomenon. At the southern Baltic coast, it was probably the only significant oscillation throughout the last 5,000 years. For other minor fluctuations, as during the Bronze Age, vague but not convincing evidence exists.

Despite the similarity of the three sea-level curves, they differ regularly in the depth–age relationship, indicating a persistent movement of the Earth's crust and thus signifying isostatic movements, with most isostasy on Rügen and least in the Wismar Bay. More information about this process can be gained only by using geophysical models or by comparing the relative curves with a curve from a nearby area believed to be tectonically stable. As the German North Sea coast (sea-level curve from Behre (2003) in Fig. 12.4) was recently identified as a subsiding region (Vink et al. 2007) and most of the Baltic coasts are influenced by uplift (sea-level curve for S-Sweden in Fig. 12.4 from Yu (2003)), the nearest curve suitable for comparison is located at the Belgian North Sea coast (Fig. 12.4; Denys and Baeteman 1995; Kíden et al. 2002). A sea-level development comparable to what was observed at the German Baltic coast can be expected for the Polish coast (Fig. 12.4, Uścinowicz 2003) as the tectonic and isostatic conditions are similar (Garetsky et al. 2001). Due to the data available from the south Baltic coastal area, the comparison was restricted to the period since 8,000 year cal BP (Fig. 12.4).

From the correctness of all curves provided, three conclusions can be drawn: (i) the intense sea-level fluctuations found by Yu (2003) and Behre (2003) are not confirmed by other investigators, although minor variations are definitely not excluded when sea-level error bands are considered. (ii) The curve from the German North Sea coast is related to mean high water level and all other curves to mean sea level, i.e. the altitude of the German North Sea curve is usually lower than all others. (iii) The assumption of a tectonically stable Belgian coast allows inferences onto the

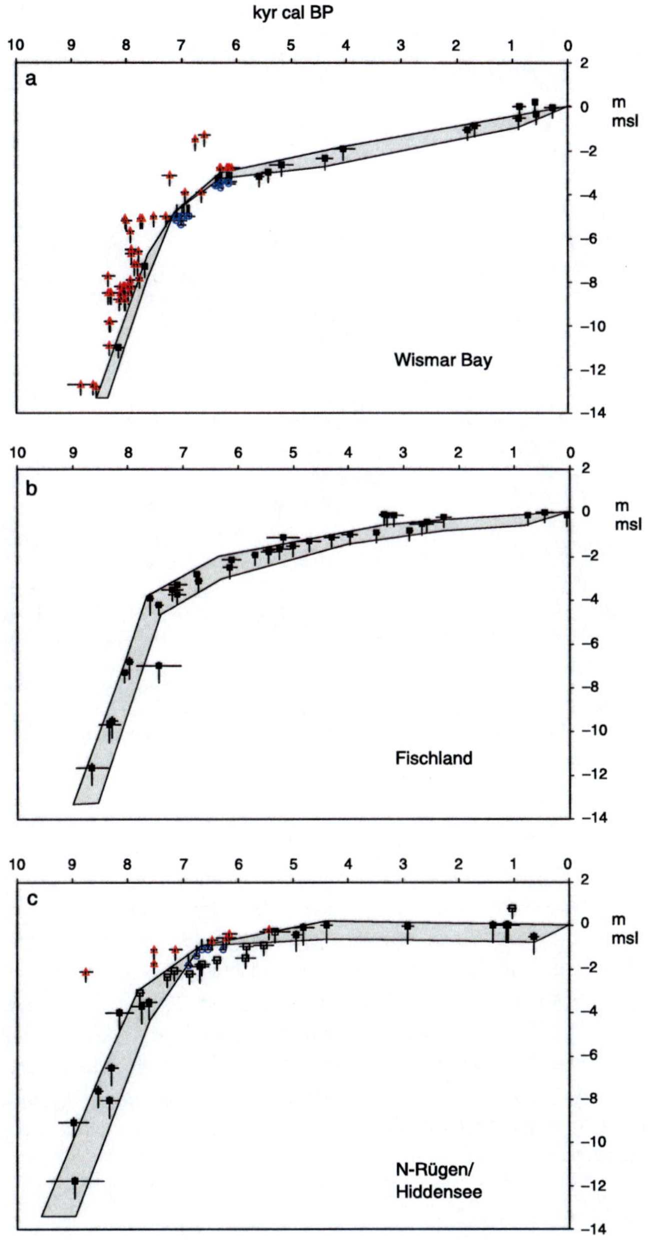

Fig. 12.3 Relative sea level (RSL) curves for the Wismar Bay (**a**), Fischland (**b**) and north Rügen/Hiddensee (**c**). *Red triangles* designate data from terrestrial deposits, *green diamonds* designate telmatic deposits and *blue circles* designate archaeological finds in marine nearshore deposits. *Horizontal bars* represent double standard deviation (2σ), *vertical bars* indicate estimated altitude error of sea-level position. The sea-level curves are depicted as error envelopes within which the sea-level position most probably was located

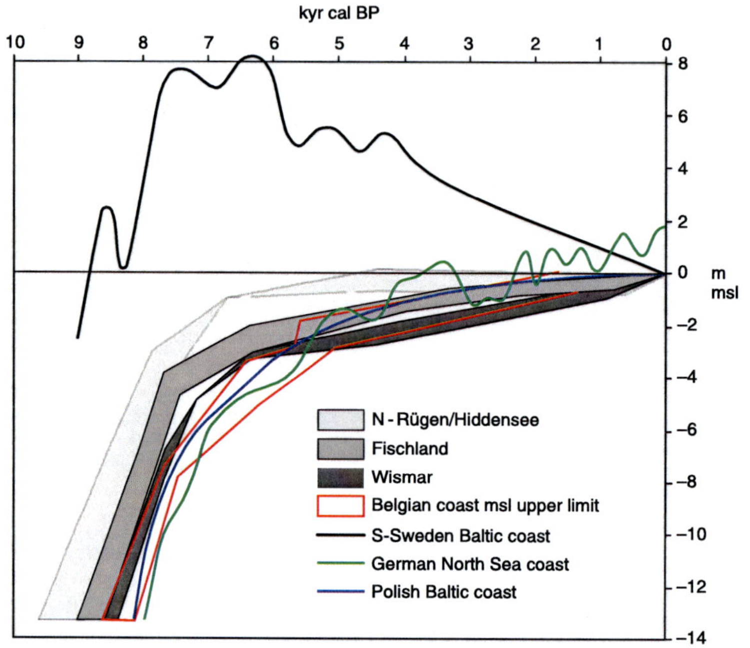

Fig. 12.4 Error bands of the RSL curves for Wismar Bay, Fischland and N-Rügen/Hiddensee. For comparison the RSL error band of the Belgian coast (Kíden et al. 2002) and the RSL curves of S-Sweden (Yu 2003), the Polish Baltic coast (Uścinowicz 2003) and the German North Sea coast (Behre 2003) are shown

isostatic movement of the other coastal areas. Higher sea-level curves are influenced by uplift, while the lower ones are influenced by subsidence. The north-eastern German coast, therefore, belongs to the outermost edge of the Scandinavian uplift, where the isostatic upheaval, or unloading effect, fades out. In the areas around Wismar and Fischland, the isostatic emergence already ceased more or less but it probably continues on Rügen. More recently, for the Wismar area, even a slight subsidence seems possible. These conclusions are in line with the results of the gauge investigations. Also the Polish coast seems to be stable. However, this curve and the German North Sea curve were constructed from regionally wider distributed data and, therefore, differences in movement between single coastal sections were possibly not detected.

12.4.2 Relief Prior to Transgression

From Fig. 12.1, differences regarding the extent of the barriers along the north-eastern German Baltic coast can be deduced. Obviously, more barriers occur eastwards of the Fischland than westwards and are wider, longer and probably more voluminous. To investigate the causes of this barrier size distribution, numerous drillings and GPR surveys on the barriers, vibrocoring and SES surveys offshore and in the lagoons were carried out. The extensive fieldwork resulted in an in-depth knowledge regarding the internal structure and facies distribution of the sediments from which the relief prior to the transgression can be deduced (Fig. 12.5 shows examples). In the coastal area from Usedom Island in the east to the Fischland in the west, the Pleistocene uplands consist predominantly of glacio-fluvial/lacustrine sand with some till beds and are characterized by a highly undulating relief with elevations up to +60 m and interjacent depressions down to –20 m. Up to a level of c. –12 to –8 m the depressions are filled with slightly carbonate-bearing fine-to-medium sand, containing diatoms and molluscs, indicating cold freshwater environments (Fig. 12.5a). The AMS radiocarbon dates imply that they are of Late Glacial Age, i.e. older than 11,700 year cal BP (cf. Walker et al. 2009). The surface of the Late Glacial sand dips very slightly to –15 to –18 m northwards towards the proper Baltic basin where a steeper decline occurs (Fig. 12.1). In the surface, numerous dead ice depressions are indented and completely filled with interbedded fine sand and silt (Fig. 12.5b). Also, some palaeo-channels intersect the surface. The most prominent one is the wide Oder palaeo-valley, located east of Rügen (Fig. 12.1). The surface

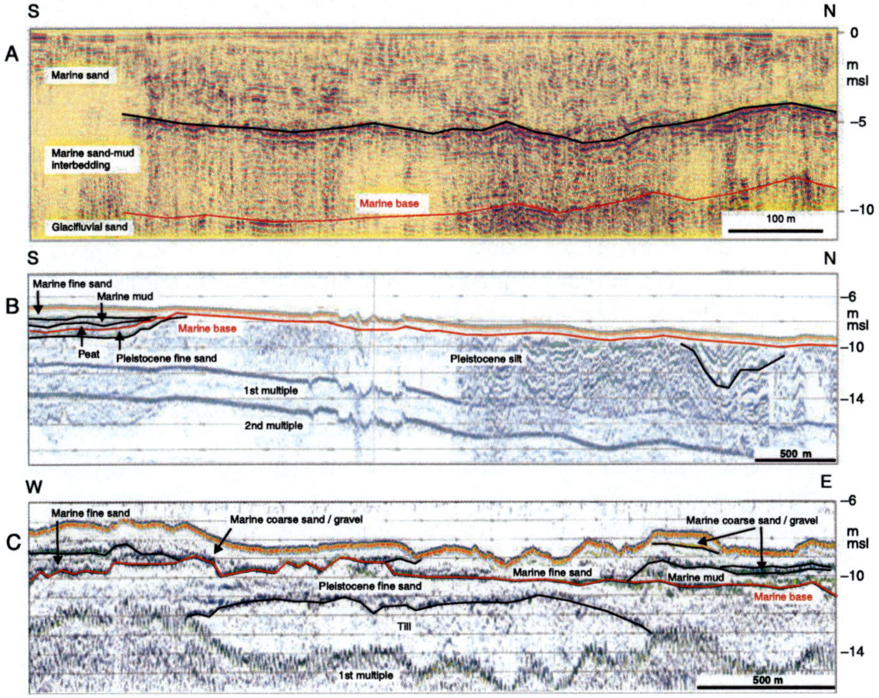

Fig. 12.5 **a** Example of a GPR record from Hiddensee Island; for location, see line 3 in Fig. 12.6. Depth is calculated as 0.053 m/ns. **b** and **c** Examples of SES records from the Falster–Rügen plane; for locations, see lines 1 and 2 in Fig. 12.6. Depth is calculated as 1,500 m/s

of the sandy deposits undulates only slightly and points to a connected Late Glacial fluvio-lacustrine system in the Pomeranian Bight, the Rügen lagoons and the Darss-Zingst area (Lampe 2005). Shallow water-filled depressions, which remained after the final drainage of this system, accumulated freshwater mud or lake marl during the Early Holocene and mostly silted up in the mid-Holocene.

By contrast, in the area west of the Fischland, the coastal relief is widely characterized by higher ground with long till cliff sections. The abrasion platforms in front of the cliffs are covered mostly by lag sediments and show a steeper inclination towards the Baltic basin. Sandy sediments are rare. Only in the Wismar Bay a more complex palaeo-valley system determines the relief, incised in the loamy ground moraine (Harff et al. 2007). The main differences between the coastal sections located east and west from the Fischland, therefore, are (i) the existence of depressions reaching far below the recent sea level, (ii) the availability of sandy material from both offshore and onshore sources and (iii) the inclination of the palaeo-relief towards the proper Baltic basins.

12.4.3 Structure and Volume of Coastal Barriers

The rising Baltic Sea caused groundwater rise in the adjacent coastal mainland and thus favoured peat accumulation upon the land surface. This 'basal peat', mostly some centimetres to two decimetres thick, was later inundated due to the landward-migrating shoreline. The transgression contact is often marked by a hiatus of several hundred years due to the erosion of the peat top layer. Usually, a clear nearshore/beach facies cannot be observed or consists only of 1-cm-thick sandy layer. These deposits are overlain by muddy sediments consisting of silt with different admixtures of fine sand and up to 25% organic matter. The lowermost section is strikingly enriched with shells of *Hydrobia* sp., *Cerastoderma* sp. and *Scrobicularia* sp., indicating an evolutionary stage where the barriers not yet existed and the present lagoon areas were still bays of the Baltic. The mud gradually changes into sand which built the main base of the barrier. At the seaside of the barriers the grain size is coarser, sometimes gravelly, while at the lagoon side, mud–sand interlayerings are observed, which change upwards into fine sand (Fig. 12.6).

Peat is never intercalated in the siliciclastic sequence; only allochthonous floral detritus layers occur. This is an important difference from the coastal sediment sequence described for the southern North Sea (Behre 2003, Streif 2004) and underlines the statement that no significant sea-level fluctuation occurred during its accumulation. At the sea coast the surface of the marine sand package is covered by progradational beach ridges and dunes, and on the lagoon side the barrier surface is flat and covered with fenland peat ('cover peat'). Radiocarbon datings show that the cover peat accretion began at about 800–1,200 year cal BP (Jeschke and Lange 1992, Lampe and Janke 2004).

The dense net of boreholes in combination with about 80-km GPR tracks, which facilitate interpolation between the drilling profiles, allows for modelling of the base

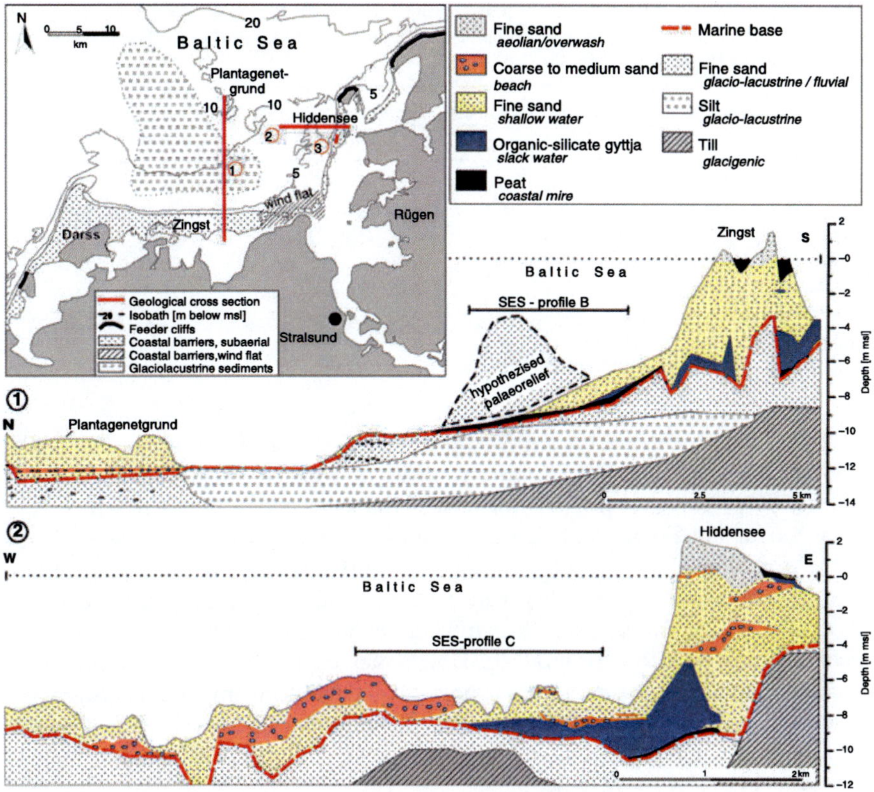

Fig. 12.6 Cross sections of the eastern Zingst peninsula (*1*), Hiddensee island (*2*) and the offshore areas. The GPR transect marked with (*3*) is shown in Fig. 12.5 as example A and the SES profiles are shown in Fig. 12.5 as B and C

and the surface of the coastal sediments and calculation of the volume of the barriers. The sediment volume located at the shoreface was neglected, due to the mostly uncertain distribution and thickness in the near-coastal zone. The estimated volumes are shown in Table 12.1. Two arguments allow relating the barrier volume with the retreat of the neighbouring cliffs:

(a) In the study areas, no rivers are located, which would deliver significant amounts of sandy material having the potential for nearshore accumulation.
(b) Due to the many offshore finds of Preboreal, Boreal and Early Atlantic lakes, mires, forests and archaeological sites, we can exclude any significant erosion and landward sediment transport to build up the barriers.

The above arguments imply that the predominant part of the barrier sand volume must have been provided by cliff abrasion, and maybe also to a minor extent by shoreface abrasion. Considering the height and length of the feeding cliffs, the

Table 12.1 Volume of selected coastal barriers and corresponding retreat of feeder cliffs at the southern Baltic coast (Barthel 2002, Hoffmann 2004, Naumann 2006) during the last 8,000 years. For barrier and cliff locations, see Fig. 12.1

Coastal barrier (no. in Fig. 12.1)	Estimated volume (mill m^3)	Mean length/height of feeding cliffs (m)	Calculated cliff retreat (m)
Kieler Ort (2)	11	4,000/4	690
Zingst (6)	450	No cliff	?
Hiddensee (7)	270	4,500/40	1,500
Bug (8)	66	8,000/8	1,030
Schaabe (9)	103	17,300/35	170
Peenemünde (13)	420	10,500/30	1,810
Pudagla (14)	150		

above-mentioned volumes correspond to a mean cliff retreat during the last 8,000 years as listed in Table 12.1. Similar estimations were published by Uścinowicz (2003, 2006), who assumes that the Polish coast has receded 1,000–1,500 m.

Except for the Zingst study area, all barriers can be explained to be built up mainly from eroded sediment from the nearby cliff sections. For the Zingst peninsula, the provenance of the sandy material is more difficult to explain; an obvious feeder cliff does not exist today. The offshore area is widely covered by silty sediment of glacio-lacustrine origin (Fig. 12.1) which cannot provide barrier building sand. The only possible sources are hypothetical glacio-lacustrine/fluvial sand bodies scattered offshore in the vicinity of today's peninsula. Analogue sediment bodies were found at the base of the coastal barriers building the backbone of the recent peninsula (Fig. 12.6). These sediments probably built the glacial lake shore and fringing small deltas and were fluvially intersected after the lake level fell. The form and extent of these sediment bodies cannot be reconstructed because they are now completely eroded. The eroded material built spits which became permanently reshaped and transgressed with the rising sea level to the recent position of the peninsula. Behind the slowly moving spits, slack water areas occurred temporarily. Here, small-sized thin lagoonal mud layers accumulated which are located today some hundred to thousand meters offshore of Zingst and Hiddensee and are the only remaining traces of the transgressive proto-barriers.

12.5 Discussion and Conclusions

The internal structure of the barriers shows some critical aspects. The lagoonal mud underlying the barrier sand was deposited under calm and sheltered conditions, provided by seaward islands, spits or barriers. Further, the occurrence of mud implies a barrier transgression over the lagoonal sediments and to the limited extent of the barrier shift (Hurtig 1954, Kliewe and Janke 1982, Lampe 2005). As evident from ship-based offshore investigations (vibrocoring and SES surveys), non-compacted Late Glacial and Early Holocene lake and peat deposits exist seawards of the present barriers, but relict estuarine mud are not found there, with the exception of the offshore Zingst and Hiddensee areas. That means that no barriers developed during the early transgression phase and only narrow, flat, exiguous beach ridges transgressed rapidly over the very gently dipping surface (0.05 ... 0.15°, inherited from Late Glacial lakes) driven by the rising sea. Some few overstepped beach ridges occur only where coarse gravels crop out (Gromoll 1994). Although much sand has been available for sediment transport, the shape and character of the surface left by the Late Glacial drainage system and the modest current and wave energy of the Baltic, further reduced by bottom friction effects, prevented barrier formation. Hence, we can confirm the results of barrier translation models (Roy et al. 1994, Stolper et al. 2005) that on a flat substrate, increasing friction and decreasing wave power led to a reduction in barrier size.

The transgressive beach ridges finally stranded at the toes of the Pleistocene elevations interspersed mostly seawards between the Late Glacial lake sediments. Due to the steeper substrate slope, the shoreline recession decelerated and erosion possibly started. At that point the ratio between sediment supply and accumulation space became critical and determined whether the volume of the beach ridges grew and developed into spits and barriers or the elevations were finally eroded and drowned and the embryonic spits were eroded. Under this perspective it becomes clear that small elevations may play a special role in the process of beach ridge stranding. They fix the migrating beach ridges/barriers but can maintain them for a longer time only if (i) they will not get drowned by the rising sea and (ii) the sediment supply is big enough to fill the still growing accommodation space. Therefore, the present barriers are all connected to viable feeder cliffs and in all of them, cores from Pleistocene sediment can be found. All lower elevations located farther seawards became eroded, drowned and today, build shoals and reefs.

After the sea-level rise ceased – and hence the accommodation space started to shrink – the stranded spits grew faster. Where the distance between spit anchors (i.e. the accommodation space) was adequately small in relation to sand supply, spit ends grew together and progradation started. This process led to the development of bay barriers with wide beach ridge plains and dune fields (Fig. 12.7). Where the distance between the spit anchors was large compared to the sediment supply, the bays were not completely cut off from the open sea. In fact, these spits are still growing, thereby receding landwards and traversing lagoon sediments (Hiddensee, Zingst, Bug, Rustwerder, Kieler Ort, see Fig. 12.1).

Large areas behind the beach ridges or dune belts on the barriers developed as wind flats, whose surface was levelled due to frequent floodings occurring through shallow inlets. Vertical sand accumulation in the flats kept pace with the moderate sea-level rise until the inlets became truncated by a beach ridge and peat accumulation started (Fig. 12.8; Hoffmann et al. 2005). The preservation of wide flats behind the beach ridges or dune belts is generally considered to be evidence of shoreline stability or only slight retreat which is in accordance with the finding that the lagoonal sediments usually do not crop out at the sea shoreface.

Where the headlands or islands which provided anchor or hinge points to the barriers were not high or voluminous enough to survive, they drowned or were abraded,

Fig. 12.7 Aerial image of the barrier Schaabe; view is towards east and the Baltic is located on the left side (for location, see Fig. 12.1, no. 9). Two beach ridge systems are evident: an older system (*A*), which consists of short, low-lying ridges, bent into the former bay and a second system of higher elevated dune ridges (*B*), which cut off the bay from the sea and show rapid progradation. This system is covered by dense pine forest (Photo: R. Lampe 2007)

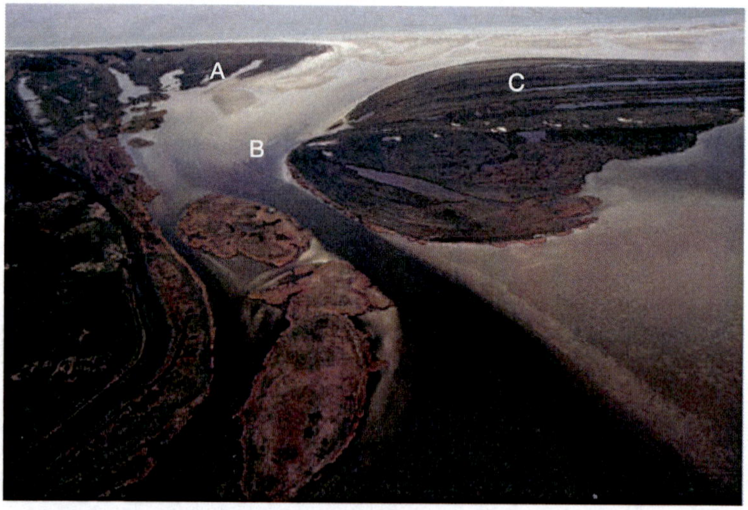

Fig. 12.8 Aerial image of the eastern tip of the Zingst peninsula and the offshore sand flats and islands; view is northwards (for location, see Fig. 12.1, no. 6). The inlet is progressively truncated by a spit (*A*) growing from west (*left*) to east (*right*). Sand flats (*B*) with incised flooding channels spread between the islands. In the northern part of the offshore island, progradational beach ridges (*C*) are visible (Photo: R. Lampe 2007)

thereby opening new sediment transport passages. The former barrier eroded and the sediment was incorporated in a new spit/barrier structure evolving further landwards at higher elevations. Finally, where the sediment supply exceeded accommodation, the bay was filled and the prograding beach matched the neighbouring coastal cell (Hoffmann and Barnasch 2005). Many permutations are possible between these evolutionary types. It can be concluded that the barriers and spits transgressed only to a minor extent and only when hinge points were drowned or abraded or when space between anchor points was too large. The main processes in shaping of the present coast have been stranding, progradation and elongation and are significantly controlled by the inherited relief.

The rough volume calculation emphasizes the assumptions that the cliffs provided enough material to build up the barriers and that they receded between 1 and 2 km since the Littorina transgression reached the present coastal area. Therefore, the anchor points of the spits experienced approximately the same dislocation. When the sea-level rise decreased at 7,800 year cal BP and, hence, accommodation space grew slower, the main phase of barrier building started. In the subsequent 1000 years or so, the sediment supply from cliff erosion still continued due to ongoing coastal re-equilibration but exceeded the steadily shrinking accommodation, thus causing fast barrier building. However, this process was different in the three study areas. While on Rügen the spits were rapidly closed to prograding barriers, this process took much more time in the Darss-Zingst area and occurred in the Wismar Bight to a very limited extent. The difference is caused by both factors: sediment starvation in the Wismar area where cliffs providing sufficient sediment are rare and accommodation space, which decreased much faster on Rügen than at Wismar due to the more rapid isostatic uplift.

The slowing down of sea-level rise between 6,000 and 1,200 year cal BP led gradually to a decrease in sediment supply and – in sections – to cliff stabilization, which must have been most pronounced on Rügen. Since c. 1,200 year cal BP, coastal dynamics has increased again as is evident by the occurrence of transgressive dunes. From the younger historical record, fast elongation of spits and impending barrier breaching is known. The increased dynamics can be related to the onset of the post-Littorina (Late Subatlantic) transgression, the timing of which is evident from the accumulation of the cover peat.

12.6 Summary

To study the interplay between sea-level evolution, crustal movement, accommodation and sediment supply, new and detailed investigations of the evolution of the southern Baltic coast were conducted based on intensive drilling and geophysical surveys both onshore and offshore. An important result of the project was the identification of three local relative sea-level curves which clearly show that a fading crustal upheaval occurred which is still in progress on Rügen and – to a minor extent – on the Fischland area, while the movement ceased or changed to a slight subsidence in the Wismar Bay. Because the new RSL curves are well proven by both archaeological and geological evidence and based on a consistently calibrated ^{14}C-AMS data set, much clearer conclusions can be drawn about sea-level fluctuations than before.

Backed by a comprehensive borehole data set, the existing ideas about the Holocene coastal evolution in the southern Baltic were critically evaluated. It could be shown that the sediments underlying the barriers belong to a complex fluvio-lacustrine drainage system of Late Glacial Age. Also, the large flat sand/silt deposits of Usedom and Zingst were primarily built in the Late Glacial period and were reshaped only slightly during the main Littorina transgression phase. This underlines the importance of the sea-level rise rate for coastal erosion processes. Higher rates cause flatter shoreface profiles which deflect from a Bruun-like equilibrium. When the rise rate decreases, the shoreface profiles tend to rebuild an equilibrium which leads primarily to higher erosion due to deeper profile moulding, but to decreasing erosion (and sediment supply) with ongoing approach to equilibrium.

All data from the Fischland in the west to the Usedom Island in the east point to the existence of an interconnected Late Glacial fluvio-lacustrine system which can be related to a level of –8 to –12 m found below all West Pomeranian barriers. During the Littorina transgression the flat surface of these deposits was inundated very rapidly and its small inclination is assumed to be the reason that less voluminous beach ridges/barriers developed and migrated with the transgressing sea. Only at locations of steeper ground could small coastal features like spits grow, but they were eroded or overstepped when the hinge points drowned. After the sea-level rise became slower at c. 7,800 year cal BP, the sediment supply exceeded the growth of the accommodation space. Large beach ridge systems began to accumulate at higher elevated islands and promontories and developed into barriers. Progradation started and in bays with closed sedimentary systems, barriers were shaped to perfectly swash aligned beaches. The sediment volume finally incorporated in the barriers corresponded to a 1–2-km retreat of the feeding cliffs. Shore profiles became equilibrated due to the very slow rise rate and the cliff retreat rates decreased.

An important alteration in the sediment dynamics occurred with the onset of the Late Subatlantic transgression which started at about 1,200 year cal BP. Former progradation changed into retrogradation which was connected with sediment mobilization and faster shoreline erosion leading to elongation of spits, more frequent inundation and overwash and the development of transgressive dune fields. On the back barriers, peat accumulation started. After an interruption of the rise due to the Little Ice Age, the higher dynamic mode in coastal behaviour continues after about 1850.

Acknowledgements This study was possible due to the financial support provided by the Deutsche Forschungsgemeinschaft, which is gratefully acknowledged (FO 488/1). We thank all members of the SINCOS Research Group for valuable data and discussions, and students and staff from Greifswald University for their help in the field and laboratories. We acknowledge the recommendations of three anonymous reviewers and the editors which helped to improve the chapter.

References

Bankwitz P (1971) Geologische Auswertung von geodätisch ermittelten rezenten Krustenbewegungen im Gebiet der DDR. Petermanns Geographische Mitteilungen 115(2):130–140

Barthel A (2002) Aufbau und Entwicklung der holozänen Sedimente der Insel Hiddensee und deren Modellierung mit geostatistischen Methoden. Thesis, Universität Greifswald

Behre K-E (2003) Eine neue Meeresspiegelkurve für die südliche Nordsee. Probleme der Küstenforschung im südlichen Nordseegebiet 28:9–63

Bennike O, Jensen JB, Lemke W, Kuijpers A, Lomholt S (2004) Late- and postglacial history of the Great Belt, Denmark. Boreas 33:18–33

Björck S (2008) The late Quaternary development of the Baltic Sea basin. In: The BACC Author Team (eds) Assessment of climate change for the Baltic Sea basin. Springer, Berlin, Heidelberg, pp 398–407

Danzeglocke U, Jöris O, Weninger B (2007) CalPal-2007 online. http://www.calpal-online.de/

Denys L, Baeteman C (1995) Holocene evolution of relative sea level and local mean high water spring tides in Belgium – a first assessment. Marine Geology 124:1–19

Dietrich R, Liebsch G (2000) Zur Variabilität des Meeresspiegels an der Küste von Mecklenburg-Vorpommern. Zeitschrift für Geologische Wissenschaften 28(6):615–624

Ekman M (1996) A consistent map of the postglacial uplift of Fennoscandia. Terra Nova 8(2): 158–165

Endtmann E (2005) Erste Ergebnisse der neuen paläobotanischen Untersuchungen am mesolithischen Fundplatz von Lietzow-Buddelin auf Rügen. Bodendenkmalpflege in Mecklenburg-Vorpommern, Jahrbuch 2004 52:197–209

Fjeldskaar W (1994) The amplitude and decay of the glacial forebulge in Fennoscandia. Norsk Geologisk Tidsskrift 74:2–8

Freund H, Streif H (1999) Natürliche Pegelmarken für Meeresspiegelschwankungen der letzten 2000 Jahre im Bereich der Insel Juist. Petermanns Geographische Mitteilungen 143(Pilotheft 2000):34–45

Garetsky RG, Ludwig AO, Schwab G, Stackebrandt W (2001) Neogeodynamics of the Baltic Sea depression and adjacent areas – results of IGCP project 346. Brandenburgische Geowissenschaftliche Beiträge 8(1):1–47

Gehrels WR, Szkornik K, Bartholdy J, Kirby JR, Bradley SL, Marshall WA, Heinemeier J, Pedersen JBT (2006) Late Holocene sea-level changes and isostasy in western Denmark. Quaternary Research 66(2):288–302

Gromoll L (1994) Fossile Küstenlinien als Wasserstandsmarken in der südlichen Ostsee – Effekte von Isostasie, Eustasie und Neotektonik. Zeitschrift für geologische Wissenschaften 22(3): 287–304

Harff J, Lemke W, Lampe R, Lüth F, Lübke H, Meyer M, Tauber F, Schmölcke U (2007) The Baltic Sea coast – a model of interrelations among geosphere, climate, and anthroposphere. The Geological Society of America Special Paper 426:133–142

Hoffmann G (2004) Rekonstruktion und Modellierung der Küstenevolution im Bereich der Pommerschen Bucht in Abhängigkeit von holozänen Meeresspiegelschwankungen. PhD thesis, Greifswald University

Hoffmann G, Barnasch J (2005) Late Glacial to Holocene coastal changes of SE Rügen Island (Baltic Sea, NE Germany). Aquatic Science 67:132–141

Hoffmann G, Lampe R, Barnasch J (2005) Postglacial evolution of barriers along the West Pomeranian coast, NE Germany. Quaternary International 133/134:47–59

Hurtig Th (1954) Die mecklenburgische Boddenlandschaft und ihre entwicklungsgeschichtlichen Probleme. Ein Beitrag zur Küstengeschichte der Ostsee. Deutscher Verlag der Wissenschaften, Berlin

Jensen JB, Bennike O, Witkowski A, Lemke W, Kuijpers A (1997) The Baltic ice lake in the southwestern Baltic: sequence-, chrono- and biostratigraphy. Boreas 26:217–236

Jensen JB, Bennike O, Lemke W, Kuijpers A (2005) The Storebælt gateway to the Baltic. Geological Survey of Denmark and Greenland Bulletin 7:45–48
Jeschke L, Lange E (1992) Zur Genese der Küstenüberflutungsmoore im Bereich der vorpommerschen Boddenküste. In: Billwitz K, Jäger K-D, Janke W (eds) Jungquartäre Landschaftsräume – aktuelle Forschungen zwischen Atlantik und Tienschan. Springer, Berlin
Jol HM, Lawton DC, Smith DG (2003) Ground penetrating radar: 2-D and 3-D subsurface imaging of a coastal barrier spit, Long Beach, WA, USA. Geomorphology 53(1–2):165–181
Keckler D (1997) SURFER® for Windows user's guide. Contouring and 3D surface mapping, Version 6. Golden Software Corporation
Kíden P, Denys L, Johnston P (2002) Late Quaternary sea-level change and isostatic and tectonic land movements along the Belgian–Dutch North Sea coast: geological data and model results. Journal of Quaternary Science 17:535–546
Kliewe H, Janke W (1982) Der holozäne Wasserspiegelanstieg der Ostsee im nordöstlichen Küstengebiet der DDR. Petermanns Geographische Mitteilungen 126(2):65–74
Kolp O (1982) Eustatische und isostatische Veränderungen des südlichen Ostseeraumes im Holozän. Petermanns Geographische Mitteilungen 123(3):177–187
Lampe R (2004) Reliefgenese und Faziesdifferenzierung am steinzeitlichen Fundplatz Saiser Fpl. 1 (Lietzow-Buddelin), Kr. Rügen. Bodendenkmalpflege in Mecklenburg-Vorpommern, Jahrbuch 2004 52:185–195
Lampe R (2005) Late Glacial and Holocene water-level variations along the NE German Baltic Sea coast: review and new results. Quaternary International 133/134:121–136
Lampe R, Janke W (2004) The Holocene sea-level rise in the southern Baltic as reflected in coastal peat sequences. Polish Geological Institute Special Papers 11:19–30
Lampe R, Ziekur R, Schuricht R, Hoffmann G (2004) Georadaruntersuchungen zur Erkundung der Faziesarchitektur im Küstenholozän NE-Deutschlands unter Einbeziehung von Rammkernsondierungen und geoelektrischen Messungen. Zeitschrift für Geologische Wissenschaften 32(2–4):97–112
Lampe R, Endtmann E, Janke W, Meyer H, Lübke H, Harff J, Lemke W (2005): A new relative sea-level curve for the Wismar Bay, NE-German Baltic coast. Meyniana 57:5–35
Lampe R, Meyer H, Janke W, Ziekur R, Endtmann E, Schmedemann N (2007) Holocene evolution of an irregularly sinking coast – interplay of eustasy, crustal movement and sediment supply. Bericht der Römisch-Germanischen Kommission 88:15–46
Lange W, Menke B (1967) Beiträge zur frühpostglazialen erd- und vegetationsgeschichtlichen Entwicklung im Eidergebiet, insbesondere zur Flußgeschichte und zur Genese des sogenannten Basistorfes. Meyniana 17:29–44
Liebsch G (1997) Aufbereitung und Nutzung von Pegelmessungen für geodätische und geodynamische Zielstellungen. Deutsche Geodätische Kommission bei der Bayerischen Akademie der Wissenschaften, Reihe C(485)
Ludwig AO (2001) Vertical movements since the beginning of Rupelian stage (map 1). Brandenburgische Geowissenschaftliche Beiträge 8(1):5–12
Mandelkow E, Frenzel P, Lampe R, Kaute P, Schindler G (2005) Paläontologische Untersuchungen von Sedimentprofilen der achäologischen Grabung Stralsund-Mischwasserspeicher. Bodendenkmalpflege in Mecklenburg-Vorpommern, Jahrbuch 2004 52:263–281
Ministerium für Bau, Landesentwicklung und Umwelt (1994) Generalplan Küsten- und Hochwasserschutz Mecklenburg-Vorpommern. Schwerin
Montag H (1967) Bestimmung rezenter Niveauverschiebungen aus langjährigen Wasserstandsbeobachtungen der südlichen Ostseeküste. Arbeiten aus dem Geodätischen Institut Postdam 15:1–139
Mörner N-A (1980) The northwest European "sea-level laboratory" and regional Holocene eustasy. Palaeogeography, Palaeoclimatology, Palaeoecology 29:281–300
Naumann M (2006) Faziesdifferenzierung und holozäne Küstenentwicklung der Halbinsel Bug/Rügen. Thesis, Universität Greifswald
Neal A (2004) Ground-penetrating radar and its use in sedimentology: principles, problems and progress. Earth-Science Reviews 66(3–4):261–330

Nocquet JM, Calais E, Parsons B (2005) Geodetic constraints on glacial isostatic adjustment in Europe. Geophysical Research Letters 32(6):1–5
Posamentier HW, Allen GP (1999) Siliciclastic sequence stratigraphy: concepts and applications. SEPM Concepts in Sedimentology and Paleontology 7:1–210
Roy PS, Cowell PJ, Ferland MA, Thom BG (1994) Wave-dominated coasts. In: Carter RWG, Woodroffe CD (eds) Coastal evolution – Late Quaternary shoreline morphodynamics. Cambridge University Press, Cambridge
Stigge H-J (2003) Beobachtete Wasserstandsvariationen an der deutschen Ostseeküste im 19. und 20. Jahrhundert. Küste 66:79–102
Stolper D, List JH, Thieler ER (2005) Simulating the evolution of coastal morphology and stratigraphy with a new morphological–coastal behaviour model (GEOMBEST). Marine Geology 218:17–36
Streif H (2004) Sedimentary record of Pleistocene and Holocene marine inundations along the North Sea coast of Lower Saxony, Germany. Quaternary International 112:3–28
Uścinowicz S (2003) Relative sea level changes, glacio-isostatic rebound and shoreline displacement in the southern Baltic. Polish Geological Institute Special Papers 10:1–80
Uścinowicz S (2006) A relative sea-level curve for the Polish Southern Baltic Sea. Quaternary International 145/146:86–105
van Heteren S, Fitzgerald DM, McKinlay PA, Buynevich IV (1999) Radar facies of paraglacial barrier systems: coastal New England, USA. Sedimentology 45:181–200
Vink A, Steffen H, Reinhardt L, Kaufmann G (2007) Holocene relative sea-level change, isostatic subsidence and the radial viscosity structure of the mantle of northwest Europe (Belgium, the Netherlands, Germany, southern North Sea). Quaternary Science Reviews 26(25–28): 3249–3275
Walker M, Johnsen S, Olander Rasmussen S, Popp T, Steffensen J-P, Gibbard P, Hoek W, Lowe J, Andrews J, Björck S, Cwynar LC, Hughen K, Kershaw P, Kromer B, Litt T, Lowe DE, Nakagawa T, Newnham R, Schwander J (2009) Formal definition and dating of the GSSP (Global Stratotype Section and Point) for the base of the Holocene using the Greenland NGRIP ice core, and selected auxiliary records. Journal of Quaternary Science 24/1:3–17
Winn K, Erlenkeuser H, Nordberg K, Gustafsson M (1998) Paleohydrography of the Great Belt, Denmark, during the Litorina transgression: the isotope signal. Meyniana 50:237–251
Yu SY (2003) The Litorina transgression in southeastern Sweden and its relation to mid-Holocene climate variability. LUNDQUA thesis 51, 22 p + 6 apps, Lund
Yu SY, Berglund BE, Sandgren P, Lambeck K (2007) Evidence for a rapid sea-level rise 7600 years ago. Geology 35:891–894

Part V
Sediment Dynamics

Chapter 13
On the Dynamics of "Almost Equilibrium" Beaches in Semi-sheltered Bays Along the Southern Coast of the Gulf of Finland

Tarmo Soomere and Terry Healy†

Abstract Beaches along the northern coast of Estonia form an interesting class of almost equilibrium, bayhead beaches located in bays deeply cut into the mainland in an essentially non-tidal, highly compartmentalised coastal landscape, and that develop mostly under the influence of wave action. These beaches, although often suffering from a certain sediment deficit, are stabilised by the postglacial land uplift. We describe the basic features of their appearance and functioning from the viewpoint of sediment transport processes. Wave action normally impacts a relatively narrow nearshore band and additionally stabilises the beaches through littoral drift of sandy sediment and gravel towards the bayheads. Eolian transport and fluvial sediment supply have typically very modest magnitude. Such beaches, in general, evolve quite slowly and may represent an almost equilibrium stage, even when the active sand mass is very limited. The concept of the equilibrium beach profile is an adequate tool for their analysis. As an example, its parameters and longshore transport patterns are evaluated for Pirita Beach based on a granulometric survey and long-term simulation of wave climate. It is demonstrated that net sand changes for such beaches can be estimated directly from the properties of the equilibrium profile, land uplift rate, and loss or gain of the dry beach area. Another type of highly dynamic equilibrium exists owing to interplay of the effects of river flow and wave action at the mouths of large rivers such as the Narva River.

Keywords Beaches · Gulf of Finland · Baltic Sea · North Estonian coast · Sediment transport · Almost equilibrium beaches · Equilibrium beach profile · Wave climate · Sediment loss

T. Soomere (✉)
Institute of Cybernetics, Tallinn University of Technology, 12618 Tallinn, Estonia
e-mail: soomere@cs.ioc.ee

†Terry Healy passed away in July 2010.

J. Harff et al. (eds.), *The Baltic Sea Basin*, Central and Eastern European Development Studies (CEEDES), DOI 10.1007/978-3-642-17220-5_13,

13.1 Beaches Along the North Estonian Coast

The complexity of the dynamics of the Baltic Sea (Fig. 13.1) extends far beyond the typical features of water bodies of comparable size. Pronounced salinity gradients and rich mesoscale dynamics distinguish this basin from large lakes and create a similarity of its basic processes with those occurring in the open ocean (Alenius et al. 1998). The regular presence of sea ice plays a considerable role in its functioning (Kawamura et al. 2001, Granskog et al. 2004). Marine meteorological conditions reveal remarkable anisotropy and non-homogeneous patterns of wind and wave fields (Myrberg 1997, Soomere and Keevallik 2003, Soomere 2003).

The most interesting basin in this respect is the Gulf of Finland (Fig. 13.2, Soomere et al. 2008c). Its hydrodynamical fields reveal highly interesting patterns of currents (Andrejev et al. 2004) and its small size is the basis of its high susceptibility with respect to (changes of) the external forcing factors. Dominant winds blow obliquely with respect to the axis of the gulf, giving rise to wave systems with a specific orientation (Kahma and Pettersson 1994, Pettersson et al. 2010) that frequently differs from the wind direction. A short "memory" of wave fields (Soomere 2005) combined with highly intermittent local wave regime (Soomere 2008) makes it frequently possible to identify the impact of single storm or wind event in the coastal landscape. While large parts of the Baltic Sea coasts express relatively simple geomorphic and lithodynamic features (e.g. the almost straight eastern coast

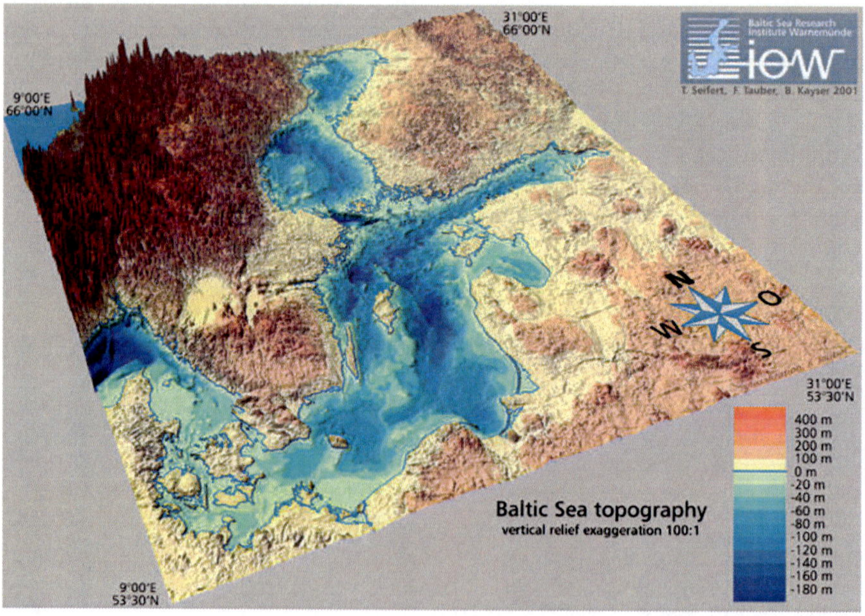

Fig. 13.1 Location and bathymetry of the Baltic Sea. From Seifert et al. (2001) by kind permission of T. Seifert

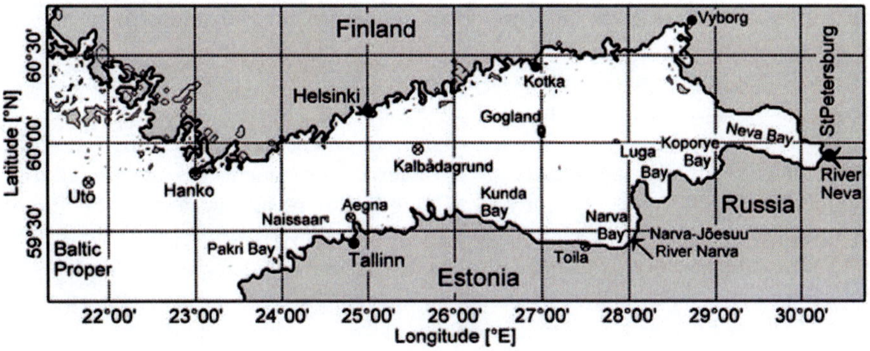

Fig. 13.2 Scheme of the Gulf of Finland

from Poland up to Latvia or bedrock-based, extremely stable archipelago areas along the Swedish and Finnish coasts), understanding of physics and dynamics of lithohydrodynamical processes along the coasts of the Gulf of Finland is still a challenge. The most interesting are the southern and eastern coasts of the Gulf of Finland that belong to a rare type of young, relatively rapidly uplifting beaches. This chapter makes an attempt to depict the basic factors jointly governing their evolution and to make use of their closeness to the equilibrium for practical estimates of sand loss or gain.

Different from several downlifting coasts in southern Sweden or in Denmark that are largely open to substantial hydrodynamic loads and are rapidly developing (Hanson and Larson 2008), beaches in the Gulf of Finland area are stabilised by relatively rapid postglacial uplift, the magnitude of which ranges from about 1 mm/year in the eastern part of Estonia near Narva up to about 2.8 mm/year in the northwestern part of the coast (Vallner et al. 1988, Miidel and Jantunen 1992). This uplift combined with relatively low hydrodynamic activity and limited supply of sand has led to a specific type of "almost equilibrium" beaches that develop relatively slowly. Such a slow net development is not specific to the Baltic Sea and in many cases the evolution of beaches is governed by highly dynamic processes (such as littoral drift that generally carries sediments to the bayheads, supply of sediments by rivers, the mouths of which are located at the bayheads, and the above-mentioned land uplift) that jointly keep the beach in an equilibrium state.

The above suggests that the beaches in question eventually are extremely sensitive to changes in external factors. For example, an increase of the global sea level, increased discharge during more pronounced spring floods in the climate of the future (The BACC Author Team 2008), construction of a dam to regulate the river flow (Velegrakis et al. 2008), or a new coastal engineering structure blocking the littoral drift (Soomere et al. 2007) may easily distort the balance. An important issue for sustainable management of such beaches is establishing the parameters of their equilibrium regime, the magnitude of the sediment supplies, and the basic patterns of the natural sediment transport processes. Based on this information, well-justified decisions can be made for their protection, controlled modification (for example, managed retreat, see Healy and Soomere 2008), or reconstruction.

Another type of equilibrium exists at the mouths of large rivers such as the Narva River where seasonal variations of the river flow and wave intensity give rise to interesting seasonal changes of the sand bar (Laanearu et al. 2007). Cyclic variations in the height of the bar apparently are an inherent component of a specific kind of equilibrium in such areas. A highly interesting feature of such systems is the role of stratification of water masses in the vicinity of the sill.

The central goal of this chapter is to summarise the characteristic features of the appearance and the dynamics of almost equilibrium beaches on the southern coast of the Gulf of Finland and to present the recently developed applications for rapid estimates of their basic parameters based on a few relatively easily measurable or computable parameters. The presentation is mostly based on two examples. Pirita Beach, located at the head of Tallinn Bay, is supported by a multitude of factors and is a typical example of a beach whose evolution is generally slow and has been largely controlled by development works. Narva-Jõesuu Beach in the Narva River mouth area represents a more or less straight, widely open beach, with its highly interesting interplay of processes forced by the littoral drift and the voluminous river inflow.

The chapter is organised as follows. First, we describe the basic properties of the beaches, major drivers governing the evolution of the beaches, and main features of sediment transport. These aspects are discussed in more detail for Pirita and Narva-Jõesuu beaches, with emphasis on estimates of the parameters of the classical (Dean's) equilibrium profile for Pirita Beach. Further on, the sediment budget for Pirita Beach and its changes in the recent past are discussed together with the potential of a recently developed application for express estimates of the net gain or loss of sediment based on inversion of the Bruun Rule. Finally, the nature of seasonal variations in the dynamic equilibrium caused by the interplay of littoral drift and river discharge is analysed for Narva-Jõesuu Beach.

13.2 Forcing Factors of Sediment Transport Processes

Sediment transport processes in the littoral system are driven by a large number of external processes such as oscillatory wave motions, wind-induced transport, coastal currents and wave-induced longshore flows, variations of water level, sea ice. Equally important are the local factors such as the geometry of the coast, the sediment textural characteristics, and the availability of mobile sediments.

13.2.1 Internal Properties of Beaches

The beaches of the southern and the northern coasts of the Gulf of Finland (Fig. 13.2) are completely different. The northern coast is characterised mostly by "skären"-type beaches, the evolution of which is weakly affected by hydrodynamic factors (Granö and Roto 1989). The coast largely consists of extremely stable and very irregular bedrock formations that are mostly stripped of finer sediment. The presence of such formations gives rise to highly irregular bathymetry, extensive archipelago areas, and extremely complex geometry of the coastline.

In contrast, the eastern and southern coasts of this gulf were formed and developed predominantly under the effect of wave action (Orviku and Granö 1992). The coasts obtained their contemporary shape only a few millennia ago (Raukas and Hyvärinen 1992). The volume of sediment and the magnitude of littoral drift are modest. The most common type of coasts here are the embayed coasts which are straightening with sediment accumulation. The most stable beaches are located in deeply indented bays.

Most of the sandy beaches along the North Estonian coast overlie ancient dunes and river deltas. An overview of geology and geological history of the entire Gulf of Finland area is presented in the collection by Raukas and Hyvärinen (1992). The descriptions of general properties of the beaches at its southern coast are mostly published in Russian (Orviku 1974, Orviku and Granö 1992). A shorter review of the relevant knowledge is given by Soomere et al. (2007).

Two distinctive subsections can be distinguished along the North Estonian coast. Deeply embayed beaches along the northern coast of Estonia (including Pirita Beach) westwards from the longitude 27°E reveal many properties of bay beaches for which waves are generated under the effective fetch distances <50 km (Nordstrom 2005). They are mostly geometrically sheltered from high waves coming from a large part of the potential directions of strong winds (Soomere 2005). As a consequence, their local wave climate is relatively mild compared with that in the open part of the Gulf of Finland or in the adjacent sea areas. For example, the annual mean significant wave height is as low as 0.29–0.32 m in different sections of Pirita Beach (Soomere et al. 2007). On the other hand, wave fields at widely open sections of the coast eastwards from the longitude 27°E are almost totally governed by the properties of wind in the open parts of the Gulf of Finland (Laanearu et al. 2007).

The western part of the North Estonian coast can be divided into many small sedimentary compartments and isolated beaches of length frequently <1 km (Soomere et al. 2007) separated by rocky peninsulas and headlands. Viimsi Peninsula located next to Tallinn (Fig. 13.3) divides the embayed beaches of the northern coast of Estonia into two subsets. The features of the coast eastwards from this peninsula are mainly related to glacial and fluvioglacial formations and deposits of the foreklint lowland while the bays westwards (including Tallinn Bay) are mostly associated with structural blocks and ancient erosional valleys cut into the bedrock (Orviku and Granö 1992). This division is immaterial from the viewpoint of this chapter.

The composition of the upper layers of the sand mass in several North Estonian beaches reveals typical features of bay beaches formulated by Nordstrom (2005). Studies of drill cores extending to a depth of 2.1 m into the sea floor near Pirita show that in deeper areas (down to depths of 15–20 m), the sampled layer consists entirely of relatively well-sorted material. In contrast, several thin medium-

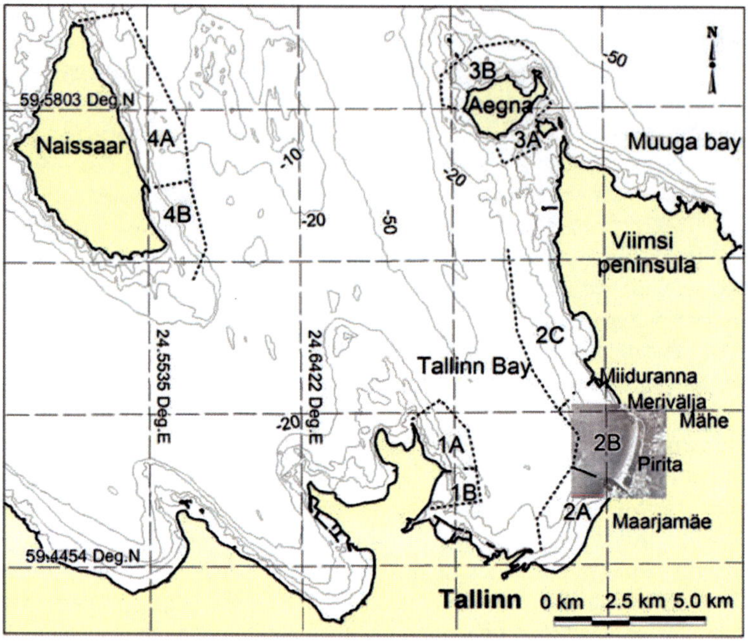

Fig. 13.3 Location scheme of Pirita Beach and adjacent coastal engineering structures. *Dotted lines* show cells of sediment transport according to Soomere et al. (2007): cell 1 along the western coast of the inner part of Tallinn Bay (divided into subcells 1A and 1B by Katariina jetty); cell 2 along the eastern coast of Tallinn Bay (divided into subcells 2A, 2B, and 2C by Pirita Harbour and the Port of Miiduranna); cell 3 surrounding the island of Aegna (subcells 3A and 3B at the southern and northern coasts, respectively), and cell 4 along the eastern coast of the island of Naissaar (divided into subcells 4A and 4B by Naissaar Harbour). Isobaths of –2, –5, –10, –20, and –50 m are shown based on information from Estonian Land Board. The step of the co-coordinate grid is 5 km. Graphics by A. Kask. Reprinted with permission from the Estonian Academy Publishers

and coarse-grained sand bodies were detected at water depths between 2 and 10 m. Similar depositional sequences have been detected in other sandy areas adjacent to several North Estonian river mouths (Lutt 1992).

The transition between the fine and coarse sand bodies is quite sharp whereas the transition between fine sand and coarse silt is generally gradational. Coarser sand bodies are poorly sorted at Pirita and contain a number of different fractions, none of which dominates, whereas the fine sand bodies are usually well sorted (Lutt 1992). This property is also frequent for the North Estonian sandy areas (Lutt 1985, Lutt and Tammik 1992, Kask et al. 2003a). Most probably, it reflects the generic property of bay beaches where the depth of mobilisation of sediments frequently is fairly shallow and the active beach may be only a thin veneer of unconsolidated material overlying an immobile layer of sediments (Nordstrom 2005).

13.2.2 Forcing Factors

The effect of ice (which is usually present 60–80 days annually, Sooäär and Jaagus 2007) is mostly indirect and consists in either moving of boulders, damage to dune forest, or reducing the wave loads during the ice season. For cliffed coasts and exposed till and dune bluffs, the effect of frost heave and spring melt on the sand, and layered cliffs contribute sediment to the littoral system. The tidal range is 0.01–0.02 m in the area in question. The tidal currents are hardly distinguishable from the other motions. Water level at the beaches is mainly controlled by hydrometeorological factors. The range of its monthly mean variations is 0.2–0.3 m (Soomere et al. 2008c) but its short-term deviations from the long-term average are larger and frequently reach several tens of centimetres. Water levels exceeding the long-term mean more than 1 m are rare. The highest measured level at the Port of Tallinn is 1.52 m on 09.01.2005 (Suursaar et al. 2006) and the lowest is –0.95 m. Even larger variations of the extreme water level occur in the eastern part of the Gulf of Finland, for example, the historical highest water level has been 4.21 m in Sankt Petersburg and 2.02 m in Narva-Jõesuu (Soomere et al. 2008c).

Coastal currents induced by large-scale circulation patterns are modest in the whole Gulf of Finland (Alenius et al. 1998). Their speed is typically 0.1–0.2 m/s and only in exceptional cases exceeds 0.3 m/s. In the bayheads, such as the nearshore of Pirita Beach, current speeds apparently are even smaller. Although there is some evidence about a relatively stable pattern of coastal currents in Narva Bay (Andrejev et al. 2004), the current speed is usually modest there. Local currents are at times also highly persistent in the coastal zone next to Pirita Beach (Erm et al. 2008) and may provide appreciable intensity of transport of finer fractions of sand that are suspended in the water column even though the typical settling time of these fractions is only a few minutes.

The magnitude of wave-induced bedload transport greatly exceeds that of the current-induced transport even at relatively large depths (8–10 m) of open-sea areas (see Soomere et al. 2007 and references therein). Therefore, wave action in the surf zone evidently plays the decisive role in functioning of the beaches at the southern coast of the Gulf of Finland, as is typical for beaches located in microtidal seas. Exceptions form the mouths of relatively large rivers where seasonal variation of a sill height is jointly governed by a similar variation of the magnitude of river outflow (that erodes the sand bar) and the longshore wave-induced sediment transport (that increases the sill height) (Laanearu et al. 2007).

The wave climate of the Gulf of Finland matches that in the open Baltic Sea. It is generally mild, with the annual mean significant wave height well below 1 m and the typical wave periods usually not exceeding 7–8 s (Soomere 2005). The average and, in particular, the maximum wave heights in the gulf are much smaller than those in the Baltic Proper (Soomere 2008). Yet very rough seas with the significant wave height >4 m occur in the open parts of the gulf approximately once in a decade (Soomere et al. 2008c). The wave activity has a strong seasonal variation in the Gulf of Finland with the highest wave loads usually occurring during the autumn and early winter (September–December) whereas the late spring and summer months are relatively calm (Zaitseva-Pärnaste et al. 2009).

For long and high waves excited by strong southwestern storms, the geometry of the northern Baltic Sea and the Gulf of Finland suggests that full geometric blocking (Caliskan and Valle-Levinson 2008) should occur. Most of the waves affecting beaches on the southern coast of the Gulf of Finland thus originate from the gulf itself. Under specific conditions, however, western winds may still bring appreciable amounts of wave energy stemming from the northern sector of the Baltic Proper to certain coastal sections of the gulf (that are open to the west). If this happens, very high and extremely long waves may penetrate deep into the gulf (Soomere et al. 2008a). Such wave systems have much longer periods ($T = 12$–14 s) and impact the nearshore at much greater depths compared to local wind seas. As such events are usually accompanied by high water levels, major damage may occur to beaches that are widely open to the Gulf of Finland such as Narva-Jõesuu Beach.

The "memory" of wave fields is relatively short and the changes in the wind field are fast reflected in the wave pattern. As a consequence, the instantaneous wave fields in smaller sub-basins (such as Tallinn Bay or Narva Bay) rapidly mimic the changes of the open-sea winds (Soomere 2005, Laanearu et al. 2007).

A specific feature of the northern coast of Estonia is the largely intermittent nature of the local wave climate. As different from the classical examples of bay beaches, the bayhead beaches here are only partially sheltered from intense waves. Very high waves occasionally penetrate into such bays and cause intense erosion of their coasts (Kask et al. 2003b). For example, the significant wave height in Tallinn Bay usually exceeds 2 m each year, may reach 4 m in NNW and western storms (Soomere 2005), and may overshoot 2.5 m in the nearshore of Pirita Beach during NNW storms (Soomere et al. 2007). Such storms usually cause littoral transport towards the bayhead beaches, but may severely damage coastal sections in their neighbourhood.

13.2.3 Local Sediment Transport

As contemporary rivers in North Estonia are fairly small (except for the Narva River) and cross mostly areas overlying limestone, they bring relatively little amounts of sediments into the sea. Moreover, sand forms only a very little fraction of the fluvial sediments. For example, the Pirita River provides about 400 m^3 of suspended matter annually. Most of its fluvial sediments (74%) possess a grain size from 0.01 to 0.05 mm and about a quarter has a size from 0.0025 to 0.01 mm (Lutt and Kask 1992, pp. 149–152). Only a few rivers cross sandstone layers and fairly small sandstone sections of the North Estonian coast are open to wave action.

The largest source of sand are bluffs cut into glacial deposits (including eskers, Estonia lies in the periphery of the esker distribution area of the Scandinavian glaciation, Karukäpp 2005) and ancient beach ridges which are open to the wave action along some sections of the coast. Their sand content is relatively small, usually well below 25%. The material contains many cobbles, pebbles, and boulders which frequently form a protective pavement in the surf zone. Coastal erosion thus also insignificantly feeds the system with coarse sedimentary material. It is therefore not unexpected that the entire North Estonian coast suffers from beach sediment deficit and even the healthiest sections of coast (that usually show clear accumulation features) at the bayheads are from time to time subject to erosion.

Wave- and current-induced sand recycling is concentrated in a relatively narrow nearshore band that extends from the seaward border of the surf zone to the area influenced by the runup of largest waves. Its location is relatively stable, which is in contrast to open coast tide-dominated environments where the position of this band frequently varies with respect to the dune toe. Only in (infrequent) high water conditions does the band extend to the dry beach. Owing to the mild overall wave climate, its width is a small fraction of that for the open ocean, high-energy beaches (Wright and Short 1984). This feature together with the generic properties of bay-head, low-energy beaches allows the existence of persistent beaches consisting of a relatively small volume of sand. Such beaches are, however, extremely vulnerable with respect to new forcing factors such as waves from fast ferries (Soomere et al. 2009).

Fine sand that frequently dominates in the North Estonian beaches (Fig. 13.4) can easily undergo eolian transport when dry; however, winds sufficient for extensive eolian transport are infrequent. The shoreline of many beaches (including Pirita Beach) is more or less parallel to the predominant southwestern and western winds (Soomere and Keevallik 2003). Strong onshore (northwestern) winds typically occur either during the late stage of storms or during the autumn months when sand is wet. Although wind may even carry a certain amount of wet sand to the dunes, the overall intensity of dune building is modest. The height of the existing semi-active dunes, which are already largely vegetated, is a few metres. Only sand on the exposed side, seawards from the faceted dune face, actively undergoes eolian transport. The role of eolian transport apparently has been larger in the past (Raukas and Teedumäe 1997, Soomere et al. 2007). A similar situation occurs at Narva-Jõesuu Beach, the

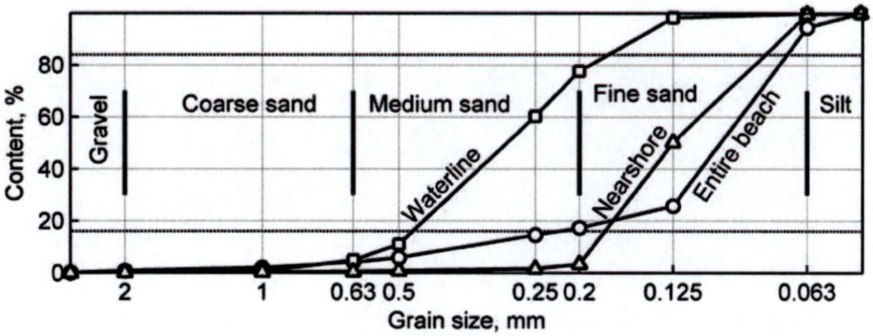

Fig. 13.4 The cumulative distribution of different grain size fractions at Pirita based on samples collected in 2005 and 2007. The *horizontal dashed lines* represent the 16 and 84% of the sand mass (Soomere et al. 2007; reprinted with permission from the Estonian Academy Publishers)

eastern part of which is wide enough to support relatively intense eolian transport along the shoreline (Laanearu et al. 2007).

13.3 Features of Pirita Beach and Narva-Jõesuu Beach

Owing to historical reasons, there exists a very limited amount of detailed studies into properties of North Estonian nearshore and beaches starting from the 1940s (Suuroja et al. 2007). The reason is that this coast was mostly a border zone in the USSR and thus mostly closed to public and almost inaccessible also for scientific research. Only a few sections of the coast (such as sandy beaches at Pirita and Narva-Jõesuu) were open for recreational purposes and data about their evolution cover to some extent also the time interval from the 1940s to the end of 1980s.

Both beaches in question have been subject to considerable anthropogenic impact for several decades. Prior to the mid-twentieth century, Pirita Beach (Fig. 13.3) was apparently stabilised by the postglacial uplift and natural sediment supplies. During recent decades, however, a gradual decrease of the dry beach width, rapid recession of the till cliff at the northern end of the beach, and extensive storm damage to the dunes (Fig. 13.5) have occurred despite the postglacial uplift and attempts to refill the beach with material dredged from a neighbouring harbour or transported from mainland quarries. The main reason behind the gradual beach degradation is the human intervention that has cut down the major natural sand supplies to the beach (Soomere et al. 2007).

Fig. 13.5 Erosion and loss of pine forest on the low dune during a strong storm in January 2005 (above; photo by I. Kask; from Soomere et al. (2007) with permission from the Estonian Academy Publishers)

Natural supplies of sand to the beach originate from the Pirita River, littoral transport along the western coast of Viimsi Peninsula, and sporadic erosion of sand from a glacial till scarp at the northern end of the sandy sector as well as from the dunes of the middle and the northern part of the beach (Soomere et al. 2007). All these sources have undergone major changes within the twentieth century. The construction of a small jetty out to about 3 m water depth in 1925–1927 just to the north of the beach diminished the transport of coarser sediments. The quays of Miiduranna Port, commenced in the 1970s, extend out to the natural depth of 6–8 m and almost completely block wave-induced alongshore sediment transport. At the turn of the millennium the depth of its fairway was dredged to 13 m. This blocking subsequently leads to sediment deficit and relatively fast erosion southwards from these constructions.

The supply of fluvial sediments by the Pirita River (about 400 m^3/year in the past, Lutt and Kask 1992) is entirely blocked by the Olympic sailing harbour that was built in the mid-1970s and today acts as a settling basin. A revetment from granite stones was constructed along the dune toe in the 1980s to protect dunes in the northern sections of Pirita Beach against erosion. The till scarp that was subjected to direct wave action under storm surge conditions and also supplied a certain amount of sand was protected by a new seawall in 2006–2007 (Soomere et al. 2007, 2008b). There have been several attempts to increase the active sand mass of the beach starting from the late 1950s by pumping sediments from the river mouth to different sections of the beach. The joint effect of all the human activities led not only to blocking of the major supplies of coarser sand of relatively high recreational value but also to a gradual decrease of the beach dominant grain size owing to the potential misbalance of the supply of different fractions.

On the other hand, Pirita Harbour blocks the lateral sand loss from the beach. The beach profile, therefore, should be relatively stable and the concept of the equilibrium beach profile is accordingly an appropriate tool for its analysis.

The vicinity of the Narva River mouth that also hosts Narva-Jõesuu Beach (Fig. 13.6) serves as an example of a littoral system that is completely open to the Gulf of Finland. This river is the largest in Estonia. Its long-term mean volume discharge is around 400 m^3/s (Protasjeva and Eipre 1972) whereas considerably (up to two times as large) larger monthly values occur during spring floods. The bay is mostly sandy at the coast and is open to the dominating winds (Laanearu and Lips 2003). The river mouth area therefore is an interaction zone between wave- and current-induced sand motions.

According to the classification of Wright (1985), the Narva River mouth belongs to the Senegal type characterised by high wave energy, strong littoral drift, and relatively steep offshore shoaling slope. The presence of littoral drift commonly results in the formation of a submerged sand bar (sill) that gradually moves in the prevailing direction of the littoral drift. The bar forces the river flow to bend in the same direction and occasionally (for example, when the river flow is weak) it creates inconvenience by initiating flooding of the river delta or hindering navigation (Carter 2002). Indeed, the sand bar in the Narva River mouth presents a highly

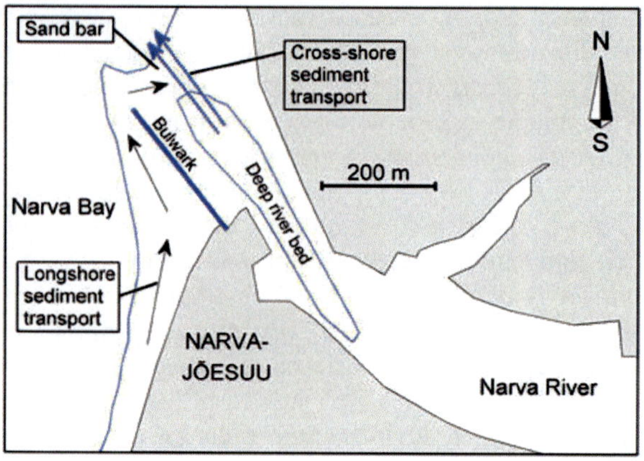

Fig. 13.6 Scheme of geometry, bathymetry, and patterns of sediment transport at the Narva River mouth. Redrawn based on an image from Laanearu et al. (2007)

inconvenient obstacle to ship traffic during relatively low water-level conditions (Laanearu et al. 2007).

Littoral processes near the Narva River mouth have also been strongly modified by the presence of the Narva-Jõesuu breakwater. It is a concrete structure about 300 m in length and built in the late 1980s (i) to facilitate navigation between the harbour and sea and (ii) to prevent the extensive beach erosion observed in the 1970s. The structure is built approximately perpendicular to the beach, westwards from the river mouth. It effectively traps sand that generally is transported eastwards along the coast. At the western side of the bulwark, the width of the dry beach has

Fig. 13.7 Damages to the Narva-Jõesuu bulwark in 2005

increased considerably whereas strong erosional features are evident at its eastern side (Fig. 13.7). While it probably prevented the forming of the sand bar for some time, the bar has re-emerged at the seaward end of the bulwark. The entire system seems to have reached a more or less equilibrium stage in terms of long-term variations. This equilibrium at the river mouth is, however, highly dynamic in the sense that considerable seasonal variations of the height of the sand bar may occur in the system (Laanearu et al. 2007).

13.4 Equilibrium Profiles and Transport Patterns

Although a particular beach profile may undergo substantial changes, an average of the instantaneous profiles over a long period usually preserves a relatively constant shape called the equilibrium beach profile (EBP, Dean 1991). The temporal and spatial resolution of available surveys at the southern coast of the Gulf of Finland is too low for adequate estimate of properties of the EBPs. For that reason the relevant studies (Soomere et al. 2007, 2008b) rely on theoretical estimates of the shape of the EBP based upon the concept of uniform wave energy dissipation per unit water volume in the surf zone (Dean and Dalrymple 2002, Chap. 7).

The water depth $h(y)$ along such profiles at a distance y from the waterline is $h(y) = Ay^{2/3}$, where the profile scale factor A depends on the grain size of the bottom sediments. As Narva-Jõesuu Beach is an example of large, high-energy beach in the Gulf of Finland context, it is natural to assume that its grain size is also largely homogeneous. This is largely the case also for Pirita Beach where in 2005–2007 bathymetry and sediment texture were mapped in the nearshore between the waterline and the 11 m depth contour along an about 2.5-km-long section of the coast (Soomere et al. 2007). The average grain size in the nearshore of Pirita Beach is close to 0.12 mm. Although the mean grain size does vary to some extent along the beach, the corresponding variations of the factor A are fairly small: it is approximately 0.07–0.08 for the northern and about 0.063 for the southern part of the beach (Soomere et al. 2008b). Therefore, it is adequate to use a fixed value of the factor A=0.07 that corresponds to the overall average grain size for Pirita (Dean et al. 2001).

Several variations of the mean grain size and the content of the fractions should occur naturally within the beaches in question. Since waves sort the sediments and the finer fractions are gradually transported offshore, deeper areas usually host the finest sediments and the coarser material is concentrated in the vicinity of the breaker line and at the waterline (Dean and Dalrymple 2002). Coarser-grained sand is found along the waterline (where the maximum content of medium sand is up to 84%) and finer components in deeper areas of Pirita Beach indeed (Soomere et al. 2007). The mean grain size along the waterline is much larger than in the rest of the study area (Fig. 13.4).

The proportion of coarser sand generally decreases offshore. This is evidently related to the above-discussed highly intermittent nature of wave activity: the breaker line is poorly defined and the relevant band of relatively coarse sand is not always apparent. This feature, not entirely typical but also not very surprising (Dean and Dalrymple 2002, Chap. 2.3.2), may play an important role in planning of beach nourishment activities because material with the grain size much smaller than the one at the waterline may be lost relatively fast. Moreover, relatively coarse and well-sorted sand is perceived to be of the largest recreational value. In other words, beach fill with fine sand would lead to a decrease in the beach quality.

Another basic parameter of the equilibrium beach profile is the depth of closure h^* at which repeated survey profiles pinch out to a common line (Kraus 1992). It represents the maximum depth at which the breaking waves effectively adjust the surf zone profile. Seawards from the closure depth, waves may occasionally move bottom sediments but they are not able to maintain a specific profile. The closure depth may be different for different sections of the beach and generally should be treated as a function $h^*(x)$ of the distance x along the shoreline.

Several authors have suggested simple empirical expressions for h^* based on certain integral measures of the wave activity. A specific feature of wave climate in the entire Baltic Sea is that the average wave conditions are mild, but very rough seas may occur episodically in long-lasting, strong storms (Soomere 2008). Waves in such storms are much higher than one would estimate from the average wave conditions. Moreover, the strongest storms in the Gulf of Finland tend to blow from directions from which winds are not very frequent (Soomere and Keevallik 2003, Soomere 2005). As a result, the simplified estimates based on the annual mean wave parameters substantially underestimate the closure depth (Soomere et al. 2007, 2008b). More elaborate estimates that additionally account for the duration of the strongest storms and for the wave periods in such storms lead to adequate results. For example, for Pirita an acceptable approximation for h^* is (Birkemeier 1985)

$$h^* = p_1 H_{s,0.137} - p_2 \frac{H_{s,0.137}^2}{gT_s^2}, \quad p_1 = 1.75, p_2 = 57.9, \tag{1}$$

where $H_{s,0.137}$ is the threshold of the significant wave height that occurs 12 h a year, that is, the wave height that is exceeded with a probability of 0.137%, and T_s is the peak period in such wave conditions (Soomere et al. 2007, 2008b).

Although only two parameters are necessary to adequately estimate the closure depth, the relevant information generally is not provided, nor is it in the wave atlases or able to be extracted from the existing wave measurements (Kahma et al. 2003, Pettersson 2001) in the northern Baltic Sea and in the Gulf of Finland. Similar problems are frequently encountered in many regions of the world and long-term numerical simulations are a feasible way to approach them.

The wave climate in the vicinity of Pirita Beach and in Narva Bay was estimated on the basis of a simplified scheme for long-term wave hindcast with the use of a triple-nested version of the WAM model (Laanearu et al. 2007, Soomere et al. 2008b). This model, although constructed for open ocean conditions and for relatively deep water (Komen et al. 1994), gives good results in the Baltic Sea, provided the model resolution is appropriate and the wind information is correct (Soomere 2005). Since waves are relatively short in the Gulf of Finland, the innermost models

(24 evenly spaced directions, grid step of about 1/4 nautical miles, 24 frequencies from 0.042 to 0.41 Hz with an increment of 1.1 for Narva Bay, 42 frequencies up to 2.08 Hz for Tallinn Bay) allow description of wave properties in the coastal zone, up to depth of about 5 m and as close to the coast as about 200–300 m (Soomere 2005).

The model was forced with wind data from Kalbådagrund (Fig. 13.2, 59°59′°N, 25°36′°E). This is the only measurement site in the Gulf of Finland that correctly represents marine wind conditions. The presence of ice is ignored. The computed annual mean parameters of wind waves are, therefore, somewhat overestimated and represent average wave properties during the years with no extensive ice cover. The model was used for calculation of long-term statistics for Tallinn Bay (Soomere et al. 2007, 2008b) and for time series of wave properties in 2002 for Narva Bay (Laanearu et al. 2007).

Detailed calculations have been performed for sections with a length of about 0.5 km relating to the nearshore off Pirita for 1981–2002. The threshold for the significant wave height occurring with a probability of 0.137% varies between 1.45 and 1.58 m along the beach. The typical peak period T_s in such storms is about 7 s. Expression (1) gives reasonable values of 2.36–2.57 m for the closure depth that match the bathymetric survey data (Soomere et al. 2007). These values are apparently typical for many bay beaches along the northern coast of Estonia. Only a few more exposed sections may be subject to larger wave loads (in terms of both the threshold of the significant wave height that occurs 12 h a year and the peak period in such wave conditions) and host equilibrium profiles extending to somewhat greater depths. For the case of Pirita, given the approximate value of the scale factor $A = 0.07$, the width of the equilibrium profile is expected to be about 250 m and its mean slope approximately 1:100.

13.5 Applications for "Almost Equilibrium" Beaches

One of the basic properties characterising the beaches is the magnitude of sediment transport. Its properties and spatial patterns of longshore transport can be relatively easily calculated for the almost equilibrium beaches under consideration. It is convenient to estimate the intensity of alongshore sediment transport in terms of its potential rate Q_t (Coastal Engineering Manual 2002). An equivalent measure is the potential immersed weight transport rate

$$I_t = (\rho_s - \rho)\, g\, (1 - p)\, Q_t, \tag{2}$$

which accounts for voids between sediment particles and the specific weight of the sediment components. Here ρ_s and ρ are the densities of sediment particles and seawater, respectively; $g = 9.81$ m/s^2 is the acceleration due to gravity; and p is the porosity coefficient. Both these measures express the volume of sediments carried through a cross-section of the beach in ideal conditions within a unit of time. The factual magnitude of the transport is much less along the North Estonian beaches

(Soomere et al. 2008b). The difference is particularly large (up to several orders of magnitude) when the sediment layer is not continuous (as it is northwards of Pirita Beach) or has a limited thickness. For this reason, the calculated rates must be used with caution. For example, the difference between the estimates for different sections of the beach carries the key information about their vulnerability with respect to changes of sediment transport processes. Another key quantity is the ratio of net and bulk transport rates that characterises the intensity of sediment transit at a particular site.

13.5.1 Sediment Balance at Pirita Beach

Detailed calculations of the potential transport rate for Pirita Beach have been performed by Soomere et al. (2008b) based on numerical simulation of the local wave properties and the CERC (*Coastal Engineering Research Center*) method. The latter is based on the assumption that the potential immersed weight transport rate I_t is proportional to the rate of beaching of wave energy flux (wave power) per unit of the coastline P_t. The latter quantity depends on the wave height, period, and approach angle. The relevant expression $I_t = KP_t$ is usually referred to as the CERC formula. The non-dimensional proportionality coefficient, K, is frequently expressed as a certain function of the wave approach angle, the maximum orbital velocity in breaking waves, and the sediment fall velocity in the surf zone, the latter dependence implicitly expressing the properties of sediments (Coastal Engineering Manual 2002, part III-1). Such model set-ups have been widely used in the southern Baltic Sea conditions (Kuhrts et al. 2004, Fröhle and Dimke 2007).

The calculations revealed that the potential transport rate (consequently, also the overall functioning of the sedimentary system) at Pirita is almost independent of the grain size for a fairly wide range (from 0.063 to 0.2 mm) of the mean size. Longshore sediment motions at Pirita are thus almost entirely governed by the match of the wave propagation direction and the geometry of the coast. This feature suggests that potential changes of the transport patterns when the grain size is modified (for example, through beach refill) are fairly modest.

The calculated transport rate patterns generally coincide with different geomorphic features such as sections of intense sediment transit, areas of erosion and accretion, the presence and orientation of sand bars, or sections with extensive retreat of the dune toe during strong storms. The discrepancies between numerically simulated transport patterns and the appearance of the beach are fairly minor and become evident only in limited sections. This match shows that the quality and resolution of the wave and sediment transport models in use, the quality of the information about the granulometry, and the quality of the atmospheric forcing are sufficient for resolving the basic features of functioning of typical beaches along the North Estonian coast.

The performed simulations made it possible to derive an estimate of the anthropogenic changes of the magnitude of the littoral drift from the North to Pirita Beach. The beach is fed by the flux of relatively fine sediments from the North (say, with a magnitude R, which usually is a small fraction of the transport rate I_t) and by unsorted material abraded from the cliff at the northern end of the sandy strip. The erosion of the cliff has been mapped by several topographic surveys. The cliff sediments comprise roughly 1/3 of sand and gravel. If the amount of M is abraded from the cliff, the beach receives about $\frac{1}{3}M$ of material. Also, at times a certain amount of sand S is eroded from the dune scarp and berm along the sandy beach. The latter quantity has been estimated as $S \approx 400\ \text{m}^3$/year in 1997–2006 from the results of subsequent topographic surveys (Soomere et al. 2007). The earlier observations suggest that the sand volume of the beach was more or less unchanged before the 1970s (Soomere et al. 2007). The balance equation for the sand volume was thus

$$\Delta Q = R + \frac{1}{3}M + S - D = 0, \tag{3}$$

where D is the net loss of sand volume to the deeper areas. There are no lateral loss terms in Eq. (2), because (i) the Pirita Harbour completely stops the littoral drift and (ii) the southwards drift overwhelmingly dominates at the northern border of the beach. Assuming that the beach was in equilibrium in the past, this balance equation can be used to calculate the flux R in the past provided the contemporary average rate D of net sand loss from the beach to offshore is known.

13.5.2 Sediment Loss from Almost Equilibrium Beaches

To obtain an accurate estimate of the net sand loss normally requires long-term measurements of sediment transport, or sediment trapped at a groin, or historical geomorphic and bathymetric changes, and thus is time-consuming and costly. A simple method is proposed by Kask et al. (2009) for rapid estimation of this quantity for beaches where the sediment loss or gain is almost balanced by the land uplift or downsinking. The method consists of inverting the Bruun Rule (Bruun 1962). The sediment loss or gain is expressed in terms of the changes of the dry land area, the width of the equilibrium beach profile, and the uplift or downsinking rate. The method essentially relies on the existence of a more or less persistent beach profile.

Usually, the Bruun Rule is expressed as the linear relation $\Delta y = -\Delta S/\tan\theta$ between the shift Δy of the shoreline and the relative water level rise ΔS, where the proportionality coefficient is the inverse mean slope $\tan\theta$ of the equilibrium profile. This relation is valid for any shape of the equilibrium profile with the mean slope $\tan\theta$.

Consider now a situation in which a certain loss of sand has occurred from the equilibrium profile and the entire profile has been shifted shoreward (Fig. 13.8). For small changes of the shoreline position the slope of the dry beach can be ignored. The curved regions ABD and 0EC are obviously identical. To a first approximation, the cross-section of the entire profile has been shifted to the left and the volume of lost sand is $\Delta V \approx h^* \Delta y$, where h^* is the closure depth.

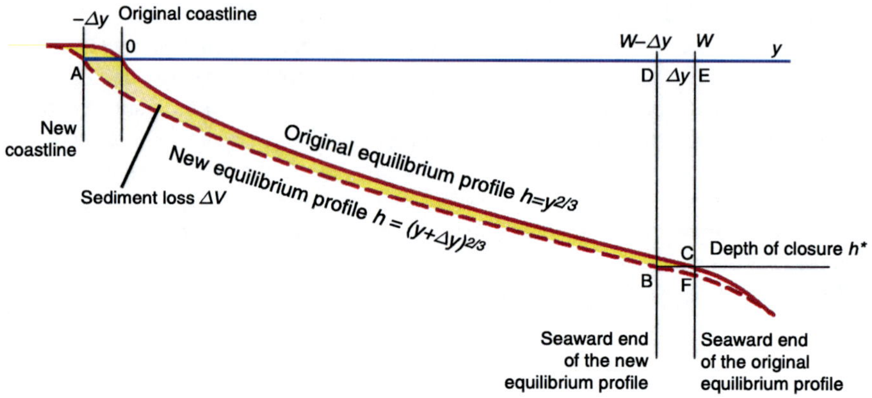

Fig. 13.8 Scheme of the calculation of the change of sand volume for small changes of the position of the coastline

The problem of calculating sand loss for a small section of coastline has therefore been reduced to determination of the shift of the shoreline and the closure depth. This approach neglects (i) the amount of sediment in the subaerial beach and (ii) a part of sediment located between the original and the new seaward end of the equilibrium beach profile. The first constituent of the error is small when the subaerial beach is gently sloping. For a perfectly equilibrium profile, the second constituent, equivalently, the error of this estimate, is smaller than $\frac{1}{2}\Delta y \tan\theta$ and obviously can be neglected for small coastline changes.

The resulting sand loss over a longer section of a homogeneous beach (along which the closure depth is constant) only depends on the changes of the area of the dry land:

$$\Delta V_{\Sigma} = h^{*} \int \Delta y \, dx. \tag{4}$$

Details of the derivation of Eq. (4) and further discussion of the applications of the method are presented in Kask et al. (2009).

Given the calculated typical value of the closure depth for Pirita $h^{*} \approx 2.5$ m, Eq. (4) predicts that each square meter of gain or loss of the dry land at Pirita corresponds to the change of the volume of sand by $\Delta V \approx 2.5\ \text{m}^3$ per each meter of the beach. Realistic values representing long-term gain or loss obviously can only be obtained for beach sections of considerable length, along which the integral in Eq. (4) is calculated.

The accuracy of the resulting estimate with the use of Eq. (4) is the best for beaches for which the different sand supplies and losses are almost balanced. The relative change of the water level is equivalent to an extra loss or gain of sediment. The relevant modification of the above estimate is presented for Pirita Beach in Kask et al. (2009). It is necessary to first calculate the mean slope $\tan\theta$ of the equilibrium profile, for example, using the typical grain size (defining the parameter A in the relevant EBP) and the closure depth. The potential gain or loss of the dry land area is then calculated under an assumption of zero net loss from the Bruun Rule. This potential gain is then compared with the factually measured gain or loss.

Kask et al. (2009) illustrate the method based on two examples. The coastline changes at Pirita Beach were first quantified from topographical maps based on measurements with a time lag of about 15 years (1986 and about 2000). The uplift rate at Pirita is about 2.5 mm/year (Vallner et al. 1988). If the sand volume were constant at Pirita, the expected coastline shift within approximately 15 years would have been about 4 m and the gain of dry land in the entire sandy beach with a length of about 2 km to about 8,000 m^2. In reality, the total gain of land is about 3,000 m^2, which corresponds to a mean coastline shift of about 1.5 m seawards. Consequently, the net loss of sand from the beach is about 5,000 $m^2 \times 2.5$ m=12,500 m^3. The net annual loss of sand is thus of the order of 1,000 m^3.

Another example of a similar estimate is obtained from the comparison of the results of two high-resolution surveys from 1997 and 2006, during which the area of dry beach remained practically unchanged. The expected coastline shift within 10 years, however, would have been about 2.5 m and would have resulted in the gain of about 5,000 m^2 of dry land. Therefore, the net loss of sand during these years is also about 12,500 m^3. The net loss of sand from the beach is thus about 1,250 m^3/year during this decade.

13.5.3 Interplay of Littoral Transport and River Flow at Narva-Jõesuu

While there is fairly weak net longshore transport in the middle sections of bay-head beaches (Soomere et al. 2008b), the situation is completely different in the eastern section of the North Estonian coast. The dominant wave approach direction along a long section of the almost straight coast is from the northwest. Although at times waves generated by easterly winds cause westward sediment drift, the basic geomorphic features reflect the overall intense sediment transport to the east. This transport leads to the formation of sand bars across river mouths.

The intensity of wind waves has a pronounced seasonal cycle in the entire Baltic Sea (Soomere 2005, Broman et al. 2006, Soomere and Zaitseva 2007, Räämet and Soomere 2010). The monthly mean wave height at Pakri (the only wave observation point, in the western part of the Gulf of Finland) varies from 0.38 m during spring and early summer (April–June) to 0.75 m in late autumn and early winter. The seasonal cycle is also clearly visible in the most typical wave conditions, dominant wave periods, and higher percentiles of observed wave heights (Zaitseva-Pärnaste et al. 2009). The similar cycle in the intensity and direction of littoral transport is much more strongly pronounced, because (i) westerly winds dominate during the later autumn and (ii) relatively strong easterly winds usually occur in early spring when much of the near-coastal wave activity is damped by the presence of ice.

The interplay of seasonal variation of wave intensity and river discharge leads to an interesting pattern of seasonal variation of the river mouth bar or sill height at

Narva-Jõesuu. The water depth is modest (about 2–4 m) in the vicinity of the seaward end of the breakwater, where the sand bar is highest. The water depth increases considerably upstream, reaching 8–10 m at the natural river mouth, and increases seawards to 10 m about 2 km offshore. The flow in the mouth of the Narva River can thus be classified as "sill flow" (Baines 1998). The cross-shore or diabathic transport over the sill is mostly driven by the discharging river, and the longshore transport by waves. The comparatively slow changes in sill height can be treated, as a first approximation, as a nearly balanced situation between the cross- and longshore transport.

The observed hydrological conditions and the estimated hydraulic parameters suggest that the straightforward, one-layer hydraulic model fails to adequately describe changes in the bottom topography of the river mouth. Hence, a two-layer exchange flow approach is adopted by Laanearu et al. (2007) to incorporate the observed stratification in the river mouth area. The observed spatial salinity distribution confirms that the river plume extends far into the bay during the spring months when the river discharge is comparatively large and the flow has a two-layer nature. During the autumn and winter months the river mouth is mainly stratified.

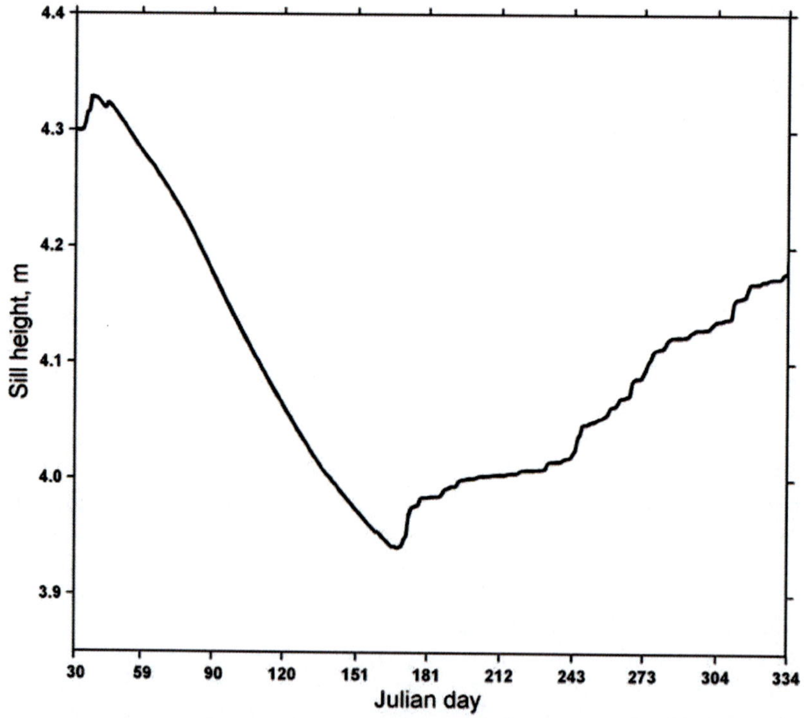

Fig. 13.9 Modelled variation of the sill height at the seaward end of the Narva-Jõesuu breakwater. Image by J. Laanearu, based on (Laanearu et al. 2007)

The combination of the seasonality of wave fields and this interesting switch between the two regimes has important consequences on the sediment transport in the sill area. While wave-induced increase in the sill height dominates during the autumn and winter season (when the river flow is modest and the lower layer is usually arrested), the mostly one-layer, voluminous outflow causes intense seaward transport of sediments in the sill area and a rapid reduction of the sill height during the spring flood (Fig. 13.9). A simple hydraulic model of such a two-layer flow combined with a model of cross-shore transport adequately reproduced the seasonal variation of the sill height, the magnitude of which is about 1 m (Laanearu et al. 2007). As the properties of stratification of seawater play an important role in the switch of the river flow between the one-layer and two-layer flows, any changes in the large-scale hydrophysical properties (for example, an increase in the precipitation and river discharge, or a decrease in the intensity of salt water inflow into the Baltic Sea) may affect this balance.

13.6 Discussion and Conclusions

Many beaches along the northern coast of Estonia belong to an interesting class of almost equilibrium, bayhead beaches located in an essentially non-tidal, highly compartmentalised coastal landscape. They develop mostly under the influence of wave action which is normally active in a relatively narrow nearshore band, but occasionally (under meteorologically forced high water conditions) may reach a certain sediment deficit. Their development is largely stabilised by the littoral drift or finer sediment and gravel (usually eroded from till bluffs in the neighbourhood of the beaches) towards the bayheads and by the postglacial uplift with a rate between 1 and 2.5 mm/year. Eolian transport and fluvial sediment supply have typically very modest magnitude. As is typical for bayhead beaches, there is no lateral loss of sediments towards the entrance of the bays. The combination of the listed features suggests that such beaches, in general, develop quite slowly and may be in an almost equilibrium stage even when the active sand mass is very limited.

Owing to a specific combination of the geometry of the coastline and dominant wind directions in strong storms, wave conditions along such beaches are highly intermittent. While the overall wave climate, estimated in terms of average wave properties, is usually very mild and wave periods are comparatively small, at times ferocious storms blowing from specific directions generate high and long waves that directly attack the beach. The development of such beaches, therefore, is step-like: many years of very slow development under low wave conditions approaching from a fixed direction are interspersed with large changes occurring infrequently during a strong storm.

Another class of beach in the area in question forms sections of the coast located eastwards from the longitude of 27°E. The coast is almost straight from this longitude, with a few headlands of moderate size around Kunda. This part of the coast is significantly exposed to wave approach from the Gulf of Finland. Its dynamics largely mimics that of an open ocean, high-energy coast and develops under the action of predominant wave approach from the west. Consequently, the sediment compartments are relatively large and there is an active littoral drift to the east. Under such conditions, the classical dynamic balance of open ocean beaches hosting considerable sediment transit is combined here with a seasonally varying balance between the wave-induced longshore transport and river-flow-induced cross-shore transport at river mouths (Laanearu et al. 2007).

The basic advantage of the analysis of such beaches is that various concepts applicable for equilibrium systems can be used to forecast their properties as well as their reaction to human intervention, either in the form of various coastal engineering structures that disturb the flow of natural processes or of coastal protection measures. One has, of course, to account for the intermittency of wave climate when calculating basic properties of equilibrium profiles such as the closure depth (Soomere et al. 2008b). Also, one has to critically evaluate many results of calculations of sediment transport. For example, the formal estimates of the potential transport rate are frequently overestimated by several orders of magnitude simply because the active layer of sediments may be very thin and/or be present only at specific places.

One of the largest advantages, however, can be achieved by combination of the theory of (basically one-point) equilibrium beach profile with the almost equilibrium state of the entire beach. In such cases, greatly simplified methods, based on a few parameters of the beach and the local wave climate, can be used for estimation of such necessary parameters for coastal management as the overall net sand loss.

The above analysis has also shown that the equilibrium of the beaches in question is largely based on a specific long-term balance of the sediment properties, geometry of the coast, and the forcing conditions. In this respect, the beaches are apparently very sensitive with respect to changes in the external forcing. Numerous changes in the forcing conditions (such as an increase in the average wind speed along the northern coast of the Gulf of Finland (Soomere and Keevallik 2003) or rapid decrease in the length of the ice season (Sooäär and Jaagus 2007)) and in the reaction of the water masses of the Gulf (such as an increase in the variability of sea level (Johansson et al. 2001)) have been identified during the latter decade; see The BACC Author Team (2008) for more examples. Moreover, the trends of the average and of extreme values of certain properties are different. This feature has been recently identified, among other processes, for wave conditions (Soomere and Healy 2008). Both instrumental wave data from Almagrundet (Broman et al. 2006) and visual wave data from Vilsandi (Soomere and Zaitseva 2007) suggest that during the 1980s there was an increase in the annual mean wave height in the northern Baltic Proper but a drastic decrease in the wave activity has occurred since 1997. At the same time in December 1999 (Kahma et al. 2003) and at the turn of 2004/2005 (Soomere et al. 2008a) extremely rough seas occurred. The beaches in question may be used for early detection of consequences of such changes.

Acknowledgements The chapter is based largely on two presentations to the 33rd International Geological Congress, Oslo, 6–14 August 2008: "Sediment transport patterns and rapid estimates of net loss of sediments for "almost equilibrium" beaches of tideless embayed coasts" by T. Soomere, A. Kask, and T. Healy, and "Formation of sand deposits in Estonian coastal sea" by A. Kask,

J. Kask, and T. Soomere. Financial support from the Estonian Science Foundation (Grant 7413), targeted financing by the Estonian Ministry of Education and Research (grants SF0140077s08 and SF0140007s11), and Tallinn University of Technology towards participation of TS and AK in the Congress is gratefully acknowledged. A large part of the chapter was written during the visits of one of the authors (TS) to the Centre of Mathematics for Applications, University of Oslo, within the framework of the MC TK project CENS-CMA (MC-TK-013909).

References

Alenius P, Myrberg K, Nekrasov A (1998) Physical oceanography of the Gulf of Finland: a review. Boreal Environment Research 3:97–125

Andrejev O, Myrberg K, Alenius P, Lundberg PA (2004) Mean circulation and water exchange in the Gulf of Finland – a study based on three-dimensional modeling. Boreal Environment Research 9:1–16

BACC (The BALTEX Assessment of Climate Change for the Baltic Sea Basin) Author Team (2008) Assessment of climate change for the Baltic Sea basin. Springer, Berlin, Heidelberg

Baines PG (1998) Topographic effects in stratified flows. Cambridge University Press, Cambridge

Birkemeier WA (1985) Field data on seaward limit of profile change. Journal of Waterway Port Coastal Ocean Engineering ASCE 111:598–602

Broman B, Hammarklint T, Rannat K, Soomere T, Valdmann A (2006) Trends and extremes of wave fields in the north–eastern part of the Baltic Proper. Oceanologia 48(S):165–184

Bruun P (1962) Sea level rise as a cause of erosion. Journal of Waterway Harbors Division ASCE 88:117–133

Caliskan H, Valle-Levinson A (2008) Wind-wave transformations in an elongated bay. Continental Shelf Research 28:1702–1710

Carter RWG (2002) Coastal environments. Academic Press, London

Coastal Engineering Manual (2002) Department of the Army. U.S. Army Corps of Engineers. Manual No. 1110-2-1100

Dean RG (1991) Equilibrium beach profiles: characteristics and applications. Journal of Coastal Research 7:53–84

Dean RG, Walton TL, Kriebel DL (2001) Cross-shore sediment transport. Coastal Engineering Manual, U.S. Army Coastal and Hydrodynamics Laboratory (CD)

Dean RG, Dalrymple RA (2002) Coastal processes with engineering applications. Cambridge University Press, Cambridge

Erm A, Alari V, Kõuts T (2008) Transport of sediments resuspended by ferries. Proceedings of the 2008 IEEE/OES US/EU-Baltic international symposium, Tallinn, Estonia, pp 392–397

Fröhle P, Dimke S (2007) Analysis of potential longshore sediment transport at the coast of Mecklenburg-Vorpommen. Baltic Sea Science Congress, Rostock 2007, March 19–22. Abstract volume, Part 1:37

Granö O, Roto M (1989) The duration of shore exposure along the emerging Finnish coast. Journal of Coastal Research 5:49–55

Granskog M, Leppäranta M, Ehn J, Kawamura T, Shirasawa K (2004) Sea ice structure and properties in Santala Bay, Baltic Sea. Journal of Geophysical Research 109:C02020

Hanson H, Larson M (2008) Implications of extreme waves and water levels in the southern Baltic Sea. Journal of Hydraulic Research 46:292–302

Healy T, Soomere T (2008) Managed retreat – is it really an option for mitigation of chronic erosion and storm surge flooding. In: Wallendorf L, Ewing L, Jones C, Jaffe B (eds) Solutions to coastal disasters 2008. American Society of Civil Engineers, Hawaii, pp 456–462

Johansson M, Boman H, Kahma KK, Launiainen J (2001) Trends in sea level variability in the Baltic Sea. Boreal Environment Research 6:159–179

Kahma K, Pettersson H (1994) Wave growth in a narrow fetch geometry. The Global Atmosphere and Ocean System 2:253–263

Kahma K, Pettersson H, Tuomi L (2003) Scatter diagram wave statistics from the northern Baltic Sea. MERI – Report Series of the Finnish Institute of Marine Research 49:15–32
Karukäpp R (2005) Eskers in the periphery of their distribution in North Estonia. Proceedings of the Estonian Academy of Sciences. Geology 54:26–39
Kask J, Kalberg T, Talpas A, Schwarzer K, Wegner C (2003a) Sea floor deposits in Keibu Bay and Nõva Bay (northwestern Estonia) and the adjacent offshore area. Proceedings of the Estonian Academy of Sciences. Geology 52:42–62
Kask J, Talpas A, Kask A, Schwarzer K (2003b) Geological setting of areas endangered by waves generated by fast ferries in Tallinn Bay. Proceedings of the Estonian Academy of Sciences. Engineering 9:185–208
Kask A, Soomere T, Healy T, Delpeche N (2009) Sediment transport patterns and rapid estimates of net loss of sediments for "almost equilibrium" beaches. Journal of Coastal Research Special Issue 56, II:971–975
Kawamura T, Shirasawa K, Ishikawa N, Lindfors A, Rasmus K, Ehn J, Leppäranta M, Martma T, Vaikmäe R (2001) A time series of the sea ice structure in the Baltic Sea. Annals of Glaciology 33:1–4
Komen GJ, Cavaleri L, Donelan M, Hasselmann K, Hasselmann S, Janssen PAEM (1994) Dynamics and modelling of ocean waves. Cambridge University Press, Cambridge
Kraus NC (1992) Engineering approaches to cross-shore sediment processes. In: Lamberti A (ed) (Proceedings of the short course on) Design and reliability of coastal structures. Proceedings of the 23rd International conference on Coastal engineering (ICCE), Venice, 1–3 October 1992, pp 175–209
Kuhrts C, Fennel W, Seifert T (2004) Model studies of transport of sedimentary material in the western Baltic. Journal of Marine Systems 52:167–190
Laanearu J, Lips U (2003) Observed thermohaline fields and low-frequency currents in the Narva Bay. Proceedings of the Estonian Academy of Sciences. Engineering 9:99–106
Laanearu J, Koppel T, Soomere T, Davies PA (2007) Joint influence of river stream, water level and wind waves on the height of sand bar in a river mouth. Nordic Hydrology 38:287–302
Lutt J (1985) Bottom sediments of Väinameri sea area. Valgus, Tallinn (in Russian)
Lutt J (1992) Bottom deposits. In: Raukas A, Hyvärinen H (eds) Geology of the Gulf of Finland. Valgus, Tallinn, pp 158–218 (in Russian)
Lutt J, Kask J (1992) Sources of sedimentation. In: Raukas A, Hyvärinen H (eds) Geology of the Gulf of Finland. Valgus, Tallinn, pp 137–157 (in Russian)
Lutt J, Tammik P (1992) Bottom sediments of Tallinn Bay. Proceedings of the Estonian Academy of Sciences. Geology 41:81–86 (in Estonian)
Miidel A, Jantunen T (1992) Neotectonics and contemporary movements. In: Raukas A, Hyvärinen H (eds) Geology of the Gulf of Finland. Valgus, Tallinn, pp 87–98 (in Russian)
Myrberg K (1997) Sensitivity tests of a two-layer hydrodynamic model in the Gulf of Finland with different atmospheric forcings. Geophysica 33:69–98
Nordstrom KF (2005) Bay beaches. In: Schwartz ML (ed) Encyclopedia of coastal science. Springer, Dordrecht, The Netherlands, pp 129–130
Orviku K (1974) Estonian coasts. Tallinn (in Russian)
Orviku K, Granö O (1992) Contemporary coasts. In: Raukas A, Hyvärinen H (eds) Geology of the Gulf of Finland. Valgus, Tallinn, pp 219–238 (in Russian)
Pettersson H (2001) Directional wave statistics from the Gulf of Finland. MERI – Report Series of the Finnish Institute of Marine Research 44:1–37
Pettersson H, Kahma KK, Tuomi L (2010) Predicting wave directions in a narrow bay. Journal of Physical Oceanography 40:155–169
Protasjeva M, Eipre T (eds) (1972) Resources of the surface water of the USSR. Baltic region. Estonia. Gidrometeoizdat, Leningrad (in Russian)
Räämet A, Soomere T (2010) The wave climate and its seasonal variability in the northeastern Baltic Sea. Estonian Journal of Earth Sciences 59:100–113
Raukas A, Hyvärinen H (eds) (1992) Geology of the Gulf of Finland. Valgus, Tallinn (in Russian)
Raukas A, Teedumäe A (1997) Geology and mineral resources of Estonia. Estonian Academy Publishers, Tallinn

Seifert T, Tauber F, Kayser B (2001) A high resolution spherical grid topography of the Baltic Sea, 2nd edn. Baltic Sea Science Congress, Stockholm, 25–29 November 2001, Poster #147, www.io-warnemuende.de/iowtopo

Sooäär J, Jaagus J (2007) Long-term variability and changes in the sea ice regime in the Baltic Sea near the Estonian coast. Proceedings of the Estonian Academy of Sciences. Engineering 13:189–200

Soomere T (2003) Anisotropy of wind and wave regimes in the Baltic Proper. Journal of Sea Research 49:305–316

Soomere T, Keevallik S (2003) Directional and extreme wind properties in the Gulf of Finland. Proceedings of the Estonian Academy of Sciences. Engineering 9:73–90

Soomere T (2005) Wind wave statistics in Tallinn Bay. Boreal Environment Research 10:103–118

Soomere T, Zaitseva I (2007) Estimates of wave climate in the northern Baltic Proper derived from visual wave observations at Vilsandi. Proceedings of the Estonian Academy of Sciences. Engineering 13:48–64

Soomere T, Kask A, Kask J, Nerman R (2007) Transport and distribution of bottom sediments at Pirita Beach. Estonian Journal of Earth Sciences 56:233–254

Soomere T (2008) Extremes and decadal variations of the northern Baltic Sea wave conditions. In: Pelinovsky E, Kharif C (eds) Extreme ocean waves. Springer, Berlin, pp 139–157

Soomere T, Healy T (2008) Escalating extremes over descending trends of the northern Baltic Sea wave fields. In: Wallendorf L, Ewing L, Jones C, Jaffe B (eds) Solutions to coastal disasters 2008. American Society of Civil Engineers, Hawaii, pp 129–138

Soomere T, Behrens A, Tuomi L, Nielsen JW (2008a) Wave conditions in the Baltic Proper and in the Gulf of Finland during windstorm Gudrun. Natural Hazards and Earth System Science 8:37–46

Soomere T, Kask A, Kask J, Healy T (2008b) Modelling of wave climate and sediment transport patterns at a tideless embayed beach, Pirita Beach, Estonia. Journal of Marine Systems 74(Supplement):S133–S146

Soomere T, Myrberg K, Leppäranta M, Nekrasov A (2008c) The progress in knowledge of physical oceanography of the Gulf of Finland: a review for 1997–2007. Oceanologia 50:287–362

Soomere T, Didenkulova I, Parnell KE (2009) Implications of fast-ferry wakes for semi-sheltered beaches: a case study at Aegna Island, Baltic Sea. Journal of Coastal Research Special Issue 56, I:128–132

Suuroja S, Suuroja K, Talpas A (2007) Seafloor geological survey in Estonian shelf area. Baltica 20:53–58

Suursaar Ü, Kullas K, Otsmann M, Saaremäe I, Kuik J, Merilain M (2006) Cyclone Gudrun in January 2005 and modelling its hydrodynamic consequences in the Estonian coastal waters. Boreal Environment Research 11:143–159

Vallner L, Sildvee H, Torim A (1988) Recent crustal movements in Estonia. Journal of Geodynamics 9:215–223

Velegrakis AF, Vousdoukas MI, Andreadis O, Adamakis G, Pasakalidou E, Meligonitis R, Kokolatos G (2008) Influence of dams on downstream beaches: Eressos, Lesbos, Eastern Mediterranean. Marine Georesources Geotechnology 26:350–371

Wright LD, Short AD (1984) Morphodynamic variability of surf zones and beaches – a synthesis. Marine Geology 56:93–118

Wright LD (1985) River deltas. In: Davis JrRA (ed) Coastal sedimentary environments. Springer, New York, NY, pp 1–76

Zaitseva-Pärnaste I, Suursaar Ü, Kullas T, Soomere T (2009) Seasonal and long-term variations of wave conditions in the northern Baltic Sea. Journal of Coastal Research Special Issue 56, I:277–281

Chapter 14
Modelling Coastline Change of the Darss-Zingst Peninsula with Sedsim

Michael Meyer, Jan Harff, and Chris Dyt

Abstract Coastlines do not change because of sea level variation alone. Instead, the changes are the result of a complex interaction between climate and geologically controlled processes. Especially on a local scale, sedimentary dynamics play an important role. Even with a rising sea level, concurrent sediment accumulation may prevent coastline retreat. On the other hand, erosion may accelerate marine transgressions remarkably. The southern coast of the Baltic Sea is an impressive example for the impact of erosion, transport, and accumulation of sediments to coastline change during the Holocene. Since the end of the Littorina transgression the coastline morphology has been shaped here mainly by longshore sediment transport controlled by the geological situation and glacioisostatic influence. The longshore sediment transport is driven by wind and consequently waves shaping young Holocene structures like the Darss-Zingst peninsula. In order to model these processes, Sedsim (SEDimentary Basin SIMulation), a stratigraphic forward modelling software, has been applied for the Darss-Zingst peninsula on a centennial time scale. In Sedsim, the sedimentary dynamics are modelled by an approximation to the Navier–Stokes equation. Using high-resolution digital elevation data, information about the local wave characteristics, geology, estimates of sea level rise, and experimental scenarios for the development of the Darss-Zingst peninsula through the coming 840 years are presented. The results of the experiments show possible implications to the area of investigation and may serve as a basis for decision makers in coastal zone management.

Keywords Coast line change · Sediment transport modelling · Sedsim · Southern Baltic Sea · Darss-Zingst peninsula

---

M. Meyer (✉)
Leibniz Institute for Baltic Sea Research Warnemünde, D-18119 Rostock, Germany;
Institute for Environmental Engineering, University Rostock, 18057 Rostock, Germany
e-mail: michael.meyer@uni-rostock.de

J. Harff et al. (eds.), *The Baltic Sea Basin*, Central and Eastern European Development Studies (CEEDES), DOI 10.1007/978-3-642-17220-5_14,

14.1 Introduction

Coastlines are changing through time and space, controlled by climate and geological processes. Climate steers the hydrography, either by wind and storms triggering waves and surges or by the thermal-adjusted mass balance between continental ice and marine water. This overlaps with geological parameters, because in times of transitions between glacials and interglacials a change of this balance results in high magnitudes of glacioisostasy (Miettinen 2004). Tectonics caused by processes within the earth's interior are a morphogenetic factor too, influencing the coastline shift. A quantification of coastline changes requires the combination of data about isostasy and eustasy with digital elevation models. In a first step, regional scenarios for the Baltic Sea, presented by Meyer (2003) and Rosentau et al. (2007), take this into consideration. However, on a local scale sedimentary dynamics like erosion, transport, and accumulation play an important role too (Lehfeldt and Milbradt 2000, Harff et al. 2009). Therefore an extended approach for modelling coastline change is required, integrating sedimentary dynamics with eustasy and isostasy.

Here, this integration was accomplished by the modelling software package Sedsim (Tetzlaff and Harbaugh 1989, Martinez and Harbaugh 1993). The program simulates the behaviour of coastal sediments with respect to eustasy and isostasy during geological and short time periods (Li et al. 2004). Sedsim is a forward modelling tool, depending on defined initial conditions. Before the implementation of experiments for the geological past, in a first stage Sedsim was used for validation experiments on the basis of recent initial conditions. With assumptions about the development of the future sea level during the next 840 years (Voß et al. 1997) and different parameter set-ups, various coastline scenarios have been calculated with Sedsim. These experiments are located at the southern Baltic Sea coast, in the area of the Darss-Zingst peninsula. This structure, shaped by longshore sediment transport, is an excellent example for modelling as it is typical for many young Holocene formations along the southern and southeastern Baltic Sea. The results of the modelling are not predictive, instead they have to be considered as case scenarios. A reasonable evaluation of the simulations is a precondition for modelling the past on longer geological periods, based on reconstructed palaeo data sets.

14.2 Area of Investigation

The Darss-Zingst peninsula is located at the southern coast of the Baltic Sea (Fig. 14.1). It is part of Mecklenburg-Vorpommern, the most northeastern state of Germany. The peninsula has an area of about 160 km^2. The distance from the Fischland in the west to the Bock Island in the east is 40 km on average, while the north–south length is approximately 20 km. The elevation is very low with maxima of 15 m in the Altdarss and Fischland areas. Generally, elevations of 0.5–2 m are common. The surrounding water is also shallow with water depths not exceeding 4 m in the lagoons sheltered by the peninsula.

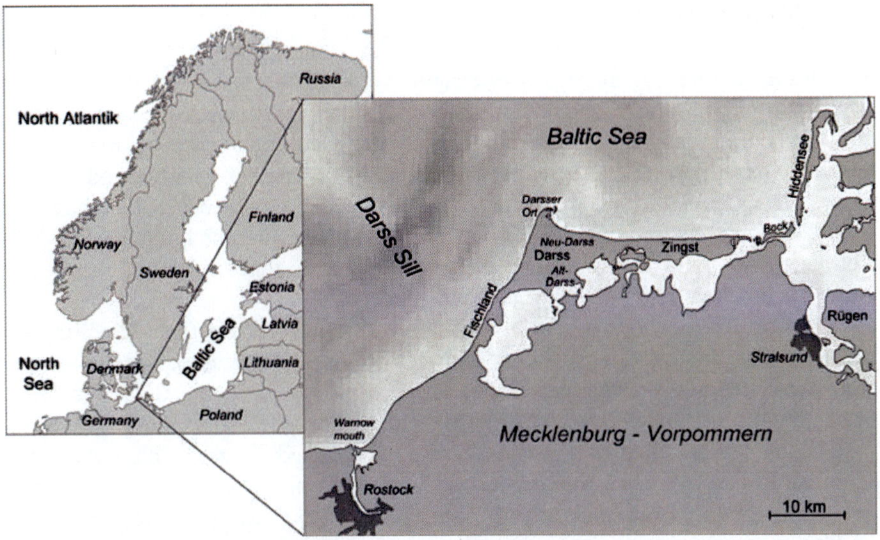

Fig. 14.1 Area of investigation. The peninsula, located in the southern Baltic Sea, between the cities Rostock and Stralsund, shelters a chain of lagoons from the open sea

The shape of the peninsula today is the result of Holocene sediment transport processes. During the initial phase of the Littorina transgression at 8,000 ^{14}C years before present, the land simply drowned without remarkable coastal erosion. The sea level rise was very fast, approximately 1 cm/year, documented quite well by relative sea level curves (e.g. Lampe et al. 2005). Starting around 4,000 ^{14}C years before present this speed slowed down rapidly, causing a straightening of the coast (Meyer et al. 2008). Large parts of the southern Baltic Sea coast were reshaped by erosion, transport, and accumulation of sediment, forming spits and lagoons. In the Baltic Sea, these lagoons are called "Bodden" or "Haff". For the evolution of the Darss-Zingst Bodden chain various scenarios do exist (Schumacher 2000, Lampe 2002). The main concept is the erosion of glacial till complexes with a transport of the resulting silt to deeper water depths and the accumulation of the remaining coarser sediments along the coast. A prominent example for a glacial till complex is the Fischland cliff, while the Neudarss area evolved step by step by the accumulation of sand (Schumacher 2000). The main direction of the longshore transport of this sediment is west–east aligned, indicated by the shape of the peninsula. In addition, an accretion by landward over-washes is discussed by Lampe (2002). These regimes are still valid at recent time, with sediment sources at cliff regions in the Fischland area, and accumulation mainly eastwards at the Darsser Ort and in the very shallow waters around the Bock Island.

14.3 Methodology

As outlined before, the modelling of coastline changes requires data about surface elevation, vertical crustal movement, and climatically driven sea level change. In addition, on a local scale, erosion, transport, and accumulation play a major role. These morphodynamic processes are controlled by sediment properties and driving forces. They are simulated by the stratigraphic modelling software Sedsim, originally developed by Tetzlaff and Harbaugh (1989) and maintained today by CSIRO Petroleum Australia (CSIRO 2004). Sedsim calculates sediment budget changes in time as a function of the depositional environment.

The structure of the program is modular with distinct algorithms handling the different physical processes which effect the sediment distribution. Sedsim is controlled by a text file, with each of the separate processes having its own section which can be selectively used. An orthogonal regular grid is used to describe the surface and the cumulative deposition and erosion of sediments on that surface are recorded at user-specified time intervals. Four types of user-specified siliciclastic sediments are allowed as well as two carbonate types and two organic types.

Fluvial processes are controlled by a marker in cell technique, which flows Lagrangian fluid elements over the imported digital elevation model (DEM) grid surface, depositing or eroding sediment depending upon its current transport capacity. Each fluid source is specified by location, initial velocity, volume, and initial sediment composition, with fluvial, hypopycnal, hyperpycnal, and debris flows capable of being modelled.

Vertical movements of the earths crust are controlled either by specifying the tectonic movement of each surface point directly or with the ISOSTACY module which determines the flexure of a rigid plate in response to loading (Li et al. 2004). In addition to surface movement, sea level fluctuations are implemented via a simple input file.

The influence of waves can be incorporated in a range of different waves depending upon the detail of data available for input. From a general wave direction and height which lasts for the entire simulation, to a time-varying direction and height, through to a complete wave field detailing the changes at each grid point. Wave refraction into shallow water can also be selectively used. The module works by calculating the sediment mobilized by wave impact on the coastline and determines the amount of that sediment transported alongshore depending on the wave incident angle, wave height, depth of mobile bed, and wave base.

Storm surges effects are included either by specifying known storm data including the time, incident angle, wave height, and time duration or by creating synthetic storms by listing the mean storm return time, direction, deviation in direction, wave height, and storm duration. The storm module calculates the storm erosional impact above storm wave base and moves the sediment offshore to below storm wave base perpendicular to the coastline.

Sedsim also offers other modules for modelling slope failure through oversteepening of sediments, the ability to grow organics and carbonates through a system of fuzzy rules, compaction due to loading of overlying sediments, a cellular automata-based aeolian module, as well as the ability to include contour currents

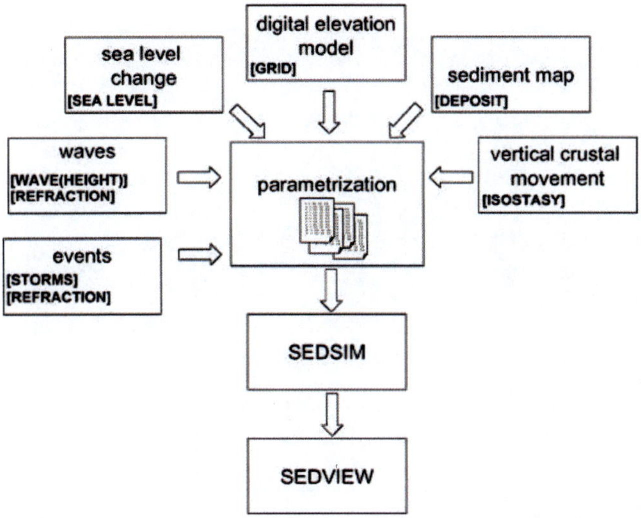

Fig. 14.2 Schematic flow chart and parameters used for the modelling. Sedsim modules used are listed in *brackets*

or input oceanic circulation patterns. However, these modules were not used in this study (Fig. 14.2).

14.4 Data

14.4.1 Digital Elevation Model

Elevation models for the Darss-Zingst peninsula are available from the Land Survey Administration Mecklenburg-Vorpommern, Germany, in different scales. However, these data sets comprise only terrestrial elevations. For the modelling with Sedsim a combined data set is required, which couples bathymetric data with land elevations. A first, but rough approach is the integration of the data set provided by Seifert et al. (2001). This data set comprises land and sea bottom elevation for the area of investigation with a spatial resolution of approximately 1 km^2 per grid cell. In a second step, the terrestrial part for this data was replaced by the DEM25 data set from the Land Survey Administration Mecklenburg-Vorpommern (2006). This data set has a spatial resolution of 25 m. For a high-resolution bathymetry, over 5 million bathymetric measurements covering the German Baltic coast provided by the Federal Maritime and Hydrographic Agency of Germany were used (Meyer et al. 2008). Finally, the resolution of the resulting elevation model was set to 50 m. However, initial experiments showed that the amount of data due to this high resolution is too much to calculate scenarios with Sedsim in a reasonable time. Therefore, the DEM was resampled to a 150 m resolution, covering the area between 12.11° east to 13.17° east and 54.07° north to 54.76° north (Fig. 14.3). The coordinate system used is Gauss Krüger, meridian stripe #4, on Bessel ellipsoid.

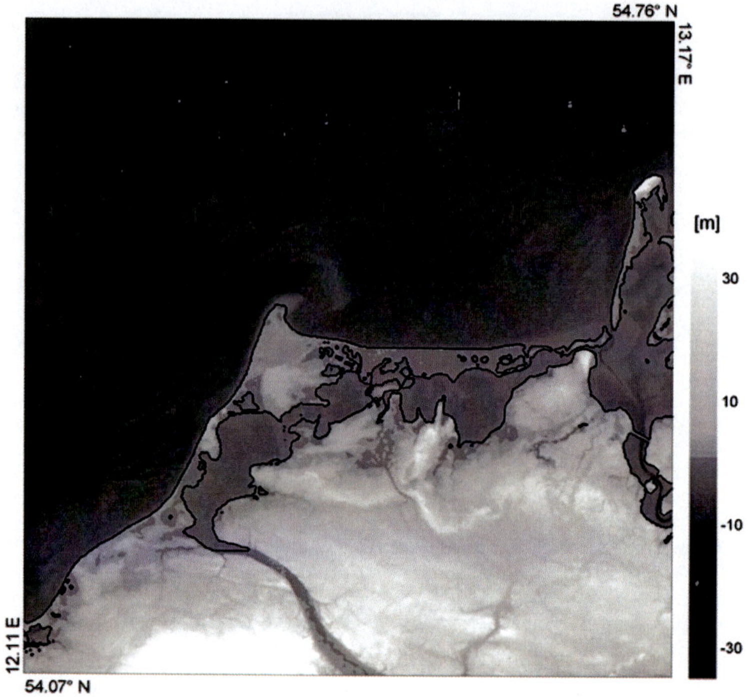

Fig. 14.3 Digital elevation model for the Darss-Zingst peninsula region. Recent coastline is indicated by the *black line*

14.4.2 Sediment Map

A consistent sediment distribution for the Darss-Zingst peninsula is given by Heck et al. (1957). The details on land are mapped very accurately and can be considered as standard still today; however, the sediments at the sea floor are not included. Again, a compilation of onshore and offshore models is required. Since the beginning of the 1990s, the sediments for the offshore area of the German Baltic Sea have been mapped by the Federal Maritime and Hydrographic Agency. The most recent maps are published by Tauber and Lemke (1995) and Tauber et al. (1999). For the area of investigations, relevant parts of these two maps have been digitized. Together with a digital version of the map by Heck et al. (1957), a geological surface model was assembled. There is a high diversity in sediment types, not to mention different nomenclatures for terrestrial and marine data. Therefore, for modelling purposes the classification of the sediments was simplified into three major types. These types are mud plus very fine sand, sand, and glacial till. Figure 14.4 shows the distribution of these sediments in the area of investigation. The spatial parameters for this grid are congruent to the DEM.

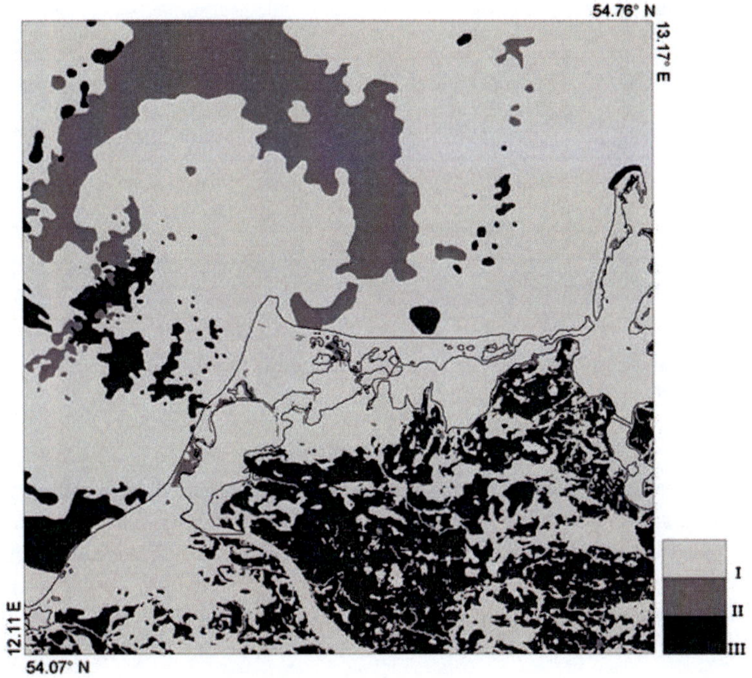

Fig. 14.4 Simplified sedimentological surface model of the Darss-Zingst region. Recent coastline is marked by the *solid black line*. Sediment types: (I) sand, (II) very fine sand and mud, (III) glacial till

Internally, Sedsim requires information about the sediment composition for each model cell, based on a system of four grain size classes. These are coarse, medium, fine, and silt classes with parameters adjustable in the Sedsim command file. The grain properties influence the results in several ways. Denser and larger particles are harder to erode, so that once finer particles have been removed from the top layer of a particular grid cell, the coarser particles may shield the cell from further erosion. Coarser particles are also able to reside at a much steeper submarine angle than the fine particles (a user-defined parameter). The result of this is that fine material tends to be eroded easier and transported further. This results in a more sand-rich coastline and a more mud-rich deep marine environment.

A conversion of the sediment map according to the scheme shown in Table 14.1 was performed. The percentages used are validated by Hoffmann et al. (2004) for the Usedom peninsula, ca. 100 km east of the Darss-Zingst region, but genetically similar.

Such a sediment distribution can only be acquired for the terrain surface and detailed data about the vertical distribution of sediments are rather rare. For the Darss-Zingst region no consistent 3D model of the sediment structure with a satisfying spatial resolution is known. Therefore, as a general presumption the thickness of

Table 14.1 Translation scheme of the simplified sedimentological surface model (Fig. 14.4) to a Sedsim-compatible sediment composition

	Coarse (%) Ø: 0.75 mm ρ: 2,650 kg/m^3	Medium (%) Ø: 0.375 mm ρ: 2,650 kg/m^3	Fine (%) Ø: 0.15 mm ρ: 2,650 kg/m^3	Silt (%) Ø: 0.03 mm ρ: 2,550 kg/m^3	Porosity (%) Ø: 0.75 mm ρ: 2,650 kg/m^3
I	8	28	54	10	40
II	0	0	0	100	80
III	12	22	35	31	20

Ø: grain size diameter, ρ: density

the outcropping sediments was set to a homogeneous value of 10 m, parameterized in Sedsim by the DEPOSIT module.

Considering the borders of the study area, additional sediment input can be neglected. To the south, the mainland forms a natural barrier, while to the north the open sea with larger water depths serves as a discharge area. In the west the area is bordered by the river Warnow. This river's mouth is very important for sea traffic and constantly dredged, therefore, preventing additional sediment input from the west. Because the longshore transport proceeds from west to east, the area bordering to the east is excluded as a sediment source, too.

14.4.3 Vertical Movement of the Earth's Crust

The Darss-Zingst region is located on the southern border of Scandinavia, which is lifting up because of glacioisostatic adjustment (Björck 1995). The maximum magnitude of this uplift is about 9 mm/year in the Gulf of Bothnia, while there are negative values in a surrounding subsiding area (Harff et al. 2001, Rosentau et al. 2007). Although these maps are considered to show the vertical movement of the earth's crust, it has to be noted that they are constructed from gauge measurements and include the eustatic signal. Only a removal of this parameter reveals the true pattern of vertical movement of the earth's crust (Harff and Meyer 2007), which is required for a Sedsim input parameterization. For the western Baltic Sea, a eustatic sea level rise of approximately 1.0 mm/year during the last century is postulated by Hupfer et al. (2003). This value was used by Harff and Meyer (2007) for the calculation of a revised model of the vertical movement of earth's crust.

On a first glance, these movements are rather small in the study area (Fig. 14.5). Largest uplift for the peninsula is calculated with 0.4 mm/year in the Zingst region, whereas in the southwest parts the crust is actually sinking. On a scale of a millennium this will sum up to 0.4 m uplift, respectively, 0.2 m subsidence. These are magnitudes that have to be taken into account when modelling coastlines, and the module TECTONICS was parameterized using these data.

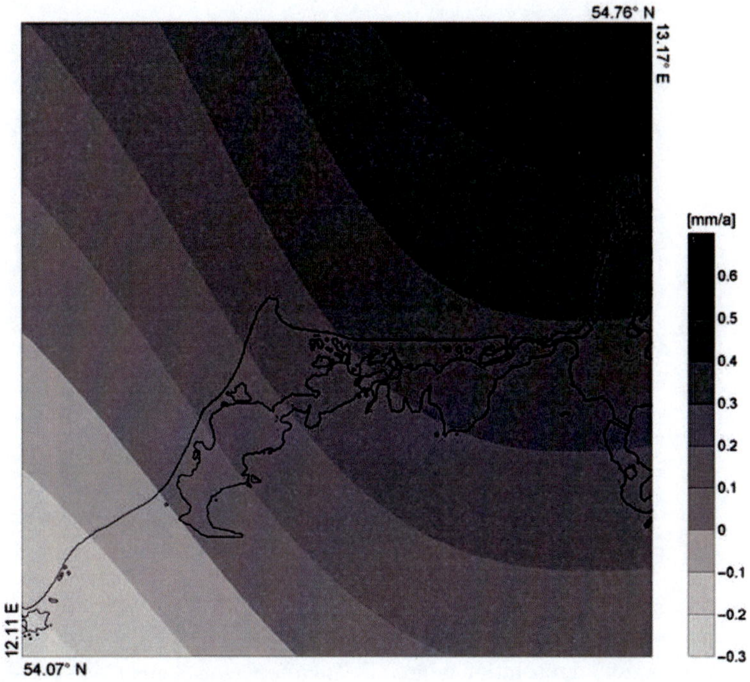

Fig. 14.5 Vertical movements of the earth's crust. Data source: Harff and Meyer (2007). The pattern results from data sets for relative sea level change corrected by the removal of a eustatic factor

14.4.4 Sea Level Change

During the last century, the eustatic sea level in the Darss-Zingst region was rising with a magnitude of 1 mm/year (Hupfer et al. 2003). According to IPCC projections the speed will increase; for the next 100 years a rise of more then 200 mm is supposed (Metz et al. 2007). Therefore a simple linear extrapolation of the value given by Hupfer et al. (2003) for the next millennium is not possible. Voß et al. (1997) calculated with the global atmosphere–ocean circulation model ECHAM/LSG (Roeckner et al. 1996) a long-term time series of sea level development for the next 840 years, based on IPCC scenario A (Houghton et al. 1990). This scenario assumes an increase of the atmospheric CO_2-concentration up to a specific limit with interconnected global warming effects. After reaching the limit, the concentration is considered to stay constant. Sea level change models are available for a CO_2 doubling during the first 60 model years, respectively, a CO_2 quadrupling within the first 120 model years. Unfortunately, the LSG model scales only on a very

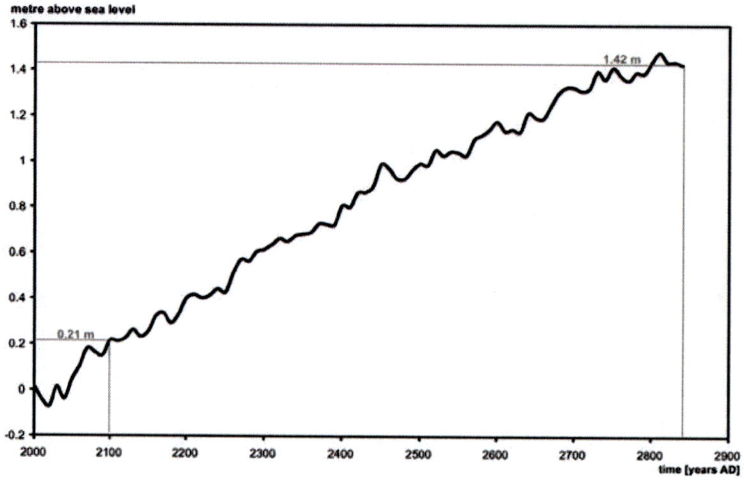

Fig. 14.6 Sea level model for the study area for the next 840 years. *Data source*: Voß et al. (1997)

rough spatial horizontal resolution with 5.6°. Therefore Voß et al. (1997) do not provide values for the Baltic Sea itself. The average of bordering model cells from the North Sea and the Northern Atlantic is used as a sea level model for the study area (see Fig. 14.6). In this study we focused on the CO_2 quadrupling scenario with a modelled rise of the sea level for the next century of 21 cm. More recent global sea level rise scenarios (IPCC 2007) suggest a range in sea level rise for the next 100 years from 18 to 59 cm. In comparison, the data we used have to be considered as a "best-case scenario", with 21 cm at the lower border of the range. This value is also within the range of the scenarios proposed by Meier et al. (2004) for the Baltic Sea. They take sea level rise values of 9, 48 cm, or, in a worst-case scenario, 88 cm as basis for modelling towards the end of the century.

The speed of our sea level rise scenario is about 1.7 mm/year and does not accelerate. This seems to be in contradiction to recent findings, e.g. Hammarklint (2009) who postulates an increase of the speed of the sea level rise during the last 30 years from 1.5 up to 3 mm/year. In the frame of a 840-year-long time series a time span of 30 years covers only a small interval and such an increase can be considered as a fluctuation. Although in summary linearly, the sea level rise is always superimposed by an oscillation with times of increase and decrease, reflecting irregularities in the behaviour of a complex natural system.

The sea level time series is parameterized by the Sedsim module SEA LEVEL. The projected sea level rise scenario for the next 840 years is shown in Fig. 14.6. The sea level rise continues even after the end of the increase in CO_2 concentration after 120 years. This is caused by long-term global oceanic circulations responsible for the heat transfer between atmosphere and ocean.

14.4.5 Waves

Waves are the driving force for sediment transport simulation in Sedsim. Here, they are parameterized by the module WAVE (HEIGHT). This module requires a time series with significant heights and directions for waves approaching the study area. The behaviour of the waves adjusted by morphological conditions is calculated in Sedsim internally, with refraction simulated by the module WAVE REFRACTION. A consistent time series of waves between 1958 and 2002 for the study area is available, modelled for a gauging station site at 54.69° north, 12.69° east, approximately 27 km to the north of the Zingst peninsula. Data source is the coastDat database, hosted by the GKSS-Research Centre Geesthacht, Germany (Weisse et al. 2009). The data set records a mean significant wave height of 50 cm (Fig. 14.7). Most of the waves travel either from the northeast, between 50° and 80°, or from the west, with a more dispersed orientation between 220° and 300°. This modelled data set provides values in hourly intervals resulting in a very large data volume. For the application in Sedsim, the data set was averaged into semi-yearly time slices. The winter season lasts from October to April, while the summer season covers the rest of the year. According to the time span to be covered by the modelling (see previous chapter), a time series for 840 years is required; therefore, the 44-year data set was prolonged with a simple line-up. Although it seems obvious that waves will change in the future because of global warming, the linear character of the

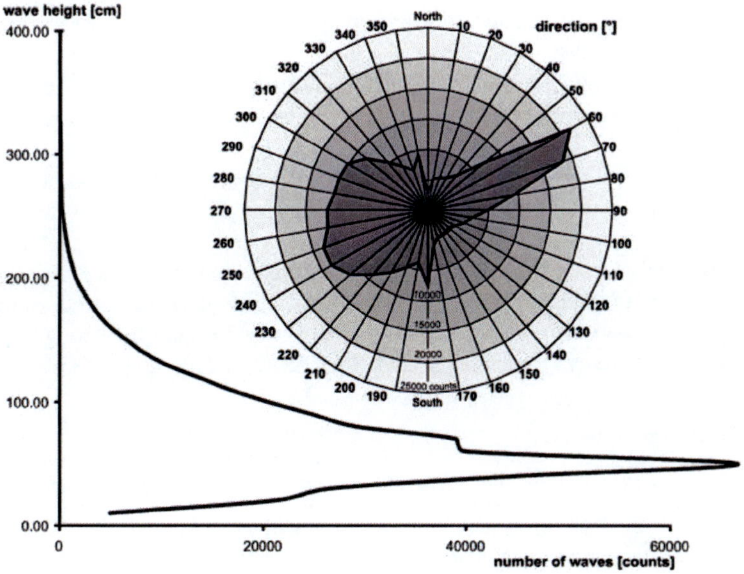

Fig. 14.7 Wave statistics for the Darss-Zingst peninsula region. *Data source*: coastDat database (Weisse et al. 2009)

eustatic scenario proposed within this study suggests otherwise. There is no evidence that the wave system will change because of such a stationary process. We assume the wave regime to be stationary, too. This is in contradiction to model results from Grabemann and Weisse (2008) who detected a significant increase of the wave heights because of future climate change. However, these results have been acquired on behalf of the North Sea as a model region, and there is no equivalent application for our area of interest.

14.4.6 Events

The wave data described in the previous chapter contains information about wave heights, but because of the seasonal averaging extreme wave heights accompanied with short-time storm events have to be added separately. The surges caused by storms are crucial for the coastline development. Suddenly, previously safe onshore areas are exposed to the forces of rising sea water, often resulting in dramatic coastal erosion. For the coast of the study area the statistics in Table 14.2 clarifies return frequencies for different surge types.

For the experiments with Sedsim, the STORMS module was set up with a distribution of storms occurring every second year with an increase of the sea level by 1 m and an additional significant wave height of 2 m. Actually, there is no overall agreement whether frequency or intensity will change in the future (BACC 2008). According to Weisse and Storch (2009) studies on this matter always have a strong regional form and may not be generalized. For the area of investigation no detailed study about changes in surge statistics is applicable; therefore, a linear extrapolation seems reasonable at least.

With the rise of the sea level during storms, onshore sediments are included in the modelling of the erosion, transport, and accumulation. Also, a corresponding shift of the wave base is taken into consideration.

The most devastating storm surge at the coast of Mecklenburg-Vorpommern ever recorded and measured in detail was in 1872, during the night from the 12th to the 13th of November. Locally, the height of the sea level reached over 3 m above normal sea level. Today, this surge is considered as a benchmark for determining the defence level for coastal protection. A reconstruction of the behaviour of the sea level caused by this event was modelled by the Federal Office for Navigation and Hydrography, Germany (Rosenhagen and Bork 2009). In the area of investigation

Table 14.2 Return frequency for different storm surge types at the coast of Mecklenburg-Vorpommern. Modified after Hupfer et al. (2003)

Surge type	Sea level height above normal sea level (m)	Return frequency (years)
Light	1.00–1.25	1–2
Medium	1.25–1.50	5–10
Heavy	1.50–2.00	5–20
Very heavy	> 2.00	50–100

the sea level rose about 2.5 m. This information was used for the simulation of a local worst-case sea level scenario.

14.5 Results and Discussion

The natural development of the Darss-Zingst peninsula for the next 840 years was modelled with various set-ups in order to develop an understanding about the relevance and influence of the different input data and providing sensitivity tests for the Sedsim model. The structure of Sedsim allows an easy toggle of the different modules. In experiment A, the WAVE (HEIGHT) module was switched on, together with biannual storms declared in STORMS module. The resulting shape of the model area after 840 years is depicted in Fig. 14.8a.

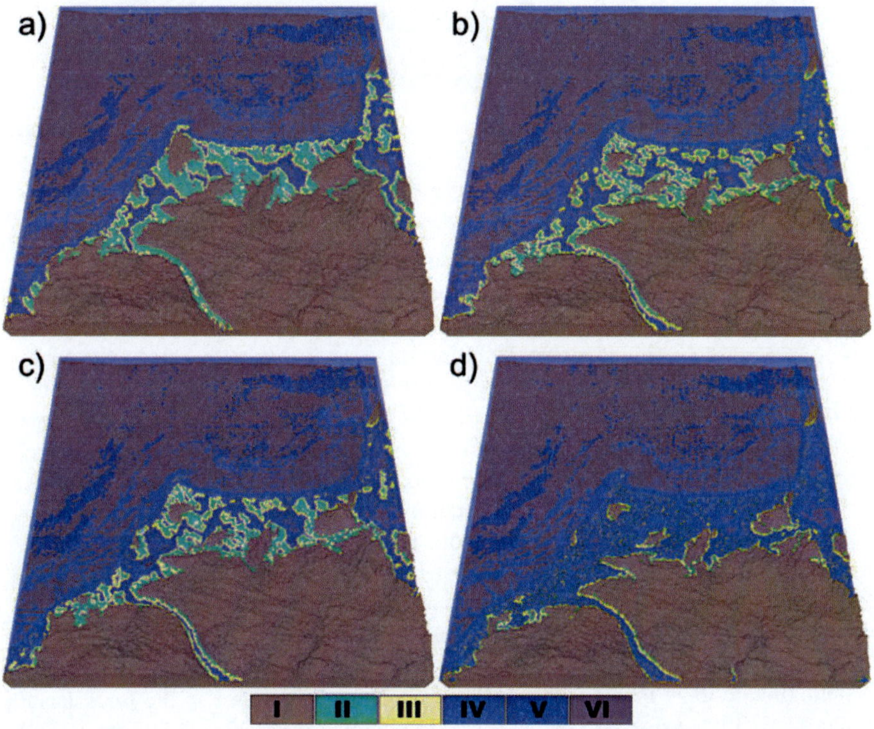

Fig. 14.8 Modelling results after 840 years. Distance from shore classification: (I) unchanged inland, (II) modified inland, (III) shoreline, (IV) marine but near shore, (V) open marine, (VI) unchanged open marine. **a** Experiment A, active modules: WAVE (HEIGHT) and STORMS. **b** Experiment B, active modules: WAVE (HEIGHT), STORMS, and SEA LEVEL. **c** Experiment C, active modules: WAVE (HEIGHT), STORMS, SEA LEVEL, and TECTONICS. **d** Experiment D, active modules: WAVE (HEIGHT), STORMS, SEA LEVEL, TECTONICS. In addition, the sea level height was set to the height of the storm flood in 1872

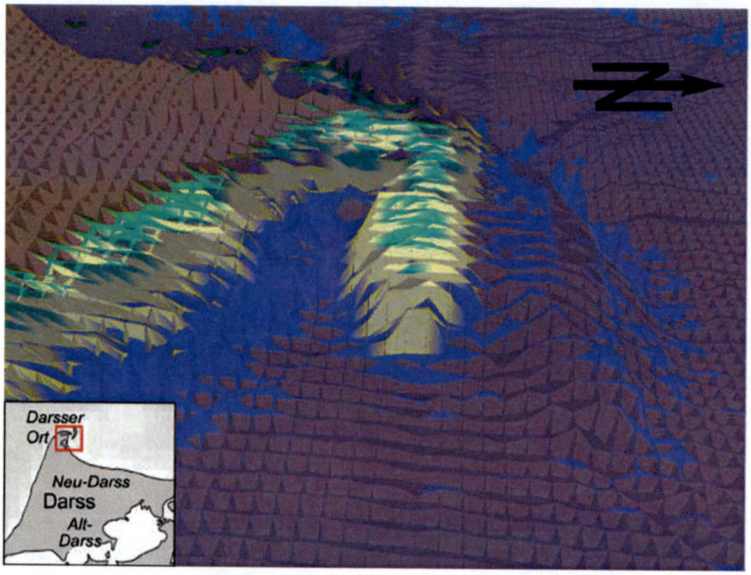

Fig. 14.9 The Darsser Ort after 840 years (experiment A). A west–east-directed spit has been accumulated with two distinguishable ridges

The difference to the recent coastline (Fig. 14.3) is apparent. Looking at the sediment source area of Fischland, erosion is visible. There are even three inlets between the open sea and the Bodden chain. In contradiction, the most northern site of the Darss, the Darsser Ort, experienced accumulation. A typical spit is forming here with a west–east elongated shape, common for coastal formations along the southern Baltic Sea. A closer look to this structure (Fig. 14.9) suggests a sequence-like composition. A first sequence is located right in front of the main peninsula, separated only by a narrow inlet. This is the main spit body that is located above sea level. A second sequence with a similar shape follows to the north, but still below the sea level. This scheme fits well with the structure of the Neu-Darss, that is composed by a pattern of barrier beaches with intermediate depressions (Janke and Lampe 1998).

Going further eastwards, coastal regions along northern Zingst, exposed to the open sea, have been eroded with the result of small bay-like structures, while the Bodden chain was filled up with sediments. The results of this experiment agree with the general understanding about erosion, transport, and accumulation for this system, though there is rather minimal accumulation in the area of the Bock Island that is generally considered as a major accumulation zone (Janke and Lampe 1998).

The overall set-up of the second experiment B is comparable to the first simulation, however, with the SEA LEVEL module switched on. As shown in Fig. 14.6, the sea level rise after 840 years is about 1.4 m. This additional parameter causes major changes in the model results (Fig. 14.8b). The difference to experiment A is striking. The Zingst is no longer a coherent peninsula but rather a chain of small islands. There are big channels between the inner Bodden and the open sea. This significant feature is also common in the Fischland area. An intense water exchange can be expected with remarkable effects for hydrography and biology. The rising sea level also prevents the formation of a spit at the Darsser Ort. There is no sediment accumulation visible, but a retreat of the coast. The northern forefront is smoothed.

The vertical movement of the earth's crust acts contrary to the sea level rise in the study area, mostly. With a maximum value of 0.4 mm/year (Fig. 14.5) this process has a low influence on the system. Summed up for the next 840 years, this culminates in a maximum uplift of 0.34 m, valid for the most northeastern parts. The results for experiment C, that includes this uplift, is shown in Fig. 14.8c.

In experiment D (Fig. 14.8d) the extreme conditions of the storm surge from 1872 are added to the set of inputs so far. Now, the recent coastline cannot be recognized any longer onshore but is visible as a sharp submarine ridge. From the peninsula, only some islands remain. The Alt-Darss and a part of Fischland are still above sea level as well as the area south of the Bock.

The comparison of the different experimental results points out a major controlling effect of sea level rise for the development of the coastline, even on the local scale. A shift of sea level not only has a direct impact because of morphological adjustments but also influences the sediment transport system by changing the exposure of terrain to waves.

14.6 Summary

For the area of the Darss-Zingst peninsula, experimental scenarios for the development of the coastline during the next 840 years have been calculated with the sediment transport modelling software Sedsim. Climate-driven parameters taken into account are sea level change, wave regime, and extreme storm events. The parameterization of the sea level aligns to IPCC CO_2 concentration scenarios, while for the waves data from past decades have been extrapolated. On the geological side, vertical movement of the earth's crust and the distribution of different sediment types are included into the modelling. These parameters have been adjusted according to most recent maps and investigations from the study area. A digital elevation model with a spatial resolution of 150 m serves as the structural frame.

Different parameterization set-ups have been tested. In a first approach, the impact of waves in combination with biannual storms has been simulated. The results of this experiment show a continuation of the recent sedimentation regime with sediment longshore transport from west to east. This pattern changes drastically if the module responsible for the simulation of sea level rise is activated. A formation of the spit at the Darsser Ort cannot be recognized anymore, and the consolidated coast breaks apart into some small islands. Now, the former isolated Bodden chain is connected to the open sea by channels. By the integration of the module for vertical movement of the earth's crust, these model results are not modified significantly. In a worst-case scenario the sea level height from the storm surge in 1872 is superimposed. Only some areas remain above sea level, while the majority of the land is flooded. All these experiments do not take coastal protection into account. Instead, they can be used to identify locations and areas where such activities may be necessary and worthwhile.

Altogether, Sedsim proved to be an appropriate tool for the modelling of Holocene coastal structures at the southern Baltic Sea. The model results for the next 840 years are geologically plausible. Therefore, in order to investigate the long-term coastal evolution model runs for the geological past are proposed.

Acknowledgements This chapter is a result of the project SINCOS (Sinking Coasts – Geosphere, Ecosphere and Anthroposphere of the Holocene Southern Baltic Sea) which was funded by the German Research Foundation.

The compilation of digital elevation data, provided by the Land Survey Administration Mecklenburg-Vorpommern and the Federal Maritime and Hydrographic Agency, was prepared by Mayya Gogina, Leibniz Institute for Baltic Sea Research Warnemünde, Germany. Anke Barthel, PhD student at the Ernst-Moritz-Arndt University Greifswald, Germany, digitized the terrestrial sediment distribution map. Prof. Dr. Cedric Griffiths, CSIRO Australia, granted access to the SEDSIM simulation software and the incorporated hardware resources.

References

BACC Author Team (2008) Assessment of climate change for the Baltic Sea Basin. Regional Climate Studies, 474 p

Björck S (1995) A review of the history of the Baltic Sea, 13.0–8.0 ka BP. Quaternary International 27:19–40

Commonwealth Science and Industrial Research Organisation (CSIRO) Petroleum, Australia (2004) SEDSIM demonstration manual. PC demonstration 2004. http://strata.geol.sc.edu/PDF-Files/Simulations/SedsimManual2004.pdf: 23 pp. Accessed 30.09.2008

Grabemann I, Weisse R (2008) Climate change impact on extreme wave conditions in the North Sea: an ensemble study. Ocean Dynamics 58:199–212. doi:10.1007/s10236-008-0141-x

Hammarklint T (2009) Swedish Sea level series – a climate indicator. Swedish Meteorological and Hydrological Institute, 5p

Harff J, Frischbutter A, Lampe R, Meyer M (2001) Sea level change in the Baltic Sea – interrelation of climatic and geological processes. In: Gerhard J, Harrison WE, Hanson BM (eds) Geological perspectives of climate change. American Association of Petroleum Geologists Bulletin Special Publication, Tulsa, Oklahoma, pp 231–250

Harff J, Bobertz B, Graf G (2009) Dynamics of natural and anthropogenic sedimentation (DYNAS). Journal of Marine Systems 75(3–4):315–316

Harff J, Meyer M (2007) Changing Holocene coastal zones of the Baltic Sea – a modeling approach. In: Harff J, Lüth F (eds) Sinking coasts-geosphere, ecosphere and anthroposphere of the Holocene Southern Baltic Sea. Berichte der Römisch-Germanischen Kommission, vol 88, pp 241–266

Heck H-L, Breitbach J, Büttner K, Groba E, König G, Stahff U, Tattenberg P, Vollbrecht K (1957) Geologische Karte des Norddeutschen Flachlandes, 1:1000000. Geologische Karte, Einheitsblatt 10:2 sheets, Berlin

Hoffmann G, Musolff A, Meyer T, Schafmeister M-Th (2004) Der geologische Aufbau des oberflächennahen Grundwasserstockwerkes im Nordosten der Halbinsel Gnitz (Usedom/Mecklenburg-Vorpommern). Rostocker Meeresbiologische Beiträge 12:9–21

Houghton JT, Jenkins GJ, Ephraums JJ (eds) (1990) Climate change: the IPCC scientific assessment. Cambridge University Press, Cambridge

Hupfer P, Harff J, Sterr H, Stigge HJ (eds) (2003) Der Wasserstand an der Transgressionsküste der südwestlichen Ostsee. Entwicklung – Sturmfluten – Klimawandel. Die Küste 66:311p

IPCC (2007) Summary for policymakers. In: Solomon S, Qin D, Manning M, Chen Z, Marquis M, Averyt KB, Tignor M, Miller HL (eds) Climate change 2007: the physical science basis. Contribution of Working Group I to the IVth assessment report of the Intergovernmental panel on climate change. Cambridge University Press, Cambridge and New York, NY

Janke W, Lampe R (1998) Die Entwicklung der Nehrung Fischland – Darß – Zingst und ihres Umlandes seit der Litorina-Transgression und die Rekonstruktion ihrer subrezenten Dynamik mittels historischer Karten. Z Geomorph N F, Suppl-Bd 112:177–194. Berlin – Stuttgart

Lampe R (2002) Holocene evolution and coastal dynamics of the Fischland-Darss-Zingst peninsula. Greifswalder Geographische Arbeiten 27:155–164

Lampe R, Endtmann E, Janke W, Meyer H, Lübke H, Harff J, Lemke W (2005) A new relative sea-level curve for the Wismar Bay, N-German Baltic coast (Eine neue relative Meeresspiegelkurve für die Wismarbucht, norddeutsche Ostseeküste). Meyniana 57:5–35

Land Survey Administration Mecklenburg-Vorpommern (2006) Digital elevation model 25 – DGM25. Digital elevation model, grid size 25 m. Schwerin, Germany

Li F, Dyt C, Griffiths C (2004) 3D modelling of the isostatic flexural deformation. Computers & Geosciences 30:1105–1115

Lehfeldt R, Milbradt P (2000) Longshore sediment transport modeling in 1 and 2 dimensions. Advances in Hydro-Science and Engineering. Proceedings of the 4th international conference on Hydro-science and engineering, Seoul. Abstract Volume 262

Martinez PA, Harbaugh JW (1993) Simulating nearshore environments. Pergamon Press, New York, 265p

Meier HEM, Broman B, Kjellstrom E (2004): Simulated sea level in past and future climates of the Baltic Sea. Climate Research 27:59–75

Metz B, Davidson O, Bosch P, Dave R, Meyer L (ed) (2007) Contribution of Working Group III to the IVth assessment report of the Intergovernmental panel on climate change. Cambridge University Press, Cambridge, 851p

Meyer M (2003) Modelling prognostic coast line scenarios for the southern Baltic Sea. Baltica 16:21–30

Meyer M, Harff J, Gogina M, Barthel A (2008) Coastline changes of the Darss-Zingst Peninsula – a modelling approach. Journal of Marine Systems 74:S147–S154

Miettinen A (2004) Holocene sea-level changes and glacio-isostasy in the Gulf of Finland, Baltic Sea. Quaternary International 120:91–104

Roeckner E, Arpe K, Bengtsson L, Christoph M, Claussen M, Dümenil L, Esch M, Giorgetta M, Schlese U, Schulzweida U (1996) The atmospheric general circulation model ECHAM-4: model description and simulation of present-day climate. Max-Planck Institute for Meteorology, Hamburg, Germany, Report No. 218:90 pp

Rosenhagen G, Bork I (2009) The extreme storm surge at the German coasts of the Baltic Sea in November 1872 – reanalysis of the wind fields for coastal purposes. In: Witkowski A, Harff J, Isemer H-J (eds) International conference on climate change. The environmental and socio-economic response in the southern Baltic region, University of Szczecin, Poland, 25–28 May

Rosentau A, Meyer M, Harff J, Dietrich R, Richter A (2007) Relative sea level change in the Baltic Sea since the Litorina Transgression. Zeitschrift für Geologische Wissenschaften 35(1/2):3–16

Schumacher W (2000) Zur geomorphologischen Entwicklung des Darsses – ein Beitrag zur Küstendynamik und zum Küstenschutz an der südlichen Ostseeküste. Zeitschrift für Geologische Wissenschaften 28:601–613

Seifert T, Tauber F, Kayser B (2001) A high resolution spherical grid topography of the Baltic Sea, 2nd edn. Baltic Sea Science Congress, Stockholm, Poster #147

Tauber F, Lemke W (1995) Map of sediment distribution in the western Baltic Sea. Deutsche Hydrographische Zeitschrift 47(3):171–178

Tauber F, Lemke W, Endler R (1999) Map of sediment distribution in the Western Baltic Sea (1:100 000), sheet Falster – Møn. Deutsche Hydrographische Zeitschrift 51(1):5–32

Tetzlaff DM, Harbaugh JW (1989) Simulating clastic sedimentation. Computer methods in the geosciences. Van Nostrand Reinhold, New York, 196p

Voß R, Mikolajewicz U, Cubasch U (1997) Langfristige Klimaänderungen durch den Anstieg der CO_2-Konzentration in einem gekoppelten Atmosphäre-Ozean-Modell. Annalen der Meteorologie 34:3–4

Weisse R, Storch Hv (2009) Marine climate and climate change: storms, wind waves and storm surges. Springer-Praxis books in Environmental sciences, Springer, Berlin; Chichester, UK, 219p

Weisse R, Storch Hv, Callies U, Chrastansky A, Feser F, Grabemann I, Guenther H, Pluess A, Stoye Th, Tellkamp J, Winterfeldt J, Woth K (2009) Regional meteo-marine reanalyses and climate change projections: results for Northern Europe and potentials for coastal and offshore applications. Bulletin of the American Metrological Society 90:849–860. http://dx.doi.org/10.1175/2008BAMS2713.1

Part VI
Interactions Between a Changing Environment and Society

Chapter 15
Settlement Development in the Shadow of Coastal Changes – Case Studies from the Baltic Rim

Hauke Jöns

Abstract The maritime zone of the Baltic basin, in all the phases of its settlement history, was of special importance to the people living there. Only there did they have access to marine resources and to the transportation and communication routes. The Baltic shore was therefore utilized, occupied, settled and even modified by humans, despite the unstable environmental conditions due to the isostatic rebound and the eustatic rise in sea level, which made it necessary to constantly adapt to a changing environment. Changes in sea level and the shoreline are generally investigated by the earth sciences. The resulting data form the base for the calculation of sea-level curves and shore-displacement models. Especially in areas with high rates of shore displacement, the data and models can then be used to reconstruct environmental conditions and to date prehistoric coastal sites. Conversely, well-excavated and dated archaeological sites that were originally located on the shore can provide detailed information about the sea level at the time of their occupation and can be used as sea-level index points. In this chapter, the opportunities and problems arising from the use of shore-displacement models for the interpretation of archaeological sites are discussed, as is the utilization of data extracted from archaeological investigations. Both models and sites are introduced in case studies that represent not only the different areas and localities but also the different stages in the development of the Baltic Sea.

Keywords Archaeology · Settlement development · Baltic Sea · Sea level index-point · Shore displacement · Coastal changes · Seafaring

15.1 Introduction

The settlement of the Baltic Sea area started a few centuries after the end of the last glaciation 15,000 years ago and has continued without any notable interruption until today. Our knowledge of the development of settlement in this area is almost

H. Jöns (✉)
Lower Saxony Institute for Historical Coastal Research, D-26382 Wilhelmshaven, Germany
e-mail: joens@nihk.de

J. Harff et al. (eds.), *The Baltic Sea Basin*, Central and Eastern European Development Studies (CEEDES), DOI 10.1007/978-3-642-17220-5_15,

entirely based on the archaeological remains of earlier cultures because contemporary written sources are not available for the period before the gradual introduction of Christianity from the ninth to the thirteenth century AD when priests and monks came to the coast of the Baltic Sea and wrote down their observations – mostly about political and military events. Thus the historiography of the Baltic area began. However, it is now agreed that even for this later phase in mankind's history the archaeological record must also be taken into consideration if there is to be a comprehensive reconstruction of living conditions in the past.

The use of archaeological and historical methods makes it possible to obtain spatially and chronologically differentiated information about the cultural characteristics of former societies as expressed, for example, in house-building traditions, costume fashions or burial customs. However, if one also wants to analyse more general living conditions, such as the climatic and environmental conditions or the available resources, historical research must be supplemented by the scientific information provided by disciplines such as botany, zoology and the geosciences. The results of these investigations permit the reconstruction of the biosphere and geosphere that gave rise to the environment of the former settlement area and they are, therefore, essential for understanding settlement behaviour and thus the anthroposphere.

This applies in principle to all landscapes that are used as a source of food or are occupied, settled and even modified by humans, but it is especially applicable to the coastal area of the Baltic Sea. The communities living there since deglaciation not only had to constantly adapt to the ever-changing composition of the flora and fauna – both on shore and in the sea – but also, in some periods, had to face and react to dramatic changes in the shoreline caused by isostatic rebound and land uplift on the one hand and the constant eustatic rise in sea level on the other. The removal of inundated settlements to more secure spots, the abandonment of graveyards in flooded areas and the relocation of silted-up landing and harbour sites or dried-out fishing fences in areas of land uplift are all evidence of the reaction of ancient communities to the changing environmental and living conditions.

To sum up, the people living on the Baltic rim in the past were continually forced to adapt their economic strategies to a changing environment. Consequently, their remains – preserved in the soil ever since they abandoned their homes – are, today, considered to be an important record not only of settlement history but also of coastal development. Especially since the beginning of the 1990s, when absolute-chronological classification by ^{14}C dating was supplemented by the AMS method, which also enabled the dating of very small samples of organic material, an increasing number of archaeological sites that were originally on the coast have been investigated not only to answer archaeological questions but also to obtain data about the sea level at the time of their occupation (Fischer 1996, Åkerlund et al. 1997).

Since then, there has been very intensive and fruitful interdisciplinary cooperation between archaeologists, geoscientists and modelling specialists in many parts of the Baltic Sea area. An example to be mentioned here is the DFG research unit SINCOS, which aims to obtain new data on the changes in the coastal landscape over the last 8,000 years in the German part of the south-western Baltic coast by close cooperation between geologists, geophysicists, geographers, geodesists, botanists, zoologists, dendrochronologists and archaeologists (Harff et al. 2007).

This chapter has to be considered against the background of this new approach to research. It presents some general reflections on methodological preconditions and several case studies from different time periods and regions of the Baltic rim that show how the multidisciplinary approach has improved our knowledge of the continuous displacement of the shoreline and the development of settlement in the shadow of such coastal changes.

To avoid any misunderstanding as far as chronology is concerned, it must be mentioned that all the dates discussed in this chapter should be understood as calendar years (calibrated ^{14}C years BC/AD), calibrated using the Calpal program by O. Jöris and B. Weninger (see Manual Calpal or www.calpal.de).

15.2 Methodology

Due to the melting of the ice masses in the glaciers of the Fennoscandian ice sheet, the global sea level started to rise rapidly at the end of the Weichselian period. During this process, the shape of the present Baltic Sea became subject to constant change, with regional differences in the dynamics and extent of this change. Today, the Baltic Sea basin is filled with brackish water but, in the past, its salinity and the proportion of freshwater changed repeatedly. In post-glacial times, the whole Baltic area experienced a period of regionally variable glacio-isostatic land uplift with its centre in the northern part of the Gulf of Bothnia. Although the rate of uplift has declined since the end of deglaciation, an on-going uplift of 9 cm/century is still being recorded today in this area (Rosentau et al. 2007, Meyer et al. Chap. 14, this book). Even though the sea level rose more or less continuously, the coastal landscape in this region was mainly shaped by a permanent regression. As a result, the coastline in central Sweden has advanced by about 300 km since 7,000 cal. BC.

The land uplift in the eastern Baltic area is much smaller and has recently averaged only 1–2 cm/century so that it is more or less balanced by the rise in sea level of 1.8 mm/year. On the other hand, in the south-western Baltic area, the uplift of about 1 cm/century is considerably less than the average rise in sea level so that, relatively, the coast is sinking and land is gradually being lost. Consequently, the coastline of the southern Baltic rim around 7,000 cal. BC was situated up to 70 km north of its present location.

Despite these changes in coastline and landscape, the contact zone between land and water around the Baltic Sea has always been an area of special importance for human communities. Only this ecosystem provides access to marine resources such as fish, mussels and oysters and an opportunity to hunt brants and ducks, walruses and seals as well as other sea birds and mammals. Together with lakes and rivers, the sea was the most important transportation system up until the Middle Ages, so the maritime landscape was also of great importance for travellers, migration, communication and the exchange of goods (Westerdahl 2000). Detailed analyses of the settlement pattern on the Danish island of Fyn show that during this period the coastal zone formed a cultural landscape of unique character and function (Crumlin-Pedersen et al. 1996). There can be no doubt that this situation can be generalized, at least for the larger islands in the Baltic Sea during the first and second millennia AD, but presumably also for the coastal zones on parts of the mainland.

Even though the reasons for and the intensity of the use of the landscape and settlements in the coastal zone changed through the ages, it was common to all the communities living there that they modified the utilized or occupied parts of the landscape, e.g. by building houses, fishing fences or boat-landing facilities – and also by leaving their refuse and rubbish on the sites (Jöns 2002).

Given the above-mentioned continuous displacement of the shoreline as a result of the changing sea level and isostatic rebound, the people had to leave their coastal settlements and move to other spots that presumably offered better conditions for the future. Most of the abandoned sites fell into oblivion and were never occupied again because the specific attractiveness that originally led to their utilization was lost as a result of the changing environment.

Today, all the traces and remains of earlier activity on these deserted sites have become an archaeological archive, full of information about a specific – locally and chronologically limited – part of the history of mankind and the environment.

15.2.1 Shore-Displacement Models as a Base for Dating Prehistoric Sites

Most of the deserted coastal settlements have been eroded through the ages, by the current in the case of inundated sites or by wind, frost, sun and rain in the case of sites on dry land. In particular, structures, tools and refuse of organic material such as wood, bone, antler and leather have often disappeared completely so that the archaeological record of these sites consists almost entirely of inorganic finds made of stone or ceramics. Due to the absence of organic material, a chronological classification of these sites is only possible by means of a typological comparison of the artefacts recovered from the site with those from better preserved sites. In the case of sites with no diagnostic artefacts, it is often not possible to date them or even assign them to an archaeological culture.

Especially in those parts of the Baltic Sea area where the glacio-isostatic rebound has led to permanent land uplift, the information available on changes in the sea level has traditionally been used by archaeologists to date such sites (Ling 2004 with further references). In central and northern Sweden (Linden et al. 2006, Berglund 2004, 2008), Norway (Fuglestvedt 2008, Grimm 2006, Gustafson 1999) and Finland (Siiriäinen 1982, Jussila 1995) analyses of sea-level curves and shore-displacement models are of great importance for the dating of archaeological sites: the rule of

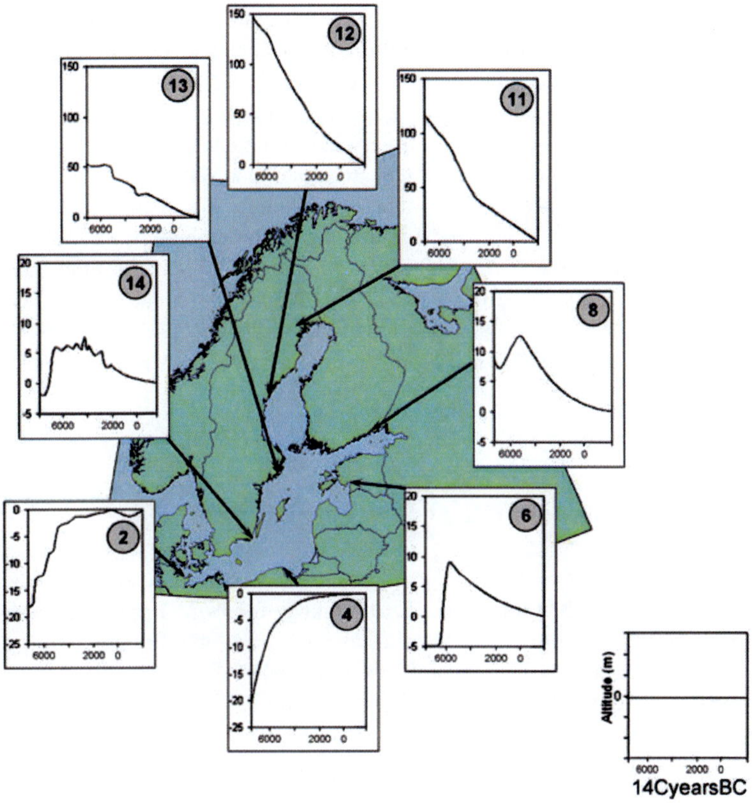

Fig. 15.1 An assortment of relative sea-level curves around the Baltic Sea during the Phase of the Littorina Sea (7,000 BC – after Rosentau et al. 2007, fig. 4)

thumb especially in the northernmost parts of this countries is 'the higher the level, the older the site' (Figs. 15.1 and 15.6).

The growing amount of relevant data obtained from new investigations over the last few decades, especially in Norway and Sweden, has led to a large number of regionally valid sea-level curves, which permit the generation of shoreline-displacement models of increasingly high quality (Rosentau et al. 2007). This new information about shoreline developments has, in some cases, already had important consequences for the archaeology-based reconstruction of the settlement history and led to changes in the research strategy.

The history of Stone Age settlement in northern Sweden can be mentioned as an example (Hörnberg et al. 2005). The glacio-isostatic land uplift there created a dynamic landscape that experienced a great deal of substantial environmental change during the Holocene such as the relocation of lake shorelines and modifications of the water flow in rivers. Consequently, the present landscape is completely different from that in the past. For a long while, only a few Stone Age sites were known from the inland parts of northern Sweden so that, until the 1970s, the region was thought to be of little archaeological significance. The situation changed recently when a model simulating early Holocene land uplift was created. This meant that the positions of contemporaneous lake shorelines could be reconstructed. These were used on field surveys and finally led to the discovery of a large number of Mesolithic settlements (Olofsson 2003, Bergman et al. 2003).

In the south-western part of the Baltic Sea, where the maritime landscape and settlement areas have been inundated and now lie on the seabed, shore-displacement models can be used to obtain an initial idea of the chronological classification of the submerged sites: here, the rule of thumb is 'the deeper the site below sea level, the older the site'. However, unlike the areas with decreasing relative sea levels, they are only of limited value as a starting point for underwater surveys in the search for new sites.

15.2.2 Archaeological Sites as Sea-Level Index Points

The quality of the sea-level curves and shore-displacement models greatly depends on the data used to compile them. Traditionally, they are based on geological and palynological investigations of stratified sequences of sediments from different deposits in the coastal area, which are analysed and interpreted. However, in some cases, data from archaeological sites are also integrated – or at least referred to – usually in order to prove the quality of the models (Lübke 2002).

The remains of settlements that were originally on the coast can be used as fossil sea-level index points, provided that the relevant parts of the sites were originally situated near the shore or constructed with specific reference to sea level (Fig. 15.2; for a summary see Behre 2004, 2007). In such cases, it can be assumed that the settlement facilities on the site, e.g. houses, hearths and pits, were above the mean sea level and, in general, secure from inundation by storm surges. In addition, it has to be ensured that the dated material represents the lowest parts of the site (Olsson and Risberg 1995). On the other hand, fish traps, fishing fences or the foundations of piers and other harbour facilities must originally have been under water.

Thus, the reliable dating of these settlement remains by archaeological methods, dendrochronology or radiocarbon analysis can help us to reconstruct the sea level at a specific point in time. This is especially valid in the case of archaeological sites that were flooded during storm surges, when their remains were covered with sediments that preserved them from erosion and conserved them in some fortunate cases for thousands of years, until today. Almost without exception, such favourable conditions only exist on sites that were inundated immediately after they were abandoned or while they were still occupied. Most of these are still under water today. This is first and foremost true of those areas in the south-western part of the Baltic Sea that were rapidly inundated during the Littorina transgression and were not affected by the glacio-isostatic uplift so that the drowned landscapes and sites

Fig. 15.2 An artist's view of a typical Mesolithic coastal settlement at the Baltic rim (Graphic: F. Bau, Århus)

remained below sea level. Although access to these sites is only possible with the methods developed by underwater archaeology, they are of great scientific value because they not only offer an opportunity to recover artefacts made of organic material but also enable information to be obtained on the dynamics of rising sea levels.

15.3 Case Studies – The Baltic Rim as a Prehistoric Anthroposphere and an Archive of Coastal Change

As already pointed out above, the settlement history of the Baltic area goes back to the climate amelioration after deglaciation and has seen the rise and fall of numerous cultures and societies. The study of their remains is the task of archaeologists and historians in the respective countries and cannot be summarized in this chapter. Instead, I will only refer to archaeological cultures and sites to the extent necessary to understand the case studies presented here and their significance for the questions of settlement behaviour and economic strategies under discussion. Moreover, the numerous climatic and environmental changes that occurred in the Baltic basin during this period will only be referred to when they are essential in order to understand the relative coastal changes. An exhaustive survey is not possible, nor is it attempted.

The case studies will be discussed in chronological order, in blocks of time that reflect the respective stages in the development of the Baltic Sea as described by Björck (1995) and Lemke (2004).

15.3.1 Late Palaeolithic Reindeer Hunters Around the Baltic Ice Lake and Yoldia Sea

The deglaciation of the Baltic area was a long drawn-out process, reaching the various regions at different times. While the south-western and south-eastern parts of the present Baltic rim were already free of ice around 15,000 cal. BC (Clausen 1997, Eriksen 2002) and 12,000–10,000 cal. BC (Zagorska 1999, Ukkonen et al. 2006), respectively, the central and northern Baltic areas became ice-free not earlier than 8,500 cal. BC (Linden et al. 2006, Berglund 2008). Deglaciation was followed by a remarkable rise in temperature that permitted the emergence of tundra vegetation characterized by bushes, low dwarf-birches and pine trees (Fig. 15.3). This new landscape offered favourable conditions for the reindeer herds that subsequently migrated into the whole area around the Baltic Ice Lake. Radiocarbon dates for fossil bones and antlers (not found in a human settlement context and therefore presumably not hunted game) indicate that the animals were very resistant and could even survive the climatic conditions of the late Glacial and early Holocene (Ukkonen et al. 2006). The oldest finds of reindeer remains are from Lithuania, Estonia and Latvia; these are dated to 14,180–11,280 cal. BC. In Denmark and western Norway the species was present around 12,800 cal. BC, in southern Sweden around 11,600 cal. BC, whereas north-western Russia and Finland first attracted reindeer herds around 6,500 cal. BC.

Several centuries later, the herds were presumably followed by hunters who specialized in hunting reindeer while the animals were crossing rivers (Terberger 2006a). Archaeological evidence of this first phase of human presence in the Baltic area is only known from northern Germany and southern Denmark (Grimm and Weber 2008) and indicates that these communities set up their camps and settlements along the reindeer migration routes. They belonged to the so-called Hamburgian group, which came to the region during the Meiendorf interstadial around 12,700 cal. BC, or to the Havelte group that developed from the former group after 12,300 cal. BC.

The landscape changed considerably during the Allerød interstadial when the temperature again rose remarkably by a total of more than 5°C (Clausen 1997). This climate change permitted the growth of birch, aspen, rowan and pine trees in the southern Baltic area. The area provided a habitat for elk as well as giant deer and wild horses: there is evidence of the existence of open woodland that lasted for 1,200 years, from 11,900 to 10,700 cal. BC (Eriksen 2002, Terberger 2006a). In this period, communities belonging to the Federmesser culture and the Brommian culture inhabited the south-western part of the Baltic rim. Especially from Denmark and Scania loads of Brommian finds and – less well represented – Federmesser finds are known (Eriksen 2002, Andersson and Cronberg 2007).

Throughout this whole period, the Baltic basin was gradually filling up with meltwater as a consequence of deglaciation. A constantly expanding freshwater lake developed – the Baltic Ice Lake (Björck 1995, Lemke 2004). For more than 3,500 years (13,000–9,500 cal. BC), this lake remained covered by ice for most of the year. The Baltic Ice Lake was not connected to the North Sea, so its water level

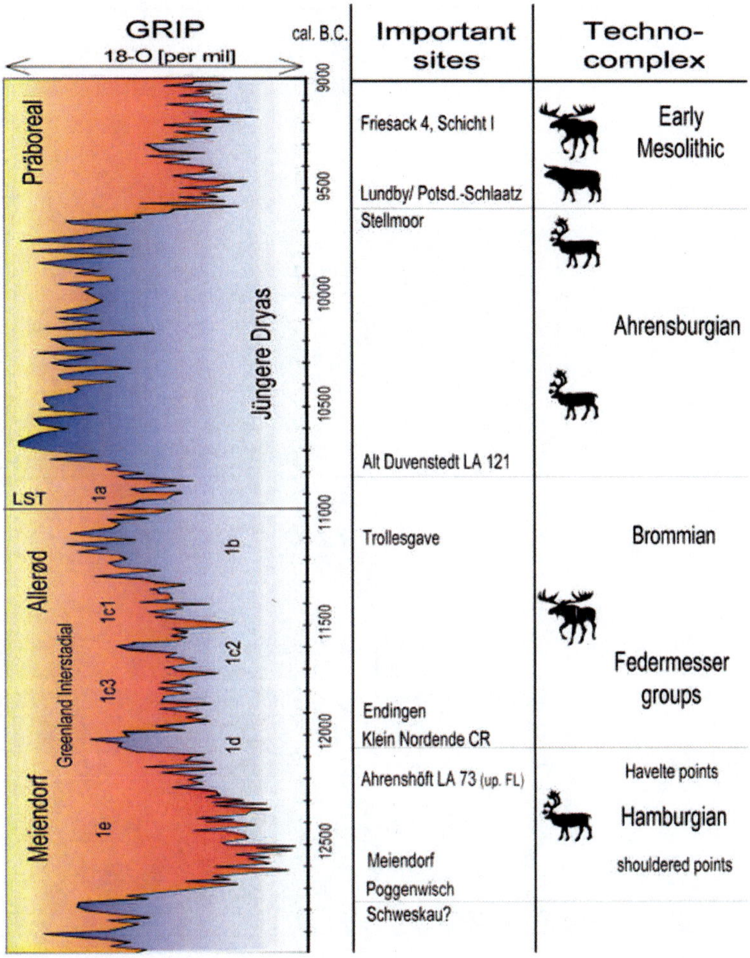

Fig. 15.3 Chronostratigraphical sequence of the Late Glacial in relation to isotope curve of the GRIP ice core (LST: Laacher See Tephra), archaeologically defined groups and cultures of Northern Germany/Southern Scandinavia, important sites and typical faunal elements (after Terberber 2006a, fig. 6)

rose to more than 25 m above the seawater level (Fig. 15.4). While the shoreline of the south-western part of the Baltic Ice Lake was still far to the north and east of the present shore, large parts of what are now the territories of the Baltic States were under water (Zagorska 1999). Only during a short period, between 10,200 and 9,700 cal. BC, a rapid regression of the level of the Baltic Ice Lake has been recorded, when the Baltic waters could flow out into the North Sea via a strait running through what is today central Sweden. But this strait closed again during the Younger Dryas period and the level of the Baltic Ice Lake again rose rapidly, causing

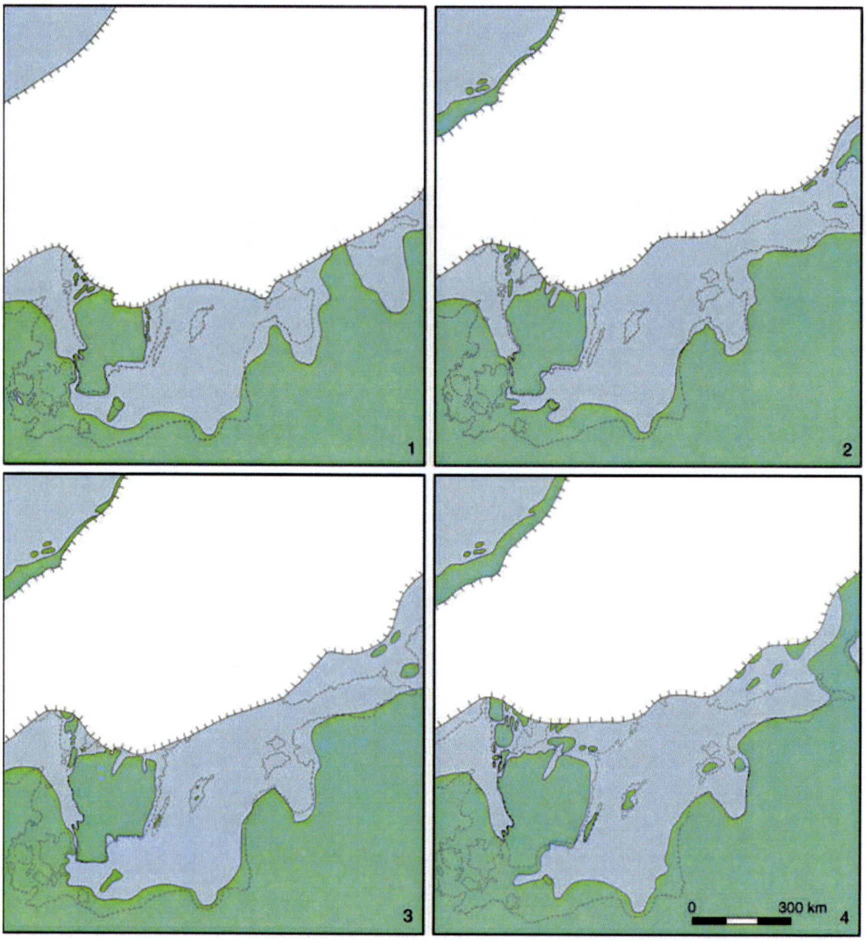

Fig. 15.4 History of the Baltic Basin during the Saltic Ice lake and the Yoldia Sea (13,000–9,000 cal. BC). *1* The Baltic Ice Lake around 12,500 cal. BC in its completely up-dammed stage. *2* The Baltic Ice Lake around 11,000 cal. BC connected by a subglacial drainage or by an open strait to the Kattegat. *3* The Baltic Ice Lake around 9,800 cal. BC, just prior to the final drainage. *4* The Yoldia Sea 9,400 cal. BC with several outlets in central Sweden, north of a large land bridge (after Lemke 2004, fig. 1, 1–4, modified by the author)

flooding over large areas along the south-western shore of the lake. As a result of the isostatic uplift in the eastern part of the Baltic area, the shoreline was continuously displaced towards the sea and new land emerged.

As far as we know today, all the settlements of the Hamburgian group, as well as those of the following Brommian and Federmesser groups, were located inland, far from the coast, so that these communities were presumably not affected by the changes in the level of the Baltic Ice Lake.

The transition from the Baltic Ice Lake to its successor, the Yoldia Sea (9,700–8,800 cal. BC), was marked by a rapid fall in the water level, about 25 m, caused by a newly opened strait through central Sweden to the Kattegat that allowed Baltic water to flow out into the North Sea. For a short period around 9,600 cal. BC, salt water also entered the Baltic through this strait and created a brackish environment. As a consequence of the decline in the water level the environmental conditions changed rapidly, especially in the southern and the western parts of the Baltic. It also led to an extension of the land bridge connecting northern Germany, Denmark and Sweden.

The transition from the Baltic Ice Lake to the Yoldia Sea occurred during the Younger Dryas stadial, which was characterized by a radical fall in the average annual temperature throughout the whole southern Baltic area, to a level even lower than that during the Meiendorf interstadial: the ice margin, which had moved far to the north during the Allerød, was now located in the area that is now central Sweden and southern Norway (Eriksen 2002). This climate change turned the open woodland into wide-spread tundra again: the reindeer herds, and with them the hunter communities, therefore returned to the southern parts of the Baltic area for the period from 10,800 to 9,600 cal. BC (Terberger 2006a). They ranged over an area stretching from Russia to England and central Germany (Eriksen 2002). A few sites with typical Ahrensburgian artefacts from central Sweden and western Norway indicate (Fuglestvedt 2008) that in that period a few communities also moved to the north. According to Terberger et al. (2004), this expansion can be seen as evidence of a climatic amelioration already before the periglacial climate finally ended during the Preboreal.

The similarities in some of their equipment and in their hunting techniques indicate that the communities living in this wide-spread area were culturally closely related but, due to the differences in their material culture, they are terminologically divided into different regional groups, e.g. the Ahrensburgian group in north-western Germany and southern Scandinavia (Eriksen 2002), the Fosna-Hensbacka in south-western Norway (Fuglestvedt 1999, 2008) and western Sweden (Kindgren 1996, Schmitt et al. 2006, 2009), and the Swiderian or Eastern Ahrensburgian group in Poland and the eastern Baltic (Zagorska 1999).

While the shores of the last phase of the Baltic Ice Lake and Yoldia Sea in the south-western part of the Baltic rim were still far from the present coastline, the eastern part experienced the emergence of new land as a result of the strong isostatic land uplift caused by deglaciation. Recently, the ancient shoreline of the Baltic Ice Lake on the territory of Latvia was observed at 55–12 m above the present sea level (Eberhards and Zagorska 2002). Palynological investigations have shown that this development took place in five phases during the Older Dryas, Allerød and Younger Dryas periods. As a result of this development, former ice-margin basins developed into lakes and the beds of the Daugava and Lielupe rivers – originally formed by glacial meltwater during the early stage of the Baltic Ice Lake – became the drainage system for the area (Zagorska 2007a). The subsequent banks of these rivers were partly formed as a vertical sequence of wide terraces. Specific artefacts such as harpoons made of bone and antler, silex tanged points and scrapers as well as numerous reindeer bones indicate that they were occupied during the younger Dryas period by reindeer hunters of the Swiderian group with connections to the Ahrensburgian group, who came to the region from their base camps in Lithuania or East Prussia as they followed the summer migration of the reindeer herds (Zagorska 1999, 2007b).

A similar preference for the occupation of river terraces on three different levels by the reindeer-hunter communities of the Swiderian, Brommian and Ahrensburgian groups during the Allerød and Younger Dryas has also been observed in Lithuania (Rimantienė 1994, 1998).

Compared to the relatively moderate environmental changes that the reindeer-hunter communities experienced during the Younger Dryas in the eastern Baltic, the first colonizers of the territories of western Sweden and south-eastern Norway had to face truly dramatic changes in their habitation area. At the beginning of the Holocene at 9,500 cal. BC, this region was still largely covered by glacial ice: according to Boaz (1999), final deglaciation occurred here during the Preboreal, which means that the landscape was ice-free around 8,200 cal. BC. The swift retreat of the Pleistocene ice mass during this phase led to extremely dynamic and short-lived environments with a particularly high sea level. Large parts of the present landscape became inundated during this period. However, due to the rapid isostatic land uplift following deglaciation, the highest points of the moraines emerged from the sea and formed a fast-growing archipelago, which provided favourable conditions for the communities that were especially adapted to life in a maritime landscape as hunters of marine mammals and fishermen (Fischer 1996, Kindgren 1996).

How these communities depended on the environment and their adaptation to the specific challenges of changing coastlines can be clearly seen at a site that was discovered at Stunner, Ski district, in the vicinity of Oslo (Gustafson 1999). Although there has not yet been any excavation, more than 700 artefacts made of silex and quartz were salvaged as surface finds on a dry stony surface. The site is flanked by two hills and lies on ground that is 165 m above the present sea level (Fig. 15.5). Fuglestvedt's (1999) analysis of the artefacts showed that the site was inhabited by the Fosna-Hensbacka group during the Preboreal.

More precise dating was possible with the help of the regional sea-level curve and the shore-displacement model developed from it (Sørensen 1979, Gustafson 1999). These prove that the regional sea level fell rapidly after deglaciation – calculated at a rate of 10 m/100 years between 10,000 and 7,000 cal. BC (Fig. 15.6).

The topography of the Stunner site indicates that it would only have been attractive for habitation when the sea level was between 160 and 162 m above the present level. Only then would the settlement have been located on an island in the archipelago and thus exposed to the sea. It would have been secure from storm floods and had access to a small strait and a bay, which could be used as a safe harbour for boats. Already at a sea level of 150 m above the present level, access to the strait would have been closed and the shore more than 1 km away.

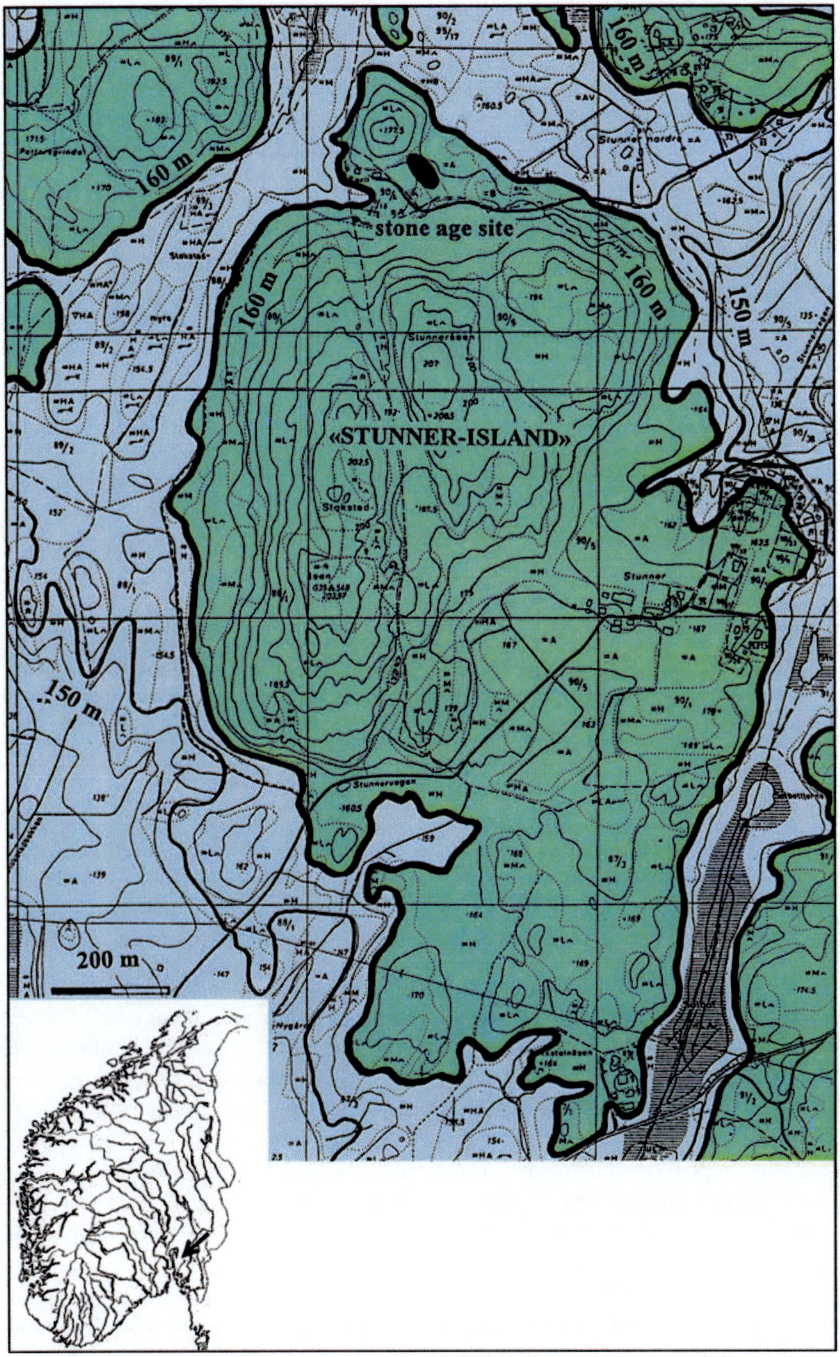

Fig. 15.5 Map of the 'Stunner Island'. The Stone Age site and the curves of 160 and 150 m are marked (after Gustafson 1999, fig. 2)

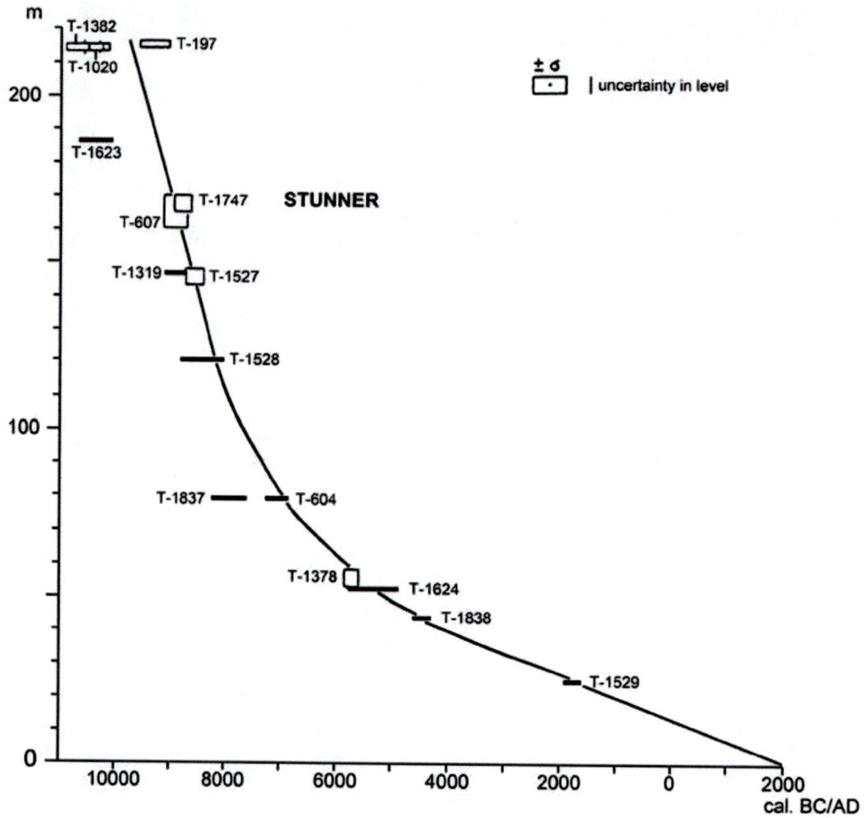

Fig. 15.6 Shore-displacement curve from the Ski district, Norway (after Gustafson 1999, fig. 3, modified by the author; graphics by R. Kiepe, NIhK)

Based on the sea-level curve, Gustafson (1999) has shown that with a sea level 160 m higher than today Stunner island can be placed chronologically at 8,900 cal. BC. This dating fits well with the typological classification of the artefacts. The blade technique used, in particular, indicates a close cultural connection between the people of Stunner and the late Ahrensburgian group (Fuglestvedt 1999).

Whether the people from Stunner and the other Fosna-Hensbacka group sites indeed travelled over the Yoldia Sea to personally visit the camps of the late Ahrensburgian communities on the present territory of Denmark and Germany, or whether they have to be regarded as their successors – well adapted to the new maritime landscape – is still a point of scientific discussion (Eriksen 2002, Fuglestvedt 2008), but there can be no doubt that they experienced more impressive shore displacements and changes in the landscape and their environment than most other communities in the history of the world.

15.3.2 Mesolithic and Early Neolithic Hunter-Gatherers and Fishermen on the Shores of the Ancylus Lake and Littorina Sea

The ninth millennium BC was again a period of intense change in the development of the Baltic basin. Due to the continuous isostatic uplift of its northern part, the strait that had connected the Baltic area with the Kattegat for centuries gradually closed and new land emerged in the area of present-day central Sweden (Fig. 15.7). The exchange of water was severely reduced, which led around 8,800 cal. BC to a change in the character of the Baltic from a brackish to a freshwater environment. In addition, the water level rose as a result of the damming of the so-called Ancylus Lake. While the containment of this lake only influenced the shoreline in the northern part of the Baltic to a small extent, because the on-going uplift compensated for the rising water level, the consequences for the southern Baltic area were dramatic. According to Björck (1995), the sea level here rose up to 10 m/century, so that vast areas were successively inundated. This situation changed again around 8,400 cal. BC, when a rapid regression has been recorded. For the following 200 years, the water level in the Ancylus Lake was equal to that of the ocean. It is assumed that there must have been a connection between the Baltic and the North Sea that permitted the outflow of Baltic water, although there is no evidence of inflowing salt water. Several possibilities for the location of this connection are still being discussed and the clarification of this question is, therefore, a field for future research. However, the connection between the Kattegat and the Ancylus Lake must have remained open until 7,000 cal. BC as the environmental conditions along its shores were more or

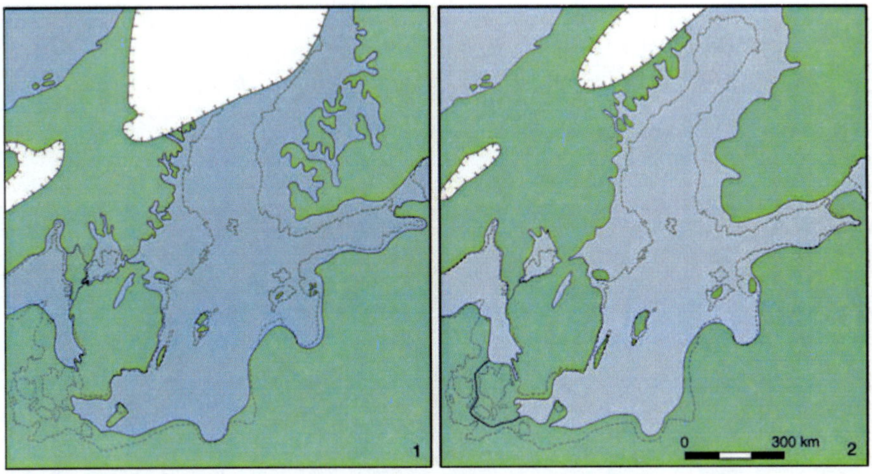

Fig. 15.7 The Baltic basin during the phase of the Ancylus Lake (9,000–8,000 cal. BC). *1*. Stadium around 8,500 cal. BC at the maximum of its transgression, *2* Stadium around 8,100 cal. BC when the former difference in level between the ocean and the Ancylus Lake was balanced (after Lemke 2004, fig. 1, 5–6, modified by the author)

less stable until then. After the end of that phase, a slight change from a freshwater to a brackish environment has been detected for some areas of the Ancylus Lake, indicating that salt water from the North Sea had started to flow into the Baltic basin (Lemke 2004).

The following phase, from 6,700 to 6,100 cal. BC, was characterized by constant uplift in the northern part of the Baltic area and a strong salt water transgression in its southern part (Lemke 2004). According to Rößler (2006) the central part of the Mecklenburgian Bight was not earlier affected by this development as recently as 6,100 cal. BC. It was caused by a rapid rise in the level of the North Sea and first led to the inundation of the Danish belts and the Öresund and thus to a new connection between the Baltic and the North Sea. In this phase, the Baltic Sea is called the Littorina Sea; other than the substantial salt water transgression, it is characterized by a brackish environment (Fig. 15.8).

For the south-western Baltic coastal area, Kliewe and Janke (1982) estimated a sea level rise of 2.5 cm/year. This development created a more structured coastline with a constantly changing topography consisting of numerous small islands and sea inlets. Not earlier than around 4,000 cal. BC, when the sea level had already reached a level only 1 m lower than today, the rising sea level lost its force and slowed down to 0.3 cm/century. The coastal landscape was consolidated and the first compensatory processes set in motion (Schmölcke et al. 2006).

The environmental conditions, too, changed considerably after the end of deglaciation: the temperature rose rapidly during the Preboreal and caused changes in the climate, the vegetation and the landscape (Schmölcke et al. 2006). Forests dominated by Scots pine and birch trees expanded more and more. Hazel quickly spread into the southern Baltic area after 8,600 cal. BC and, finally, the elm and oak arrived. These forests became the habitat of red and roe deer as well as wild boar, moose and aurochs. Brown bears, the European otter, beavers and foxes were also present. During the following Boreal millennia, lime and ash trees arrived in the southern Baltic area to complete the deciduous forests of oak, elm, hazel and birches.

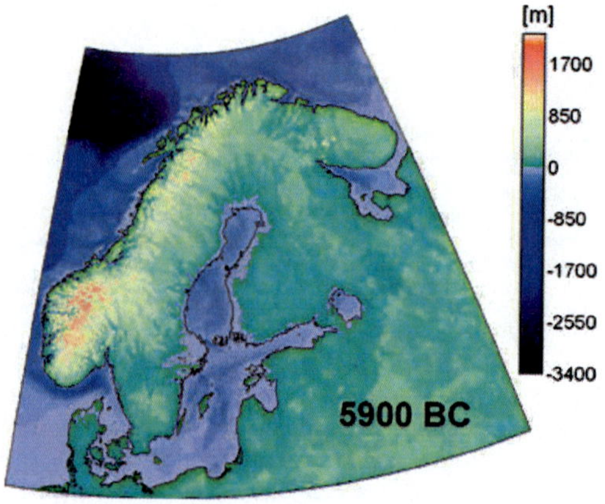

Fig. 15.8 Model of the Littorina Sea around 5,900 cal. BC (after Harff and Meyer 2007, fig. 11)

To sum up, environmental conditions in the Holocene were completely different from those in the late Pleistocene. The people no longer depended almost entirely on hunting one species – reindeer – but were also able to take advantage of the rich fish resources in lakes and in the sea, to gather vegetables – especially roots and fruit – and to hunt the locally available game and wild birds. This gave the Mesolithic communities living around the Baltic rim new economic opportunities but also forced them to develop new subsistence strategies (Terberger 2006b). The result is well known, thanks to the archaeological record for several inland and coastal sites that date from the ninth to the fifth millennium cal. BC. Unlike the highly specialized late Palaeolithic reindeer hunters, the Mesolithic people were generalists who adapted their economic strategies to the exploitation of the many different kinds of resources available in their vicinity.

The introduction of agriculture and animal husbandry during the Neolithic period at the transition from the fifth to the fourth millennium BC led to a fundamental change in the economic system that also affected the occupation of the coastal landscape. Settlements were moved to arable farmland away from the coast and the demand for meat was increasingly covered by domestic animals. Although the economic importance of game and marine food resources decreased considerably, hunting, fishing and gathering were never abandoned. In fact, they were still very much part of economic life (Hartz et al. 2007). Most of the traditional hunting and fishing methods continued to be used and known coastal locations were occupied on a seasonal basis.

As has already been pointed out, the coastal zone of the south-western Baltic rim was very much affected by the changes that occurred. In particular, the dynamic rise in sea level during the Littorina transgression, around 15 m in 600 years, led to the flooding of whole landscapes since the second half of the seventh millennium: this makes it an extraordinarily interesting area of research as far as the relationship between the geo-system, eco-system, climate and socioeconomic system is concerned (Harff et al. 2007). This was also the reason for the establishment of the multi-disciplinary research project SINCOS (Sinking Coasts: Geosphere, Ecosphere and Anthroposphere of the Holocene Southern Baltic Sea), which aims to reconstruct the coastal morphogenesis, the palaeoclimatic and ecological conditions as well as the settlement history of the area between the Oldenburger Graben and the Oder estuary during the Littorina transgression. Within this project, archaeological investigations were undertaken to obtain information on whether and how the ancient human communities reacted in the face of coastal decline and the enormous changes in their natural environment. A special focus is on whether they adapted their economic systems, social structures and/or their communication networks in reaction to these changes. The SINCOS research is concentrated in two areas, to the west and east of the Darss Sill: the Wismar Bight as part of the Mecklenburgian Bight in the west and Rügen Island in the east (Jöns et al. 2007).

In the discussion of the suitability of archaeological sites as sea-level index points, particular attention is paid to a group of more than 20 submerged settlements, today located at the bottom of the Wismar Bight at depths between 2.5 and 11 m below the present sea level (Fig. 15.9). Most of these were discovered during

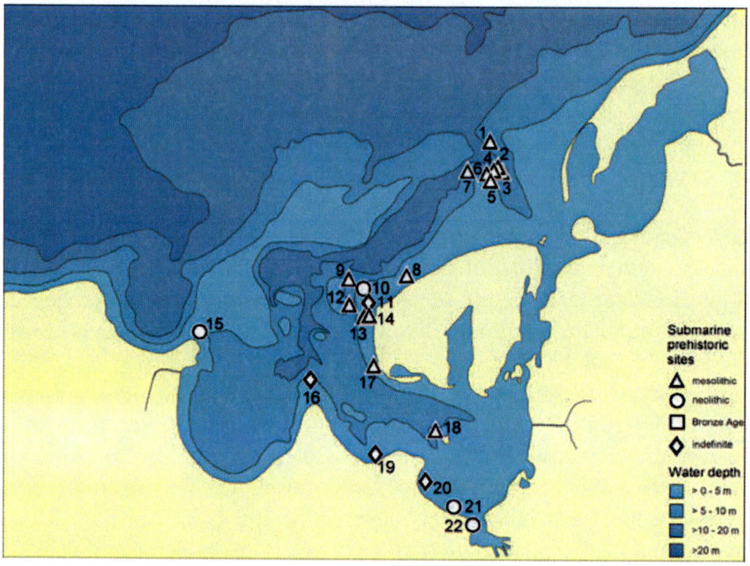

Fig. 15.9 Distribution of submerged Mesolithic and Early Neolithic sites in the Wismar Bight (after Jöns et al. 2007, fig. 2)

geophysical and Hyball surveys and, in a second step, partly excavated underwater. As well as seeking answers to several questions about the settlement pattern and chronology of the respective sites, a further aim is to gather data about ancient coastlines and the dynamics of the rise in sea levels.

Of special importance is the Jäckelberg-Huk site, located on the edge of the Jäckelberg at a depth of 8.5 m below the present sea level, because it is one of the oldest known submarine sites in the waters of the Wismar Bight. Radiocarbon analyses date the site to the period between 6,400 and 6,000 cal. BC (Fig. 15.10). The fish remains found on the site consist only of pike, perch and eel, which indicate a freshwater environment; the settlement must therefore have been situated in immediate proximity to a freshwater lake. The salvaged artefacts include trapezes, rhombic arrowheads and also a few very small longish triangular microliths, which prove that the people from this settlement were closely related to the initial phase of the southern Scandinavian Kongemose culture (Sørensen 1996).

In the same area, the late Mesolithic Jäckelgrund-Orth site was discovered at a depth of 7–8 m below the present sea level. ^{14}C dates for tree stumps from this site indicate that it was occupied from 6,000 to 5,700 cal. BC (Jöns et al. 2007).

Also of supra-regional importance is the Timmendorf-Nordmole II site. Here, parts of a fishing fence were excavated at a depth of 5 m below the present sea level, which had blocked the end of a small brook. The preservation conditions for organic material on the site were excellent; wooden artefacts such as several leister prongs and parts of a fish trap were recovered. Analysis of the find material

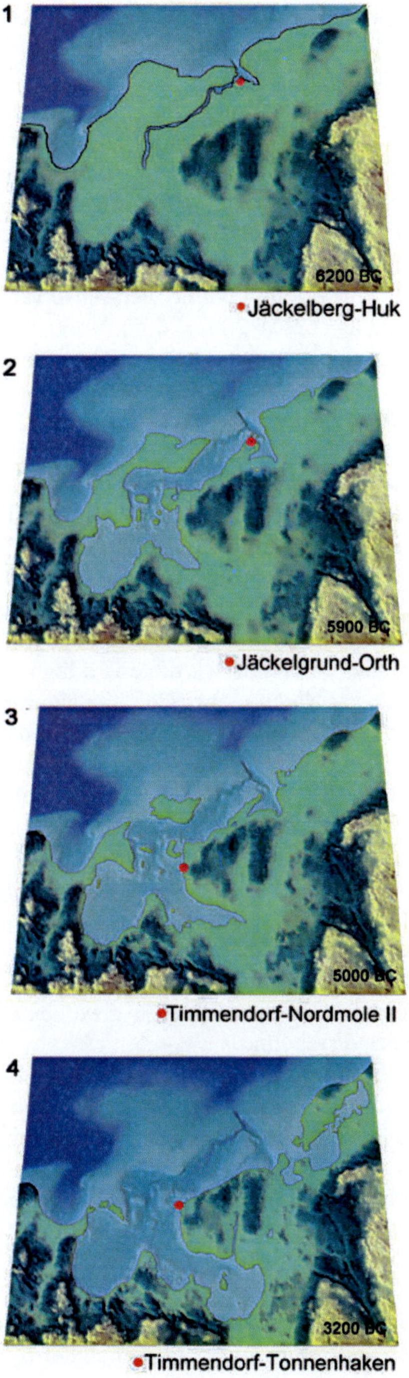

Fig. 15.10 Scenario of different stages of the Wismar Bight during the Littorina transgression (1. designed by the author, 2–4 after Harff and Meyer 2007, fig. 6)

indicated that the site belongs to an aceramic phase of the Ertebølle culture; a series of ^{14}C dates places the site in the period between 5,100 and 4,800 cal. BC (Hartz and Lübke 2006).

The neighbouring site, Timmendorf-Nordmole I, is also of great scientific value: settlement remains of the late Ertebølle culture were investigated at a depth of 2.5–3.5 m below the present sea level. They were radiocarbon dated to the period between 4,400 and 4,100 cal. BC. On this site, a pit was excavated that was covered with a number of long logs and poles that could originally have been a roof or covering for some structure. In the heterogeneous sediment that filled the pit, a truncated blade was found with a well-preserved handle made of hazel wood and lime-baste binding (Lübke 2003, 2005).

Another site was discovered at Timmendorf-Tonnenhaken, 2 m below the present sea level (Lübke 2002). It is situated on a former peninsula and has a cultural layer with well-preserved artefacts made of stone, bone and antler. Potsherds were also found here, which prove that this site was occupied by people of the early Neolithic Funnel Beaker culture. This chronology is confirmed by ^{14}C dates between 3,200 and 2,700 cal. BC and by the fact that all the bone material is from domesticated animals, e.g. cattle or pigs.

A further objective of the SINCOS project was the calculation of a new sea-level curve based mainly on geological and palynological data but also using ^{14}C dates from well-stratified archaeological sites (Lampe et al. 2005). When all these data are plotted on this curve, there is a high degree of concordance between the different sources, which emphasizes the significance of archaeology-based data from sites that were occupied for only a short time (Fig. 15.11).

The situation mentioned above for the south-western Baltic rim is completely different from that in the central and northern parts of the Baltic area. While the late Mesolithic settlements in the Wismar Bight were flooded, the simultaneous strong isostatic uplift in the north caused new land to emerge. This is especially well documented in several studies of sea-level development in central and northern Sweden (Rosentau et al. 2007). These prove that although the shore-displacement tendency in the area is generally well known, only a detailed examination of the available records and data, on both a local and a regional scale, allows a detailed picture to be drawn of the respective developments of the eustatic rise in sea level and the isostatic rebound.

To illustrate the opportunities and challenges presented by the comparative and combined interdisciplinary investigation of coastal change, shore displacement and settlement history as discussed in this chapter, the Södertörn peninsula to the south of Stockholm can be regarded as a model region (Björck et al. 1999). The area was completely inundated by the waters of the Baltic Ice Lake and the Yoldia Sea and then re-emerged as a result of the continuous process of isostatic uplift at the time of the Ancylus Lake period. During the Mesolithic period, the area was an archipelago that offered favourable conditions for communities of hunters and fishermen (Olsson and Risberg 1995). Thanks to heritage-based rescue excavations and research programs, a large number of archaeological sites of the Mesolithic and

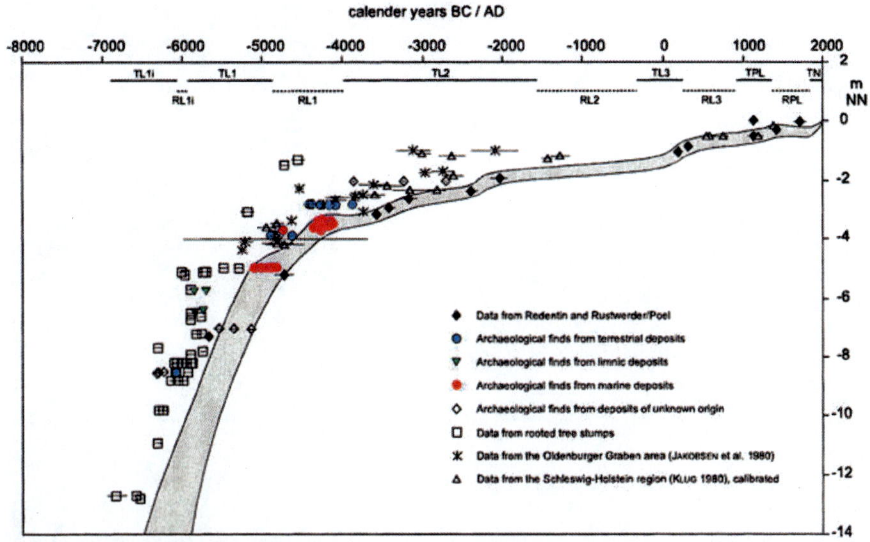

Fig. 15.11 Relative sea-level curve for the Wismar Bight as reflected by AMS-^{14}C data from peats, rooted tree trunks, archaeological finds and published data. T = transgression, R = regression (after Lampe et al. 2005, fig. 9)

Neolithic periods have been investigated and now form an excellent base for the reconstruction of the landscape and settlement patterns. The sites are located at levels between 40 and 80 m above the present sea level and span more than 4,000 years, from 8,300 to 3,900 cal. BC. Most of the settlements were originally established on large islands, in bays with access to the straits (Fig. 15.12). Over the last few years, several new sites have been discovered in the former outermost archipelago. The finds indicate that they were seasonally occupied as camps for seal hunting and fishing (Pettersson and Wikell 2006).

Several sea-level curves have been calculated for the Södertörn peninsula since the 1980s, based on data from various archives (Fig. 15.13). The Miller curve (Brunnberg et al. 1985) is based on conventional ^{14}C dates, mainly from mires, but data from archaeological sites are also included. It indicates two sub-Boreal transgressions, in the early Neolithic and the early Bronze Age. In contrast, the Risberg curve (1991) is based on radiocarbon-dated lake sediments: this does not show any transgressions, only a slowing down of the regression during the late Mesolithic period.

A third curve, published by Olsson and Risberg (1995), covers only the period from 4,700 to 3,800 cal. BC. It is mainly based on ^{14}C dates and other detailed information from 20 properly investigated and ^{14}C-dated archaeological sites. The dated material was always taken from the lowest features. In addition, the above-mentioned Miller and Risberg curves are discussed in the chapter and an attempt is made to explain the different results presented. The authors argue, on the basis

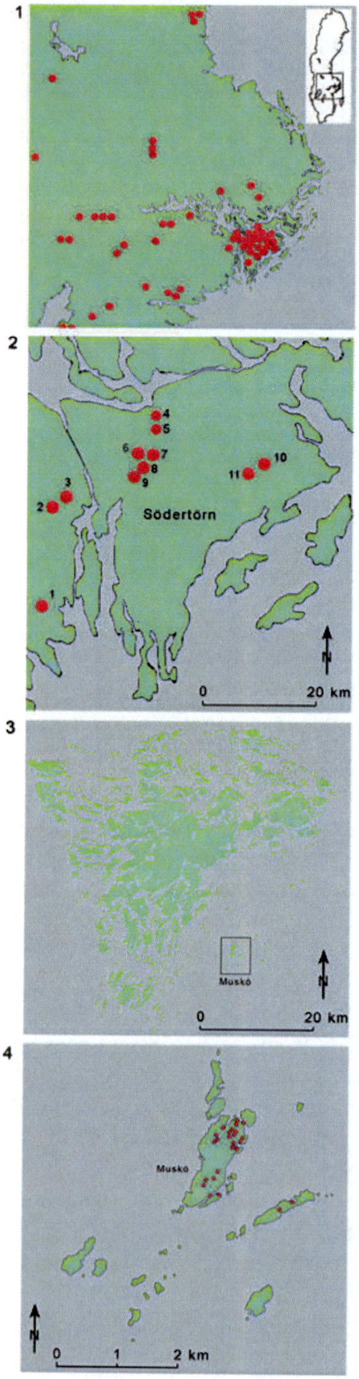

Fig. 15.12 Stone Age colonization of Eastern Sweden. 1. Distribution of Stone Age sites (after Björck et al. 1999, fig. 2). 2. Location of ^{14}C-dated Stone Age sites on the former shore of the archipelago of Södertörn peninsula: *1* Fansåker, *2* Pärlangsberget, *3* Kvedesta, *4* Masmo, *5* Häggstra, *6* Sjövreten, *7* Smällan, *8* Korsnäs, *9* Kyrktorp, *10* Söderbytorp, *11* Skogvagtartorp (after Olsson and Risberg 1995, fig. 1). 3 and 4. Models of the former archipelago on Södertörn peninsula and the island Muskö around 5,900 cal. BC with contemporaneous sites (after Pettersson and Wikell 2006, figs. 2 and 5). Graphics modified by R. Kiepe, NIhK

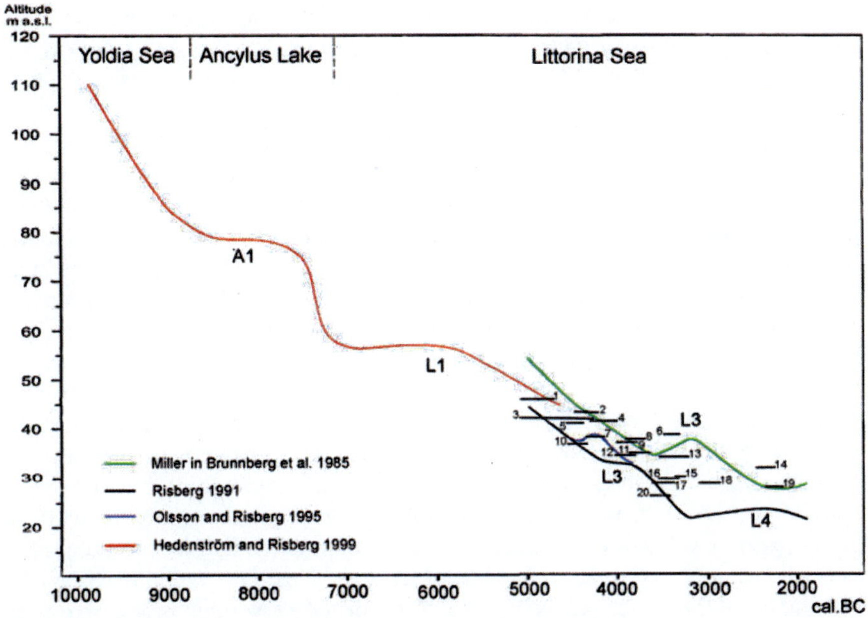

Fig. 15.13 Sea-level curves calculated for the Södertörn peninsula, based on Olsson and Risberg (1995) and Hedenström and Risberg (1999) (designed after calibration by the author). *1* Kyrktorp 9B; ph. 1, *2* Masmo l; *3* Kyrktorp 9A; *4* Häggsta l; *5* Kyrktorp 9B, ph. 2; *6* Pärlängsberget; *7* Smällan l; *8* Söderbytorp; *9* Häggsta 2; *10* Skogyaktartorp; *11* Häggsta 3; *12* Smällan 2; *13* Masmo 2; *14* Kvedesta l; *15* Korsaäs; *16* Häggsta 5; *17* Häggsta 6; *18* Kyrktorp 8 V; *19* Fänsaker; *20* Kyrktorp 8 V (for the references see Olsson and Risberg 1995)

of all the archaeological and geological evidence – especially from three of the sites – that there was a clearly detectable transgression phase with its maximum sea level around 4,000 cal. BC. During this phase, the sea level rose from 36 to 39 m above the present sea level.

Furthermore, four coastal hunter-fisher camps of the Neolithic Pitted Ware culture are analysed in the study by Olsson and Risberg (1995). These were dated by the ^{14}C method to the period between 3,800 and 2,500 cal. BC. They were located between 29 and 35 m above the present sea level.

The authors also state that the archaeological data correspond well to the Risberg curve (1991) whereas a positive correlation with the curve published by Miller (Brunnberg et al. 1985) is only visible if the dating is considered inaccurate and the curve moved back by 700 ^{14}C years (Åkerlund 1995, Åkerlund et al. 1997).

Finally, a fourth sea-level curve has been published for the Södertörn peninsula by Hedenström and Risberg (1999). This is based on diatom analysis and the radiocarbon dating of cores from sedimentary basins; archaeological data are not included. This curve covers the period between 9,300 and 4,500 cal. BC and hence cannot be used to check the above-mentioned Littorina transgression at the end of the fifth millennium. However, as far as the Ancylus Lake is concerned, it proves that around 8,000 cal. BC the regression of the sea level slowed down and there may even have been a small-scale transgression. In addition, in a very early phase of the Littorina Sea, from 7,000–5,500 cal. BC, a previously unknown transgression phase with an amplitude of ca. 2 m is assumed.

Compared with the Scandinavian rates of glacio-isostatic land uplift described above, the developments on the eastern Baltic coast have to be considered as moderate. At present, for example, the crustal movement of the area between the northern part of Lithuania and north-eastern Estonia amounts to only +1 to 2 mm/year and is thus barely able to offset the eustatic sea level rise of 1.8 mm/year (Rosentau et al. 2007). However, during most of the Stone Age this part of the Baltic rim was rising relatively so that, in principle, the shore-displacement models can also be used here for the relative dating of archaeological sites that, today, are on dry land.

In Lithuania, it has been observed that the remains of Mesolithic and early Neolithic settlements were located on river terraces and can be distinguished by their different levels (Rimantienė 1994). During the maximum of the Littorina transgression during the seventh millennium BC, the low-lying terraces I and II were flooded and habitation was only possible on the highest terraces, level III. At that time, the level of the rivers did not rise as high as in the Pleistocene but had reached a level of 5–6 m above the level I terraces. This observation regarding the topography of these sites allows a subdivision of the Mesolithic period into three

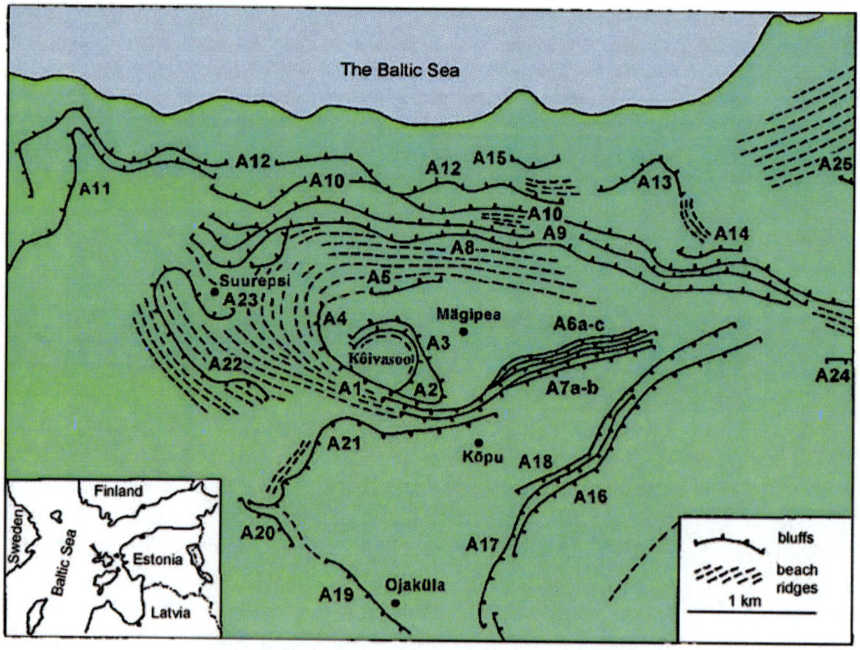

Fig. 15.14 Ancient shorelines on the Kõpu peninsula of Hiiumaa island (Raukas and Ratas 1995, fig. 2)

chronological stages: pre-transgressional, transgressional and post-transgressional (Rimantienė 1998).

In Estonia, too, the topographical location of Stone Age sites on former shores is used to determine their chronology. Sixty Stone Age sites, originally located on the Estonian Baltic coast, were recently integrated in a pilot project to prove that it is possible to apply the shore-displacement chronology in this region (Jussila and Kriiska 2004). Based on the respective levels of the sites, eight shorelines were reconstructed and dated with the help of ^{14}C dates from some of the archaeological sites to the period between 5,700 and 2,600 cal. BC. Of special importance for the study of shore displacement and settlement history was Hiiumaa island, to the west of the Estonian mainland (Fig. 15.14). On the oldest and westernmost part of the island, called Kõpu, the remains of 22 ancient shorelines are preserved on different levels, on clearly distinguishable ancient beach ridges (Raukas and Ratas 1995). On these ridges, 12 Stone Age sites were identified that dated from the late Mesolithic to the early Neolithic (Kriiska and Lõugas 1999). Although only partly excavated so far, these sites provide an excellent basis for the study of regional shore displacements and settlement history (Lõugas et al. 1996).

15.3.3 Seamen and Traders – The Post-Littorina and Limnaea Seas as a Transportation and Communication Zone

As has already been pointed out, the global sea-level around 4,000 cal. BC was 1 m below the present sea level (Lampe et al. 2005). Since then, the more or less continuous, but minor, rise in sea level has to be considered as having been periodically interrupted by phases of falling sea level (Kliewe and Schwarzer 2002). Nevertheless, the coastal landscape of the Baltic area experienced considerable changes during the last six millennia, mostly caused by further substantial isostatic rebound. The northern coast of the so-called Post-Littorina Sea, in particular, was affected by a continuous land uplift of up to 90 cm/century, which means that the former shores are now located more than 150 km from today's coastline and up to 280 m above the present sea level (Berglund 2008). In comparison, the shoreline changes in the eastern part of the southern Baltic area can be regarded as very moderate; on the territories of Lithuania, Latvia and Estonia there was a slight uplift of 10–20 cm/century. By contrast, the south-western Baltic coast has been slowly but continuously sinking, especially in the area of the Mecklenburgian Bight where it amounts to –10 cm/century (Harff and Meyer 2007, Rosentau et al. 2007).

For the period from the turn of the eras until the 1500 AD, the exchange of water between the North Sea and Baltic Sea has no longer reached the northern part of the Baltic Sea. Consequently, there was again a shift from a brackish to a freshwater environment that caused a migration of the freshwater snail *Limnaea*, which is why this stage in the development of the Baltic Sea is called the *Limnaea* Sea by some scholars (Kliewe and Schwarzer 2002).

As already mentioned, the coastal zone lost its dominant importance for the inhabitants' nutrition as a consequence of the introduction of agriculture and animal husbandry during the Neolithic period. However, hunting, fishing and gathering were never completely abandoned during the following Bronze and Iron Ages, nor during the Middle Ages, but they played a secondary role in the economy of the respective societies and communities. This is mainly proved by the archaeozoological examination of fish remains from several archaeological sites that have been investigated in detail (for a summary see Heinrich 1995, Schmölcke 2004).

Against this background, it is not surprising that archaeological evidence from coastal sites preserved in situ for the last 5,000 years is rare. Due to the isostatic land uplift in central and northern Scandinavia, the former coastal landscapes and settlements of the periods between the late Neolithic and the Middle Ages are far from the coast today, so shore-displacement models are also used to determine their chronology (Grimm 2006).

Only a few sites in the southern Baltic region have yielded data that provide information about the seashore at the time of their occupation. For example, several hearths from the Late Bronze Age (900–600 cal. BC) can be mentioned here that were exposed as a result of coastal erosion at Rerik in Mecklenburg. They show that the Baltic water level must then have been at least 1 m lower than today (Jöns et al. 2007).

A few Iron Age sites have also yielded the remains of features that are closely related to the exploitation of marine resources. Of special importance in this respect is a small group of shell middens distributed along the Baltic shore of eastern Jutland and along the Flensburgian fjord that have been affected by erosion (Harck 1973, Løkkegaard Poulsen 1978). When they emerged around the beginning of the first millennium the sea level in that region was probably only slightly lower than today (Labes 2002/03).

In addition, a fishing site dating to the Roman Iron Age should be mentioned: it was discovered during construction work in the harbour basin of Greifswald in West Pomerania (Kaute et al. 2005). Fishing fences were found here, which show that the sea level to the east of the Darss Sill in the second to fourth centuries AD must have been at least 1 m lower than today.

While the importance of the coastal zone as a source of nutrition declined from the Neolithic period onward, its significance for transportation increased. Together with rivers and lakes, the sea formed the backbone of prehistoric infrastructure right up until the late Middle Ages. As far as we know, only dugout canoes provided mobility for travellers and tradesmen before the late Neolithic period. Although this type of boat remained in use until late medieval times, better and increasingly powerful types of boats were also developed (Bill et al. 1997) that enabled voyages to be made over the Baltic Sea.

The earliest information about seagoing boats and ships is from the Bronze Age. In this period, transportation networks were established that made possible the large-scale importation of the necessary copper and tin, from eastern Central Europe and the British Isles, respectively (Harding 1999). Although no boat timbers or wrecks of that era have yet been found in the Baltic area, depictions of ships carved on rocks or bronze objects enable us to imagine how these vessels might have looked (Kaul 1998).

Most of the known rock carvings are from central Sweden, i.e. an area that experienced a strong isostatic uplift of the land. The landscape at the time when the rock carvings were made was therefore completely different from that of today. This again became obvious when a sea-level curve for the coast of Bohuslän in western Sweden was calculated recently by T. Påsse (2003) and used to generate a shore-displacement model (Ling 2004). These investigations showed that the clay-soil plains surrounding some of the well-known Bronze Age rock art sites in the coastal area around Gothenburg could not have been dry land – as had been assumed before the study – but were, in fact, at the bottom of the sea in shallow bays. Consequently, it is hypothesized that at least some of the ship carvings were originally done on or near the contemporary shore, which itself could have been a ritual landscape with special locations for cult activities.

The growing social, economic and military importance of seafaring from the first millennium AD can easily be studied from the many coastal sites along the Baltic rim. As archaeological investigations – mostly from the last three decades – have shown, maritime routes were constantly developed during that period, e.g. by the establishment of shipping barriers and channels that gave control over the waterways (Nørgård Jørgensen 2003).

The increasing importance of the long-distance water-borne transportation of wares and goods that were not available locally can also be determined from the establishment of beach markets, landing places and shipyards that were occupied on a seasonal basis (Ulriksen 1998, 2004). Most of these sites have been identified since the 1990s in Denmark, Norway and Sweden in the course of extensive surveys based on the analysis of the coastal landscape from a seaman's perspective. This new approach is also being used to an increasing degree for maritime-landscape research in Germany (Dobat 2003) and the eastern Baltic states (Mägi 2004, Ilves 2004).

Since the eighth century AD, so-called trading centres were established in many places along the Baltic coast (Fig. 15.15). They were established by the political authorities in the respective region or territory to consolidate long-distance trade and the local exchange of goods as well as to organize local handicraft production (Callmer 1994, Jöns 2008). These sites were always located in bays or on the banks of rivers to take advantage of a topographically well-protected position with direct access to the sea so that boats and ships could be loaded and unloaded safely (Crumlin-Pedersen 1999). At some of the trading centres, piers, landing bridges and other harbour facilities were built to permit the direct unloading of high-draft ships that could not be pulled onto the beach (Crumlin-Pedersen 1997). The remains of these structures are a source of information about the local sea level that, when they are well preserved and have been properly excavated and recorded, can be used as sea-level index points – or at least give an indication of the local sea level, as shown by the following examples of recently investigated sites at Haithabu and Groß Strömkendorf in the south-western part of the Baltic (Germany) and from Birka in Lake Mälaren (Sweden).

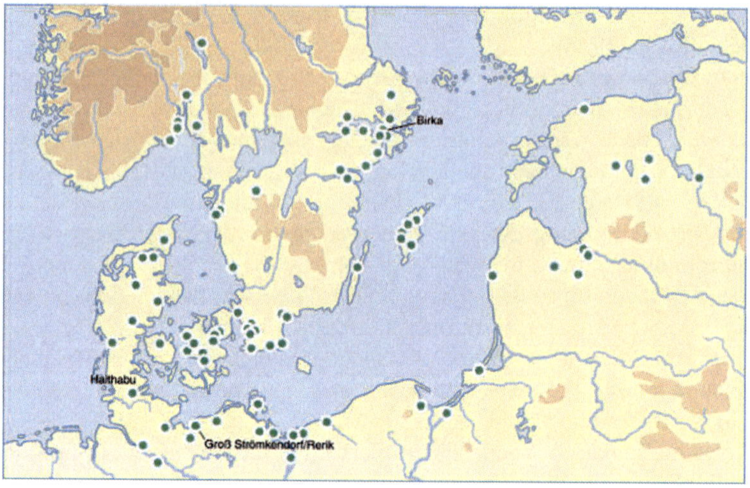

Fig. 15.15 Distribution of early trading centres of the eighth until the tenth century AD at the Baltic Rim (Graphic: H. Dieterich, Kiel)

One of the oldest trading centres in the Baltic area was at Groß Strömkendorf on the shore of the Wismar Bight, only a few kilometers south-east of the above-mentioned Mesolithic and Neolithic sites to the west of Poel island. This site was occupied from the early eighth until the beginning of the ninth century AD and is presumably identical with the *emporium reric* mentioned in the Frankish annals (Jöns 1999). The site's waterfront is of special interest in the discussion of shore displacement in the area of the Wismar Bight. Geological and geophysical investigations have proved that the harbour was located in a long stretched-out bay that had been washed out by meltwater in the deglaciation phase (Fig. 15.16). In

Fig. 15.16 Scenario of the Wismar Bight in the early Medieval period (designed by the author, based on a model by M. Meyer, IOW)

early Medieval times, the bay was still separated from the Wismar Bight by the remains of a ground moraine that formed a natural barrier. The moraine was then cut through by just a small inlet, thus connecting the Wismar Bight with the bay to create outstanding conditions for its use as a natural harbour. Given the above-mentioned rising sea level in the Wismar Bight, the ground moraine was gradually completely eroded over the last 1,200 years and the shoreline of the harbour bay displaced by about 80 m towards the coast so that the former waterfront area and harbour basin are now completely submerged. Observations made on the site indicate that the sea level in the eighth century AD was 80–100 cm lower than the present sea level. It seems possible that the gradual erosion of the ground moraine as a result of the rising sea level finally led to the loss of the harbour's natural protection, which could also be a reason for abandoning the trading centre already at the beginning of the ninth century AD.

The Haithabu trading centre can be regarded as the economic successor of *reric*/Groß Strömkendorf (Jöns 1999). It existed from the eighth to the eleventh century AD and was situated at the head of the Schlei fjord – a narrow, navigable inlet flowing into the Baltic Sea. During the ninth and tenth centuries AD, Haithabu was the most important trading centre on the southern Baltic rim and a link between the North Sea and Baltic trade routes (for a summary see Carnap-Bornheim and Hilberg 2006).

The history of the Schlei fjord has recently been reconstructed by Labes (2002/03) with regard to the sea-level changes from the Bronze Age to modern times. Her research is mainly based on a number of radiocarbon-dated tree stumps and data from several archaeological sites that were originally located on the banks of the Schlei. The data prove that the sea level around 2,500 cal. BC was approximately 2 m lower than today and that it rose during the second millennium BC up to 1 m below the present sea level. By around the beginning of the first millennium AD, the sea level is thought to have reached almost the present level. A regression to a level 1 m below the present sea level has been reconstructed for the first millennium AD, followed by repeated transgression phases during the second millennium AD until the present level was reached. These data were, in general, confirmed by the recently completed evaluation of the excavations in the harbour area of Haithabu (Kalmring 2010). In his conclusion, this author has assumed a sea level in the tenth century AD at 80 cm below the present level.

The most important trading centre in the central Baltic area was undoubtedly Birka, located on Björko island in Lake Mälaren (for a summary see Gräslund 2001). Trade and exchange, between the western and eastern Baltic as well as to the eastern Mediterranean via the Russian rivers, was organized from here between the eighth and tenth centuries AD (Noonan 1997).

The main harbour area of Birka has also been at least partially excavated and yielded evidence of several jetties at different levels between 5 and 6 m above the present sea level and at various distances from the present shore (Fig. 15.17). Considerable shore displacement and a fall in sea level during the Viking period is obvious (Ambrosiani and Clarke 1998). Numerous typologically datable artefacts, found in direct association with the jetties, could be used to adjust the regional

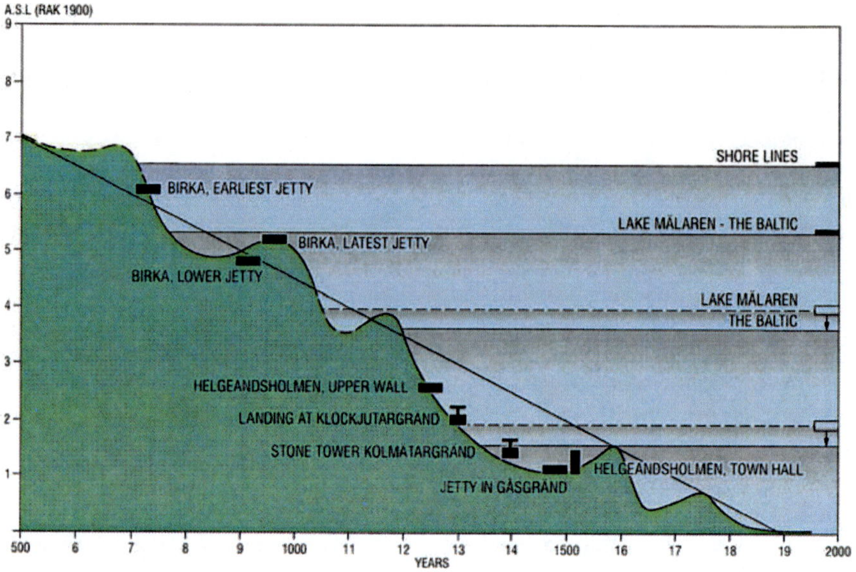

Fig. 15.17 Shore-displacement diagram for the Mälaren area for the period 500 until 2000 AD (after Amrosiani and Clarke 1998, fig. 1)

sea-level curve for Lake Mälaren. Before the excavations, it was assumed that the fall in sea level due to the isostatic land uplift should be regarded as a continuous process without fluctuations. However, together with the results of the scientific investigation of shore displacement in other parts of Lake Mälaren (Miller et al. 1997, Risberg et al. 2002), the Birka harbour stratigraphy and the finds from the site present a different picture: the relative sea level fell rapidly during the eighth and eleventh centuries AD, due to the isostatic land uplift, but rose considerably in the tenth and twelfth centuries AD. For several decades, the Birka waterfront experienced a transgression caused by a temporary fall in the eustatic sea level (Ambrosiani and Clarke 1998).

During the following centuries, the late Middle Ages and early modern times, the changes in the relative sea level on the southern coast of the Baltic Sea remained within the rate of annual fluctuation and are usually no longer detectable in the archaeological record. This is in distinct contrast to the situation in the northern part of the Baltic area, where the continuing uplift and shore displacement can still be used to determine the chronology of coastal sites in these periods.

15.4 Summary

After the ice masses in the glaciers of the Fennoscandian ice sheet had melted, the new landscape around the Baltic Sea basin began to be settled by human communities. Due to the eustatic sea-level changes and the strong isostatic land uplift, the coastal zone was an unstable habitat; in particular, it was subject to a continual displacement of the shore that forced the people living there to move their settlements, either because they were flooded or because they no longer had direct access to the sea. Given the fact that these people were highly dependent on changes in the sea level, sea-level curves and shore-displacement models can be used to determine the chronology of prehistoric sites if they were originally located on the shore. Similarly, well-preserved coastal sites can be regarded as a record of the sea level at the time of their occupation and thus used as sea-level index points. During every phase in the development of the Baltic Sea, the people living in coastal areas were – to a large extent – forced to adapt their respective economic systems to the prevailing environmental conditions. The first to arrive were groups of Palaeolithic hunter-gatherers who migrated to the late-glacial landscape as they followed the herds of reindeer. Especially in the recently deglaciated areas of central Scandinavia, they were confronted with a dramatically changing landscape with newly emerged land and a shrinking sea, which soon changed an archipelago into dry land. In the following Mesolithic and early Neolithic periods, the coastal environment provided favourable conditions not only for hunting game but also for marine food resources. While the communities inhabiting the central and northern parts of the Baltic rim experienced the continuous emergence of new land and ever-larger islands in the archipelago, a rising sea level reduced the amount of land available for habitation by the communities along the southern shore, who were apparently forced to move their settlements to more protected spots – a little higher and further inland than the old flooded sites. After agriculture and animal husbandry were introduced in the Neolithic period, the importance of marine food resources as a source of nutrition decreased. At the same time, the changes in sea level became less dramatic, although shore displacement continued. From the time of the Bronze Age, at the latest, the Baltic Sea became increasingly important for transportation and communication purposes. Landing places and beach markets as well as specialized trading centres with direct access to the sea were established during the first millennium AD almost everywhere on the Baltic rim. Most of these can provide information on sea levels at the time of their occupation and can therefore be dated by reference to shore-displacement models.

Acknowledgements This chapter was initiated by SINCOS (Sinking Coasts – Geosphere, Ecosphere and Anthroposphere of the Holocene Southern Baltic Sea), a project funded by the German Research Foundation. I wish to thank Sönke Hartz, Ulrich Schmölcke, Harald Lübke and Sven Kalmring of the Archaeological State Museum of Schleswig-Holstein as well as Thomas Terberger, University of Greifswald, Karl-Ernst Behre, NIhK Wilhelmshaven, and Ingrid Fuglestvedt, Oslo University, who gave valuable advice, especially about the Palaeolithic and Mesolithic periods but also about early Medieval period and general sea-level changes. I have also to thank Michael Meyer, Institute for Baltic Research, Warnemünde, who provided me with shore-displacement models for the Baltic area and, especially, for the Wismar Bight. Finally my thanks go to Beverley Hirschel for tidying up the English in this chapter.

References

Åkerlund A (1995) Man and Sea in the expanding coastal district of Svealand. In: Fischer A (ed) Man and sea in the Mesolithic: coastal settlement above and below present sea level. Proceedings of the international symposium, Kalundborg, Denmark 1993. Oxbow monograph 53, Oxbow, Oxford, pp 221–228

Åkerlund A, Risberg J, Miller UU, Gustafsson P (1997) On the application of the 14C method to interdisciplinary studies of shore line displacement and the settlement location. Pact 49:53–84

Ambrosiani B, Clarke H (1998) Birka and the beginning of the Viking age. In: Wesse A (ed) Studien zur Archäologie des Ostseeraumes. Von der Eisenzeit zum Mittelalter. Festschrift für M. Müller-Wille. Neumünster, pp 33–38

Andersson M, Cronberg C (2007) Moving north. The first travellers to south Scandinavia. In: Hårdh B, Jennbert K, Olausson D (eds) On the road. Studies in Honour of Lars Larsson. Acta Archaeologica Lundensia 4(26):145–150. Lund

Behre K-E (2004) Coastal development, sea-level change and settlement history during the later Holocene in the Clay District of Lower-Saxony (Niedersachsen), northern Germany. Quaternary International 112:37–53

Behre K-E (2007) A new Holocene sea-level curve for the southern North Sea. Boreas 36:82–102

Berglund M (2004) Holocene shore displacement and chronology in Ångermanland, eastern Sweden, the Scandinavian glacio-isostatic uplift centre. Boreas 33:48–60

Berglund M (2008) Time-transgressive early Holocene vegetational succession following shore displacement: a case study from central Sweden. Boreas 37:87–101

Bergman I, Påsse T, Olofsson A, Zackrisson O, Hörnberg G, Hellberg E, Bohlin E (2003) Isostatic land uplift and Mesolithic landscapes: lake-tilting, a key to the discovery of Mesolithic sites in the interior of Northern Sweden. Journal of Archaeological Science 30:1451–1458

Bill J, Poulsen B, Rieck F, Ventegodt O (1997) Fra stammebåd til skib. Dansk søfarts historie 1. Kopenhagen

Björck N, Knutsson K, Lindgren C, Hallgren F (1999) The Mesolithic in Eastern Central Sweden. In: Boaz J (ed) The Mesolithic in Eastern Central Scandinavia. Universitetets Oldsaksamlings Skrifter 22:87–123

Björck S (1995) A review of the history of the Baltic Sea 13.0–8.0 ka BP. Quaternary International 27:19–40

Boaz J (1999) Pioneers in the Mesolithic: the initial occupation. In: Boaz J (ed) The Mesolithic of Central Scandinavia. Universitetets Oldsaksamlings Skrifter 22:11–25. Oslo

Brunnberg L, Miller U, Risberg J (1985) Project Eastern Svealand. Development of the Holocene landscape. Proceedings of the 3rd Nordic conference on the Application of the scientific methods in archaeology. Iskos 5:85–91

Callmer J (1994) Urbanization in Scandinavia and the Baltic Region ca. AD 700–1100. Trading places, centres and early urban sites. In: Ambrosiani B, Clarke H (eds) Developments around the Baltic Sea in the Viking Age. Birka Studies 3:50–90. Stockholm

Carnap-Bornheim C, Hilberg V (2006) Recent archaeological research in Haithabu. In: Henning J (ed) Post-roman towns and trade in Europe, Byzantium and the Near-East, vol 1. The Heirs of the Roman West. Berlin, pp 199–219

Clausen I (1997) Neue Untersuchungen an späteiszeitlichen Fundplätzen der Hamburger Kultur bei Ahrenshöft, Kr. Nordfriesland (ein Vorbericht). Archäologische Nachrichten aus Schleswig-Holstein 8:8–49

Crumlin-Pedersen O, Porsmose E, Thrane H (1996) Atlas over Fyns kyst i jernalder, vikingetid og middelalder, Odense

Crumlin-Pedersen O (1997) Viking-age ships and shipbuilding in Hedeby/Haithabu and Schleswig. Ships & Boats of the North 2. Schleswig-Roskilde

Crumlin-Pedersen O (1999) Ships as indicators of trade in Northern Europe 600–1200. In: Bill J, Clausen BL (eds) Maritime topography and the Medieval Town. National Museum of Denmark Studies in Archaeology & History 4:11–20. Copenhagen

Dobat AS (2003) Hedeby and its maritime hinterland: the Schlei Fjord as an early medieval communication route? Bodendenkmalpflege in Mecklenburg-Vorpommern, Jahrbuch 51: 419–436

Eberhards G, Zagorska I (2002) The environment and the earliest settlement of Latvia, East Baltic. In: Eriksen BV, Bratlund B (eds) Recent studies in the final Palaeolithic of the European plain. Proceedings of a U.I.S.P.P. symposium, Stockholm, 14–17 October 1999. Jutland Archaeological Society Publications 39:85–90. Højbjerg

Eriksen BV (2002) Reconsidering the geochronological framework of Lateglacial hunter-gatherer colonization of southern Scandinavia. In: Eriksen BV, Bratlund B (eds) Recent studies in the final Palaeolithic of the European plain. Proceedings of a U.I.S.P.P. Symposium, Stockholm, 14–17 October 1999. Jutland Archaeological Society Publications 39:25–41. Højbjerg

Fischer A (1996) At the border of human habitat. The late Palaeolithic and early Mesolithic in Scandinavia. In: Larsson L (ed) The earliest settlement of Scandinavia and its relationship with neighbouring areas. Acta Archaeologica Lundensia Series 8(24):157–176. Lund

Fuglestvedt I (1999) The early Mesolithic site at Stunner, Southeast Norway: a discussion of late upper Palaeolithic/early Mesolithic. In: Boaz J (ed) The Mesolithic of Central Scandinavia. Universitetets Oldsaksamlings Skrifter 22:189–202. Oslo

Fuglestvedt I (2008) The Ahrensburgian Galta 3 site in SW Norway. Dating, technology and cultural affinity. Acta Archaeologica 78(2):87–110

Gräslund A-S (2001) Birka between West and East. Offa 58:129–140

Grimm O (2006) Großbootshaus – Zentrum und Herrschaft. Zentralplatzforschung in der nordeuropäischen Archäologie. 1–15 Jahrhundert. Ergänzungsbände zum Reallexikon der Germanischen Altertumskunde Bd. 52. Berlin

Grimm SB, Weber M-J (2008) The chronological framework of the Hamburgian in the light of old and new 14C dates. Quartär 55:17–40

Gustafson L (1999) Stunner – the "First" early Mesolithic site in Eastern Norway. In: Boaz J (ed) The Mesolithic of Central Scandinavia. Universitetets Oldsaksamlings Skrifter 22:181–187. Oslo

Harck O (1973) Eisenzeitliche Muschelhaufen an der schleswigschen Ost- und Westküste. Offa 30:40–54

Harding AF (1999) Nord-südlicher Rohstoffaustausch. In: Bundeskunsthalle Bonn (ed) Götter und Helden der Bronzezeit. Europa im Zeitalter des Odysseus. Bonn, pp 38–42

Harff J, Lemke W, Lampe R, Lüth F, Lübke H, Meyer M, Tauber F, Schmölcke U (2007) The Baltic Sea coast – a model of interrelations among geosphere, climate, and anthroposphere. In: Harff J, Hay WW, Tetzlaff DM (eds) Coastline changes: interrelation of climate and geological processes – The Geological Society of America, Special Paper 426:133–142

Harff J, Meyer M (2007) Changing Holocene coastal zones of the Baltic Sea – a modelling approach. Bericht der Römisch-Germanischen Kommission 88:241–266

Hartz S, Lübke H (2006) New Evidence for a chronostratigraphic division of the Ertebølle culture and the earliest Funnel Beaker culture on the Southern Mecklenburg Bay. In: Kind C-J (ed) After the ice age. Settlements, subsistence and social development in the Mesolithic of Central Europe. Materialhefte zur Archäologie in Baden-Württemberg 78:59–74. Stuttgart

Hartz S, Lübke H, Terberger T (2007) From fish and seal to sheep and cattle – new research into the process of neolithization in northern Germany. In: Whittle A, Cummings V (eds) Going over: the Mesolithic-Neolithic transition in north-west Europe. OUP/British Academy, London, pp 567–594

Hedenström A, Risberg J (1999) Early Holocene shore-displacement in southern central Sweden as recorded in elevated isolated basins. Boreas 28:490–504

Heinrich D (1995) Fischerei und Fischereimethoden. Reallexikon der Germanischen Altertumskunde 9:130–138

Hörnberg G, Bohlin E, Hellberg E, Bergmann I, Zackrisson O, Olofsson A, Wallin J-E, Påsse T (2005) Effects of Mesolithic hunter-gatherers on local vegetation in a non-uniform glacio-isostatic land uplift area, northern Sweden. Vegetation History and Archaeobotany 15:13–26

Ilves K (2004) The seaman's perspective in landscape archaeology. Estonian Journal of Archaeology 8:163–176
Jöns H (1999) War das "emporium Reric" der Vorläufer von Haithabu? Bodendenkmalpflege in Mecklenburg-Vorpommern, Jahrbuch 47:201–213
Jöns H (2002) Möglichkeiten und Perspektiven der Maritimen Archäologie im Ostseeraum. In: Cederlund CO, Krüger K (eds) Maritime Archäologie heute. Internationale Tagung der Maritimen Archäologie im Ostseeraum 3:19–50. Rostock
Jöns H, Lübke H, Lüth F, Terberger T (2007) Prehistoric settlements and development of the regional economic area. Archaeological investigations of the SINCOS-subproject 1.6 (Sedeco). Bericht der Römisch-Germanischen Kommission 88:149–188
Jöns H (2008) Handel und Handwerkszentren der Wikingerzeit. In: Historisches Museum der Pfalz (ed) Die Wikinger. München, pp 142–147
Jussila T (1995) Shore-level displacement of the prehistoric dwelling sites in the ancient Lake Saimaa Complex. Fennoscandia archaeological 12:39
Jussila T, Kriiska A (2004) Shore displacement chronology of the Estonian stone age. Estonian Journal of Archaeology 8:3–26
Kalmring S (in print) Der Hafen von Haithabu. Ausgrabungen in Haithabu 14. Neumünster
Kaul F (1998) Ships on bronzes. A study in Bronze age religion and iconography. National Museum of Denmark Studies in Archaeology & History, Copenhagen, 3
Kaute P, Labes S, Schindler G (2005) Ein kaiserzeitlicher Fischzaun im Greifswalder Museumshafen. Archäologische Berichte aus Mecklenburg-Vorpommern 12:24–36
Kindgren H (1996) Reindeer or Seals? Some late Paleolithic sites in Middle Bohuslän. In: Larsson L (ed) The earliest settlement of Scandinavia and its relationship with neighbouring areas. Acta Archaeologica Lundensia Series 8(24):191–205. Lund
Kliewe H, Janke W (1982) Der holozäne Wasserspiegelanstieg der Ostsee im nordöstlichen Küstengebiet der DDR. Petermanns Geographische Mitteilungen 126:65–74
Kliewe H, Schwarzer K (2002) Die deutsche Ostseeküste. In: Liedtke H, Marcinik J (eds) Physische Geographie Deutschlands. Gotha
Kriiska A, Lõugas L (1999) Late mesolithic and early neolithic seasonal settlement at Kõpu, Hiiumaa Island, Estonia. Pact 57:157–172
Labes S (2002/03) Der nacheiszeitliche Meeresspiegelanstieg der Ostsee im Bereich der Schlei. Offa 59/60:223–227
Lampe R, Endtmann E, Janke W, Meyer H, Lübke H, Harff J, Lemke W (2005) A new relative sea-level curve for the Wismar Bay, N-German Baltic coast. Meyniana 57: 5–35
Lemke W (2004) Die kurze und wechselvolle Entwicklungsgeschichte der Ostsee. Aktuelle meeresgeologische Forschungen zum Verlauf der Litorina-Transgression. Bodendenkmalpflege in Mecklenburg-Vorpommern, Jahrbuch 52:43–54
Linden M, Möller P, Björck S, Sandgren P (2006) Holocene shore displacement and deglaciation chronology in Norrbotten, Sweden. Boreas 35:1–22
Ling J (2004) Beyond transgressive lands and forgotten seas. Towards a Maritime understanding of rock art in Bohuslän. Current Swedish Archaeology 12:121–140
Løkkegaard Poulsen K (1978) Eisenzeitliche Muschelhaufen in Dänemark. Offa 35:64–85
Lõugas L, Kriiska A, Moora H (1996) Coastal adoption and marine exploitation of the Island Hiiumaa, Estonia, during the Stone Age with special emphasis on the Kõpu I site. Pact 50: 197–211
Lübke H (2002) Submarine Stone Age settlements as indicators of sea-level changes and the coastal evolution of the Wismar Bay area. Greifswalder Geographische Arbeiten 27: 202–210
Lübke H (2003) New investigations on Submarine Stone Age Settlements in the Wismar Bay Area. In: Kindgren H, Knutsson K, Larsson L, Loeffler D, Åkerlund A (eds) Mesolithic on the move. Proceedings of the 6th international conference on the Mesolithic in Europe, Stockholm 2000. Oxbow Monographs, pp 69–78

Lübke H (2005) Spät- und endmesolithische Küstensiedlungsplätze in der Wismarbucht – Neue Grabungsergebnisse zur Chronologie und Siedlungsweise. Bodendenkmalpflege in Mecklenburg-Vorpommern, Jahrbuch 52:83–110

Mägi M (2004) "... Ships are their main strength." Harbour sites, arable lands and chieftains on Saaremaa. Estonian Journal of Archaeology 8:128–154

Miller U, Ambrosiani B, Clarke H, Hackens T, Hansson A-M, Johansson B (1997) Environment and Vikings with special reference to Birka. Birka Studies 4. Pact 52:7–275

Noonan T (1997) Scandinavians in European Russia. In: Sawyer P (ed) The Oxford Illustrated History of the Vikings. Oxford University Press, New York, NY, pp 134–155

Nørgård Jørgensen A (2003) Fortifications and control of land and sea traffic in the Pre-Roman and Roman Iron Age. In: Jørgensen L, Storgaard B, Gebauer Thomsen L (eds) The spoils of victory. The North in the shadow of the Roman Empire. The National Museum, Kopenhagen, pp 194–209

Olofsson A (ed) (2003) Pioneer settlements in the Mesolithic of northern Sweden. Archaeology and environment 16. Umeå

Olsson E, Risberg J (1995) Archeological data and models of sea-level change c. 6000–3500 BP South of Stockholm, Eastern Sweden. Pact 50:219–230

Påsse T (2003) Strandlinjeförsjutning i norra Bohuslän under holocen. In: Persson P (ed) Strandlinjer och vegetationshistoria. Kvartärgeologiska undersökningar inom kust till kust projektet, 1998–2002. Gotarc ser. C. Arkeologiska skrifter 48:31–64. Göteborg

Pettersson M, Wikell R (2006) Mesolitiska boplatser i Stockholms skärgård. Fiske och säljakt på utskären under 10 000 år. Fornvännen 101:153–168

Raukas A, Ratas U (1995) Holocene evolution and paleoenvironmental conditions of Hiiumaa Island, Northwestern Estonia. Pact 50:167–174

Rimantienė R (1994) Die Steinzeit in Litauen. Bericht der Römisch-Germanischen Kommision 75:26–146

Rimantienė R (1998) The relationship between Stone Age people and their environment in Lithuania. Pact 54:143–156

Risberg J (1991) Paleoenvironment and Sea level changes during the early Holocene on the Södertörn Peninsula, Södermanland, eastern Sweden. Department of Quaternary Research, Stockholm University, Stockholm. Report 20

Risberg J, Karlsson S, Hansson A-M, Hedenström A, Heimdahl J, Miller U, Tingvall C (2002) Environmental changes and human impact as recorded in a sediment sequence offshore from a Viking Age town, Birka, southeastern Sweden. The Holocene 12:445–458

Rosentau A, Meyer M, Harff J, Dietrich R, Richter A (2007) Relative Sea Level change in the Baltic Sea since the Litorina Transgression. Zeitschrift für Geologische Wissenschaften 35(1/2): 3–16

Rößler D (2006) Reconstruction of the Litorina Transgression in the Western Baltic Sea. Meereswissenschaftliche Berichte/Marine Science Reports No. 67. Warnemünde

Schmitt L, Larsson S, Schrum C, Alekseeva I, Tomaczek M, Svedhage K (2006) 'Why they came', The colonization of the coast of western Sweden and its environmental context at the end of the last deglaciation. Oxford Journal of Archaeology 25(1):1–28

Schmitt L, Larsson S, Burdukiewicz J, Ziker J, Svedhage K, Zamon J, Steffen H (2009) Chronological insights, cultural change, and Resource exploitation on the west Coast of Sweden during the late Paleolithic/early Mesolithic Transition. Oxford Journal of Archaeology 28:1–27

Schmölcke U (2004) Nutztierhaltung, Jagd und Fischerei. Zur Nahrungsmittelwirtschaft des frühgeschichtlichen Handelsplatzes von Groß Strömkendorf, Landkreis Nordwestmecklenburg. Forschungen zu Groß Strömkendorf I. Beiträge zur Ur- und Frühgeschichte Mecklenburg-Vorpommerns, Bd. 43. Lübstorf

Schmölcke U, Endtmann E, Klooss S, Meyer M, Michaelis D, Rickert B-H, Rößler D (2006) Changes of sea level, landscape and culture: a review of the south-western Baltic area between 8800 and 4000 BC. Palaeogeography Palaeoclimatology Palaeoecology 240:423–438

Siiriäinen A (1982) Shore displacement and archaeology in Finland. Annales Academiae Scientiarium Fennicae, Ser. A III. Geologica-Geograpiae 134:173–184
Sørensen SA (1996) Kongemosenkulturen i Sydskandinavien. Færgegaarden
Sørensen R (1979) Late Weichselian deglaciation in the Oslofjord area, South Norway. Boreas 8:241–246
Terberger T, De Klerk P, Helbig H, Kaiser K, Kühn P (2004) Late Weichselian landscape development and human settlement in Mecklenburg-Vorpommern (NE Germany). Eiszeitalter und Gegenwart 54:138–175
Terberger Th (2006a) From the first humans to the Mesolithic hunters in the Northern German Lowlands. In: Moeller Hansen K, Rolland T (eds) Across the western Baltic. Symposium Vordingborg März 2003. Sydsjaellands Museums Publikationer 1:23–56. Odense
Terberger Th (2006b) The Mesolithic hunter-fisherer-gatherers on the Northern German plain. In: Moeller Hansen K, Buck Pedersen K (eds) Across the western Baltic. Symposium Vordingborg März 2003. Sydsjaellands Museums Publikationer 1:111–184. Odense
Ukkonen P, Lõugas L, Zagorska I, Lukševica L, Lukševics E, Daugnora L, Jungner H (2006) History of the reindeer (Rangifer tarandus) in the eastern Baltic region and its implications for the origin and immigration routes of the recent northern European wild reindeer populations. Boreas 35:222–230
Ulriksen J (1998) Anløbspladser. Besejling og bebyggelse i Danmark mellem 200 og 1100 e.Kr. Roskilde
Ulriksen J (2004) Danish coastal landing places and their relation to navigation and trade. In: Hines J, Lane A, Redknap M (eds) Land, Sea and Home. Proceedings of a conference on Viking-period settlement at Cardiff 2001. Society for Medieval Archaeologys´ monograph series. Maney, pp 7–26
Westerdahl C (2000) From Land to sea, from sea to Land. On transport zones, Borders and human space. In: Litwin J (ed) Down the river to the Sea. ISBSA 8:11–20. Gdansk
Zagorska I (1999) The earliest settlement of Latvia. Environmental and cultural history of the eastern Baltic region. Pact 57:131–155
Zagorska I (2007a) Earliest inhabitance of the territory of Latvia in the late glacial environment. The quaternary of Western Lithuania: from the Pleistocene glaciations to the evolution of the Baltic Sea. Proceedings of The INQUA Peribaltic group field symposium, May 27–June 02, 2007. Plateliai, pp 100–102
Zagorska I (2007b) On the road to eastern shore of the Baltic basin. In: Hårdh B, Jennbert K, Olausson D (eds) On the road. Studies in Honour of Lars Larsson. Acta Archaeologica Lundensia Series 4(26):193–198. Lund

Chapter 16
Geological Hazard Potential at the Baltic Sea and Its Coastal Zone: Examples from the Eastern Gulf of Finland and the Kaliningrad Area

Mikhail Spiridonov, Daria Ryabchuk, Vladimir Zhamoida, Alexandr Sergeev, Vadim Sivkov, and Vadim Boldyrev

Abstract Geological hazards may threaten human life, may result in serious property damage, and may significantly influence normal development of biota. They are caused by natural endogenic and exogenic driving forces or generated by anthropogenic activities. An interaction of geological processes and intense anthropogenic activities, e.g., construction of buildings, harbors, oil and gas pipelines, hydroengineering facilities, and land reclamation, has resulted in hazard potential, especially for the densely populated areas of the Russian Baltic coastal zone. These hazards may in addition be harmful for the sensitive ecosystem of the Baltic Sea. Mapping and assessment of the geological hazard potential should be the main objectives of an integrated management program for the protection of coastal zones. This study documents the first step in that process for the Russian sector of the Baltic Sea and its coastal zone. A major part of endogenic hazard potential both in the Kaliningrad area and in the eastern Gulf of Finland remains at low- or medium-risk levels, but analysis of the recent environmental conditions at the seabed of the Russian sector of the Baltic Sea and, especially, within its coastal zone shows that during the last years the activity of exogenic geological processes has increased significantly. The highest risk within both studied areas has been caused by coastal and bottom erosion. In addition, in shallow area near the shore bottom of the eastern Gulf of Finland, "avalanche" sedimentation and sediment pollution can produce hazardous situations as well.

Keywords Geological hazard potential · Coastal zone · Mapping

16.1 Introduction

Worldwide, during the last decades, several countries have suffered from increased impacts of natural hazards. Among them, the geological hazards can threaten human life and significantly impact normal development of biota. Processes that affect the

M. Spiridonov (✉)
A.P. Karpinsky Russian Research Geological Institute (VSEGEI), St. Petersburg 199106, Russia
e-mail: Michail_Spiridonov@vsegei.ru

J. Harff et al. (eds.), *The Baltic Sea Basin*, Central and Eastern European Development Studies (CEEDES), DOI 10.1007/978-3-642-17220-5_16,

coasts and marine ecosystems worldwide are caused by natural endogenic and exogenic factors or by anthropogenic activities (www.eurosion.org; www.encora.org).

Regarding endogenic processes and connected hazards, the Baltic Sea region is classified as a low seismicity area, although historical data, covering the period from 1616 to 1911, show the evidence of more than 40 strong (intensity 5–7 MSK-64 scale, 12-point scale of earthquake intensity developed by Medvedev et al. (1965)) events in the Baltic region (Nikonov and Sildvee 1991, Sliaupa et al. 1999). Since the last seismic activity maximum in 1908–1909, no hazardous seismic events have been registered here except for the Osmussaare 4.5 magnitude event in Estonia in 1976 (Sliaupa et al. 1999) and the Kaliningrad 5.0 magnitude event in September 2004 (Aptikaev et al. 2005, Assinovskaya and Karpinsky 2005). The seismotectonic activity must have been significantly higher in this region at the time of deglaciation causing extreme rates of glacial isostatic uplift (Bödvarsson et al. 2006, Mörner 2004). On the other hand, possible seismic reactivation during the Holocene has been detected in some old bedrock fracture zones in the Bothnian Sea, the Archipelago Sea, and the northern Baltic Proper (Hutri 2007). Confirming evidence comes from high-resolution, low-frequency, echo-sounding observations of disturbed sediment structures (slide and slumps, faults, debris flow and turbidite-type structures, and some gas-related structures) and their locations in the vicinity of bedrock fracture zones (Hutri 2007). In the northern Stockholm archipelago, pockmarks possibly formed by thermogenic gas seepage were also found over still active tectonic lineaments in the crystalline basement (Söderberg and Flodén 1991). Some authors have detected active tectonic faults within the Gulf of Finland (Nikonov and Sildvee 1991, Rudnik 1996, Yaduta 2003).

Marine coastal hazards in the Baltic region have recently become the focus of attention of many researchers (Valdmann et al. 2008, Pruszak and Zawadzka 2005, Zilinskas 2005).

The Workshop on Sea-Level Rise and Climate Change organized by TAIEX in April 2008 demonstrated that the problem of coastal erosion is very urgent for many Baltic countries (Workshop on Sea-Level Rise 2008), while the coasts of the northern Baltic Sea do not suffer much from coastal erosion due to geological structure, skerries type of shoreline, and tectonic uplift (Valdmann et al. 2008).

The Estonian coastal zones adjoining the Russian part of the Gulf of Finland have shown during the last 20–30 years the most marked coastal erosion events resulting from a combination of heavy storms, high sea level induced by storm surges, ice-free sea, and thawing sediments (Orviku et al. 2003).

Along the open Latvian Baltic seacoast, the recession has exceeded 50–60 m (up to 200 m) during the last 50–60 years. Only along the coast of the Gulf of Riga, coastal erosion is less prominent. In general, coastal erosion has significantly increased due to severe storms during the last 15 years. The rate of coastal erosion during any single storm has increased, averaging 3–6 m with the maximum reaching 20 m (Eberhards et al. 2009).

In Lithuania, the total annual sand balance of the coastal zones is negative. The length of accumulating zones of the Lithuanian coast decreased from 36 to 20 km between 1993 and 2003, whereas the length of eroded and stable coastal zones increased by 1.5 times (from 16 to 24 km and from 37.5 to 55 km, respectively; Bitinas et al. 2005) during the same period.

Three types of coasts are distinguished in Poland, depending on morphology and geological structure: cliffs (101 km), barriers (380 km), and coasts similar to wetlands (salt marshes) (17 km). Cliff coasts suffer from mass movements; serious risks are related to erosion of low and narrow barriers, which could be easily broken during storm surges (Uscinowicz et al. 2004). In the Pomeranian Bight, the average shoreline retreat is in the range of 0.1 m/year along Wolin Island, 0.2 m/year along Rügen Island, and 0.4 m/year along Usedom Island, despite huge areas of accumulation between the headlands of Rugen Island, at both ends of Usedom Island and the southeastern part of Wolin Island (Schwarzer et al. 2003).

The frequency of occurrence of storm surge events along the nontidal German Baltic coast shows a significant linear increase of 1–3 events/100 years for the last decades. At the same time, a simulation of the last 30 years of the twenty-first century results in only small changes in the frequency of occurrence of extreme storm events with increased high water (Baerens and Hupfer 1999).

It has been suggested that the main reason for the more intense coastal erosion is an increase of storm events, which have caused severe damages both in the Gulf of Finland and in the major part of the southern and the eastern Baltic Sea (Raukas and Huvarinen 1992, Orviku et al. 2003). In many cases, detailed analysis of the nearshore zone structure and processes gives the key to understanding of the coastal problems (Schwarzer et al. 2003).

Recent increase of construction works within the offshore areas and the coastal zones of the Baltic Sea (new harbors, oil and coal terminals, oil and gas pipelines, cables, and different hydrotechnical constructions) may transform common geological processes and phenomena into hazard potentials. Growing anthropogenic activity makes the geological hazard problem very important for spatial planning and sustainable development of the Baltic region.

16.1.1 Approaches and Methods of the Geological Hazard Classification and Typology

The European Spatial Planning Observation Network (ESPON) requested an assessment of spatial patterns and territorial trends of hazards and risks (Schmidt-Thomé 2006).

The German Advisory Council on Global Change (WBGU) suggests that a probability of hazard occurrence, the extension of its damage (with the certainty of their assessment), its location, persistency, irreversibility, delay effect, and mobilization potential are the basis for hazard classification and characterization (Fleischhauer 2006). On these grounds, it is possible to distinguish several different types of risks.

A large proportion of geological hazards and the natural processes that can provoke them are classified as the so-called Cyclops-type risk with an unknown probability and high extent of damage (earthquakes, volcanic eruptions, river floods, storm surges, tsunamis, avalanches, landslides, etc.). Phenomena like the self-reinforcing global warming and the instability of the west Antarctic ice sheets are categorized as "Pythia"-type risk with both probability and extent of damage unknown. Major parts of the technological hazards including, for example, the accidents during exploitation, transport, and storage of oil or gas as well as the operation of nuclear power stations and the storage of nuclear waste are classified as a "Damocle"-type risk with low probability but high extent of damage (Fleischhauer 2006).

The methodology of natural and technological hazard characterization and mapping was developed by P. Schmidt-Thomé (2005, 2006). Hazards and aggregated hazard maps – for instance, the risk map of Europe – can be used as an effective tool for spatial planning (Schmidt-Thomé 2006).

In Russia, there are some approaches and methods for classification of the natural (including geological) and technogenic hazards. First of all, the natural hazards are divided into two groups: *catastrophic* (threatening human life) and *unfavorable*. As a rule the catastrophic processes (events) are characterized by an unknown probability and a high rate or intensity. Among the catastrophic hazards, there are such events as meteorite impacts, earthquakes, volcanic eruptions, tsunamis, landslides, mud flows, avalanches, hurricanes, and floods (Har'kina 2000). One of the important features of the mentioned hazards is the cascade character of the processes – earthquakes can provoke landslides and tsunamis, and surges and floods cause active coastal erosion. Social vulnerability, which accompanies these catastrophic processes, depends on their intensity and rate as well as on the development level of the society. Natural hazards cannot be completely avoided, but in case of their prediction and sustainable spatial planning measures (coastal protection structures, aseismic constructions, and timely evacuation of people), it is possible to reduce the vulnerability to hazards.

Unfavorable hazardous processes can negatively influence the environment and components of human life without direct risk to the human life (Har'kina 2000). Usually, such processes have a long duration when compared with human lifetime. This group of hazard processes includes coastal erosion, sea-level changes, swamping, and karst.

According to the Russian State Standard classification, the geological hazards can be divided into two groups depending on their driving forces – endogenic processes (caused by Earth's tectonic and thermodynamic factors) and exogenic processes (controlled mostly by factors external to the lithosphere, such as the sun's energy, atmosphere, hydrosphere, and gravitation). The extent of damage of a geological hazard depends on the probability of its occurrence and intensity (duration, rate and area of source, volumes of the rock masses involved in the process, etc.) (Dzeker 1992, 1994).

The theory and methods of the probabilistic long-term prognosis of exogenic processes were established by the Russian Research Institute of Hydrogeology and Engineering Geology in 1975 and have been used since that time (Krupoderov 1994, Sheko and Krupoderov 1994, Osipov and Shoigu 2002). The exogenic processes are regarded as open multicomponent systems. The occurrence of each single process is caused by interaction of many factors which can be divided into three groups:

Table 16.1 The classification of exogenic geological potential hazards for the coastal zone of Russia (Osipov and Shoigu 2002)

	Groups	Classes	Types
I	Caused by climatic and biological factors		Weathering
II	Caused by relief and gravity	Movement of rocks without the loss of the contact with slope	Landslides
		Movement of rocks with the loss of the contact with slope	Rock collapse, slope slide
III	Caused by hydrological processes	Sediment movement as a result of wave and current impacts	Coastal erosion
			Alongshore sediment drift
			Lateral erosion by rivers and flows
			Erosion of permafrost coasts
			Sediment accumulation at high and very high rates
IV	Caused by groundwater	Sediment leakage	Erosive leakage
		Drop of groundwater level	Slumps
		Rising of groundwater level	Swamping, flooding caused by groundwater level change
		Degradation or destruction of the ground structure	Quick grounds, shifting sands
		Clayey rocks swelling	Clay swelling
V	Caused by wind		Deflation
VI	Caused by anthropogenic impact	Mineral resources mining and underground construction	Slumps and shifts of rocks
		Hydrotechnical constructions	Coastal and sea bottom erosion
		Industry, agriculture, sewage waters	Sea bottom sediment pollution

(a) permanent during the term of forecast; (b) slowly changing; and (c) rapidly changing. The "permanent" factors include geological structure and relief; they determine a possibility and degree of the hazard impact. Slowly changing processes include modern tectonic movements and stable hydrodynamic regimes. These factors control the process trend. The rapidly changing factors such as storm events and hurricanes control the rate of exogenous activities.

The classification of exogenic geological hazard potentials for the coastal zone of Russia is shown in Table 16.1. It is important to note that all listed phenomena can be classified as hazards only if they threaten people's lives, lead to essential property damage, or threaten natural environment. Stressful or even hazardous situations can result from an interaction between common natural processes and intense anthropogenic activities, i.e., on-land and underground constructions, dredging, land reclamation, construction of oil and gas pipelines, mainly in the areas with a high population density. For example, coastal erosion formed escarpments on different heights along all the coasts of the Gulf of Finland during the last 8,000 years. But it has not become a hazard potential until the rapid growth of population and industry in the coastal zones.

Although marine hazard can threaten different kinds of marine activities in the Baltic Sea area and lead to property damages, it is highly improbable that they would reach the "catastrophic" level and cause a major threat for human lives. On the other hand, the influence of anthropogenic-generated alteration in erosion–deposition processes can be harmful for the unique and sensitive Baltic ecosystem.

16.2 Materials and Methods

The Department of Marine and Environmental Geology of A.P. Karpinsky Russian Research Geological Institute (VSEGEI) has been carrying out seabed mapping and geological and environmental investigations in the eastern Gulf of Finland since 1980 (Moskalenko et al. 2004, Spiridonov et al. 1988, Spiridonov et al. 2007). During the last decade, several projects were devoted to study coastal dynamics.

In 2005–2008, VSEGEI together with the Atlantic Branch of P.P. Shirshov Institute of Oceanology (ABIO RAS) conducted multipurpose investigations within the project "Up-to-date assessment of mineral-resource potential, control over geological hazards and establishment of prediction models for the geological environment in the Baltic Sea and its coastal zone," funded by the Federal Agency on Mineral Resources of the Russian Federation. One of the tasks of the project was the mapping, analyses, and risk estimation of the geological hazard potential caused by natural and anthropogenic factors and their interaction for the Russian part of the Baltic Sea (Fig. 16.1).

Within nine key areas in the Kaliningrad region and the eastern Gulf of Finland, combined on-land and nearshore zone investigations were carried out. Repeated onshore observations included detailed description and mapping of the coast,

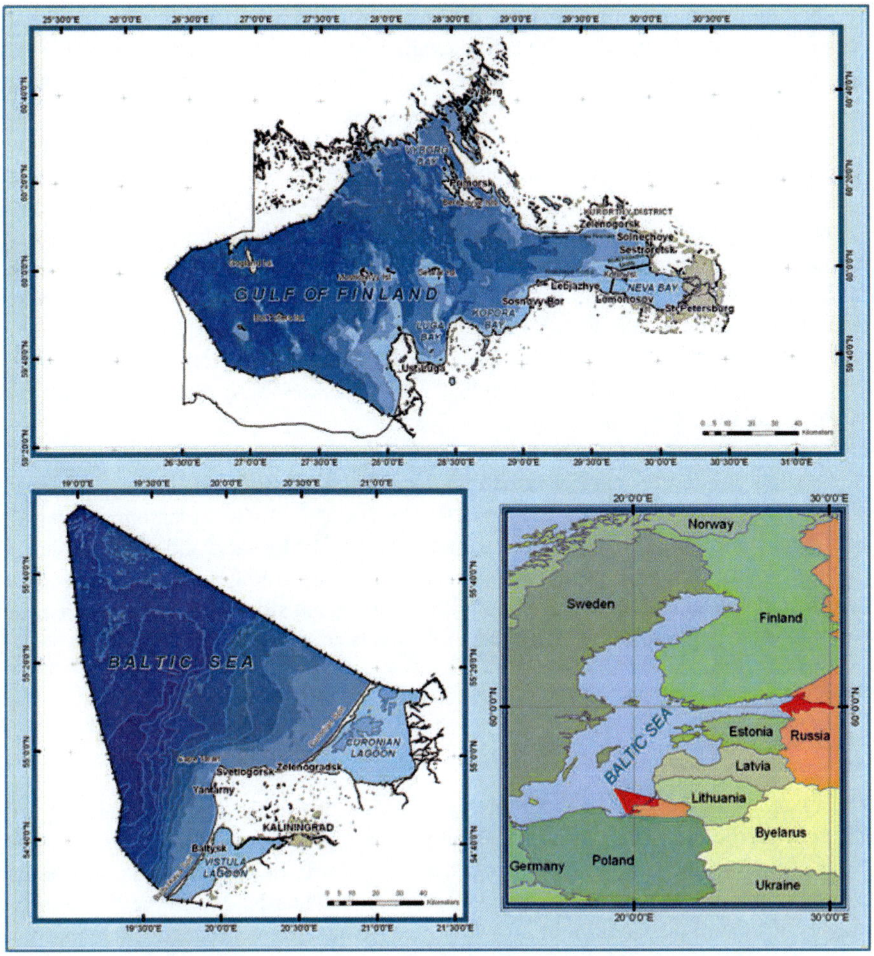

Fig. 16.1 Studied area showing two sites of investigations (*red color*)

measurement of the beach width, documentation of its specific features, description of coastal sediment composition, sediment sampling for grain-size analyses, and estimation of foredune condition. These field observations were analyzed and calibrated with remote sensing data (aerial photos of 1959–1990, resolution 0.5 m; Quick Bird space pictures of 2005, resolution 0.64 m) and with nautical and topographic charts in the scale 1:100,000–1:50,000 published in the nineteenth and twentieth centuries. Retrospective analysis of old charts and remote sensing data allowed defining shoreline recession areas, stable coastal zones, and accumulating coastal segments. Therewith, it was possible to calculate average erosion/accretion rates.

Shallow water areas of nearshore zone were studied by side-scan sonar profiling (CM2, C-MAX Ltd, UK) with search swath 100 m using a working acoustic frequency of 324 kHz. Altogether, 1,500 km of side-scan profiling perpendicular to the shoreline was done in 2005–2009 in the eastern Gulf of Finland. The distance between profiles (186 m) permitted to receive continuous acoustic images of the investigated sea bottom area as a whole, which were the basis to study the details of surface sediment-type distribution. Side-scan profiling was accompanied by echo-sounding. Repeated surveys of some nearshore zone areas and key profiles allowed exploring the development of the bottom relief and sediment dynamics through time. In the Kaliningrad area, side-scan sonar profiling was carried out along the seaward side of the Curonian and Vistula (Baltiyskaya) spits, in some areas of the Curonian and Vistula Lagoons, around the Sambian Peninsula, and along the underwater pipeline from the D6 offshore oil field ("Kravtsovskoe"). Approximately 300 km of side-scan lines were measured.

The interpretation of sonar data within both areas (the eastern Gulf of Finland and Kaliningrad area) was confirmed by sediment sampling and underwater video observations using the video-ROV Fish106M (Intershelf, St. Petersburg, Russia). Sediment sampling (230 samples in the eastern Gulf of Finland and 300 samples in the Kaliningrad area) along the side-scan sonar profiles used grab sampler and small drag sampler. The sediment sampling within the coastal slope between the coastline and the water depth of about 2.5 m was fulfilled by divers.

16.3 Results

The mapping of potential hazardous areas is one of the first steps toward hazard and risk prediction. Therefore, the geological mapping of the sea bottom and the coastal areas is the basis for the prognosis of hazards in coastal areas. There are two different approaches for mapping of geological hazards: one suggests the mapping of the actual distribution of areas affected by hazardous processes, whereas the other includes the mapping of potentially dangerous areas (the so-called method of geodynamic potential which considers the probability of the process occurrence; Krupoderov 1994, Sheko and Krupoderov 1994). As a result of our complex study, maps of geological hazard potentials for the eastern Gulf of Finland and the Kaliningrad area have been compiled (Figs. 16.2 and 16.3).

16.4 Kaliningrad Area

16.4.1 Endogenic Processes

The Baltic Sea region is traditionally characterized as an area of very low seismic activity. According to the General Seismic Zone Map of Russia, the maximal

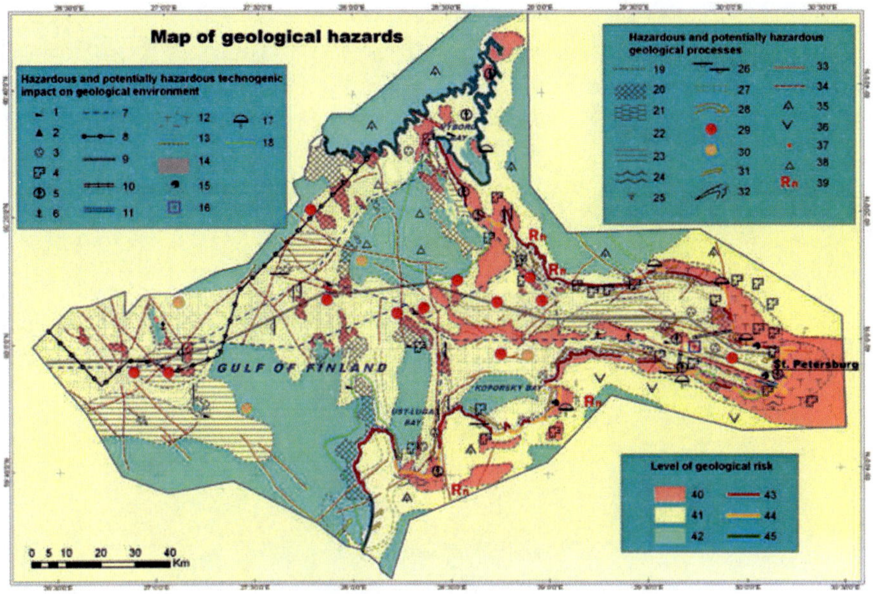

Fig. 16.2 Map of geological hazard potentials of the eastern Gulf of Finland and its coastal zone. *1,* sunken vessels; *2,* offshore oil platforms; *3,* dumps; *4,* sandpits; *5,* ports; *6,* anchorage; *7,* ship channels; *8,* oil and gas pipelines; *9,* main cables; *10,* St. Petersburg Flood Protection Facility; *11,* ship channel "Marine Channel"; *12,* areas of hazardous technogenic impact; *13,* areas of pockmark occurrence; *14,* areas of oil geological exploration; *15,* spillways; *16,* water intake point; *17,* recreation zones; *18,* nature protected areas; *19,* assumed boundaries of different geological risk areas; *20,* erosion; *21,* swamping; *22,* areas of high sedimentation rates; *23,* transit; *24,* underflooding; *25,* mud accumulation with overgrowing; *26,* landslides, landslips; *27,* buried valleys; *28,* sediment flows; *29,* geomorphic anomalies of high risk; *30,* geomorphic anomalies of medium risk; *31,* deflation; *32,* active erosion valley (incised valley); tectonic faults: *33,* fixed; *34,* assumed; *35,* tectonic uplift; *36,* tectonic subsidence; *37,* earthquakes epicenters; *38,* gas seep; *39,* Ra seep; *40,* high; *41,* medium; *42,* low; *43,* hazardous coastal erosion; *44,* potentially hazardous coastal erosion; *45,* stable coasts

seismic intensity in this area is $I \leq 5$. However, the unexpected Kaliningrad earthquake on 21 September 2004, with a main shock of 5.0 magnitude, was stronger than any other earthquake formerly instrumentally recorded within the Eastern European Platform (Aptikaev et al. 2005, Assinovskaya and Karpinsky 2005). Therefore, earthquakes should also be considered as hazard potentials around the Baltic. Slow neo and modern tectonic movements can be regarded as unfavorable geological processes. It is possible to assume that the rate of coast sinking in the Kaliningrad area reaches 1–2 mm/year (Sliaupa et al. 1999). These movements can stimulate exogenic geodynamics, which leads to hazardous erosion or, on the contrary, to silting and embankment.

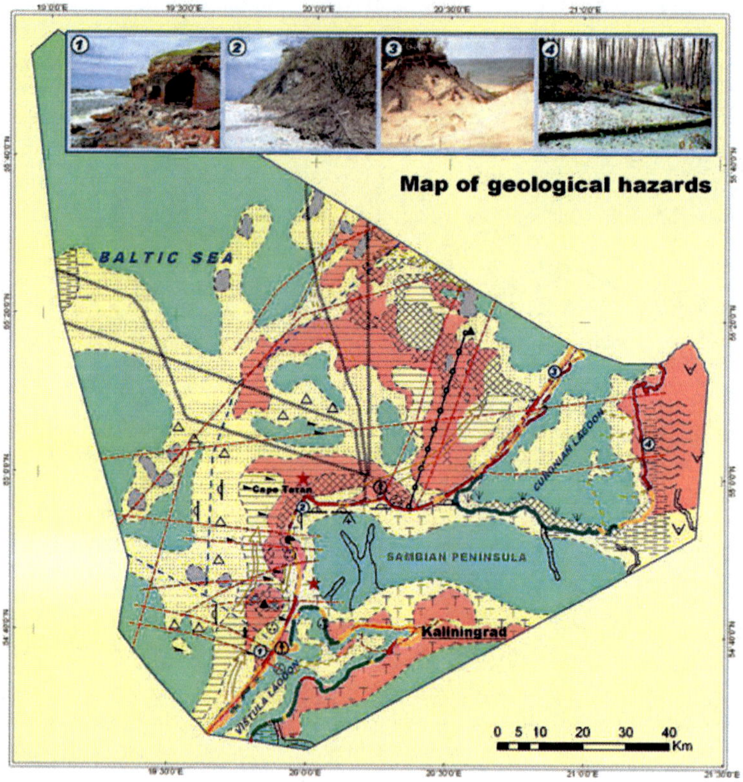

Fig. 16.3 Map of geological hazard potentials of the southeastern Baltic and its coastal zone within Kaliningrad region (legend on Fig. 16.2). Pictures: *1,* destruction of fortress (the Vistula Spit); *2,* landslide (the Sambian Peninsula) (photos by D. Ryabchuk); *3,* dune blowup (the Curonian Spit, photo by V. Boldyrev); *4,* swamping (the Neman Lowland, photo by I. Lysansky)

16.4.2 Exogenic Processes

At present, the *exogenic processes* can be considered as much more important and harmful for the Baltic Sea region due to their wide extension and activity. Among them, *coastal erosion*, caused mainly by storm surges, is one of the most intense and hazardous processes.

16.4.2.1 Coastal Erosion

Erosional processes are extremely active along the open Baltic Sea coast of the Kaliningrad area. The average rate of the cliff retreat of the Sambian Peninsula shoreline recession is 0.5–0.7 m/year. During storm surge (usually one in 5–7 years), the rate of coastal retreat increases to 4–6 m/year (Bobykina and Boldyrev 2008) (Fig. 16.4). Some significant sections of the Vistula and especially Curonian Spits'

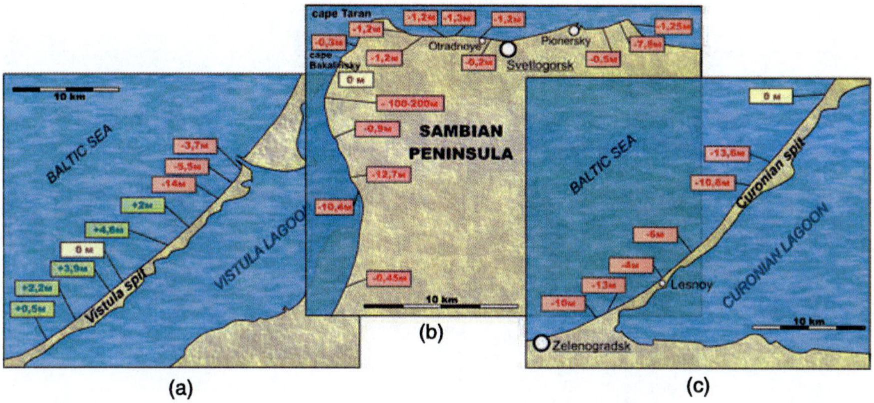

Fig. 16.4 Rate of shoreline shift in 2000–2007 according to monitoring of the ABIO RAS (Bobykina and Boldyrev 2008). **a** The Vistula Spit; **b** the Sambian Peninsula; **c** the Curonian Spit

coasts, both seaward and lagoon, are actively eroded. The most considerable impact on the unique landscapes of the Curonian Spit National Park, as well as on the freshwater environment of the Curonian Lagoon, is made by a spit breakthrough during extreme storm events. The Curonian Spit is known to have "weak points" where storm waves can break through the sand body. Between 1988 and 1996, its marine coast was threatened six times by extreme storms. During these storm events, the shoreline retreated 8–10 m along the distal part of the spit and 2–3 m close to the town of Lesnoy. In 1983, as a result of a storm event, the spit was broken through along 50–60 m of the coastline near Lesnoy (Boldyrev et al. 1990) (Fig. 16.4).

16.4.2.2 Sea Bottom Erosion

Coastal erosion is mainly caused by processes taking place at the sea bottom. The coastal slope at the Sambian Peninsula and the attached end of the Curonian Spit from Roshchino to Zelenogradsk is characterized by complicated sediment distributions. The nearshore zone along the Sambian Peninsula is a boulder bench; the amount of sand material is therefore very limited. It causes an urgent sediment deficit ("sediment starvation" effect). Areas of coarse-grained sediments (boulders and cobbles) mark outcrops of glacial till (lag deposits) and indicate active erosion processes, a sediment deficit, and low rates of tectonic subsidence. The areas of high rates of sediment removal are located mainly deeper than 10 m below sea level except for the coastal zone adjoining the Cape Taran and in the vicinity of Zelenogradsk where such areas are locally observed in immediate proximity to the coast, in a water depth less than 5 m. In areas of the coastal slope of the middle part of the Curonian Spit, elongated fields of drifting sands with ripple marks are observed (Fig. 16.5). As these fields are perpendicularly oriented to the coastline and their relative depth is 20–60 cm, it can be assumed that they serve as rivulets of water backwash and sediment outflow after storms or water run-up events. The

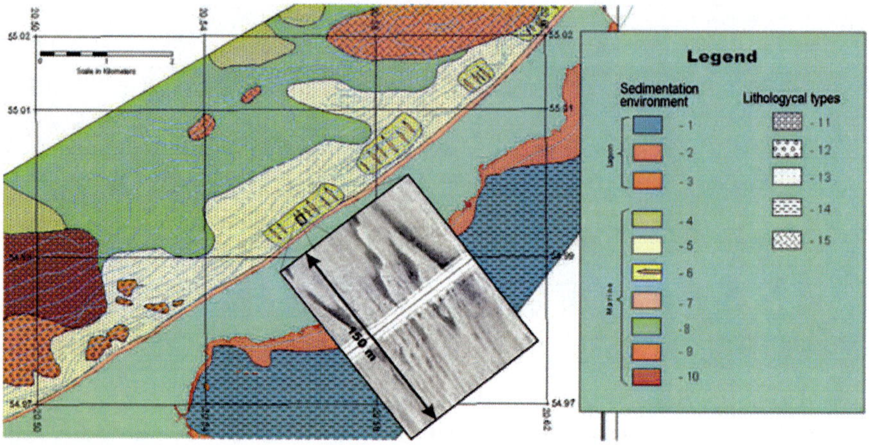

Fig. 16.5 Seabed map of the nearshore zone of the Curonian Spit. Sedimentation environment: *1–3*, lagoon: *1*, mud accumulation; *2*, wave sand accretion; *3*, erosion; *4–10* – marine: *4*, wave and current accretion; *5*, wave accretion; *6*, unstable accumulation and transit; *7*, periodic alteration of erosion/accretion processes; *8*, transit; *9*, weak erosion; *10*, intense erosion; *11–15*, lithological types of sediments: *11*, boulders, pebbles, gravel with sands; *12*, pebbles, gravel with sands; *13*, sands; *14*, silty clay mud; *15*, outcrops of dense clay deposits, partly covered by sands

lengths of some troughs exceed 100 m, with an average width of about 4–5 m (Fig. 16.5). Generally, the processes on the coastal slope of this area are dominated by longshore sediment transport. In the vicinity of Lesnoy and further to the north, there are practically no sandy sediments on the bottom surface. Boulder–pebble layer or extensive outcrops of greenish gray organic-rich laminated dense clays, partly covered by sand, were mapped offshore at depths from 5 to 15 m. Hence, erosion processes dominate within this area. Along the northern coast of the Sambian Peninsula between the Cape Taran and 25–28 km of the Curonian Spit, the value of sediment deficit is about 40 million m^3 (Boldyrev and Ryabkova 2001).

Along the western coast of the Sambian Peninsula, sediment deficit is observed between the capes Bakalinsky and Taran (Fig. 16.4). The result of this effect is an extremely high level of coastal retreat showing an annual volume of land loss by landslides and erosion of about 70,000 m^3 (Boldyrev and Ryabkova 2001).

16.4.2.3 Slope Slides

Landslides reach a very hazardous level in the Kaliningrad area. Huge landslides took place along the 35-km active cliff coast of the Sambian Peninsula (Fig. 16.3). They cause the loss of hundreds of square meters of land annually. Since 2000, the stability of the Kaliningrad coastal zone has deteriorated. In many areas, the plant cover of the slopes has been destroyed and the slope steepness of potential landslide areas has increased up to emergency levels. As a result, landslides have become more frequent. Nowadays, there are some coastal areas where a risk of building damage should be considered, for example, in the towns of Pionersky, Svetlogorsk, and Otradnoye.

16.4.2.4 Aeolian Processes

Aeolian processes have an exogenic hazard potential only within the areas of the Curonian and Vistula spits. The most hazardous are the "blowing craters" in foredunes, especially if these craters go through the entire dune system. In 2005, along the marine coast of Curonian Spit, 170 blowing craters were observed (Zhamoida et al. 2009; Fig. 16.3).

16.4.2.5 Flood and Swamping

The lowlands of the continental coasts of the Curonian and Vistula Lagoons suffer from floods and swamping (Fig. 16.3). For example, within the Neman Delta Lowland, some areas are located up to 1.5 m below sea level. The geological structure and the tectonic regime (long-time subsidence) lead to the storage of great amounts of groundwater close to the land surface and their inactive discharge. Together with the periodical sea-level rising in the Curonian Lagoon and the Neman River water supply, these factors lead to floods which sometimes cut off roads and leave the coastal villages without connection to the "mainland."

16.5 Eastern Gulf of Finland

16.5.1 Endogenic Processes

The potential hazard of endogenic processes in the eastern Gulf of Finland is questionable. According to Yaduta (2002, 2003), the area is characterized by local differentiations in trends and rates of sea bottom and land surface uplift and subsidence. Possibly, this is a result of the most recent tectonics of the so-called key type, where different hard rock blocks move in various directions under the control of a fault system. The recent study carried out by Assinovskaya and Novozhilova (2002) indicates signs of seismic activity within the Gulf of Finland and adjacent areas. These zones are traced from the territory of Finland through the Russian part of the gulf and its coastal zone. Connection to recent tectonic movements (Dvernitsky 2009) is particularly important for planned skyscriber projects in St. Petersburg, where the upper part of the geological sequence is represented by Quaternary deposits with unfavorable geotechnical properties. The other fault-related aspect of geological risk assessment is related to radon emission along those faults (Dvernitsky 2007).

VSEGEI side-scan sonar investigations carried out in the Vyborg Bay of the Gulf of Finland in 2008 detected several concentric structures (up to 10–15 m diameter) at the sea bottom in the area of known tectonic faults (Fig. 16.6). Morphology

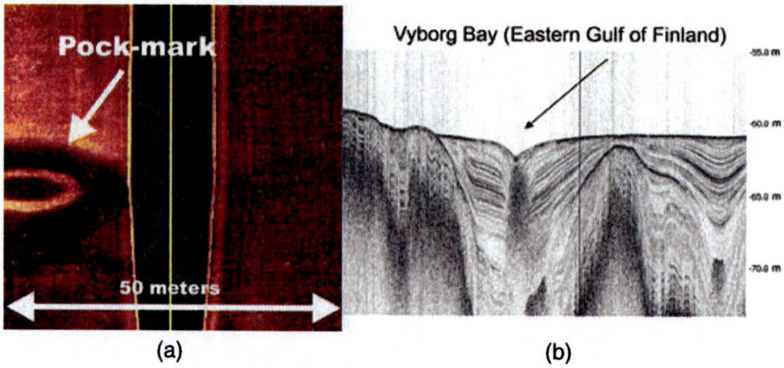

Fig. 16.6 Pockmark structure in the eastern Gulf of Finland. **a** VSEGEI side-scan sonar image; **b** fragment of echo-sounding profile, joint cruise of the Geological Survey of Finland and VSEGEI on board R/V Aranda

of these structures is similar to pockmarks found in some areas of the Baltic Sea and the Norwegian fiords (Jensen et al. 2002, Plassen and Vorren 2003, Söderberg and Flodén 1991), but the origin of these structures is debatable and needs additional detailed investigations. The pockmark distributions and their formations can be, for example, important for the construction of the Nord Stream pipeline because an interaction of natural phenomena and technological constructions can have unknown consequences.

16.5.2 Exogenic Processes

Exogenic processes are more important as potential hazards in the coastal zone of the eastern Gulf of Finland.

16.5.2.1 Coastal Erosion

The rectified length of the shoreline of the Russian part of the Gulf of Finland (skerries and islands are not counted) is about 520 km. Traditionally, the coastal zone of the easternmost (Russian) part of the Gulf of Finland was not considered as an area of active litho- and morphodynamic processes, but recent study has revealed that this area is under severe erosion. More than 40% of this coast is seasonally eroded, which causes land loss and destructions of buildings and roads.

Some parts of the coastal zone are rather stable. The northern coast (between the town of Primorsk and the Russian–Finnish border) does not suffer from wave impact due to geological structure of granite and glacial till skerries and tectonic uplift. Inner parts of the large bays (Neva Bay, Luga Bay, and Koporsky Bay) are more stable due to relatively high sediment flux of the rivers Neva, Luga, and Sista (e.g., annual suspended material flux – the Neva: 514,100 t; the Luga: 40,800 t; after Raukas and Huvarinen 1992). In the other parts of the studied coast, erosion rates reach hazardous level.

In the Neva Bay, about 70% of the northern and southern shores are eroded and under retreat. The southern coastal zone of the gulf from Lebyazhye to the St. Petersburg Flood Protection Facility is characterized by very intense coastal dynamics. According to many observations in the neighborhood of Lebyazhye and Bolshaya Izhora, between 1970 and 1980 (Raukas and Huvarinen 1992), the processes of coast destruction increased relative to previous decades. Intense erosion took place from 1975–1976 to 1989–1990 according to the analysis of aerial photo. Sandy beaches were eroded up to 30 m west of Lebyazhye and up to 70 m near Bolshaya Izhora. The comparison of the air photos from 1989 and recent high-resolution satellite images reveal that, since 1980, maximal shoreline recession in some parts of the former sand accretion areas has been more than 90 m up to now (Suslov et al. 2008).

The coastal erosion is also one of the most serious problems of the Kurortny district of St. Petersburg, which is located along the northern coast of the Gulf of Finland to the west of St. Petersburg Flood Protection Facility. This area is specifically important as a unique recreation zone of northwest Russia. The analysis of historical materials, archive of aerial photographs, and modern high-resolution satellite images has shown that the major parts of the coasts are eroded and therefore under retreat (Fig. 16.7). The average rate of shoreline retreat during the last 15

Fig. 16.7 Results of intense coastal zone erosion (the Kurortny district of St. Petersburg, the northern coast of the Gulf of Finland). *Red color* – coastal erosion with rate of shoreline retreat from 0.5 to 2.2 m/year; *green color* – stable and progradating (up to 0.5 m/year) coasts. *1,* erosion of sandy beaches; *2,* escarp in the coastal dune after the winter surge accompanied by flood (2.25 m higher than the zero water level), January 11, 2007 (photos by D. Ryabchuk); *3,* destruction of the coast protecting structures; *4,* erosion of the submarine terrace; *5,* area of submarine terrace erosion, shown in Fig. 16.9

years is about 0.5 m/year, and the maximal rate is up to 2 m/year (Ryabchuk et al. 2007). The maximal distance of shoreline retreat is observed in the vicinity of the town of Zelenogorsk (former Terijoki), where, according to the analysis of old maps and remote sensing data, the coast retreated landward at a distance of about 100 m in the course of a century (Ryabchuk et al. 2009).

16.5.2.2 Sea Bottom Erosion and Sediment Flows

Many important features of coastal dynamics are caused by nearshore processes. The eastern Gulf of Finland, for example, can be attributed as a wave- and storm-dominated shallow clastic sea (Reading 1996). According to Reading (1996), the main controls on sediment transport in such a case are (i) the frequency and intensity of storm-induced currents; (ii) nature and origin of sediment supply; and (iii) sea-level fluctuation. The important feature of the hydraulic regime of the nontidal Gulf of Finland is the nonperiodic sea-level change. The most significant sea-level variations in the eastern Gulf of Finland apparently occur as a result of combined effect of wind-induced storm surge and progressive long waves caused by cyclones moving along the Baltic Sea and the Gulf of Finland (Eremina et al. 1999). The minimal value of the sea level (–1.24 m) was registered on November 2, 1910, and the highest surge (4.21 m) occurred on November 19, 1824. About 307 floods higher than 160 cm are documented in the history of St. Petersburg during the period from 1703 to 2008. Eighty-nine percent of surges were caused by western and southwestern cyclones (Pomeranets 2005). The interaction of waves and storm-induced currents with sea-level fluctuation can be the reason for the formation of bedforms, typical for tidal seas, i.e., sand ridges, sand waves, and erosional streams.

The most significant erosion is observed within autochthonous coastal systems (Reading 1996) with minor outer sediment supply, such as the northern coast of the Gulf of Finland between capes Peschany and Dubovskoy. During fair-weather hydraulic conditions, the nearshore surface is characterized by dynamic equilibrium; storm surges break the balance and generate offshore sediment flows.

Configuration of the shoreline and dominated winds from the west and southwest resulted in sediment transport along the coast in the eastern direction (Fig. 16.8a). Therefore, sediment flux along the northern coast tends to decay in this direction. In the vicinity of Zelenogorsk, sediment flux is close to 30,000 m^3/m/year (cubic meters per meter if shoreline per year), while near the town of Repino, sediment flux is 20,000 m^3/m/year and in the vicinity of Solnechnoye, it is almost zero (Leont'yev 2008).

The eastern sediment transport along the coast discharges in front of Sestroretsk. It has changed due to coastline variations (Figs. 16.7 and 16.8a). As a result, accretion of beach sands up to 140 m wide is observed here. In the nearshore zone, there is sand accretion as well, showing a very shallow submarine terrace surface with a system of sandbars and furrows composed of fine-grained, well-sorted sands. But retrospective analyses of remote sensing data have shown that the shoreline has not aggraded here. The main reason for this phenomenon is the slope gradient (depth increase from 2 to 5 m along the distance of 100 m) of the sand terrace which is located only about 500 m from shoreline. Therefore, a big volume of sand is redeposited at the terrace foot (Fig. 16.8a).

A very special bottom relief was identified near the terrace foot. The erosion furrows (up to 30–50 cm deep) were observed at a depth of 8–12 m (Ryabchuk et al. 2007). Repeated surveys demonstrated that these forms are very stable in spite

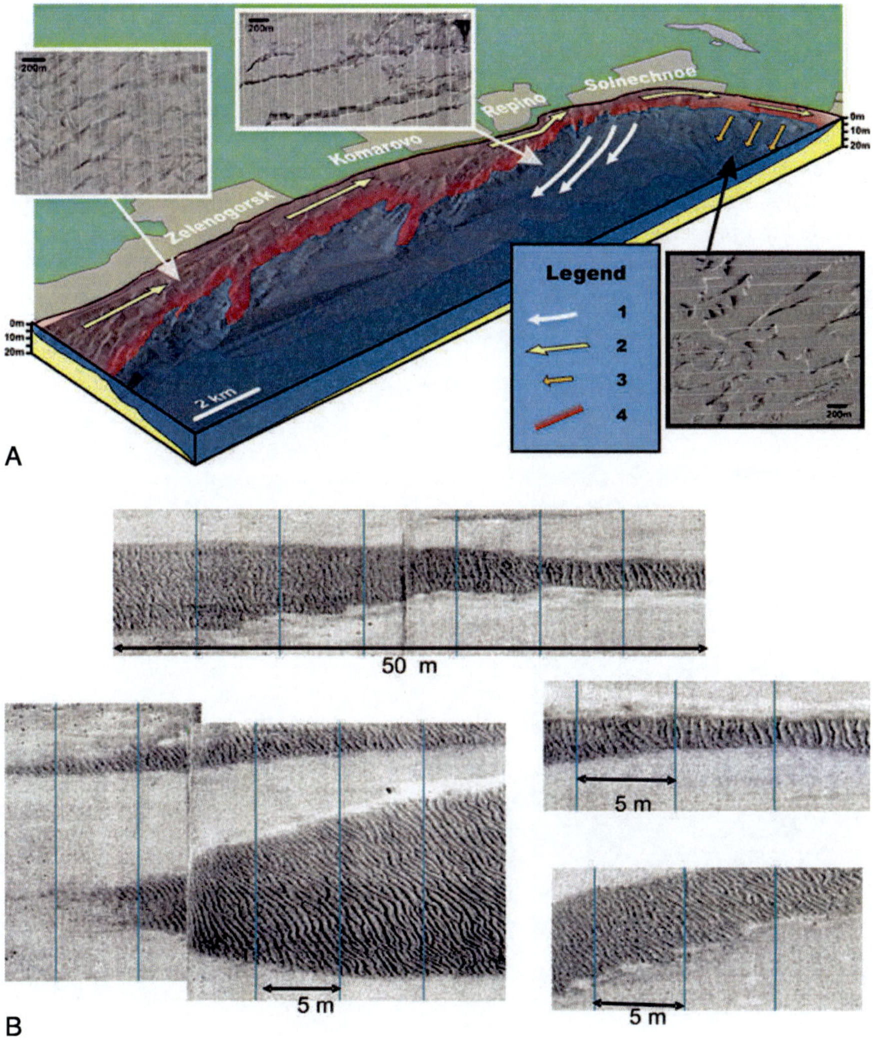

Fig. 16.8 **a** Scheme of dynamics of the northern coastal zone of eastern Gulf of Finland. *1,* Offshore sediment transport along erosion furrows; *2,* longshore sediment flow; *3,* direction of sediment transport toward the foot of submarine terrace in the area of longshore sediment flow discharge; *4,* edge of submarine terrace. **b** Side-scan sonar mosaic of erosion furrows (up to 20 m wide, 600 m long, and 0.5 m deep) with sand ripples (heights 20–30 cm; distance between crests 50–100 cm)

of their relatively small depths. Erosion furrows were traced along all coastal lines in the western direction. On the bottom surface of the erosion furrows, there are very distinct current ripples (up to 20 cm high) turned perpendicularly to the runnel direction and composed of coarse-grained sand. The distance between ripple crests ranges from 40 cm to 1 m (Fig. 16.8b). Investigation has shown that they are the ways of near-bottom currents, removing sand material offshore.

Offshore sand movement and its redeposition at depths of more than 10 m lead to sediment loss and disturbance of the dynamic equilibrium. This occurs during storms when dunes and foreshores are eroded. During these events, material is transported seaward to be deposited outside the break point, while during calmer weather, sandbars are shifted toward the coastline (Schwarzer et al. 2003).

Repeated profiling of the submarine terrace located between capes Peschany and Repino (Fig. 16.9) and the comparison of our results with old nautical maps revealed the progressive erosion of the marine edge of the terrace. As long as the terrace exists with its surface at the depth of 3–5 m, it protects the coast from erosion. Accordingly, erosion of the terrace itself is a rather dangerous process, which

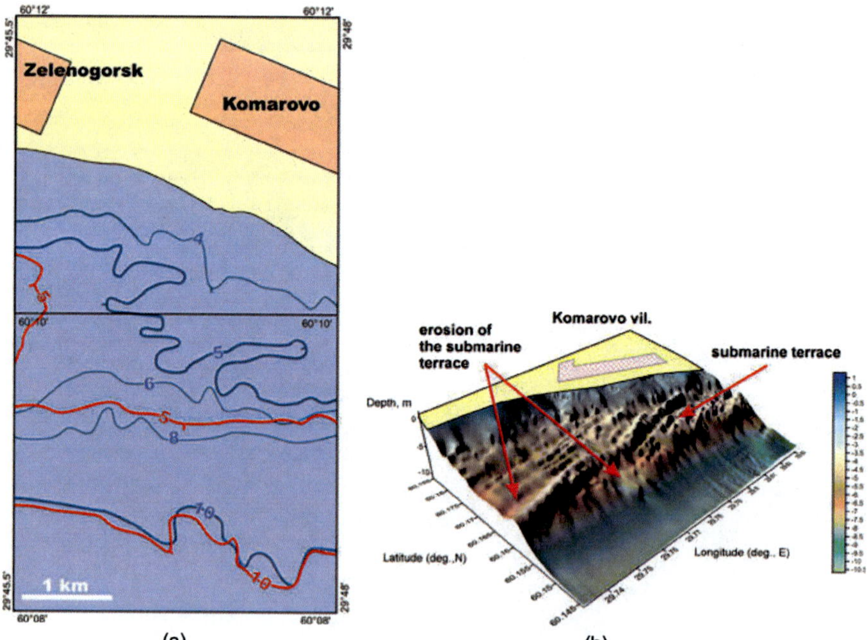

Fig. 16.9 Bottom erosion. **a** Changing of the depths in the nearshore zone as a result of the submarine terrace erosion. *Red lines* – isobaths of the navigational chart edited in 1989; *blue lines* – results of the depth measurements made by VSEGEI in 2005. **b** 3D diagram showing submarine terrace erosion

destabilizes the system. Nowadays, there is a dynamic correlation between coastal erosion and the state of the submarine terrace, i.e., the more the terrace erosion the greater the on-land problems. The situation becomes even more problematic because of the ineffective old system of coastal protection measures (Fig. 16.7) and the intense and unsustainable development of recreation infrastructure.

16.5.2.3 Ice Impact

In the eastern Gulf of Finland, ice impact is the second important factor of coastal erosion. These processes are most dangerous during winter floods when 1-m-thick ice layers can damage shallow water areas of the gulf bottom upon the depth of 30–50 cm and remove frozen blocks of sediments and even big boulders along the beach face. Ice damages trees and coastal buildings as well as coast protection structures.

16.5.2.4 Slope Slides

Landslides in the eastern Gulf of Finland are observed locally along some parts of the coast between the capes Flotsky and Peschany and in the vicinity of the town of Lebyazhye (Krasnaya Gorka fortress, southern coastal zone) where the coastal cliffs reach 25–30 m height.

The geological structure of the upper sediment layers plays an important role in the landslide processes. In the southern coastal zone, clays of Kotlin horizon of Vendian are overlapped by 5-m-thick Quaternary deposits. The intense wetting of Vendian clays and the occurrence of microfractures (especially within tectonic faults zones) lead to a decrease of property strength and can cause landslides (Auslender et al. 2002).

16.5.2.5 Sea Bottom Sediment Pollution

The problem of bottom sediment pollution was not the focus of our studies. Although it cannot be classified as a geological hazard itself, the risk assessments should keep this problem in mind for any hydroengineering activities, i.e., dredging, dumping, and pipeline construction. Within the Russian sector of the Baltic Sea, the most polluted areas are the lagoons adjacent to the big cities of St. Petersburg and Kaliningrad – the Neva Bay and the Vistula Lagoon (Spiridonov et al. 2004, Emelyanov et al. 1998). Mud of depositional basins can form a huge sink for special chemical substances. In the eastern Gulf of Finland, an extended and prolonged seafloor anoxia within local coastal depositional basins could therefore enhance the environmental problems by releasing metals and nutrients from the seafloor sediments (Kotilainen et al. 2007). The possible consequence of these pollution processes is discussed in Chapter 17.

16.6 General Classification of Geological Hazard Potential of the Eastern Gulf of Finland and the Kaliningrad Area

The results of the geological hazard potential study are summarized in the maps (Figs. 16.2 and 16.3). The analysis of interaction of natural and anthropogenic hazard potential allows estimating three levels of risk (low, medium, and high). Presented maps show that the highest level of risk corresponds mostly to the coastal zones. In the eastern Gulf of Finland, the risk level is higher in the easternmost, high populated area with developed industry.

Resulting classification of endogenic and exogenic hazard potential for the Russian Baltic and its coastal zone, based on the same principles, was developed on the basis of expert estimation and is shown in Tables 16.2, 16.3, 16.4, and 16.5.

Table 16.2 Endogenic geological hazard potentials for the coastal zone of the Russian Baltic

	Technogenic activity				Level of hazard potential	
Geological processes and phenomena	Construction	Communications, transport	Hydrotechnical constructions	New coastal territories	Kaliningrad district	Eastern Gulf of Finland
Earthquakes	+	+	+	+	Medium	Low
Vertical tectonic movements along tectonic faults	+	–	–	+	Medium	Medium
Gas flow	+	–	–	–	Low	Medium

Table 16.3 Endogenic geological hazard potentials for the sea bottom of the Russian Baltic

	Anthropogenic activity				Level of hazard potential	
Geological processes and phenomena	Marine transport, fishing	Oil and gas pipelines, cables	Dredging, dumping	Sand and gravel exploration	Kaliningrad district	Eastern Gulf of Finland
Earthquakes	–	+	–	–	Medium	Low
Vertical tectonic movements along tectonic faults	–	+	+	–	Medium	Medium
Pockmarks	–	+	–	–	Low	Low

Table 16.4 Exogenic geological hazard potentials for the coastal zone of the Russian Baltic

Geological processes and phenomena	Anthropogenic activity				Level of hazard potential	
	Construction	Communications, transport	Hydrotechnical constructions	New coastal territories	Kaliningrad district	Eastern Gulf of Finland
Seacoast erosion	+	+	+	+	High	Medium
Landslides	+	+	+	–	Medium	Low
Fluvial erosion	–	+	+	–	Medium	Medium
Aeolian processes	–	+	–	–	Medium	Low
Swamping, flooding	+	+	+	+	Low[a]	Medium
Anthropogenically induced change of the geotechnical conditions	+	+	+	+	Medium	Medium

[a]With the exception of Vistula and Curonian Lagoon coast

Table 16.5 Exogenic geological hazard potentials for the sea bottom of the Russian Baltic

Geological processes and phenomena	Anthropogenic activity				Level of hazard potential	
	Marine transport, fishing	Oil and gas pipelines, cables	Dredging, dumping	Sand and gravel exploration	Kaliningrad district	Eastern Gulf of Finland
Bottom erosion	–	+	+	+	High	High
Submarine landslides	–	+	+	–	Low	Low
"Avalanche" sedimentation, > 1 mm/year	+	+	+	+	Medium	High
Sediment anthropogenic pollution	+	–	+	+	Medium	High

16.7 Discussion and Risk Prevention

The problem of natural and anthropogenic hazards and risk prevention is attracting more and more attention of both scientists and spatial planners. Since the 1960s, both catastrophic events and insured losses have increased (Schmidt-Thomé 2006). The world coastal zones are among the most threatened and, at the same time, the most populated areas.

In the Russian Baltic and its coastal zone, the anthropogenic impact has been constantly growing since the beginning of the eighteenth century, but during the last decade this impact has drastically increased. In coastal zones (including the shallow coastal waters), anthropogenic activity has become a major factor comparable by its importance to the natural processes.

Field work and analytical data (side-scan survey, echo-sounding, sediment sampling, gamma spectrometry for ^{137}Cs, ICP-AES, and ICP-MS) as well as the investigation of archive materials have brought us to the conclusion that the sedimentation processes in the Neva Bay have completely changed during the last two centuries. Special conditions of mud accumulation have developed in the bottom depression of 5–6 m in the western part of the Neva Bay (Spiridonov et al. 2008). Significant alteration of sedimentation was caused by anthropogenic impact, i.e., defense constructions in the outer part of the bay, construction of St. Petersburg Flood Protection Facility, and hydroengineering works. The construction of a new St. Petersburg harbor in the Neva Bay, accompanied by new land creation of around 477 ha and 12–14-m-deep ship channels for cruise fairies, started in 2006. As a result of dredging and dumping processes, the concentration of suspended matter in the water was extremely high in 2007 and the trace of suspension reached Vyborg Bay (Fig. 16.10). VSEGEI study of the nearshore bottom showed that a clayey layer up to 3 cm thick had formed on the sand surface. The concentration of fine particle in beach sands of resort areas increased up to 5–7% in 2007–2008. Sedimentation system of the eastern Gulf of Finland was significantly disturbed.

Changes in sedimentary processes stressed the ecosystem of the Neva Bay and caused degradation of fish spawning and feeding areas, decrease in plankton and benthic community's productivity, and migration of marine birds and mammals (Fedorov et al. 2008). In 2006–2007, microalgal occurrence drastically decreased in the outer Neva Bay and biomass of macrozoobenthos within the bay diminished. The ecosystem started to recover after the conclusion of the active phase of hydrotechnical work.

Another big project of the eastern Gulf of Finland is the newly built Ust-Luga port complex, which is planned to be one of the world's 10 biggest ports. According to the plan, its carrying capacity of general cargo will reach 120 million tons/year (50 million tons by 2010, www.ust-luga.ru). The first part of the port complex was constructed by 2008. A large quantity of sediments (mostly sands) were removed from the coastal zone during the dredging and used for terminal construction. The new ship channels will interrupt sediment transport and act as large sediment traps. As a result, the natural sediment nourishment of the sandbars will be reduced. In the late 1970s, the area of sand accretion was 2.24 km^2, including 0.65 km^2 of the bay-head sandbar area between the Luga River and the future port. By 2003, the area of sand accretion had reduced up to 0.5 km^2. The size of the whole accretion zone decreased by about 80%, whereas the area of bay-head sandbar decreased by about 30% to 0.42 km^2 (Sergeev et al. 2009), so the bay-head sediment balance of the Luga Bay was significantly changed.

Among other large planned projects, there are two more ports: the Primorsk oil terminal, which is planned to be the biggest oil export port of northwestern Russia

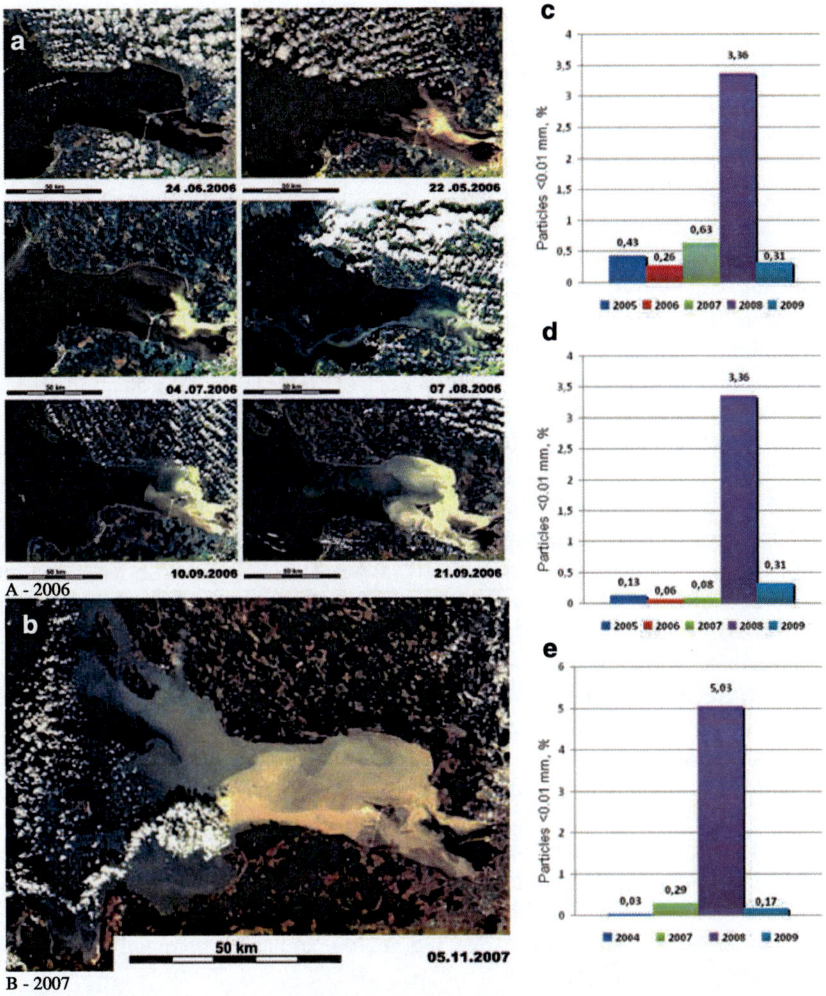

Fig. 16.10 **a, b** Increase of suspended matter concentration (*light yellow color*) in the upper water layer in 2006 and 2007 due to dredging, dumping, and new territory construction in the eastern part of the Neva Bay. **c–e** Distribution of fine particles (<0.01 mm) in the beach sands of the Kurortny district (**c** – the Repino village; **d** – the Solnechnoye village) and the Izhora Village (**e**). Satellite images analyzed by Dr. Leontina Sukhacheva (NIIKAM)

(www.mtp-primorsk.ru), and construction of 400 ha of new territories near the town of Sestroretsk. It is important to note that each separate project has environmental assessment documents, but the impact of all the activities together on the gulf ecosystem is unknown.

The coastal erosional problems have become more serious within both sectors of the Russian Baltic due to an ineffective system of coast protection and the absence of an integrated coastal zone management, both causing especially irregular measures for shore protection. These problems have resulted in extensive negative impact on the adjacent sections of the coast (Ryabchuk et al. 2009).

The modern coastal engineering methods have made it possible to estimate more or less definitely the impact of major constructions on the coastal zone. A much more complicated task is to estimate the impact of submarine sand extraction on the coastal system. For example, submarine sand mining took place in 1970–1992 in the sandy terrace area between capes Flotsky and Peschany to the west of the district of Kurortny. The total volume of extracted material was about 150 million m^3. As a result, large parts of the coastal slope at depths less than 19 m were affected (Ryabchuk et al. 2009). Keeping in mind the sediment flux volume, it is obvious that such volumes of extracted sand material significantly disturbed the coastal system.

One of the possible solutions to the problem should be avoiding new construction in the hazard coastal territories, development of modern coast protection system based on natural litho-dynamic appropriateness, and a ban against the submarine sand and gravel exploitation in the nearshore areas.

Problems of the coast erosion are becoming more and more important for regional authorities both in Kaliningrad and in St. Petersburg. Recently, State Coast Protection Programs have been developed for both regions with the participation of geologists and oceanographers from ABIO RAS, VSEGEI, and the Russian State Hydrometeorological University (St. Petersburg). Unfortunately, beginning of the programs has been delayed. The delay is due to both the lack of funding and the absence of a coastal legislation. The latter is needed for a close cooperation between regional and federal authorities, who today have different responsibilities with respect to the protection of the seacoast and the offshore part. The absence of coastal legislation leads to the increase of negative anthropogenic impacts, i.e., submarine sand exploration and unsustainable development of recreation facilities.

One of the important directions toward a future integrated coastal zone management and a risk prevention strategy is the special mapping of hazard potentials in connection with a risk assessment and analysis. The Project "Coastal zone cadastre of the Russian sector of the Baltic Sea and geological hazard potential assessment" is a first step in the right direction.

16.8 Conclusions and Future Work

Analysis of the recent environmental conditions at the seabed of the Russian sector of the Baltic Sea, especially within its coastal zone, shows that during the last years the activity of exogenic geological processes has increased significantly. In some cases, this intensification leads to the increase of negative consequences, such as coastal erosion, natural marine landscape degradation, and extreme silting. These processes support hazardous events for different types of human activities, including hydrotechnical constructions, pipelines, tourist industry, fishery, and ship navigation.

The coastal areas, located within sinking blocks of the Earth's crust, such as the Russian part of the southeastern Baltic (Kaliningrad Region), suffer the most from these changes. As a result, serious damage has been done to the beaches of the Sambian Peninsula with its famous resort areas. Furthermore, there is a real risk of Curonian Spit breakthrough during extreme storm events. This could cause the damage and vanishing of the unique landscapes of the Curonian Spit National Park and the freshwater environment of the Curonian Lagoon.

The intensity increase of exogenic geological processes and their hazardous impacts on the coastal zones of the Russian sector of the Baltic Sea are caused by a combination of several natural and anthropogenic factors. The main natural one is the increase of the frequency of disastrous storm events, which are according to several authors a possible consequence of global climate change and the resulting changes of regional hydrometeorological conditions. Our studies clearly show that the natural development trends of the Russian coastal zones are severely intensified by anthropogenic activities. Along significant parts of the Russian Baltic Sea coastal zone, especially in the Kaliningrad Region, the process of beach degradation is already irreversible. The recovery of natural coastal landscape needs strong human efforts for an environmentally friendly future development of the coasts.

The present level of knowledge about the general scheme of the coastal zone processes and especially the combination of different factors controlling hazard potentials is still insufficient. Therefore, active response from government authorities and research activities should be expanded. Special mapping of potential hazards should be started as soon as possible and a monitoring system should be in place to lead remediation and preventive action.

Acknowledgment The authors wish to thank their colleagues Elena Nesterova, Yury Kropatchev, and Svyatoslav Manuilov for their contribution to their studies. The authors gratefully thank Vasily Bukanov, Dmitry Kurennoy, and Igor Lysanskiy as well as the captain and the crew of RV "Risk" for their great contribution to the field work. The authors are thankful for critical comments and suggestion from R.O. Niedermeyer (Güstrow/Greifswald) and from an anonymous reviewer. Furthermore, the authors would like to thank Yulia Guseva and Ricardo Olea (Washington), who have kindly revised the language of the chapter.

References

Aptikaev FF, Nikonov AA, Alyoshin AS, Assinovskaya BA, Pogrebchenko VV, Erteleva OO (2005) Kaliningrad earthquake of September 21, 2004, DAMAGE. International conference on earthquake engineering in 21st century (EE-21C). Institute of Earthquake Engineering and Engineering Seismology (IZIIS-Skopje), University "Ss. Cyril and Methodius", Skopje, Republic of Macedonia, 27 August–1 September 2005

Assinovskaya BA, Novozhilova TV (2002) About a level of seismic danger of Saint-Petersburg region. Izvestiya GAO 216:394–401 (in Russian)

Assinovskaya BA, Karpinsky VV (2005) On September 21, 2004 Kaliningrad earthquake source location. In: Joeleht A (ed) The Kaliningrad earthquake September 21, 2004 workshop materials. Institute of Geology, University of Tartu. Geological Survey of Estonia, 10

Auslender VG, Yanovsky AS, Kabakov LG, Pleshivtseva ES (2002) New facts in geology of St. Petersburg. Mineral 1(4):51–58 (in Russian)

Baerens C, Hupfer P (1999) Extremwasserstände an der deutschen Ostseeküste nach Beobachtungen und in einem Treibhausgasszenario. Die Küste 61:48–72 (in German with abstract in English)

Bitinas A, Zaromskis R, Gulbinskas S, Damusyte A, Zilinskas G, Jarmalavicius D (2005) The results of integrated investigations of the Lithuanian coast of the Baltic Sea: geology, geomorphology, dynamics and human impact. Geological Quarterly 49(4):355–362

Bobykina V, Boldyrev V (2008) Tendency in shore dynamics in the Kaliningrad Oblast according to five years monitoring information. Integrated management, sustainable development indicators, spatial planning and monitoring of the South-Eastern Baltic coastal regions: materials of international conference, pp 46–47. Kaliningrad (Terra Baltica)

Bödvarsson R, Lund B, Roberts R, Slunga R (2006) Earthquake activity in Sweden. Study in connection with a proposed nuclear waste repository in Forsmark or Oskarshamn. SKB report R-06-67, Uppsala University, p 40

Boldyrev VL, Lashenkov VM, Ryabkova OI (1990) Stormy reworking of the Kaliningrad coasts of the Baltic Sea. Questions of the dynamic and paleogeography of the Baltic Sea, pp 97–127. Vilnius (in Russian)

Boldyrev VL, Ryabkova OI (2001) Coastal zone processes dynamics of the Baltic Sea (Kaliningrad region). Proceedings of Russian Geographical Society 133(5):41–48 (in Russian)

Dvernitsky BG (2007) Radon monitoring of endogenic geological processes in St. Petersburg region. Radon in geology, 16–19. Moscow

Dvernitsky BG (2009) Neotectonic risk in St. Petersburg. Proceedings of the international conference on integrating geological information in city management to prevent environmental risks (GeoInForm), St. Petersburg, pp 46–47 (in Russian)

Dzeker ES (1992) Geological hazards and risks. Environmental Geoscience (Geoekologiya) 6:3–10 (in Russian)

Dzeker ES (1994) Methodological aspects of the geological hazards and risks. Environmental Geoscience (Geoekologiya) 3:3–10 (in Russian)

Eberhards G, Grīne I, Lapinskis J, Purgalis I, Saltupe B, Torklere A (2009) Changes in Latvia's seacoast (1935–2007). Baltica 22(1):11–22

Emelyanov EM, Blazhchishin AI, Koblents-Mishke OI, Kravtsov VA, Stryuk VL, Kharin GS (1998) Ecological and geochemical situation in the eastern Baltic Sea. Problems of investigation and protection of nature at the Curonian Spit, pp 148–187. Kaliningrad (in Russian with abstract in English)

Eremina TR, Nekrasov AV, Provotorov PP (1999) Hydrophysical processes. In: Rumyantsev VA, Drabkova VG (eds) Gulf of Finland under anthropogenic impact conditions. Institute for Lake Research, Russian Academy of Sciences, Saint Petersburg, pp 5–47 (in Russian)

Fedorov MP, Shilin MB, Gorbunov NE (2008) Ecological basis of management of natural-technogenic systems. Polytechnic University Press, St. Petersburg, 506 pp (in Russian)

Fleischhauer M (2006) Spatial relevance of natural and technogenic hazards. Natural and technological hazards and risks affecting the spatial; development of European regions. Geological Survey of Finland Special Paper 42:7–16. Espoo

Har'kina MA (2000) Ecological consequences of natural disasters. Energy 1:51–56. Moscow (in Russian)

Hutri K-L (2007) An approach to palaeoseismicity in the Olkiluoto (sea) area during the early Holocene. STUK-A222, 64 pp. Helsinki

Jensen JB, Kuijpers A, Bennike O, Laier T, Werner F (2002) New geological aspects for freshwater seepage and formation in Eckernförde Bay, western Baltic. Continental Shelf Research 22:2159–2173

Kotilainen A, Vallius H, Ryabchuk D (2007) Seafloor anoxia and modern laminated sediments in coastal basins of the Gulf of Finland, Baltic Sea. Geological Survey of Finland Special Paper 45:47–60

Krupoderov VS (1994) Scientific and methodical basis of the exogenous geological processes study. Modern problems of hydrogeology, geological engineering and ecology, pp 109–129. Moscow, VSEGEINGEO (in Russian)

Leont'yev IO (2008) Sediment budget and forecast of long-term coastal changes. Oceanologia 48(N3):467–476

Medvedev SV, Shponhoyer V, Karnik V (1965) Scale of seismic intensity MSK-64. Academy of Science of USSR, 11. Moscow

Mörner N-A (2004) Active faults and paleoseismicity in Fennoscandia, especially Sweden. Primary structures and secondary effects. Tectonophysics 380:139–157

Moskalenko PE, Zhamoida VA, Manuilov SF, Spiridonov MA (2004) The geological structure, history of geological development and potential mineral resources of the eastern the Gulf of Finland. Mineral resources of the Baltic Sea – exploration, and sustainable development, pp 135–145. Hanover

Nikonov AA, Sildvee H (1991) Historical earthquakes in Estonia and their seismotectonic position. Geophysica 27(1–2):79–93

Orviku K, Jaagus J, Kont A, Ratas U, Rivis R (2003) Increasing activity of coastal processes associated with climate change in Estonia. Journal of Coastal Research 19(2): 364–375

Osipov V, Shoigu S (eds) (2002) Natural hazards of Russia, 6 volumes. KRUK, Moscow (in Russian)

Plassen L, Vorren TO (2003) Fluid flow features in fjord-fill deposits, Ullsfjorden, North Norway. Norwegian Journal of Geology 83:37–42

Pomeranets K (2005) Three centuries of Saint Petersburg floods. Saint Petersburg Art, St. Petersburg, 213 pp (in Russian)

Pruszak Z, Zawadzka E (2005) Vulnerability of Poland's coast to sea-level rise. Coastal Engineering Journal 47(2–3):131–155

Raukas A, Huvarinen H (eds) (1992) Geology of the Gulf of Finland, 422 pp. Tallinn (in Russian)

Reading HG (eds) (1996) Sedimentary environments: processes, facies and stratigraphy, 3rd edn. Blackwell Publishing Company, Hong Kong, 688p

Rudnik VA (1996) Influence of the Earth heterogeneous geological zones. Vestnik of RAS 66(8):713–719 (in Russian)

Ryabchuk D, Sukhacheva L, Spiridonov M, Zhamoida V, Kurennoy D (2009) Coastal processes in the eastern Gulf of Finland – possible driving forces and connection with the near-shore zone development. Estonian Journal of Engineering 15(3):151–167

Ryabchuk DV, Nesterova EN, Spiridonov MA, Sukhacheva LL, Zhamoida VA (2007) Modern sedimentation processes within the coastal zone of the Kurortny district of St. Petersburg (eastern Gulf of Finland). Baltica 20(1–2):5–12

Schmidt-Thomé P (2005) The spatial effects and management of natural and technological hazards in Europefinal report of the European Spatial Planning and Observation Network (ESPON) project 1.3.1. Geological Survey of Finland, 197p. Espoo

Schmidt-Thomé P (2006) Integration of natural hazards, risks and climate change into spatial planning practices. Geological Survey of Finland, pp 1–31. Espoo

Schwarzer K, Diesing M, Larson M, Neidermeyer R-O, Schumacher W, Furmanczyk K (2003) Coastline evolution at different time scales – examples from the Pomeranian Bight, southern Baltic Sea. Marine Geology 194:79–101

Sergeev AYu, Ryabchuk DV, Zhamoida VA, Nesterova EN (2009) The impact of two newly built port terminals in the eastern Gulf of Finland on sedimentation processes and coastal zone dynamics. Estonian Journal of Engineering 15(3):212–226

Sheko AI, Krupoderov VS (1994) Assessment of hazards and risks of exogenous geological processes. Geoecology 3:11–21 (in Russian)

Sliaupa S, Pacesa A, Korabliova L (1999) Indications of seismic activity in Lithuania and adjacent Baltic Sea area. Workshop: Geoindicators. Focusing on geoindicators of relevance to Eastern and Central Europe, Vilnius, Lithuania, October 11–16, pp 40–42. http://www.lgt.lt/geoin/files/vilnius_18.pdf

Söderberg P, Flodén T (1991) Pockmark developments along a deep crustal structure in the northern Stockholm Archipelago, Baltic Sea. Beiträge zur Meereskunde 62:79–102

Spiridonov MA, Rybalko AE, Butylin VP, Spiridonova EA, Zhamoida VA, Moskalenko PE (1988) Modern data, facts and views on the geological evolution of the Gulf of Finland. Geological Survey of Finland Special Paper 6:95–100

Spiridonov MA, Ryabchuk DV, Shachverdov VA, Zvezdunov SI, Nesterova EN, Suslov G, Grigoriev A (2004) The Neva Bay environmental geology, 181 pp. St. Petersburg (in Russian)

Spiridonov MA, Ryabchuk DV, Kotilainen A, Vallius H, Nesterova EN, Zhamoida VA (2007) The Quaternary deposits of the eastern Gulf of Finland. Geological Survey of Finland Special Paper 45:5–17. Espoo

Spiridonov MA, Ryabchuk DV, Kropachev YP, Nesterova EN, Zhamoida VA, Vallius H, Kotilainen A (2008) A new map of the Neva Bay bottom sediments and technogenic objects under the result of side-scan sonar profiling. Polish Geological Institute Special Papers 23: 95–107

Suslov GA, Ryabchuk DV, Nesterova EN, Fedorova EK (2008) Development of the southern coastal zone of the eastern Gulf of Finland (from Lebyazhye to the St. Petersburg flood protective dam). Polish Geological Institute Special Papers 23:109–116

Uscinowicz S, Zachowicz J, Graniczny M, Dobracki R (2004) Geological structure of the southern Baltic coast and related hazards. Polish Geological Institute Special Papers 15:95–107

Valdmann A, Kaard A, Kelpsaite L, Kurennoy D, Soomere T (2008) Marine coastal hazards for the eastern coasts of the Baltic Sea. Baltica 31(1–2):3–12

TAIEX (2008) Workshop on sea-level rise and climate change – changing processes and sustainable management of the low-lying coasts of the Baltic states, Poland and Russia – INFRA 26246 (2008) [CD-ROM], TAIEX, DG Enlargement, Palanga, Latvia

Yaduta VA (2002) The modern tectonics and stability of geological environment of the Gulf of Finland. Abstracts of all-Russian conference on the quaternary period study, 15–18. Smolensk (in Russian)

Yaduta VA (2003) Modern tectonic movements of the Earth crust in St. Petersburg and their role as geological hazards. Geologists towards 300-anniversary of St. Petersburg, pp 113–125 (in Russian)

Zhamoida VA, Ryabchuk DV, Kropatchev YP, Kurennoy D, Boldyrev VL, Sivkov VV (2009) Recent sedimentation processes in the coastal zone of the Curonian Spit (Kaliningrad region, Baltic Sea). Zeitschrift der Deutschen Gesellschaft für Geowissenschaften 160:143–157

Zilinskas G (2005) Trends in dynamic processes along the Lithuanian Baltic coast. Acta Zoologica Lituanica 15(2):204–207

Chapter 17
Seafloor Desertification – A Future Scenario for the Gulf of Finland?

Henry Vallius, Vladimir Zhamoida, Aarno Kotilainen, and Daria Ryabchuk

Abstract The Gulf of Finland is a shallow semi-enclosed sea area which due to strong anthropogenic pressure and poor water exchange is very sensitive to eutrophication. During its whole postglacial history, the seafloor of the gulf has been periodically anoxic, and anoxia below halocline can thus be seen as a natural phenomenon. During the last decades, however, this has been accompanied by a yearly repeated seasonal anoxia in the shallower basins above halocline. This yearly repeated shallower anoxia is triggered by substantial eutrophication of the sea and is a clear signal of anthropogenic pressure. The seasonal anoxia has during the last decades propagated to basins with water depths less than 20 m. The areal coverage of anoxia has thus expanded substantially. Phosphorus which is bound to oxic seafloor sediments is easily released during episodes of anoxia, which further intensifies eutrophication. It has been estimated that the concretion fields of the eastern Gulf of Finland, only, contain more than 330,000 tons of P_2O_5 which is equal to some 175,000 tons of elementary phosphorus. In case of shallowing of the area of permanent anoxia, these concretion fields would become anoxic, which would lead to rather rapid dissolution of the concretions and a release of a large amount of phosphorus together with the heavy metals which today are bound to the concretions.

Keywords Gulf of Finland · Baltic Sea · Ferromanganese Concretions · Marine sediments · Anoxia · Heavy metals · Phosphorus

17.1 Introduction

The Baltic Sea is a European epicontinental sea, one of the world's largest brackish water areas. It has connection to the Atlantic Ocean only through the narrow Danish Sounds. The water depth is moderate, only 52 m on average. There is no

H. Vallius (✉)
Geological Survey of Finland, FIN-02151 Espoo, Finland
e-mail: henry.vallius@gtk.fi

J. Harff et al. (eds.), *The Baltic Sea Basin*, Central and Eastern European Development Studies (CEEDES), DOI 10.1007/978-3-642-17220-5_17,

tide and the sea is partly ice-covered during the winter months. Eutrophication, together with chemical pollution, over-fishing, alien species and global warming are the major environmental problems threatening the Baltic Sea and its fragile ecosystem. Eutrophication is a severe environmental problem especially in the Gulf of Finland, which is a shallow brackish eastward extension of the main Baltic Sea, with an average water depth of 35 m. The Gulf of Finland is badly stressed by the population of almost 15 million people in its catchment area, the largest city being St. Petersburg with more than 4.5 million inhabitants. As eutrophication is accelerated by human activity, seasonal anoxia has turned into a more or less normal state of the seafloor of the Gulf of Finland, which brings many problems into light.

17.2 Study Area and Characteristics of the Gulf of Finland

This study deals with the Gulf of Finland, which is an eastward extension of the main Baltic Sea (Fig. 17.1), but the scenario presented in this chapter can be applied to other similar areas.

The area of the Gulf of Finland is slightly less than 30,000 km^2 and the average depth is 35 m (Vallius 1999). The salinity is very low, from close to zero in the east to a maximum of 8 PSU in the bottom waters of the western Gulf of Finland

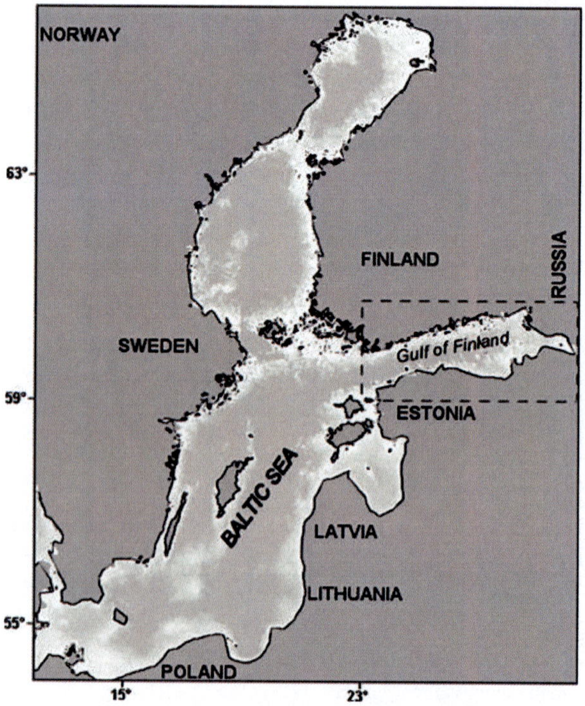

Fig. 17.1 The Baltic Sea with the Gulf of Finland indicated

(Tamsalu and Myhrberg 1995). Salinity stratification is normally strong in the gulf, with a halocline at about 60 m.

The bottom of the Gulf of Finland is made up of different kinds of sediments. The basins of modern accumulation differ in shape and size depending on which area is observed. They are separated from each other by submarine shallows made up of coarser material, such as gravel, sand, and especially glacial till which can be seen as different moraine formations. Sometimes, the bedrock penetrates the sediment as outcrops on different water depths.

It has been assumed by Kankaanpää et al. (1997) that sedimentation basins with active sedimentation would cover about 1/4 of the total area of the Gulf of Finland. Experience from echo soundings in the Gulf of Finland has revealed that the assumption is probably slightly overestimated, but it is the best guess in the absence of full coverage data. Thus, we can assume that the areal coverage of soft Holocene mud, which today is actively incorporated in binding/release of matter to/from the water phase, is about 25%. That is the area of all sediments which react very fast to changes in physico-chemical conditions, some 7,000 km^2 according to the estimate by Kankaanpää et al. (1997).

Especially in the eastern Gulf of Finland, ferromanganese concretions are rather common (Fig. 17.2). They occur on moderate depths where there is a gentle slope of clay usually ending in a basin of modern gyttja-clay accumulation. Usually the abundant concretion areas are located on the edge of the accumulation basin so that the concretions are partly covered with gyttja-clay. The conditions during growth of the concretions have to be permanently oxic. As soon as conditions change to anoxic either because of changes in the surrounding area or because of burial of the concretions into the sediment, they start to dissolve (Zhamoida et al. 2007).

One characteristic feature of the Baltic Sea as well as the Gulf of Finland is that through its entire marine postglacial history it has been occasionally anoxic (hypoxic) (Kotilainen et al. 2000, Winterhalter 2001, Zillén et al. 2008). This cyclic

Fig. 17.2 Spheroidal concretions from the eastern Gulf of Finland. Diameter of sieve is 25 cm (photo H. Vallius)

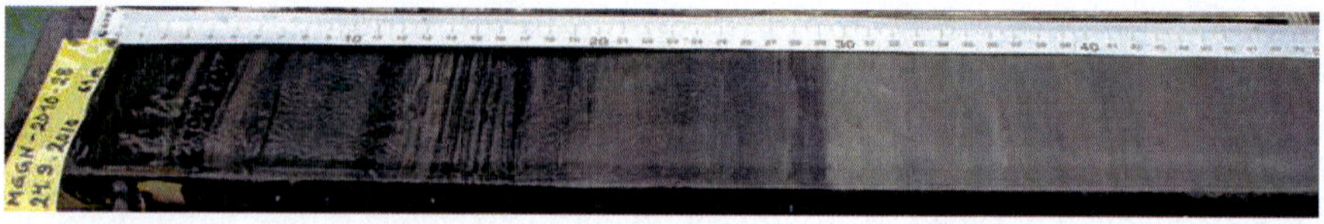

Fig. 17.3 Repeated oxic – anoxic conditions can clearly be seen in vertical sediment profiles, here core MGGN-2010-28 from the central Gulf of Finland (Photo H. Vallius/GTK)

feature is partly an indication of occasional inflows of saline water from the North Sea. Normally the hypoxic conditions between the inflows are present for some years or perhaps a decade at bottoms below the halocline. Anoxia starts after a period of stagnation, when no new saltwater inflows from the Danish Sounds have occurred for a while. The anoxia finally ends when new oxygen-rich saltwater inflows ventilate the bottom waters of the Baltic Sea, which later are pushed into the Gulf of Finland. This kind of cyclic changes can be considered normal in the Baltic Sea and are normally seen as alternating laminated and homogeneous sequences in sediment cores (Fig. 17.3). Anoxia in the Gulf of Finland can, however, be divided into two different kinds of anoxia, the more or less permanent anoxia below the halocline and the seasonal, short-term, anoxia/hypoxia which occurs more or less yearly. Both these will be explained later.

17.3 Materials and Methods

This study is based on the data of a multitude of cruises of Finnish and Russian research vessels in the Gulf of Finland arranged by the Geological Survey of Finland (GTK) and the A.P. Karpinsky Russian Research Geological Institute (VSEGEI) during the last decades. The seafloor has been surveyed using different techniques, such as echo sounding, side scan sonar, shallow seismics, and still photography, as well as observation with ROV cameras, not to mention surface sampling with different kinds of samplers. The amount of survey line kilometres, which exceeds many thousands, has been complemented with thousands of surface samples. During each survey, the seafloor has first been mapped by echo sounding after which it has been inspected by different ground truthing methods, including surface sampling. Thus, there is at present a rather good picture of the Holocene sedimentary environment of the Gulf of Finland.

17.4 Present Situation

The Gulf of Finland is unfortunately very sensitive to changes, as the water volume is small because of the shallow water depth of the gulf. Because of the rather strong freshwater inflow from east and inflow of saltier surface water from west, the water column is usually strongly stratified, which connives seafloor anoxia. Rather high organic content of the surface sediments of the soft bottoms increases anoxia through the breakdown of organic matter. Vallius and Leivuori (2003) report the total carbon content of the surface soft sediments of the eastern Gulf of Finland to be normally above 4% and occasionally up to 6% or more. New, still unpublished studies in the same area have revealed areas with surface TC concentrations of more than 10% (Vallius, in preparation). As there is virtually no oxygen exchange between the sea surface and the near-bottom water, because of the strong stratification of the water column, the periods of anoxic bottom waters remain for quite long times. Only occasionally, for example during strong winter storms, at least the shallower bottoms are periodically oxidized. In such cases, these bottoms are oxidized during spring for a short period, in order to be reduced again after the spring diatom bloom (Leivuori and Vallius 1998), when newly accumulated organic matter starts to break down. The periods of oxic conditions are usually too short for most benthic fauna to be able to colonize the bottoms. The spring blooms are usually followed by a short period of less accumulation, to be followed by cyanobacterial blooms during mid-summer weeks (Leivuori and Vallius 1998). There is normally also a third peak of accumulation in autumn; most of the organic matter is, however, accumulated during the spring bloom (Leivuori and Vallius 1998).

During the second half of the past century, the Gulf of Finland turned into a badly eutrophic water body. This has through the process explained above involved regularly repeated anoxic conditions, first in the deeper basins, but later also in shallower areas (Kotilainen et al. 2007) (Fig. 17.4). Thus the areas of seasonal seafloor anoxia have largely extended through the last decades. Seafloor anoxia is a harmful condition at least for benthic animals as they cannot survive in such conditions; only anaerobic bacteria can survive and often they build up mats like carpets on the seafloor (Fig. 17.5) (Vallius 2006). Anoxia also releases phosphorus from the nutrient-loaded seafloor sediments, which additionally increases eutrophication (i.e. internal loading). Unfortunately, the process goes on for a long period of time even if external phosphorus load is radically decreased. This has during the last decades been seen as extensive mid-summer blooms of harmful blue-green algae in the sea areas.

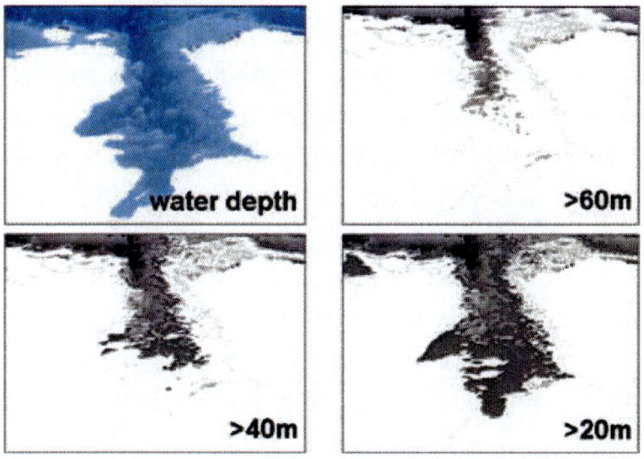

Fig. 17.4 Bathymetry of the Gulf of Finland as seen from east and areas of > 60 m, > 40 m and > 20 m water depth

Fig. 17.5 White bacterial growth on top of core MGGN-2004-1 from station E3 in the central Gulf of Finland, in autumn 2004. Bottom depth 89 m (Photo J. Hämäläinen/GTK)

17.5 Worst Scenario

In the Gulf of Finland, ferromanganese concretions are formed in rather shallow water depths. Due to global warming, permanent anoxia in the Baltic Sea might further expand in future. Thus, even the areas of Fe/Mn concretion growth might be affected, which may cause dissolution of already formed concretionary matter. This will further release more phosphorus into the water column as the concretions normally act as good phosphorus traps. The concentrations of phosphorus in the Baltic Sea Fe/Mn concretions can be up to 2–3% (Winterhalter 2004) or even up to 7.16% in the Gulf of Finland (Zhamoida et al. 1996). It has been calculated that the concretion fields of the eastern Gulf of Finland, only, contain more than 330,000 tons of P_2O_5 (Zhamoida et al. 2007), which is equal to some 175,000 tons of elementary phosphorus (Fig. 17.6). If we speculate that all concretions would be dissoluted in extremely anoxic conditions, this new phosphorus input would strongly contribute to eutrophication and a further seafloor desertification of the Gulf of Finland, a situation probably never seen before during postglacial times. Important to remember in this scenario are also the heavy metals, which normally are well trapped in the concretions in rather high concentrations. According to Emelianov (2004), it seems that especially the concretions on the shallow bottoms (27–53 m) have high heavy metal concentrations. The concentrations of most metals, except copper, are 1.5–5 times higher in the shallow Gulf of Finland concretions compared to average concentrations in the Gulf of Finland seafloor surface gyttja clays (cf. Vallius and Leivuori 1999, 2003). During extreme anoxia, the heavy metals incorporated in the dissolving concretions would be released and their concentrations would rapidly increase in the near-bottom waters. The near-bottom waters would then be overloaded with nutrients as well as thousands of tons of heavy metals (Zhamoida et al. 2004, Emelianov 2004).

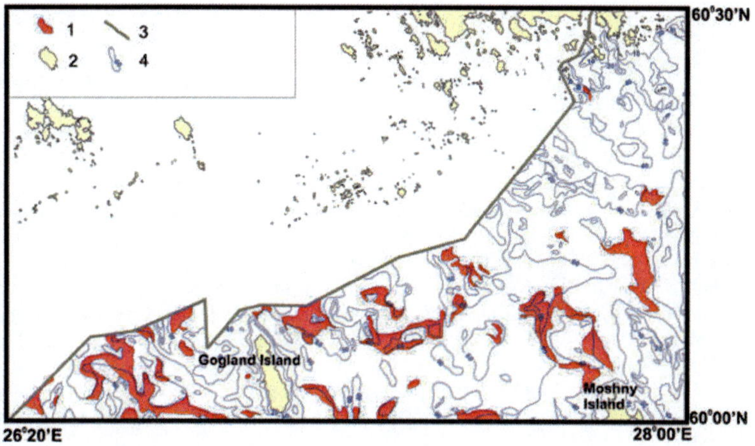

Fig. 17.6 Schematic map of concretion fields in the Russian part of the northern Gulf of Finland. 1 = concretion fields, 2 = islands, 3 = Russian–Finnish border line, 4 = isobath (from Zhamoida et al. 2007)

The scenario explained above is a worst-case scenario and might sound fictitious. But in many sea areas around the world, anoxia has been reported to intensify and extend during the last decades, even though often those areas are not so closed and sensitive as the Baltic Sea and especially the Gulf of Finland. Such episodes have been reported for example by Chan et al. (2008) from the coast of Oregon, Naqvi et al. (2000) from the western coast of India, and Scavia et al. (2003), Justić et al. (2005) and Turner et al. (2006) from the well-known Gulf of Mexico–Mississippi River delta anoxic zone.

Also, when taking into account that the seasonal anoxia in the Gulf of Finland has since the 1950s propagated to shallower bottoms from decade to decade, now reaching bottoms of less than 20 m (Kotilainen et al. 2007), it can also be speculated that the more or less permanent anoxia below the halocline, at about 60 m depth, can propagate to shallower bottoms, if conditions remain favourable for it.

17.6 Discussion

Climate change is affecting the whole human environment. In the seas, these changes are slow and often unpredictable with today's knowledge and methods. Thus it is important to speculate with different scenarios such as the one explained in this chapter. The authors understand that much of the scenario presented here is speculative and perhaps not possible. However, as long as there are such large amounts of phosphorus lying on the seafloor of the eastern Gulf of Finland, these factors should be taken into account in future Gulf of Finland management and political decision making in addition to reduction of external phosphorus loads and other measures.

References

Chan F, Barth JA, Lubchenco J, Kirincich A, Weeks H, Peterson WT, Menge BA (2008) Emergence of anoxia in the California current large marine ecosystem. Science 319:920

Emelianov EM (2004) The Baltic Sea deeps as a model for explaining iron and manganese ore formation. Zeitschrift für Angewandte Geologie, Special Issue 2:161–176

Justić D, Rabalais NN, Turner RE (2005) Coupling between climate variability and coastal eutrophication: evidence and outlook for the northern Gulf of Mexico. Journal of Sea Research 54:25–35

Kankaanpää H, Vallius H, Sandman O, Niemistö L (1997) Determination of recent sedimentation in the Gulf of Finland using ^{137}Cs. Oceanologica Acta 20:823–836

Kotilainen A, Vallius H, Ryabchuk D (2007) Seafloor anoxia and modern laminated sediments in coastal basins of the eastern Gulf of Finland, Baltic Sea. Geological Survey of Finland Special Papers 45:49–62

Kotilainen AT, Saarinen T, Winterhalter B (2000) High-resolution paleomagnetic dating of sediments deposited in the central Baltic Sea during the last 3000 years. Marine Geology 166:51–64

Leivuori M, Vallius H (1998) A case study of seasonal variation in the chemical composition of accumulating suspended sediments in the central Gulf of Finland. Chemosphere 36:2417–2435

Naqvi SWA, Jayakumar DA, Narvekar PV, Naik H, Sarma VVSS, D'Souza W, Joseph S, George MD (2000) Increased marine production of N_2O due to intensifying anoxia on the Indian continental shelf. Nature 408:346–349

Scavia D, Rabalais NN, Turner RE, Justic D, Wiseman J Jr (2003) Predicting the response of Gulf of Mexico hypoxia to variations in Mississippi River nitrogen load. Limnology and Oceanography 48:951–956

Tamsalu R, Myhrberg K (1995) Ecosystem modelling in the Gulf of Finland. I. General features and the hydrodynamic prognostic model FINEST. Estuarine Coast Shelf and Science 41:249–273

Turner RE, Rabalais NN, Justic D (2006) Predicting summer hypoxia in the northern Gulf of Mexico: Riverine N, P, and Si loading. Marine Pollution Bulletin 52:139–148

Vallius H (1999) Recent sediments of the Gulf of Finland: an environment affected by the accumulation of heavy metals. Åbo Akademi University, 111pp

Vallius H, Leivuori M (1999) The distribution of heavy metals and arsenic in recent sediments of the Gulf of Finland. Boreal Environmental Research 4:19–29

Vallius H, Leivuori M (2003) Classification of heavy metal contaminated sediments in the Gulf of Finland. Baltica 16:3–12

Vallius H (2006) Permanent seafloor anoxia in coastal basins of the northwestern Gulf of Finland, Baltic Sea. Ambio 35:105–108

Winterhalter B (2001) On sediment patchiness at the BASYS coring site, Gotland Deep, the Baltic Sea. Baltica 14:18–23

Winterhalter B (2004) Ferromanganese concretions in the Gulf of Bothnia. Mineral resources of the Baltic Sea – exploration, exploitation and sustainable development. Zeitschrift für Angewandte Geologie 2:199–212

Zillén L, Conley DJ, Andrén T, Andrén E, Björck S (2008) Past occurrences of hypoxia in the Baltic Sea and the role of climate variability, environmental change and human impact. Earth Science Review 91:77–92

Zhamoida V, Butylin WP, Glasby GP, Popova IA (1996) The nature of ferromanganese concretions from the eastern Gulf of Finland, Baltic Sea. Marine Georesources Geotechnology 14:161–176

Zhamoida V, Glasby GP, Grigoriev AG, Manuilov SF, Moskalenko PE, Spiridonov MA (2004) Distribution, morphology, composition and economic potential of ferromanganese concretions from the western Gulf of Finland. Zeitschrift für Angewandte Geologie, Specical Issue 2:213–227

Zhamoida V, Grigoriev A, Gruzdov K, Ryabchuk D (2007) The influence of ferromanganese concretions-forming processes in the eastern Gulf of Finland on the marine environment. Geological Survey of Finland Special Papers 45:21–32

Chapter 18
Sources, Dynamics and Management of Phosphorus in a Southern Baltic Estuary

Gerald Schernewski, Thomas Neumann, and Horst Behrendt

Abstract Today, phosphorus is regarded as the key nutrient for Baltic Sea eutrophication management. Major sources are large rivers like the Oder, Vistula and Daugava in the southern Baltic region. Before entering the Baltic Sea, these rivers discharge their nutrient load into coastal estuaries, bays and lagoons. The quantitative role of these coastal waters, with restricted water exchange, for Baltic Sea management is very important, but not well known. Taking the Oder/Odra estuary as an example, we analyse the long-term pollution history and the major sources for phosphorus and calculate a phosphorus budget, with special focus on anoxic phosphorus release from sediments. The budget shows that due to internal eutrophication in July 2000 the lagoon became a major temporary source of phosphorus for the Baltic Sea. A phosphorus emission reduction scenario, taking into account diffuse and point sources in the entire Oder/Odra river basin, is presented. Phosphorus load reductions have only limited effect on the eutrophic state of the lagoon. The lagoon is more sensitive to nitrogen load reductions. Therefore, both elements have to be taken into account in measures to reduce eutrophication.

Keywords Szczecin lagoon · Hypoxia · Eutrophication · Water quality · Nutrient loads · Sediment

18.1 Background and Objectives

The Baltic Sea is one of the world's largest brackish water bodies (412,000 km^2) with a water residence time of about 25–30 years, a drainage basin of 1,734,000 km^2 and a population in the drainage basin of about 85 millions. According to the Baltic Sea Action Plan (HELCOM 2007), "eutrophication is a major problem in the Baltic

G. Schernewski (✉)
Leibniz Institute for Baltic Sea Research Warnemünde, Rostock, Germany
e-mail: gerald.schernewski@io-warnemuende.de

H. Behrendt (Deceased)

J. Harff et al. (eds.), *The Baltic Sea Basin*, Central and Eastern European Development Studies (CEEDES), DOI 10.1007/978-3-642-17220-5_18,

Sea, caused by excessive inputs of nitrogen and phosphorus which mainly originate from inadequately treated sewage, agricultural run-off and airborne emissions from shipping and combustion processes. Eutrophication leads to problems such as intensified algal blooms, murky water, oxygen depletion and lifeless sea bottoms. The plan's objectives are: concentrations of nutrients close to natural levels, clear water, natural levels of algal blooms, natural oxygen levels, and natural distributions and abundance of plants and animals."

Managing eutrophication in the Baltic Sea ecosystem requires a large-scale approach, integrating watersheds, coasts and sea. This knowledge is already reflected in the European Water Framework Directive (WFD) and was adapted by the Baltic Sea Action Plan (HELCOM 2007), which asks the Baltic Sea countries to "develop national programmes, by 2010, designed to achieve the required reductions".

HELCOM (2007) assumes that for a good environmental status (clear water objective), the maximum allowable annual nutrient inputs into the Baltic Sea would be 21,000 t of phosphorus and about 600,000 t of nitrogen. Over the period of 1997–2003, average annual inputs amounted to 36,000 t of phosphorus and 737,000 t of nitrogen. Therefore, annual load reductions of 15,000 t of phosphorus and 135,000 t of nitrogen would be necessary.

Today, phosphorus is regarded as the key nutrient for Baltic Sea eutrophication management (Elmgren and Larsson 2001, Elmgren 2001, Boesch et al. 2006, Wulff et al. 2001). Unlike nitrogen, there are no processes in the Baltic Sea which can compensate phosphorus shortages and it is a potentially limiting nutrient for primary production. In detail, the discussion whether phosphorus alone or nitrogen and phosphorus together control eutrophication is more complex, still controversial, and requires a spatial and temporal in-depth analysis (Conley et al. 2009, Schindler and Hecky 2009).

Over 90% of phosphorus enters the Baltic Sea via rivers and over 50% of the loads enter along the south coast of the Baltic Sea (Helcom 2005). Therefore, rivers like the Oder (Polish: Odra), Vistula and Daugava with large river basins, draining the southern and south-eastern Baltic, are of outstanding importance for Baltic Sea management. Usually rivers do not enter the Baltic Sea directly but discharge their nutrient load into coastal estuaries, bays and lagoons. The quantitative role of these coastal waters, with restricted water exchange, for Baltic Sea management is well known. These systems serve as converters for nutrients, sinks and retention ponds and control the amount and composition of the nutrients entering the Baltic Sea (Lampe 1999, Meyer and Lampe 1999).

The Oder estuary serves as an example for our study. With a length of 854 km, a catchment of 120,000 km^2 and an average water discharge of 17 km^3 (530 m^3/s), the Oder is one of the most important rivers in the Baltic region. It contributes about 10% of the annual phosphorus load to the Baltic Sea. The Oder discharges into the Oder estuary, which consists of the Szczecin (Oder) Lagoon and the Pomeranian Bay (Fig. 18.1). The Oder Lagoon is connected to the Baltic Sea (Pomeranian Bay) via three outlets, has a surface area of 687 km^2 and has an average depth of only 3.8 m. The Oder River contributes at least 94% to the lagoon's water budget.

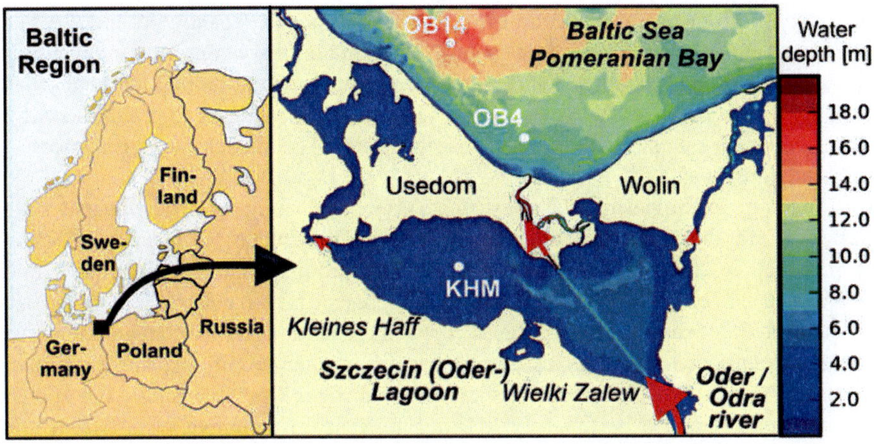

Fig. 18.1 The Oder (Szczecin) Lagoon and the Pomeranian Bay together form the Oder estuary. Indicated are the sampling stations KHM: Oder Lagoon centre, OB4: mouth of the Swina channel and OB14: Pomeranian Bay. The lagoon's maximum length is about 40 km

The objectives of this study are (a) to give an overview about the long-term phosphorus load history in the river and its effects in the estuary, (b) to reflect the present state of integrated river basin–coastal water nutrient modelling and its contribution to our understanding of the phosphorus dynamics in the estuary, (c) to analyse the different phosphorus sources and their quantitative role in the phosphorus availability of the estuary, with special focus on anoxic phosphorus release from sediments, and (d) to discuss the implications for the management of the estuary and the Baltic Sea. For this purpose, we use the long-term monitoring data in combination with a river basin nutrient load model (MONERIS) and a 3D ecosystem model of the estuary (ERGOM).

18.2 Methods and Models

German/Polish monitoring data have been provided by the Wojewódzki Inspektorat Ochrony Środowiska w Szczecinie (WIOS) and the Landesamt für Umwelt, Naturschutz und Geologie (LUNG). The regular hydro-biological and hydrochemical monitoring in the estuary started in the early 1970s. In the lagoon, altogether 12 stations are sampled on the Polish and on the German side; 4 stations exist in the Pomeranian Bay. Nowadays, the sampling frequency is 1 month and intercalibration exercises as well as harmonized sampling dates ensure reliability and comparability. We refer to three sampling stations, KHM in the Oder Lagoon centre, OB4 at the Swina channel mouth and OB14 in the coastal Pomeranian Bay. The stations are indicated in Fig. 18.1.

The ecosystem model ERGOM is an integrated biogeochemical model linked to a 3D circulation model covering the entire Baltic Sea. The circulation model is an application of the Modular Ocean Model (MOM 3) code (Pacanowski and Griffies 2000) and includes an explicit free surface, an open boundary condition to the North Sea and freshwater discharge with rivers. A horizontal resolution of 3 nautical miles was applied in the estuary. However, for time-slice experiments (e.g. 2000–2005), the Oder estuary was resolved with 1 nautical mile. The vertical layer thickness in our study area was 2 m.

The biogeochemical model consists of nine state variables. The nutrient state variables are dissolved ammonium, nitrate and phosphate. Primary production is provided by three functional phytoplankton groups: diatoms, flagellates and cyanobacteria (blue-green algae). Diatoms represent larger cells which grow fast in nutrient-rich conditions. Flagellates represent smaller cells with an advantage at lower nutrient concentrations, especially during summer conditions. The cyanobacteria are able to fix and utilize atmospheric nitrogen, and therefore, the model assumes phosphate to be the only limiting nutrient for cyanobacteria. Due to the ability of nitrogen fixation, cyanobacteria are a nitrogen source for the system. A dynamically developing bulk zooplankton variable provides grazing pressure on phytoplankton. Dead particles are accumulated in a detritus state variable. The detritus is mineralized into dissolved ammonium and phosphate during the sedimentation process. A certain amount of the detritus reaches the bottom, where it is accumulated in the sedimentary detritus. Detritus is buried in the sediment, mineralized or resuspended in the water column, depending on the velocity of near-bottom currents. The development of oxygen in the model is coupled with the biogeochemical processes via stoichiometric ratios. Oxygen concentration controls processes such as denitrification and nitrification. The biogeochemical model is coupled with the circulation model by means of advection diffusion equations for the state variables. To analyse the phosphorus dynamics, a new model version with a more detailed phosphorus and sediment module was applied. In this module, phosphate in the sediment layer and iron oxides can form iron–phosphate compounds under oxic conditions, which precipitate and accumulate in the sediment. Under anoxic conditions, iron–phosphates are reduced and dissolved and phosphates are released into the water body. Depending on the sediment thickness, a portion of the particulate iron–phosphate complexes is buried in the sediments. Neumann (2000), Neumann et al. (2002) and Neumann and Schernewski (2008) provide detailed model descriptions and validations. Weather data were taken from the ERA-40 re-analysis (grid of 50 km and 6-hourly data) for the entire period between 1960 and 2001. For later time period, data from ERA-interim were used.

The model MONERIS was applied to calculate the nutrient inputs and loads in the entire Oder river basin. The model calculates the annual nutrient load into the coastal waters, resulting from point and various diffuse sources. MONERIS is based on a geographical information system (GIS), which includes various digital maps and extensive statistical information. To be able to run the model, large amounts of spatial information had to be compiled and transferred into the GIS, including the river system, catchment and administrative borders, land use classifications, soil maps, topographical information, groundwater tables, hydro-geological and hydro-meteorological information as well as data on atmospheric deposition, river flow and water quality. Details about sources and data quality are given in Behrendt and Dannowski (2005). Point discharges from wastewater treatment plants and industries enter the river system directly, but diffuse emissions into surface waters have very different pathways and are modelled separately. Altogether six diffuse pathways are considered in MONERIS: point sources, atmospheric deposition, erosion, surface run-off, groundwater, tile drainage and paved urban areas. Along the way from the emission source into the river, many transformation, retention and loss processes have to be taken into account.

The use of a GIS allows a regional differentiated quantification of nutrient emissions into river systems. In contrast to the study of Behrendt and Dannowski (2005), the recent results on the long-term changes of the nutrient loads in the Oder have been carried out with a higher spatial resolution of the river basin. Altogether, 484 different river sub-catchments were calculated separately. Later, the data were aggregated for larger units and the entire river system. Because of data availability and funding reasons, detailed model calibrations and validations took place for the period 1993 until 1997 and 1998–2002 (Behrendt et al. 2008). For these periods, detailed and spatially high-resolved data on different phosphorus sources are available and formed the basis for the formulation of the scenario. Details about the model, processes and validations are given in Behrendt and Dannowski (2005). In this study, the output of MONERIS was used as input for ERGOM. We transformed the annual river load data from MONERIS into monthly data, by applying a typical annual dynamic of the nitrogen loads. This means that results with a temporal resolution of a month are in general reliable but do not reflect real conditions.

18.3 Long-Term Pollution History

To a huge extent, the Oder discharge controls the nutrient dynamics in the Oder Lagoon. Dissolved inorganic phosphorus (DIP) concentrations in the Oder River were stable between the 1960s and the early 1970s. They increased afterwards from around 4 to nearly 8 mmol/m^3 between the mid-1980s and the early 1990s and decreased afterwards again (Fig. 18.2). The concentrations reflect this general pattern, but the water discharge is very variable between the years and so are the loads. The concentrations during the late 1980s were not the highest but due to wet years, the loads reached the maximum during that time. In wet years, the P load discharged by the Oder can be up to twice as high as in dry years. The average total phosphorus concentrations in the lagoon showed a decline between the late 1980s (11 mmol/m^3) and the late 1990s (6 mmol/m^3). This load pattern is very well reflected in phosphorus concentrations in the lagoon. The reduction in nutrient contents observed in the early 1990s was largely an effect of the warm, dry years and cannot be attributed to anthropogenic nutrient load reductions (Schernewski and Wielgat 2001).

The DIP concentrations in the lagoon show a strong variability between the years. In some years, the model ERGON is very well able to simulate the concentrations. During the 1990s, this was, for example, true for the years 1994, 1995,

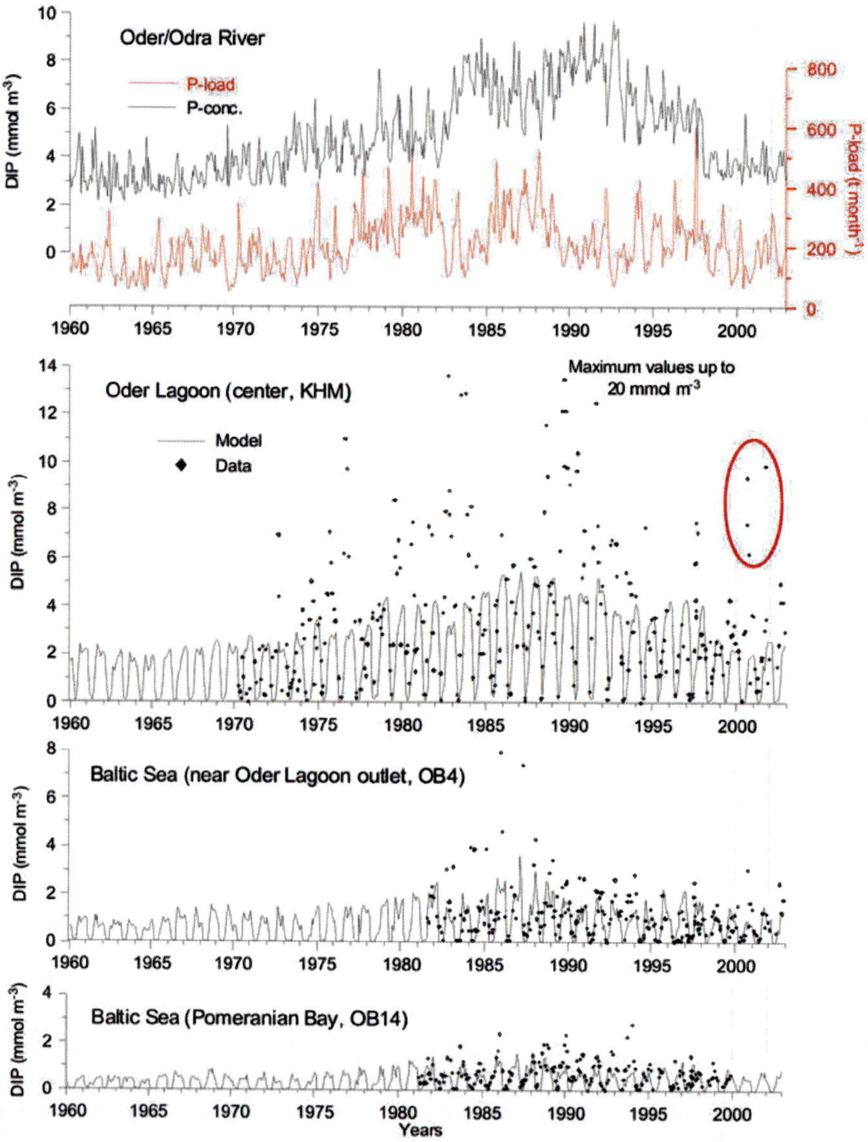

Fig. 18.2 Phosphorus loads (bio-available P) and concentrations (dissolved inorganic P, DIP) in the Oder River and in the estuary. The labels and years indicate the 1st of January. Oder river loads are based on MONERIS model simulations. In the estuary, concentrations simulated with the ERGOM model are aggregated to monthly averages, while the data represent single samplings near the water surface (data source: LUNG, Güstrow). The focus years of this study (2000 and 2001) are indicated

1996 or 1998. In autumn, winter, spring and early summer of most years, the model results are well in agreement with the data. However, between July and September, many years show much higher concentrations (up to 20 mmol/m^3) than is predicted by the model. Nearly all very high concentrations in the lagoon (station KHM) are observed during this period. In the Pomeranian Bay, a few kilometres off the Oder Lagoon outlet (OB4), DIP concentrations are in general much lower, because the bay is part of the Baltic Sea. Extremely high concentrations are rare and the model is well able to reflect the annual pattern. Unusual high concentrations above 3 mmol/m^3 are, in most cases, observed during winter and cannot be easily related to high concentrations in the lagoon. The same is true for the station OB14 in the Pomeranian Bay.

According to the model simulation, the three stations show a strong gradient with declining concentrations from the lagoon towards the outlet and the open bay. However, similarities in the long-term concentration pattern are obvious. The model clearly suggests that high riverine loads, like during the late 1980s, are causing increased DIP concentrations at all three stations in the lagoon as well as in the bay. In the data, this is not visible, but this is very likely an effect of insufficient data sampling frequencies. Further, concentrations simulated with the ERGOM model are aggregated to monthly averages, while the data represent single samplings near the water surface.

18.4 Annual Dynamics and the Role of Sediments

In the following, we focus on the years 2000 and 2001 when outstandingly high phosphorus concentrations in the lagoon have been observed. Figure 18.3 shows the weather conditions during these years and the dates with extreme total phosphorus concentrations. The green bars indicate periods with low average wind speeds of about 3 m/s. It becomes obvious that high phosphorus concentrations occur only between July and September and seem to be linked to calm periods with low wind speeds. Wind direction, temperature or cloudiness, as a measure for global radiation, seems to play only a minor role.

Detailed model results are shown in Fig. 18.4. The model separates the lagoon's water body into two horizontal layers, a bottom layer and a surface layer. Figure 18.4b shows the oxygen concentration in the bottom layer, iron–phosphates in the sediment and the dissolved inorganic phosphorus concentrations in the water body. In early spring of both years, the oxygen concentration shows a decline to values below 6 ml/l which are maintained until autumn. Anoxia in the bottom layer is not observed in the model results. Bottom oxygen concentrations are not included in the German monitoring but data from Polish stations confirm concentrations below 7 ml/l near the sediment in summer. Anoxia was found neither in the Polish monitoring data nor in the model results but has been reported for the central lagoon by scientists for the mid-1990s. As mentioned before, the model ERGON is very

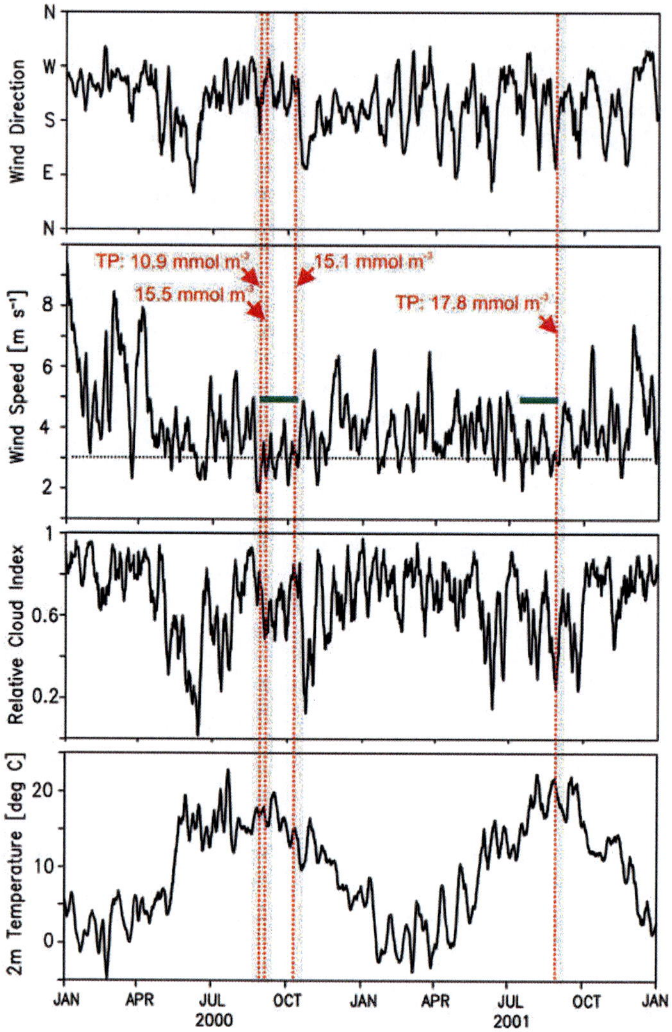

Fig. 18.3 Weather data for the Oder estuary region for the period 2000–2001 (6-hourly data based on a weather model). The data serve as input for the three-dimensional flow and ecological Baltic Sea model ERGOM. Dates with outstandingly high total phosphorus (TP) concentrations in the lagoon are indicated

well able to simulate the phosphorus concentrations in years like 1994, 1995, 1996 or 1998. However, the model fails to simulate the observed, extreme summer phosphorus concentrations above 15 mmol/m^3. Our conclusion is that the model is well able to simulate the mineralization of nutrients from organic material in the sediment and in the water body under oxic conditions. But the model does not predict the fast release of phosphorus from the sediment under anoxic conditions.

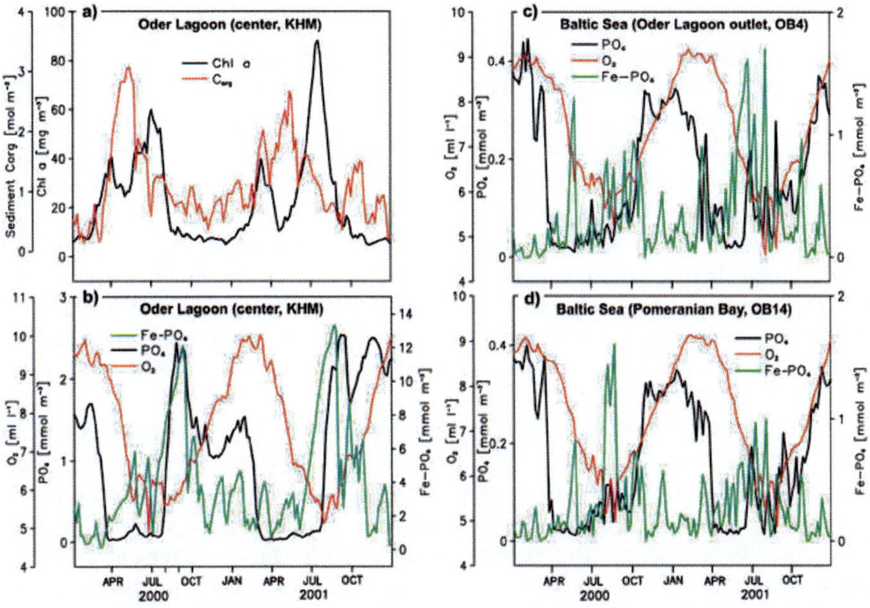

Fig. 18.4 Results of the three-dimensional flow and ecological Baltic Sea model ERGOM for three stations in the estuary during the period 2000–2001. The data were aggregated to 5-day averages a) Central lagoon: Chlorophyll-concentrations in the water and organic carbon concentrations in sediments. b) Central lagoon: Concentrations of iron-phosphate in the sediment, oxygen above the sediment and phosphate in the water body. The same parameters are shown for the Oder Lagoon outlet (c) and the Pomeranian Bay (d)

It is well known that oxygen depletion at the sediment surface can cause a large release of phosphorus from the sediment into the water body, the so-called internal eutrophication. This is especially true if large amounts of phosphorus are bound to iron. This is the case in the lagoon, where Fe concentrations between 2 and 6% were found in surface sediments (Leipe et al. 1998) and Fe redox processes can be expected to play a major role in P dynamics. Dahlke (personal communication) measured several vertical pore water profiles in the sediments in 1994 and 1995. An increase in P concentrations, from about 10 mmol P/m^3 below the sediment surface to 20–40 mmol P/m^3 in a depth of 10 cm, depending on the date, was found. These concentrations are about 2–3 times higher compared to the concentrations in the water body.

About 13 t dissolved phosphorus is stored in the pore water of the upper 10-cm sediment layer of the Kleines Haff and is generally available for a fast release. The surface sediments (0–6 cm) of the Kleines Haff contain about 0.36% P. This concentration decreases with increasing sediment depth. According to the data in Leipe et al. (1998), at least 10,000 t P is stored in the upper 10 cm of the 61% muddy sediment surface of the Kleines Haff. The opposite vertical gradients between pore water and particulate phosphorus suggest that dissolution from particulate phosphorus compounds maintains the pore water gradient and flux towards the surface. During this process the amount of particulate P in deeper sediments decreases with time. High amounts of P bound to iron and a fast P transport from deep sediments are a feature often observed in iron-rich eutrophied lakes (Schernewski 1999). Schernewski and Wielgat (2001) assume that large amounts of phosphorus in sediment are available for a fast anoxic release in the Oder Lagoon. However, experiments failed to prove an anoxic release.

The model results clearly indicate that mineralization cannot be the reason for extreme summerly phosphorus concentrations as well as the fast increase of concentrations in the water body, which has been observed in years like 2000 and 2001. External phosphorus sources cannot be the reason either. Internal eutrophication, the release of phosphorus from the sediment under anoxic conditions, seems to be the only possible process.

According to the model, the bio-available phosphorus concentrations in the water column of the lagoon are always below 0.3 mmol/m^3 between spring and summer. This is well in agreement with the data. Especially in April and May, phosphorus can become a limiting resource for phytoplankton growth. During summer, increasing mineralization causes a sufficient P supply and after June, nitrogen becomes the least abundant resource. This is the reason why in July and August neither the strong ongoing mineralization of phosphorus nor an additional release of P under anoxic conditions has any consequences for the chlorophyll concentration (Fig. 18.4a), which is an indicator of algal biomass. In the coastal Pomeranian Bay (OB4), model and data show DIP concentrations of about 0.2 mmol/m^3 between June and July. Afterwards, the concentrations increase. In September 2000, concentrations exceed 3 mmol/m^3 DIP. Even in the Pomeranian Bay, phosphorus is abundant in summer in 2000 and 2001 and algal biomass would not increase with increased P concentrations. Other years, like 1998 or 1999, are different. Here, phosphorus remains a scarce resource until July. An improved model has to show if the differences between the years in the bay are a result of anoxic processes and internal eutrophication or a result of the high P import from the lagoon. Only an improved spatial resolution of the model will be able to analyse the phosphorus dynamics in detail. Several questions remain, e.g. Are a series of short anoxic events with an accumulation of P in the water body responsible for the high concentrations or was it an anoxic period over several weeks? Can high P concentrations result from intensive short-term sediment resuspension processes and what is the role of iron–phosphate precipitation? On 29 August, 2001, for example, 17.8 mmol/m^3 total phosphorus was observed but only 3 mmol/m^3 DIP. In other years with extreme concentrations, DIP has a much higher share than did particulate fractions.

18.5 Phosphorus Budget in the Lagoon

Apart from anoxic release, the model ERGOM includes all internal and external phosphorus sources and reflects all major processes, which alter the phosphorus content in the water body. Therefore, the model can be used to estimate the anoxic release from the sediment into the lagoon, at least for a period of 1 month. This is not possible for the Pomeranian Bay because we cannot say which processes cause an increase in P concentrations, import from the lagoon or internal eutrophication. The result for the lagoon is shown for July 2000, where we observed a steep increase in phosphorus concentrations in the water column. The difference between the observed and simulated model concentrations is attributed to anoxic release from the sediment.

This budget calculation has many weaknesses and limitations but it gives an idea of the importance of different phosphorus sources. During summer, mineralization processes contribute a similar amount of phosphorus like the Oder River (Fig. 18.5). With 280 t/month, the anoxic release is, in this very special situation, about four times higher than the monthly river load. The lagoon does not serve as a sink for phosphorus anymore but becomes a significant source for the Baltic Sea. About 100 t DIP is additionally released into the Baltic Sea during July 2000.

Fig. 18.5 Budget for dissolved inorganic phosphorus (DIP) in the Oder Lagoon for August 2000. The content is the average during this month. The concentrations of organic phosphorus compounds were largely constant during this month, while DIP concentrations showed a steep increase in the lagoon. The total riverine P load into the lagoon was 246 t and the total P loss to the Baltic Sea was about 340 t. The photographs give an impression of the poor water quality in and tourism at the lagoon in summer

During short periods, the sediment can become a major source and counteract load reduction measures applied in the river basin. This is especially important because the release happens during the summer period and can potentially enhance coastal eutrophication.

18.6 Phosphorus Load Reductions in the River Basin

Compared to the total riverine P loads, internal eutrophication has only a minor importance. Further, it cannot be directly controlled by management measures. Therefore, phosphorus management has to start in the river basin. Figure 18.6 shows the total emissions and the different pathways into the Oder River for three periods, around 1960, 1990 and 2000, based on MONERIS model simulations.

In 1960, settlements, which summarize paved urban areas and sewer systems, and agriculture were of similar importance and contributed nearly 50% to the total emissions of 6,000 t total phosphorus. Until 1990, increasing population and the lack of sewage treatment plants caused a steep increase in point source emissions and a strong increase in total P emissions (>15,000 t). Around the year 2000, the

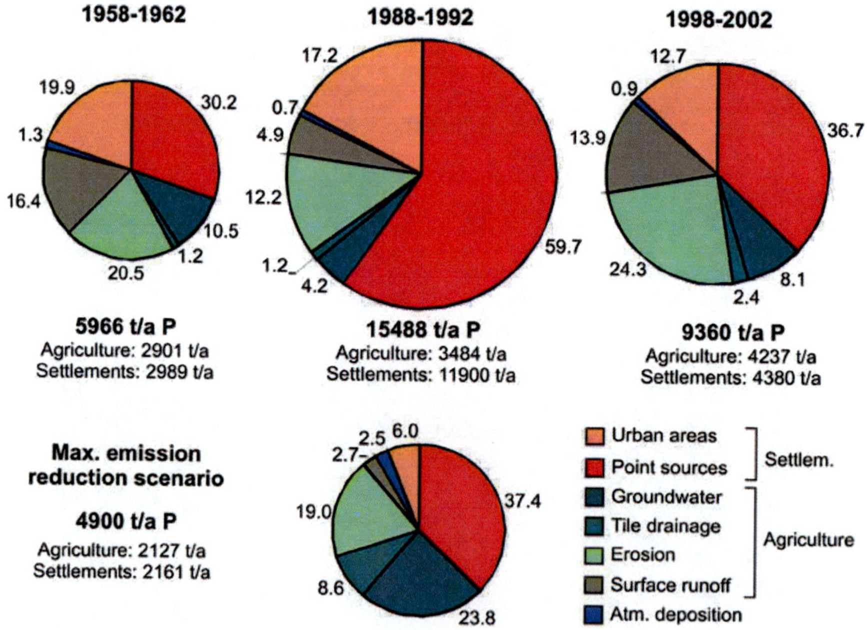

Fig. 18.6 Total phosphorus emissions into the Oder River for three periods as well as the results of a maximum emission reduction scenario. All data are based on MONERIS model calculations. Due to phosphorus retention in the river, only 39% of these emissions or 3,640 t/a (for the period 1998–2002) enter the Oder Lagoon with the river. Twenty-two percent or 2,070 t/a of the total emissions are bio-available for primary production in the lagoon

relative share of agricultural sources and settlements was similar to the 1960 level again and the total emissions decreased again to less than 10,000 t. However, the total emissions were still 50% higher compared to 1960. The Polish part of the Oder basin is responsible for 84% of the P emissions. Eleven and five percent of the P emissions can be attributed to the Czech and German parts of the Oder basin, respectively.

Due to phosphorus retention in the river, only 39% of these emissions or 3,640 t/a (for the period 1998–2002) enter the Oder Lagoon with the river. Twenty-two percent or 2,070 t/a of the total emissions are bio-available for primary production in the lagoon. In 1960, the total P load entering the Oder Lagoon was about 2,400 and increased to 6,200 t around 1990. The load around 2000 was only 60% above the level of 1960.

Behrendt and Dannowski (2005) calculated several emission reduction scenarios, which describe the effect of single management measures or sets of measures in the river basin on the nutrient loads into the Oder River. Figure 18.6 (bottom) shows the results of one scenario, which links all best-practice measures in the river basin and shows to what extent the total phosphorus input into the Oder River can be reduced. Ninety-five percent of the Oder river basin belongs to Poland and the Czech Republic. Both countries are member states of the European Community but their sewage water treatment quality does not yet comply with EC standards.

This optimal scenario is based on the following assumptions and measures: The emissions from point sources in the entire river basin meet the requirements of the Urban Waste Water Treatment Directive (91/271/EEC). The following thresholds shall not be exceeded: Biological oxygen demand (BOD) = 25 mg O_2/l, chemical oxygen demand (COD) = 125 mg O_2/l, SS = 35 mg/l, total phosphorus (TP) = 2 mg/l, total nitrogen (TN) = 15 mg/l for municipalities with a population between 10,000 and 100,000 as well as 1 mg TP/l and 10 mg TN/l for municipalities with more than 100,000 inhabitants. The use of phosphorus-free detergents is postulated in Poland and the Czech Republic.

It is assumed that best management practices on arable land are implemented to reduce the load from diffuse sources. Soil erosion is strongly reduced as well. Conservation tillage is applied on all arable land in the Oder basin. The nutrient load reduction in this scenario is realistic and the required measures could be implemented during the next two decades.

Despite the fact that only around 5% of the present loads can be regarded as natural background emissions, the potential for emission reductions is limited. For this scenario, emissions of 4,900 t TP are calculated. This means that a reduction of about 47% of the emissions in 2000 seems possible or an emission reduction of about 18% compared to 1960.

What does this mean for the water quality in the Oder Lagoon? A comparison of model results for the years 2000 and 2001 with the years 1960, 1961 and 1962, where the annual load was only about 20% above the loads in the emission reduction scenario, can give an impression of how the ecosystem would react to load reductions. However, the effect of changes in nitrogen loads and shifts in N/P ratios cannot be taken into account. In general, the DIP concentrations in April and May of the early 1960s were significantly below 0.1 mmol/m^3 DIP. In 2000 and 2001, the concentrations were higher during these months but remained below 0.1 mmol/m^3 as well. Altogether, the period with very low DIP concentrations was not longer in the 1960s and significant limiting effects on algae biomass were not obvious. However, more detailed studies are necessary to answer this question finally.

18.7 Discussion and Conclusion

The application of the 3D flow and ecosystem model ERGOM in the Oder estuary is a clear step forward compared to the box model approach in the lagoon (Wielgat and Witek 2004). The new sediment module has the potential to serve as an important tool to understand anoxic processes and the exchange between sediment and water body. The model is well able to simulate the long-term behaviour of the estuary and the impact of changing loads and can be regarded as a reliable tool. However, the model in its present state is not able to simulate short-term anoxic sediment processes. The process formulation is not the major shortcoming. Problems result from the model's horizontal and vertical resolution. A horizontal grid significantly below 500 km seems necessary to calculate the exchange between lagoon and Baltic Sea. Four to five vertical layers instead of two will be needed to simulate short-term oxygen depletion above the sediment and simultaneous inflow of Baltic water and outflow of lagoon water into the lagoon. Further, the accuracy as well as the spatial and temporal resolution of input data now becomes a limiting factor for the quality of the model performance. Wind data with hourly resolution, provided in a 5-km grid, will be necessary to simulate short-term stratifications in the water column.

The model allowed the calculation of a coarse phosphorus budget for July 2000 and the quantification of internal eutrophication. Due to a strong riverine phosphorus load reduction during the last decade, the process of internal eutrophication gained relative importance. The model clearly shows that riverine loads and internal processes in the lagoon influence the coastal Baltic Sea. The lagoon serves as a converter, sink and sometimes as a source of nutrients for the Baltic Sea. However, in the present state, the model is not able to simulate the consequences of internal eutrophication in the lagoon during summer on the ecology of the coastal sea.

The Oder example shows that nutrient management between land and sea requires a comprehensive approach, has to link external and internal management measures and has to follow guiding principles. First, the application of nutrients on terrestrial systems and their loss to the sea has to be minimized. Second, nutrient cycles have to be established and/or strengthened.

Nutrients are used as fertilizer in agriculture and are partly lost to ground and surface waters and end up in the river and finally in the sea. The application of fertilizer and agricultural practice has to be optimized to reduce the loss. Measures in the river basin can increase the retention of nutrients. Denitrification in wetlands and tile drainage systems are examples. Vegetated strips along watercourses to reduce run-off and sediment input are another example.

The results show that these measures in combination can reduce the phosphorus loads below the loads of the year 1960. However, in the early 1960s, the lagoon was already in a highly eutrophic state. It is very likely that for an efficient protection of the Baltic Sea and a significant improvement of coastal water quality, additional measures in the coastal waters themselves have to be applied. Possible additional measures are mussel farms, managed mussel beds and enlarged natural mussel beds, algal farms, increased reed belts (supported by pile rows) and extended submersed macrophyte areas and/or dredging of sediment and dumping on land. With the mussel or algal harvest, the nutrients would be removed back to the land and would end up as fertilizer in agriculture. The nutrient cycle would be closed.

The results indicate that a phosphorus load reduction has only limited effect on the eutrophic state of the lagoon. The lagoon is much more sensitive to nitrogen load reductions. We strongly support Conley et al. (2009), who consider both nitrogen and phosphorus as controlling elements for coastal and marine eutrophication and ask for measures which reduce the loads of both elements. Integrated management should always take all nutrients into account and should not merely focus on the needs of the open Baltic Sea. Our results show that coastal systems have their own dynamics, are important sinks and transformators and also act as temporary sources for nutrients. A detailed understanding of the behaviour of coastal systems is imperative for management.

Acknowedgements This chapter is dedicated to Horst Behrendt, who died, much too early, in December 2008. The work has been supported by the projects IKZM-Oder III (Federal Ministry for Education and Research; 03F0403A & 03F0465A) and BONUS+ project AMBER (Assessment and Modelling Baltic Ecosystem Response). Data have been kindly supplied by the State Agency of Environment, Protection of Nature and Geology Mecklenburg-Vorpommern (LUNG). Supercomputing power was provided by HLRN (Norddeutscher Verbund für Hoch- und Höchstleistungsrechnen).

References

Behrendt H, Dannowski R (eds) (2005) Nutrients and heavy metals in the Odra river system. Weißensee Verlag, Berlin

Behrendt H, Opitz D, Kolanek A, Korol R, Stronska M (2008) Changes of the nutrient loads of the Odra River during the last century – their causes and consequences. Journal of Water Land Development 12:127–144

Boesch D, Hecky R, O'Melia C, Schindler D, Seitzinger S (2006) Eutrophication of Swedish seas. Swedish Environmental Protection Agency, Naturvårdsverket, Stockholm, Sweden, ISBN 91-620-5509-7

Conley DJ, Paerl HW, Howarth RW, Boesch DF, Seitzinger SP, Havens KE, Lancelot C, Likens GE (2009) Controlling eutrophication: nitrogen and phosphorus. Science 323:1014–1015

Elmgren R, Larsson U (2001) Eutrophication in the Baltic Sea area. In: Bodungen B, Turner RK (eds) Science and integrated coastal management. Dahlem University Press, Berlin, pp 15–35

Elmgren R (2001) Understanding human impact on the Baltic ecosystem: changing views in recent decades. Ambio 30:222–229

Helsinki Commission (Helcom) (2005) Airborne nitrogen loads to the Baltic Sea. Report, pp 24

HELCOM (2007) Baltic Sea action plan, www.helcom.fi/BSAP/ActionPlan/en_GB/ActionPlan/. Accessed 26 November 2010

Lampe R (1999) The Odra estuary as a filter and transformation area. Acta hydrochimica et hydrobiologica 27:292–297

Leipe T, Eidam J, Lampe R, Meyer H, Neumann T, Odsadczuk A, Janke W, Puff T, Blanz T, Gingele FX, Dannenberger D, Witt G (1998) Das Oderhaff – Beiträge zur Rekonstruktion der holozänen geologischen und anthropogenen Beeinflussung des Oder-Ästuares. Meereswiss. Berichte No. 28, 61 S

Meyer H, Lampe R (1999) The restricted buffer capacity of a South Baltic estuary – the Oder estuary. Limnologica 29:242–248

Neumann T (2000) Towards a 3D-ecosystem model of the Baltic Sea. Journal of Marine System 25(3–4):405–419

Neumann T, Fennel W, Kremp C (2002) Experimental simulations with an ecosystem model of the Baltic Sea: a nutrient load reduction experiment. Global Biogeochemical Cycles 16(7-1):7–19

Neumann T, Schernewski G (2008) Eutrophication in the Baltic Sea and shifts in nitrogen fixation analyzed with a 3D ecosystem model. Journal of Marine System 74:592–602

Pacanowski RC, Griffies SM (2000) MOM 3.0 manual. Technical report, Geophysical Fluid Dynamics Laboratory

Schernewski G (1999) Der Stoffhaushalt von Seen: Bedeutung zeitlicher Variabilität und räumlicher Heterogeniät von Prozessen sowie des Betrachtungsmaßstabs. Marine Science Reports 36:275

Schernewski G, Wielgat M (2001) Eutrophication of the shallow Szczecion Lagoon (Baltic Sea): modeling, management and the impact of weather. In: Brebbia CA (ed) Coastal engineering: computer modelling of seas and coastal regions. WIT Press, Southampton, pp 87–98

Schindler DW, Hecky RE (2009) Eutrophication: more nitrogen data needed. Science 324:721

Wielgat M, Witek Z (2004) A dynamic box model of the Szczecin Lagoon nutrient cycling and its first application to the calculation of the nutrient budget. In: Schernewski G, Dolch T (eds) The Oder estuary, against the background of the Water Framework Directive. Marine Science Reports 57:99–125

Wulff F, Bonsdorff E, Gren I-M, Johansson S, Stigebrandt A (2001) Giving advice on cost effective measurements for a cleaner Baltic Sea: a challenge for science. Ambio 30:254–259

Part VII
Hydrogeological Modeling

Chapter 19
Potential Change in Groundwater Discharge as Response to Varying Climatic Conditions – An Experimental Model Study at Catchment Scale

Maria-Theresia Schafmeister and Andreas Darsow

Abstract The possible change in groundwater discharge from a medium-scale catchment to the Baltic is studied by means of a numerical groundwater flow model. The test area northeast of Wismar (Mecklenburg-Vorpommern, Germany) is built by quaternary glaciofluvial sands and intercalated tills. Today's groundwater recharge is calculated as 24% of the recent average annual precipitation of 600 mm in the test area, and its submarine groundwater discharge is modelled to 14.3% of the precipitation. Based on climate scenarios calculated by the Swedish Meteorological and Hydrological Institute (SMHI) and the Hadley Centre (HC) three sea-level scenarios in combination with four precipitation scenarios are modelled for steady-state groundwater conditions in order to assess potential response in discharge. The temporal development is observed in a simplified schematic model for transient conditions. For the given conditions the influence of sea-level rise is almost not noticeable. However, the modelled scenarios indicate that changes in groundwater recharge as a consequence of climate-induced changes in precipitation lead to notable variations of submarine groundwater discharge.

Keywords Hydrogeology · Climate change · Coastal aquifers · Submarine discharge · Freshwater resources

19.1 Introduction

Climate change will undoubtedly affect the coastal regions of the Northern hemisphere. Changes in temperature, precipitation and sea level have a strong influence on the hydrodynamics of coastal aquifers.

At the Baltic Sea region the climate change effects are superimposed on the isostatic crustal movement (Harff et al. 2005). A detailed description of the effects

M.-Th. Schafmeister (✉)
Institute for Geography and Geology, University of Greifswald, 17489 Greifswald, Germany
e-mail: schaf@uni-greifswald.de

J. Harff et al. (eds.), *The Baltic Sea Basin*, Central and Eastern European Development Studies (CEEDES), DOI 10.1007/978-3-642-17220-5_19,

to the long-term mean sea level at the Baltic Sea coast is given by Meier et al. (2004). There are three effects: (i) the isostatic uplift of the land, (ii) the eustatic sea-level rise and (iii) the water balance of the Baltic Sea (Johansson et al. 2003). In addition other effects are described (Sherif and Singh 1999), i.e. the expansion of the ocean due to a warm-up and the melting of ice sheets and glaciers will increase the total volume of the ocean, and both processes will have a big impact on sea level. However, climate change will also affect groundwater resources by the change in rainfall intensity and geographical distribution of rainfall resulting in a change in groundwater recharge (Sherif and Singh 1999).

Not all effects and resulting impacts to the geosphere are adequately understood at the moment but due to the importance of groundwater as a major resource of global water source they cannot be neglected. Groundwater discharge is a key factor controlling water table conditions, surface and groundwater quality, lake levels and baseflow of rivers and streams.

Recently the contribution of terrestrial groundwater and its potential load of nutrients and other contaminants are increasingly discussed, but little is known on the variety of discharge processes and how they depend on geological and hydrogeological conditions. However, it is expected that groundwater discharge will change quantitatively and qualitatively in response to changes in land use, groundwater recharge, groundwater management and civil engineering hydraulic activities in coastal areas induced by climate change (IPCC 2007).

Different approaches are currently used to quantify groundwater discharge; either the direct outflow is measured in situ or the submarine groundwater discharge is quantified as a component of the coastal water budget. In the first approach seepage meters are used to measure discharge rates and to collect samples in order to assess the hydrochemical composition. However, the support of this method is limited to meter scale or less. Balancing the water budget, on the other hand, yields spatially integrated values, which may be useful in order to predict total mass loads. In this study the latter approach is followed.

The northern border of the state of Mecklenburg-Vorpommern comprises 340 km of the southern Baltic coast. It shows all characteristics of a typical simplification coast with its long sandy bars and lagoons (e.g. Island Usedom) and its steep slopes of glacial tills (e.g. Island Poel). It is therefore the result of a long and highly dynamic geological history which started with the Littorina transgression (7,800 years BP) and is still ongoing. As a consequence the coastline, which builds the interface between the brackish sea water and the terrestrial fresh water (surface and subsurface), has moved landwards (Meyer 2003). This process still holds on.

The discussion on the future evolution of the sea level and its consequences is receiving more and more attention. Within the framework of the IPCC (Special Report on Emission Scenarios) 40 standard scenarios (SRES) of greenhouse gas emission have been published for the period from 1990 until 2100. The SRES scenarios are based on various estimates of demographic and socio-economic developments. Four groups of scenarios, A1, A2, B1 and B2, are distinguished (Nakicenovic et al. 2000) and transferred to the regional conditions in the Baltic Sea region (Meier et al. 2004).

The Swedish Meteorological and Hydrological Institute (SMHI) has calculated four future climatic scenarios and three sea-level heights up to the year 2100 for the Baltic Sea region based on two driving global models from the Hadley Centre (HC) and the Max Planck Institute for Meteorology (MPI) which are combined with the two IPCC emission scenarios (A2 and B2). Accordingly three sea-level scenarios for the southern Baltic, 0.24, 0.41 and 0.82 m, and four precipitation scenarios, +1% (MPIA2), +10% (MPIB2), +5% (HCA2) and –1% (HCB2), of recent precipitation rate are predicted.

The German coast of the Baltic has received much attention with respect to the dynamics of sea levels (Harff et al. 2005). However, the impact of climate change on the water budget of coastal aquifers in this region has not yet been attempted. The general mechanisms of aquifer response to changing climatic conditions at a typical till-dominated coastline still need to be investigated. An existing medium-scale groundwater flow model at the Wismar Bay (Darsow 2004) serves as an example site. The intention of this study is to provide a general understanding of the aquifer response at the southern Baltic. Here the relative importance of sea-level rise and changing groundwater recharge conditions on the amount of coastal groundwater discharge is of major interest.

Based on an existing calibrated groundwater flow model (FE model 'Catchment', Darsow 2004) the specific goals are (i) to quantify groundwater discharge for a typical medium-scale catchment area at the southern Baltic border for today's climatic conditions, (ii) to calculate groundwater discharge in the same catchment for different combinations of the predicted sea levels and precipitation rates and (iii) to analyse the change in groundwater discharge during a period of 100 years of increasing precipitation rates and/or rising sea level by means of transient simulation for a simplified model (FD model 'Simple').

19.2 Materials and Methods

A simple balance concept is used to derive the long-term submarine groundwater discharge rate from the hydrologic budget equation. This equation is applied to a coastal catchment at the border of the Baltic Sea. A groundwater flow model is calibrated for recent climatic and hydrologic conditions with respect to sea level and aerially varying groundwater recharge. Based on this, steady-state conditions of predicted climatic scenarios are calculated and the change in submarine groundwater discharge is determined. Finally a simplified FD model is used to analyse the development of submarine groundwater discharge for transient conditions.

19.2.1 Balance Concept

The global long-term water budget is given by Eqs. (1) and (2)

$$P = \mathrm{ET} + Q \tag{1}$$

and

$$Q = Q_S + Q_{GW} \tag{2}$$

with P precipitation, ET evapotranspiration, Q discharge, Q_S surface runoff and Q_{GW} groundwater discharge. For coastal areas we can assume that the discharge Q drains directly into the sea. Assuming that (i) the surface and subsurface catchment areas are equal, (ii) overland flow can be neglected and rivers are fed by groundwater only and (iii) no other sources or sinks exist, we can reformulate that the total discharge equals the amount of groundwater recharge R_{GW} (3):

$$P - \mathrm{ET} = Q = R_{GW} \tag{3}$$

Then the submarine groundwater discharge Q_{SGD} is the difference between recharge and surface runoff Q_S (4):

$$Q = Q_S + Q_{SGD} \tag{4}$$

Compared to diffuse submarine groundwater discharge, the runoff in streams can be measured with relative ease and accuracy.

19.2.2 Groundwater Recharge Assessment

Groundwater recharge is a key parameter in groundwater budgets but is difficult to assess. Scanlon et al. (2002) offer an excellent review of approaches to quantify groundwater recharge. Methods of groundwater recharge assessment based on budget considerations provide integral results and thus lack desirable accuracy with respect to spatial and temporal resolution. Direct measurements, e.g. by lysimeters, are point supported and require appropriate regionalization if larger catchments are considered.

The northern part of Germany is characterized by a generally flat topography, unconsolidated glacial sediments, predominant agricultural land use and humid climate conditions. Groundwater recharge for these conditions was found to be well estimated by areal differentiation methods which are based on empirical regression equations and which can easily be implemented in GIS systems. The method of Renger and Wessolek (1990) has proven to provide reliable results of groundwater recharge as difference between annual precipitation and real evapotranspiration.

Renger and Wessolek (1990) estimate the annual real evapotranspiration as follows:

$$\mathrm{ET_r} = a \cdot P_S + b \cdot P_W + c \cdot \log W_{Pl} + d \cdot \mathrm{ET_p} + e \tag{5}$$

$\mathrm{ET_r}$ – real evapotranspiration
$\mathrm{ET_p}$ – potential evapotranspiration
P_S – precipitation summer
P_W – precipitation winter
W_{Pl} – amount of water useable for plants stored in soil.

This empirical relation is a function of summer and winter precipitation, soil moisture and potential evapotranspiration after Haude (1955). The coefficients (*a–e*) differ depending on land use and soil types. Relevant data are provided by weather services and publicly available databases, e.g. Corine Landcover (2000) or Global Landcover (2000). Equation (5) is implemented in a GIS scheme thus combining the input data and resulting in aerially differentiated maps of groundwater recharge (Meyer and Tesmer 2000). Since groundwater recharge is calculated on an annual basis seasonal variations of groundwater recharge are not considered in this study.

19.2.3 Numerical Groundwater Models Feflow and Modflow

Groundwater recharge, as a function of precipitation, is the key parameter when considering submarine groundwater discharge. Groundwater discharge to the sea is estimated by means of two different groundwater flow modelling codes.

By means of the finite-element code Feflow Wasy® (Diersch 2005) a three-dimensional high-resolution model of a coastal catchment was established for recent conditions by Darsow (2004). This model (‘Catchment’) serves as the base model, which is then run for different groundwater recharge and sea-level conditions.

The FE model ‘Catchment’ is run for steady-state conditions only, thus neglecting the temporal response to varying boundary conditions, namely groundwater recharge change due to precipitation variation and sea-level rise.

The temporal response is therefore analysed by means of a simplified finite-difference model (‘Simple’) using the Modflow code (Harbaugh et al. 2000). Here a rectangular catchment area is represented by one layer only (two-dimensional) and subdivided into finite-difference cells. The aquifer is assumed to be homogeneous in space. The hydraulic impact of groundwater recharge and surface water bodies is modelled by second and third kind boundary conditions, respectively. However, sea level is modelled by a modified prescribed head boundary, which allows for temporal variation of a first kind boundary condition. Thus the simplified model is able to reflect transient conditions.

19.3 Test Site: Subcatchment at Wismar Bay

The test area is located at the Wismar Bay, at the southwestern coast of the Baltic Sea, just across Poel island. The total area is 140 km^2. The coastline extends about 23 km from SW to NE and is mostly straight, being only interrupted by very few small creeks (Fig. 19.1). The area is generally flat, ranging from 0 m asl at the coast up to 101.6 m asl in the southeast.

The catchment is drained by three gaining streams that flow towards northwest to the coast with a total length of 38.5 km (Fig. 19.1). A reservoir of 1.2×10^6 m^3 is situated in the centre of the area. The climate is humid with an average temperature of 8.4°C. The annual precipitation is 600 mm (non-corrected), of which 275 mm occurs in winter and 325 mm in the summer half year, respectively. The potential

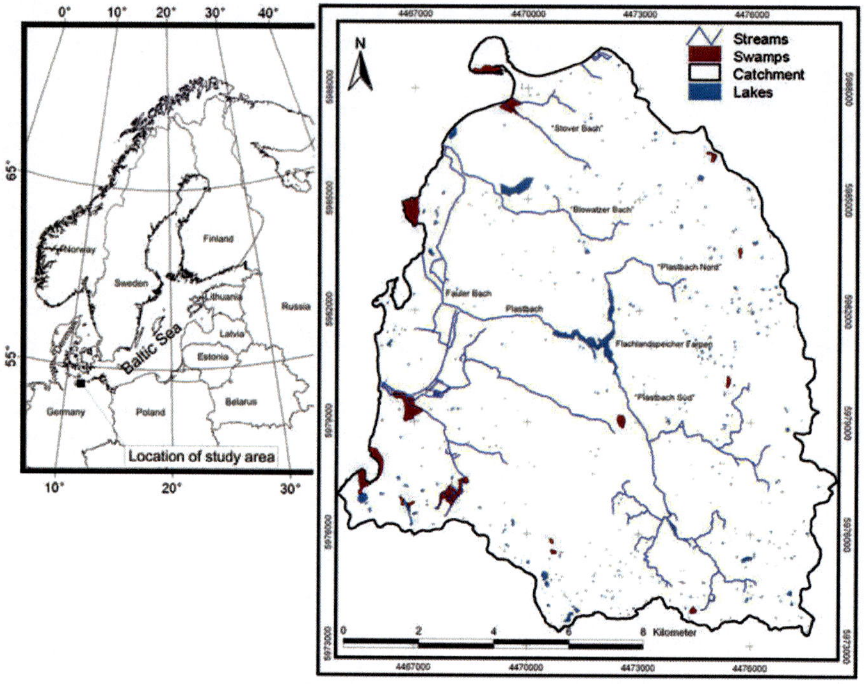

Fig. 19.1 Catchment area at Wismar Bay

evapotranspiration (Haude 1955) is calculated as 575 mm/a. The land use is crop farming on loamy soils and forestry on sands.

Three approximately 10 m thick aquifers are built by glaciofluvial sands of Saalian and Weichselian age. They are separated by up to 20 m thick glacial till layers which locally pinch out. The uppermost aquifer is partly phreatic. Hydraulic gradients of up to 0.5% are slightly steeper than reported for northeast German conditions (Jordan and Weder 1995).

19.4 Model Assumptions and Results

19.4.1 Groundwater Recharge

Given today's climatic conditions, groundwater recharge is estimated as 145 mm/a, i.e. 24% of the average annual precipitation rate of 600 mm (non-corrected). This value is representative for the area which is dominated by arable land and forests on loamy and sandy soils. Assuming that land use will not change, groundwater recharge will vary according to changes in precipitation. Based on the predictions

Table 19.1 Calculated (Renger and Wessolek 1990) annual groundwater recharge

Predicted change in precipitation	–1%	Today	+1%	+5%	+10%
Climate scenario	HCB2	–	MPIA2	HCA2	MPIB2
P (mm/a)	594	600	606	630	660
ET_r (mm/a)	453	455	457	464	473
R_{GW} (mm/a)	141	145	150	166	187
Change in estimated groundwater recharge	–3%	–	+3%	+14%	+29%

of –1, +1, +5 and +10%, respectively, groundwater recharge was estimated to change accordingly (Table 19.1).

For the given conditions of soils and land use, an increase in precipitation by 10% leads to an increase in groundwater recharge of 30% relative to today. Similarly any percentage change in precipitation results in a threefold change in percentual groundwater recharge. Note, however, that the calculation of real evapotranspiration in Renger and Wessolek (1990) is based on estimates of potential evapotranspiration not corrected for temperature rise.

19.4.2 Finite-Element Model: 'Catchment'

Recent and predicted groundwater flow conditions are modelled for steady-state conditions by means of Feflow Wasy®.

The area is subdivided into triangular prisms (finite elements) and the steady-state flow equation is calculated at each node. Groundwater recharge is implemented as an areally varying second kind boundary condition (constant flux Neumann condition, special case flux=0, no flow), whereas the sea level is modelled as a first kind (prescribed head Dirichlet condition). The water exchange between surface water bodies and groundwater is modelled as leakage (third kind Cauchy condition), i.e. the exchange rate depends on the head gradient and distance between surface water body and groundwater table and on the hydraulic conductivity of the bottom layer of the surface water body.

The model consists of about 65,075 elements and 40,368 nodes. The model is three-dimensional comprising five layers, i.e. three aquifers separated by two confining layers, which locally pinch out, thus providing direct hydraulic contact between the aquifers. Groundwater recharge is incorporated as given in Table 19.1. Hydraulic conductivity and effective porosity is derived from lithological information from drilling logs (Table 19.2). The average total thickness of the modelled aquifer system is 100 m; the three sandy aquifers comprise about 25–30% of the total thickness.

The base model (Fig. 19.2) reflects the recent spatially varying hydrogeological conditions, e.g. groundwater recharge, hydraulic conductivity and surface water bodies, reasonably well. The modelled hydraulic heads have been calibrated

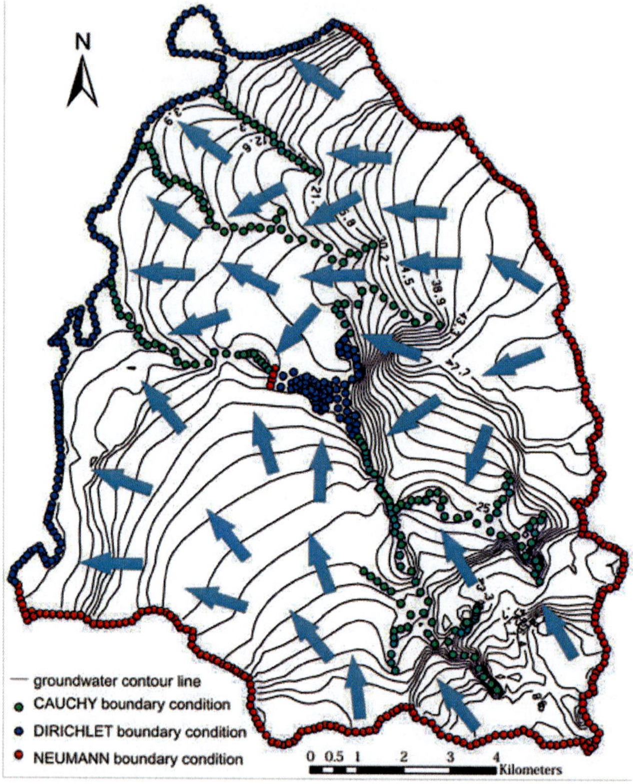

Fig. 19.2 Boundary conditions and calibrated groundwater contours of the model 'Catchment' for today's climatic conditions. The special case of Neumann boundary condition ('no-flow') was used along the watershed to outline the catchments area. *Blue* arrows represent the groundwater flow direction

($R = 0.98$) to 20 groundwater measurement points at different depths by variation of hydraulic conductivity (Darsow 2004). Groundwater recharge, as one of the key parameters in this study, is kept as evaluated by Eq. (5).

Since sink terms other than discharge are not modelled, the total discharge equals groundwater recharge, i.e. 56,000 m^3/day or 146 mm/a. Almost 59%

Table 19.2 Horizontal hydraulic conductivity K and effective porosity n_e used in the groundwater flow model 'Catchment'

	Layer 1 Upper aquifer	Layer 2 Confining layer	Layer 3 Middle aquifer	Layer 4 Confining layer	Layer 5 Lower aquifer
K (m/s)[a]	1.8×10^{-4} -7.2×10^{-4}	3.7×10^{-7} -1.2×10^{-5}	3.4×10^{-4} -5.0×10^{-4}	2.0×10^{-8}	3×10^{-5} -2.5×10^{-4}
n_e	0.2	0.1	0.2	0.1	0.2

[a]Vertical hydraulic conductivity is assumed to be 1/10 of horizontal hydraulic conductivity

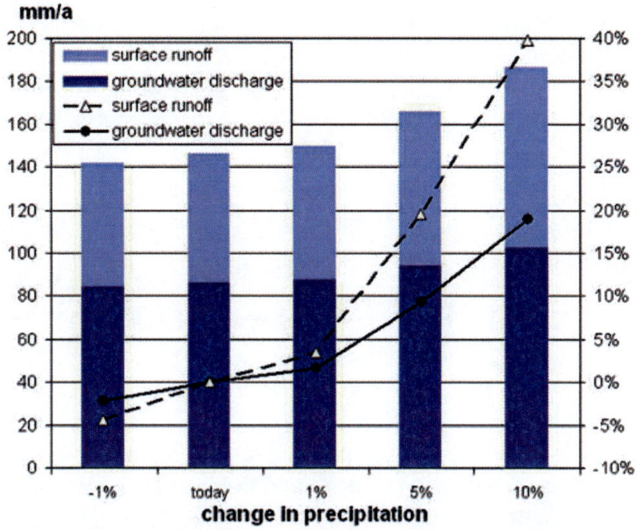

Fig. 19.3 Surface runoff (Q_S) and submarine groundwater discharge (Q_{SGD}) for today's conditions and four scenarios of change in precipitation. *Lines* illustrate relative change in percent as compared to today's conditions

(86 mm/a = 14.3% of precipitation) drain as groundwater discharge directly to the sea, in terms of coast length that is 868 m^3/day/km.

Figure 19.3 illustrates the response of surface and groundwater discharge, respectively, when the precipitation rates change. For all scenarios groundwater discharge Q_{SGD} exceeds surface runoff Q_S, i.e. 55–60% of groundwater recharge. As mentioned earlier, any percentage change in precipitation results in a threefold change in percentual groundwater recharge relative to today. Since no other sink terms are considered here, the total discharge reflects the variation of groundwater recharge. When precipitation increases, so do the discharge rates. However, of the total change in discharge, two-thirds go into submarine groundwater discharge and one-third drains through the surface water bodies.

The impact of sea-level rise was modelled for three additional stages, 0.24, 0.41 and 0.82 m, respectively. Rising sea levels flatten the hydraulic gradient of the groundwater table, which in turn leads to decreasing groundwater discharge at the coast. However, this effect is minimal (Fig. 19.4) compared to the effect of changes in groundwater recharge since it affects only the near-coastal areas.

19.4.3 Simplified Finite-Difference Model 'Simple' for Transient Simulation of Sea-Level Rise

In order to analyse the temporal response of submarine groundwater recharge to sea-level rise and increasing groundwater recharge a simplified conceptual model is designed and implemented into the Modflow finite-difference code. This code was

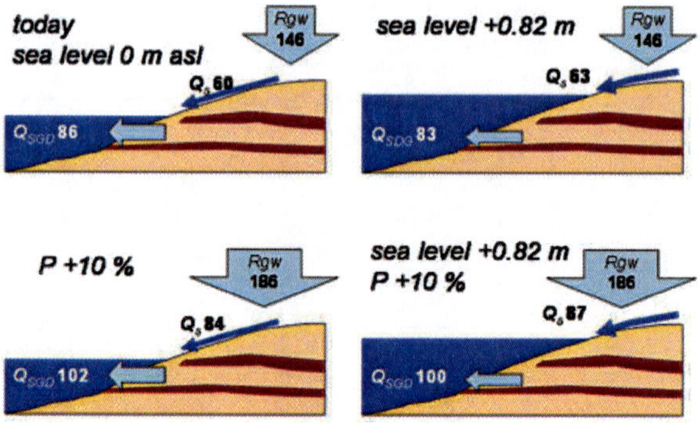

Fig. 19.4 Impact of sea-level rise on submarine groundwater discharge (Schematic cross section of the model area. *P* precipitation, R_{GW} groundwater recharge, Q_{SGD} submarine groundwater discharge, Q_S surface runoff, mass balance terms in [mm/a])

chosen because Modflow allows for time-variant specified heads (first kind boundary condition) within stress periods (Leake and Prudic 1991). The model geometry is simplified because the focus of this study is to understand the general mechanisms rather than the specific test site at Wismar Bay.

A 100 year period is simulated during which (i) the sea level rises from 0 m asl up to 1 m asl, (ii) the average annual groundwater recharge increases 30%, i.e. from 150 mm/a up to 195 mm/a and finally (iii) both scenarios are combined.

The model area is of the same size as the true catchment (140 km^2), subdivided into 100 by 140 finite-difference cells, each 100 m by 100 m. The top row (North) is defined as time-variant specified head boundary (first kind) representing a 10 km long coastline with sea-level rise of 1 m in 100 years. The Modflow code allows linearly varying head values during transient simulation. Surface runoff Q_S is represented by one central river of 10 km length, which is simulated by a third kind boundary condition.

The aquifer is represented by one single unconfined layer with a homogeneous hydraulic conductivity of $1 \times \cdot 10^{-4}$ m/s. The bottom of this layer is assumed to be at –10 m asl.

Figure 19.5 illustrates the model design and the resulting hydraulic heads after 100 years simulation period. The heads range from 78 m asl in the south to sea level in the north. According to Eq. (4) and similar to the steady-state FE model, the amount of groundwater which drains into the central river is accounted for as surface runoff (Q_S). The model is adjusted to starting conditions which reflect today's conditions as calculated with the steady-state FE model, i.e. surface runoff amounts to 40% of total discharge.

The transient simulation of the response of submarine groundwater discharge to changes in groundwater recharge and/or sea-level rise confirms the steady-state results. Given the predicted scenarios, a sea-level rise of 1 m does not affect groundwater discharge significantly. Conversely, the rise in net water inflow, i.e. groundwater recharge, leads to a significant increase of both surface runoff and submarine groundwater discharge. However, the absolute increase slightly favours the submarine groundwater discharge (Fig. 19.6).

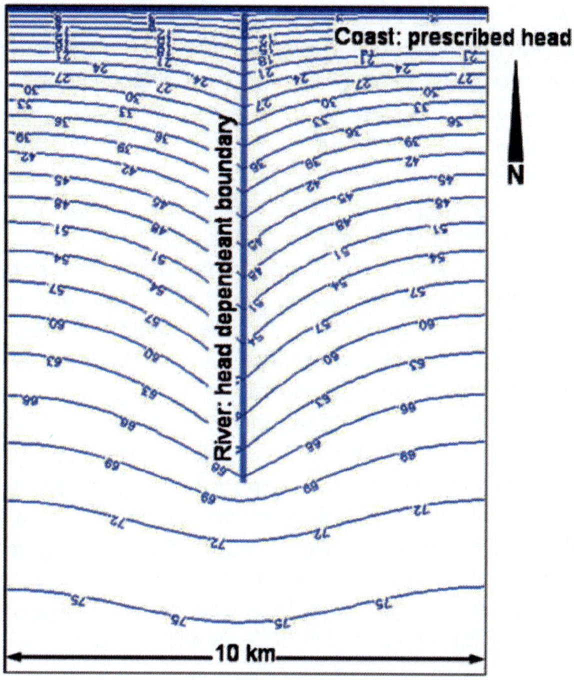

Fig. 19.5 Model design of simplified finite-difference model ('Simple'). Contours represent simulated hydraulic heads

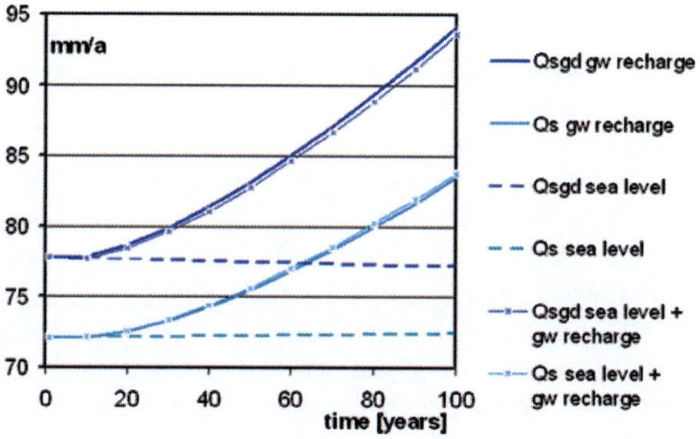

Fig. 19.6 Temporal development of submarine groundwater discharge (Q_{SGD}, *dark blue*) and surface runoff (Q_S, *light blue*) in response to a sea-level rise of 1 m (*dashed lines*), increase of groundwater recharge of 30% (*bold lines*) and both scenarios combined (*lines with markers*)

19.5 Discussion

The recent submarine groundwater discharge for the investigated catchment (140 km^2) is calculated as 86 mm/a or 12·× 10^6 m^3/a. This value compares well with measurements of Schlüter et al. (2004). By means of measurements of ^{222}Rn he determined that the direct groundwater discharge in the Eckernförde Bay/Schleswig-Holstein is between 4×10^6 m^3/a and $57 \cdot \times 10^6$ m^3/a. The geological conditions and the land use of the catchment of Eckernförde Bay are similar to those in our area of study. However, the mean annual precipitation for Eckernförde Bay, which is situated 100 km to the west of Wismar Bay, is higher. At Kiel, which is close to Eckernförde Bay, 757 mm is measured (Hillebrandt 2007). Thus the groundwater discharge in the west is higher too (236 mm/a for the state Schleswig-Holstein, AG Angewandte Geologie/Hydrogeologie 2004).

Given that the calculated recent submarine groundwater discharge is realistic, future scenarios can be simulated. The predicted worst-case scenario with a sea-level rise of 0.82 m for the southern Baltic does not have a significant impact on the groundwater discharge for the given conditions at the test site. The observed changes in groundwater discharge are in the range of a few millimetres per year, which is within the error range of quantification of groundwater recharge in this region (Meyer and Tesmer 2000). One possible reason is that the observed hydraulic gradient of 0.5% in the test area is not representative for most aquifers at the Baltic southwestern coast (Jordan and Weder 1995). In general the potentiometric surface is flatter, which in turn will result in less groundwater discharge. However, any slight change in sea level will then affect the amount of groundwater discharge more distinctively. Any rise of sea level which leads to a rise of groundwater table in the coastal region must be considered. This may result in higher soil moisture, thus provoking landslides and engineering problems.

Changes in precipitation lead to changes in groundwater recharge, i.e. the positive budget term (inflow) is directly affected. Thus the total discharge and its two components, surface and subsurface flow, respectively, respond to precipitation changes. For the considered region the submarine groundwater discharge exceeds the surface runoff. This relation becomes even more pronounced when the groundwater recharge increases.

In this study groundwater recharge is calculated for a net increase of precipitation rates. However, only the annual rates have been increased according to the predicted scenarios. Any seasonal balance shifts, where the winter precipitation increase exceeds the summer increase, are still to be analysed. It should be noted that changes in seasonal precipitation will affect evapotranspiration and surface runoff. Also an increase of the mean annual air temperature and the resulting increase of evaporation have not yet been considered. It is also obvious that climate change will result in a change in land use. However, this effect could not be predicted for this study.

The schematic model finally can only illustrate trends, which still require verification by field data. Many unknowns are still to be investigated. The temporal response strongly depends on information about the storativity of the considered aquifers. In addition the exchange rates for riverbed layers are not yet well investigated.

19.6 Conclusion

The assessment of coastal groundwater discharge can be estimated based on a budget approach if groundwater recharge can be considered equal to total discharge, i.e. no other sink terms exist. These assumptions are valid for most flat coastal catchment areas of medium size at the southern Baltic coast. Then only the surface runoff which drains through the surface water bodies (rivers) must be distinguished from the direct exfiltration from the aquifers to the sea.

For the catchment (140 km^2) in the vicinity of Wismar the groundwater recharge is calculated as 146 mm/a which corresponds to 24.3% of the mean annual precipitation of 600 mm/a; 14.3% exfiltrate as direct groundwater discharge, i.e. $12 \cdot \times 10^6$ m^3/a. Similar values have been reported for Eckenförde Bay in the west (Schlüter 2004). The residual 10% of groundwater recharge drains through the rivers to the Baltic.

Small changes in precipitation result in a more pronounced change in groundwater recharge, i.e. a 10% precipitation increase results in a 29% increase in groundwater recharge. However, seasonal shifts of precipitation change, possible alteration of land use and changes in evapotranspiration rates are still to be considered.

The predicted sea-level rise in the Baltic will have a major impact on many aspects, e.g. socio-economy. However, groundwater discharge does not show significant response to sea-level rise for the given hydraulic conditions at the test site.

Acknowledgements The authors are grateful for critical comments by Daniel M. Tetzlaff. Comments provided by two reviewers and the editors helped to improve the manuscript.

References

AG Angewandte Geologie/Hydrogeologie (2004) Grundwasserneubildungsberechnungen für das Bundesland Schleswig-Holstein. Technical report, University of Greifswald, unpublished

CLC (2000) Corine Land Cover 2000, Coordination of Information on the Environment der Europäischen Union

Darsow A (2004) Mesoskalige Modellierung eines küstennahen Grundwasserleiters nordöstlich der Hansestadt Wismar unter Verwendung von Feflow, „Modeling an coastal aquifer by the use of Feflow (Wismar, Mecklenburg-Western Pommerania). Unpublished Diploma thesis at Institute for Geography and Geology, University of Greifswald

Diersch HJ (2005) Finite element subsurface flow and transport simulation system. Reference manual, Feflow, Version 5.3, Wasy GmbH, Berlin

GLC (2000) Global Land Cover 2000, Institute of Environment and Sustainability (IES), European Joint Research Centre

Harbaugh AW, Banta ER, Hill MC, McDonald MG (2000) MODFLOW-2000, the U.S. Geological survey modular ground-water model – User guide to modularization concepts and the ground water flow process and USGS Open-File Report 00-92, Reston, VA

Harff J, Lampe R, Lemke W, Lübke H, Lüth F, Meyer M, Tauber F (2005) The Baltic Sea – A model ocean to study interrelations of geosphere, ecosphere, and anthroposphere in the coastal zone. Journal of Coastal Research 21(3):441–446

Haude W (1955) Zur Bestimmung der Verdunstung auf möglichst einfache Weise. Mii deutsch Wetterd 2, 11, 24 pp, Bad Kissingen

Hillebrandt O (2007) Quantifizierung des direkten Grundwasserabflusses von Schleswig-Holstein in die Ostsee. Bachelor thesis (unpublished), 29pp, Greifswald University

Interngovernmental Panel on Climate Change (IPCC) (2007) Intergovernmental panel on climate change. In: Pachauri RK, Reisinger A (eds) Fourth assessment report-climate change 2001. Synthesis report. http://www.Ipcc.ch/pub/reports.html

Johansson MM, Kahma KK, Boman H (2003) An improved estimate for the long-term mean sea-level on the Finnish coast. Geophysica 39(1–2):51–73

Jordan H, Weder HJ (1995) Hydrogeologie – Grundlagen und Methoden und Regionale Hydrogeologie: Mecklenburg-Vorpommern, Brandenburg und Berlin, Sachsen-Anhalt, Sachsen, Thüringen. Enke Verlag, Stuttgart, pp 603

Leake SA, Prudic DE (1991) Documentation of a computer program to simulate aquifer-system compaction using the modular finite-difference ground-water flow model. U.S. Geological Survey Techniques of Water-Resources Investigations, book 6, chap. A2, 68p

Meier M, Broman B, Kjellström E (2004) Simulated sea level in past and future climates of the Baltic Sea. Climate Research 27:59–75

Meyer M (2003) Modelling prognostic coastline scenarios for the southern Baltic Sea. Baltica 16:21–32

Meyer T, Tesmer M (2000) Ermittlung der flächendifferenzierten Grundwasserneubildung in Südost-Holstein nach verschiedenen Verfahren unter Verwendung eines Geoinformationssystems. PhD thesis, Freie Universität Berlin, Verlag im Internet, Berlin

Nakicenovic N, Alcamo J, Davis G, de Vries B, Fenhann J, Gaffin S, Gregory K, Grübler A (2000) Special report on emissions scenarios, Working Group III of the Intergovernmental panel on climate change, IPCC. Cambridge University Press, Cambridge, 595pp

Renger M, Wessolek G (1990) Auswirkungen von Grundwasserabsenkung auf die Grundwasserneubildung. Mitteilungen des Instituts für Wasserwesen 386:295–307; Universität der Bundeswehr Munich

Scanlon BR, Christman M, Reedy RC, Porro I, Simunek J, Flechinger GN (2002) Intercode comparison for simulating water balance of surficial sediments in semiarid regions. Water Resources Research 38(12):591–596

Schlüter M, Sauter E-J, Andersen C-E, Dahlgaard H, Dando P-R (2004) Spatial distribution and budget for submarine groundwater discharge in Eckernförde Bay (Western Baltic Sea). Limnology and Oceanography 49(1):157–167

Sherif MM, Singh VP (1999) Effects of climate change on sea water intrusion in the coastal aquifers. Hydrological Processes 13:1277–1287

Part VIII
Monitoring

Chapter 20
Monitoring the Bio-optical State of the Baltic Sea Ecosystem with Remote Sensing and Autonomous In Situ Techniques

Susanne Kratzer, Kerstin Ebert, and Kai Sørensen

Abstract This chapter focuses on recent advances in water quality monitoring of the Baltic Sea using remote sensing techniques in combination with optical in situ measurements. Here the Baltic Sea ecosystem is observed through its bio-optical properties, which are defined by the concentration of optical in-water constituents governing the spectral attenuation of light. In the introduction, typical geographical patterns and seasonal variations of optical properties and the cause of the mass occurrence of cyanobacterial blooms in summer are discussed. The optical characteristic of Baltic Sea waters is clearly dominated by a relatively high load of dissolved organic matter and, during the productive season, by phytoplankton growth, stimulated by nutrients mostly originating from land. In the coastal zone, inorganic suspended matter also has a significant effect on the light attenuation, which increases with proximity to land. The ecological status of the coastal zone may be synthesized using a bio-optical model, summarizing important ecosystem state variables such as terrestrial runoff and phytoplankton production. The optical properties can also be observed with visible, spectral satellite remote sensing providing repetitive and optically consistent data for the whole Baltic Sea basin. Such observations have already significantly influenced our understanding of Baltic Sea dynamics and provide us with a new look into this brackish ecosystem. The focus in this chapter is on the ocean colour sensor 'Medium Resolution Imaging Spectrometer' (MERIS), which since 2002 is flying continuously onboard the European Environmental Satellite ENVISAT, developed by the European Space Agency (ESA). The advantage of MERIS data is its good spatial resolution of 300 m allowing the analysis of coastal features and bays of the Baltic Sea. Algorithms to retrieve bio-optical parameters from MERIS data are continuously improved and extended. The MERIS mission will be continued with the Ocean and Land Colour Instrument (OLCI), an optically similar sensor which will be flown on SENTINEL-3, scheduled to be operational until 2023, to assure

S. Kratzer (✉)
Department of Systems Ecology, Stockholm University, 106 91 Stockholm, Sweden
e-mail: Suse@ecology.su.se

J. Harff et al. (eds.), *The Baltic Sea Basin*, Central and Eastern European Development Studies (CEEDES), DOI 10.1007/978-3-642-17220-5_20,

long-term monitoring of water quality from space. The wide aerial coverage, the frequent repetition and continuity of the satellite observations, the consistency of the measured data, and a relative cost-effectiveness clearly respond to the demands of a modern operational monitoring system, and the requirements of effective Baltic Sea management. An overview of existing monitoring approaches is given, and operational online systems that combine remote sensing and autonomous in situ measurements are discussed.

Keywords Baltic Sea · Optically complex waters · Bio-optical monitoring · Remote Sensing · MERIS · validation

20.1 Introduction

20.1.1 The Baltic Sea from an Optical Perspective

The Baltic Sea basin may be regarded as an extended fjord of the Atlantic Ocean or as a large estuary with relatively weak tides of less than 5 cm and with broad shallow margins. The geology of Scandinavia and the northern Baltic Sea is characterized by rifts of old rigid rocks, whereas the shoreline of the southern Baltic consists mostly of younger, more easily eroding rocks leading to sandy beaches and sand flats along the German and Polish coast, and northwards to the Estonian coast (Milliman 2001). The Baltic Sea water is brackish in nature due to its restricted water exchange with the North Sea and a high freshwater input from rivers (Voipio 1981). The water column is characterized by a permanent density stratification with a brackish surface layer and heavier bottom water of higher salinity originating from the North Sea. The permanent halocline ranges between 40 and 70 m depth. A seasonal thermocline develops during spring and summer at depths between 15 and 20 m in most parts of the Baltic Sea, providing another density barrier for vertical exchange. Apart from vertical density stratification, the high fluvial input from the north and the saline input of water from the North Sea produce a horizontal salinity gradient across the whole Baltic Sea basin. The surface salinity decreases progressively from 8–6 in the Baltic Sea Proper, to 6–5 in the Bothnian Sea, down to 3–2 in the Bothnian Bay. The salinity in the surface mixed layer is hence very low compared to other semi-enclosed seas such as, e.g., the Mediterranean Sea with 38.

The dominance of freshwater from river discharge is associated with a high content of humic substances, consisting of humic and fulvic acids. The main part of humic substances can be measured optically and is also termed coloured dissolved organic matter (CDOM). Another common term for CDOM is yellow substance, as it is yellow in colour because of its high absorption in the blue part of the visible spectrum.

Salinity is inversely related to CDOM: the higher the freshwater influence, the lower the salinity but the higher the CDOM concentration (Williams et al. 1996, Kratzer et al. 2003). Consequently, there are marked parallel horizontal gradients in both, surface salinity and CDOM, across the whole Baltic Sea basin. In the Bothnian

Bay the water is visibly coloured brown due to the high runoff of humic substances originating from bogs, lakes, and rivers (Kirk 2010).

Besides the CDOM content, the phytoplankton biomass, given as the concentration of the main phytoplankton pigment chlorophyll-*a*, and the load of total suspended matter (TSM) characterize a water body optically.[1] In coastal areas, terrestrial and river runoff in addition to wind-driven re-suspension of sediments leads to high TSM loads.

20.1.2 Seasonal Variations in Optical Properties

Major seasons of high runoff are the thawing period in spring and when the annual precipitation reaches its maximum in summer (Voipio 1981). Along with terrestrial runoff, precipitation increases the input of nutrients and dissolved and particulate matter, but decreases the salinity (Meier and Kauker 2003).

According to Voipio (1981), the spring bloom in the southern and central parts of the Baltic Proper occurs generally in the second half of April, whereas in the northern Baltic Sea it occurs in early May, somewhat later in the Bothnian Sea, and in the Northern Bothnian Bay not until June, which is partially related to a later development of the seasonal thermocline. However, the spring bloom has shifted forward since the 1980s and may be already observed between March and April in the southern Baltic, in early April in the Northern Baltic, in mid-April in the Gulf of Finland (Fleming and Kaitala 2007), and in the central Gotland Sea in May (Siegel and Gerth 2008). From the end of June or early July onwards plankton blooms of nitrogen-fixing, filamentous cyanobacteria start to occur in the Baltic Proper consisting mostly of *Nodularia spumigena, Aphanizomenon* sp., and – in low-salinity areas – of several *Anabaena species.* This annual summer phytoplankton bloom is exceptional in its intensity, extent, and duration. Until early autumn, the extensive blooms rise to the surface during calm and stable weather conditions. Scientific reports of filamentous cyanobacteria in the Baltic Sea date back to the middle of the nineteenth century. Paleo-oceanographic studies indicate their occurrence as long as the Baltic has been a brackish sea (Bianchi et al. 2000).

20.1.3 Eutrophication in the Baltic Sea

Eutrophication has been identified as the main environmental problem of the Baltic Sea ecosystem (HELCOM 2007). It is caused by a combination of increased nutrient input from land and atmosphere, which increases phytoplankton biomass. The increased organic production leads to organic matter enrichment in the bottom sediments after the spring bloom, which in turn leads to a higher consumption of

[1]The CDOM content is measured in terms of absorption at 440 nm; unit: [m^{-1}]), chlorophyll-*a* concentration is measured in units of [μg/l]) and TSM load is measured in units of [g/m^3]).

oxygen when the organic matter is being degraded by bacteria. Eventually this can lead to anoxia in bottom waters and to so-called dead-zones, causing fauna mortality. The Baltic Sea is considered to be the largest anthropogenic dead zone in the world (Diaz and Rosenberg 2008). Anoxic bottom areas and bottom waters with low oxygen content result in the release of phosphorus from the sediment. In the deep basins of the Baltic Sea the permanent stratification of the brackish water body and the slow and irregular exchange of bottom waters enhance the build-up of stagnant conditions. This causes oxygen depletion, which in turn leads to a decrease in inorganic nitrogen reserves by denitrification. It also causes the trapping of inorganic phosphorus in the bottom water and a phosphorus flux from the sediment, resulting in a decrease in the ratio of dissolved inorganic nitrogen (DIN) to dissolved inorganic phosphorus (DIP). The DIN:DIP ratio is especially low in the open Baltic Sea, which means that the open Baltic Sea is nitrogen limited in summer, which is one of the key factors for the occurrence of blooms of filamentous nitrogen-fixing cyanobacteria. Because of their ability to fix nitrogen they are more competitive under these conditions than other phytoplankton.

20.1.4 Baltic Sea Ecology Observed from Space

Figure 20.1 shows such a cyanobacterial bloom observed in July 2005 by the MERIS ocean colour sensor. Filamentous cyanobacteria contain gas vacuoles causing positive buoyancy and allowing them to position themselves close to the surface

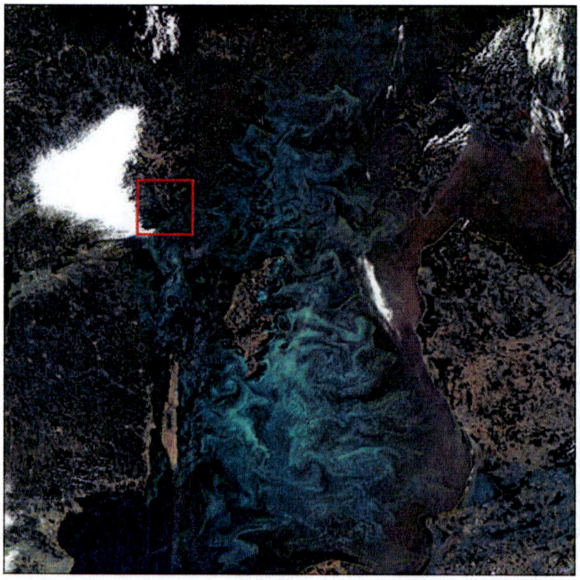

Fig. 20.1 Cyanobacteria bloom in the Baltic Sea on 13 July 2005. RGB-composite from MERIS full resolution data on ENVISAT. The *red rectangle* shows Himmerfjärden, a north-south facing bay approximately 40–70 km south of Stockholm in the northwestern Baltic Sea. Himmerfjärden is also shown in Fig. 20.2

to improve access to light, subsequently building up as surface accumulations. Furthermore, the gas vacuoles of the cyanobacteria reflect strongly, which makes them visible in satellite imagery. The image also demonstrates how cyanobacteria blooms act as visible tracers of Baltic Sea dynamics. The horizontal surface current field in the Baltic Sea has a weak cyclonic pattern with anticlockwise rotation (Kullenberg 1981, Stigebrandt 2001). Residual currents from the north occur along the Swedish coast and from the south along the Finnish coast (Kahru et al. 1995, Victorov 1996). The meso-scale features of the cyanobacterial bloom seen in Fig. 20.1 indicate horizontal eddies and fronts.

Kahru (1997) discussed a potential increase of cyanobacteria blooms in the 1990s based on sea surface temperature (SST) satellite data from 1982 to 1994. Siegel et al. (2006) showed an increase in summer temperature in the Baltic Sea during the 1990s and early this century and could connect it to the mass occurrence of cyanobacteria (Siegel and Gerth 2008). Foremost the toxic cyanobacteria species *Nodularia spumigena* is favoured by higher water temperatures in its growth (Kononen and Leppänen 1997).

In summer, strong thermal stratification can be studied in the Baltic Sea, along with local wind-driven coastal upwelling of colder, subsurface water at 5–20 km

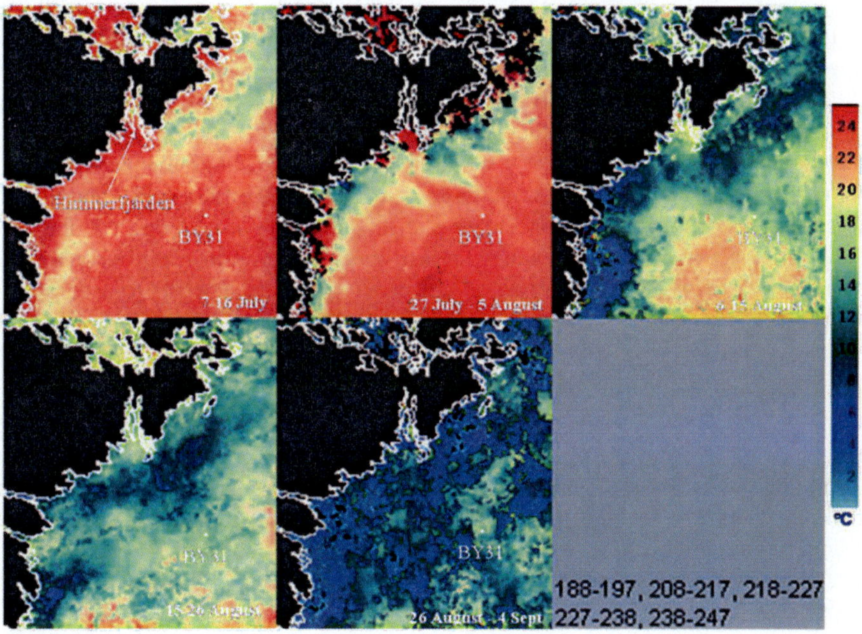

Fig. 20.2 Sea Surface Temperature (SST) in the northwestern Baltic Sea, derived from NOAA/AVHRR data of the period 7 July–4 September 2001, binned into composite images of 10 days each. The 10-day composites reveal temperature differences close to the coast, most probably caused by coastal upwelling. Station BY 31 is Landsort Deep, the deepest part of the Baltic Sea with 459 m depth (from: Kratzer and Tett 2009)

offshore (Gidhagen 1987, Lehmann and Myrberg 2008). Figure 20.2 illustrates the pattern of sea surface temperature in the northwestern Baltic Sea. The sequence of binned images (i.e. 10-daily averages) shows the development and the break-up of the seasonal summer thermocline from July to September 2001 (Kratzer and Tett 2009). The colder filaments shown on the second plate of Fig. 20.2 are typical features caused by coastal upwelling. Because of the permanent salinity stratification with a brackish, lighter layer at the top, upwelling plays an important role in keeping a cyanobacteria bloom alive. Upwelling may mix phosphorus-rich water from below the thermocline with the nutrient-depleted upper layers, thus sustaining cyanobacterial growth (Leppänen et al. 1988). Up-welling seems to stimulate mostly the production of *Aphanizomenon* sp. However, there is a lag of 2–3 weeks before this effect takes place (Vahtera et al. 2005, Lehmann and Myrberg 2008), and the initial effect is a decrease of cyanobacteria biomass due to dilution and decrease in water temperature, which has a strong negative effect on the growth of *N. spumigena* as this species has a growth optimum at somewhat higher temperatures than *Aphanizomenon* sp. (Kononen and Leppänen 1997).

20.1.5 Bio-optical Properties of Natural Waters

In 1961, Preisendorfer introduced a system which separated optical properties into two categories – inherent and apparent (Kirk 2010). Apparent optical properties (AOPs) of a water body are all the properties depending on the geometry of the light field, e.g. the radiant quantities radiance (I) and irradiance (E). Inherent optical properties (IOPs) are independent of changes in the radiance distribution and depend only on the optical substances within the aquatic medium. Examples for IOPs are the absorption coefficient, the scattering coefficient, or the backscattering coefficient. Changes in ocean colour are changes in the spectral variation of the sea surface reflectance, R. R is strongly correlated to the ratio of backscattering to absorption coefficient:

$$R \approx f b_b (a + b_b)^{-1},$$

where $f = 0.33$ (Morel and Prieur 1977). This means that both the scattering and the absorption properties of the in-water optical constituents, as well of the water determine the spectral reflectance, and therefore the colour emerging from the sea. The main scatterers in a natural water body are pure seawater and TSM. Total absorption is made up of the sum of absorption by water, phytoplankton pigments, TSM, as well as CDOM.

Clear ocean waters belong to so-called optical Case-1 waters. In these waters, the optical signal is dominated only by the sea water itself, by chlorophyll-*a*, and co-varying CDOM (Morel and Prieur 1977). Chlorophyll-*a* has two absorption peaks, a major peak in the blue part of the electromagnetic spectrum around 440 nm and one in the red part around 670 nm. In the green part of the spectrum, there is the lowest chlorophyll absorption. Remote sensing algorithms for retrieval of chlorophyll-*a* are based on empirical relationships between changes in phytoplankton pigment concentration and ratios of water-leaving radiance at different wavelengths, e.g. the ratio of blue to green radiances, or reflectances (Gordon and Morel 1983, Sathyendranath et al. 1994, Aiken et al. 1995). For these waters, chlorophyll can be determined from space with high accuracy.

In coastal, optically complex waters, such as the Baltic Sea, the optical properties are influenced not only by water itself and by phytoplankton but also by varying concentrations of CDOM and total suspended matter (TSM).[2] These waters are referred to as optical Case-2 waters (Morel and Prieur 1977). CDOM absorption in the Baltic Sea is high compared to other seas (Siegel et al. 1999, Darecki et al. 2003) and CDOM is normally the dominant optical signal in the Baltic Sea (Kowalczuk et al. 2006, Kratzer 2000, Kratzer and Tett 2009). It absorbs strongly in the blue spectral region and the absorption decreases exponentially with increasing wavelength. The exponent (slope factor) for CDOM in the Baltic differs from other seas. Schwarz et al. (2002) showed that the mean exponent measured in the Baltic Sea is relatively high: 0.0193 (±0.0024), whereas in other marine areas it is 0.0165 (±0.0035). CDOM absorption is still significant in the green spectral region around 550 nm; hence standard band ratio algorithms based on the blue to green band ratio tend to overestimate the concentration of chlorophyll-*a* in the Baltic Sea (Jorgensen 1999). Coastal waters with high concentrations of suspended sediments, e.g. waters highly influenced by tidal action, have a relatively high backscatter because inorganic sediments increase the backscatter of light from the water body, and therefore the reflection. However, the waters of the open Baltic Sea are dominated by CDOM absorption and therefore reflect relatively little because of the high CDOM absorption. The open Baltic Sea appears therefore much darker from space than, e.g., the North Sea.

20.1.6 Historical Trends in Water Quality Assessment

Secchi depth is one of the oldest methods used in oceanography. It originated with Angelo Secchi (1818–1878), who was requested to measure the transparency in the Mediterranean Sea. A white circular disk of 30 cm diameter is lowered into the water (Fig. 20.3) until the observer loses sight of it (Preisendorfer 1986). The observer notes the depth at which the disk vanishes; the deeper the Secchi depth, the clearer the water. In the Baltic Sea it is common to use Secchi depth as an indicator for eutrophication (Kautsky et al. 1986, Sandén and Håkansson 1996, HELCOM 2007). In the open Baltic the Secchi depth varies from a couple of meters during strong blooms to almost 20 during winter in the southern Baltic. Long time series indicate that the Secchi depth has decreased about 0.05 m/year since 1910. This decrease has

[2]TSM is also referred to as suspended particulate matter (SPM) and consists of an organic and an inorganic fraction (organic and inorganic SPM).

Fig. 20.3 Secchi depth measurement (Photo: Susanne Kratzer, Irish Sea 1997)

been explained by the increase in phytoplankton biomass (Sandén and Håkansson 1996) and therefore Secchi depth has been used as an indicator for eutrophication. In the coastal areas and the Stockholm archipelago, a recovery of the Secchi depth has been observed over the last decade, whereas the decreasing trend seems to continue for the open Baltic Sea (Bernes 2006).

Another way to measure water transparency is by measuring the rate of decrease of light with depth. Light energy which enters the water from above and is transmitted downwards is known as downwelling irradiance, E_d. In the case of monochromatic light with uniform angular distribution E_d diminishes in an approximately exponential manner with depth:

$$E_d(z) = E_d(0)\, e^{-K_d z} \text{ (Beer's Law)}$$

where E_d (0) and E_d (z) are the values of downward irradiance just below the surface and at depth z, respectively (Kirk 2010). K_d is the average value of the diffuse attenuation coefficient for the downwelling light field over any defined depth interval; note that K_d, the rate of light decrease, is wavelength dependent. The diffuse attenuation coefficient and Secchi depth inversely correlated (Kratzer et al. 2003, Kratzer and Tett 2009). They are influenced not only by phytoplankton or Chlorophyll-*a* concentration but also by CDOM and TSM load. Secchi depth can therefore only be used as an indicator for eutrophication where phytoplankton clearly dominates water attenuation (Wasmund et al. 2001) or where there is little variability in any of the two other optical components.

Chemical parameters such as the nutrient concentrations cannot be retrieved optically as they do not sufficiently interact with light. However, they may be retrieved indirectly, e.g. by relating Secchi depth to the total nitrogen concentration (Tett et al. 2003).

20.1.7 A Multiscale Approach to Monitoring

The EC Water Framework Directive (WFD) requires the assessment of the ecological status of European waters (C.E.C. 2000). In transitional and coastal waters the quality elements that must be determined according to the WFD include phytoplankton biomass, the amount of dissolved carbon, the frequency, and intensity of blooms and water transparency, all of which can be assessed by combining ocean colour remote sensing with in situ observational techniques.

Traditional monitoring programs in oceanography consist of water samples taken regularly at defined sites with standardized methods. The monitoring sites are usually sampled at different depths in order to get vertically resolved information. In order to protect the marine environment, strong emphasis is now placed on the complex biological–ecological status instead of on purely physiochemical parameters. The conceptual model in Fig. 20.4 illustrates a multiscale monitoring approach by applying a combination of techniques (Kratzer et al. 2003). Data derived from ocean colour remote sensing provide synoptic information of the whole Baltic Sea basin. Optical in situ data, autonomous in situ measurements on moorings or on light houses, as well as on ships-of-opportunities (ferries) are combined with optical models to interpret and validate the information from remote sensing data. Dedicated sea-truthing campaigns provide the in situ data for the development of local algorithms and retrieval methods regionally adapted to the Baltic Sea and for the validation of satellite data. Remotely measured radiances are interpreted with specific retrieval algorithms, and bio-optical parameters are derived. For local point

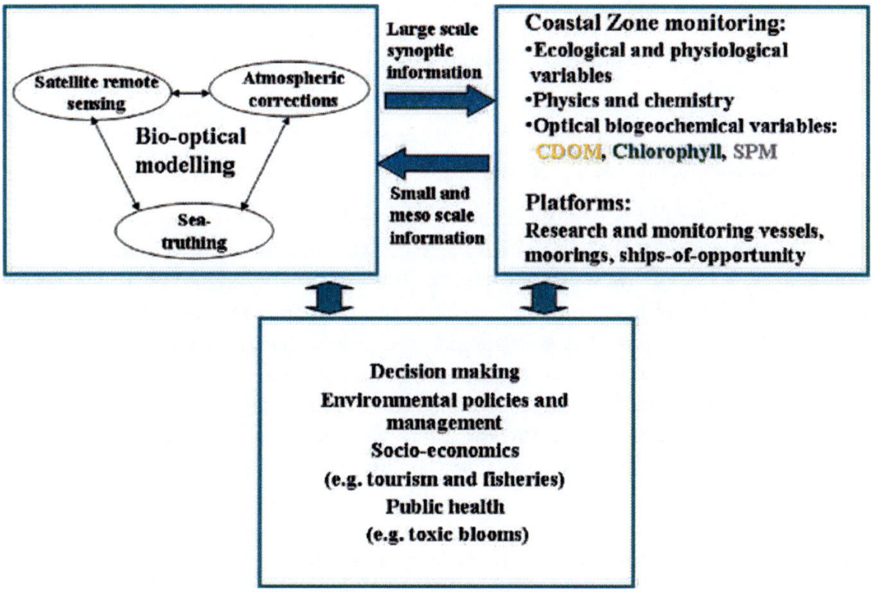

Fig. 20.4 A multiscale approach to monitoring (Kratzer et al. 2003)

measurements, ship-based data may provide complex bio-geochemical parameters including the information on vertical distribution. Horizontal transects from ships-of-opportunity provide improved temporal and spatial resolution using automatic measurements.

20.2 Methods Applied

20.2.1 Remote Sensing Methods

20.2.1.1 Background

In order to monitor an aquatic ecosystem, various biological, chemical, and physical parameters, indicating its state, are required. Marine satellite remote sensing uses a wide range of measurement techniques to derive a set of important bio-geophysical parameters. The satellite sensors work in active or passive mode. The parts of the electromagnetic spectrum used are in the microwave, infrared, and visible/near-infrared range. Active microwave sensors, i.e. radars, are applied to derive information on sea surface height, wave height, wind surface fields, and the detection of specific events such as oil spills (Brekke and Solberg 2005). Passive microwave radiometry is used to detect sea-ice zones and ice parameters, temperature, and wind (Askne and Dierking 2008). This technique has also been used to detect surface accumulations of cyanobacteria in the Baltic Sea (Subramaniam et al. 2000). Passive sensors working in the thermal-infrared spectral range are used to derive sea surface temperature (Robinson 2004), an example of which is shown in Fig. 20.2. The new sensor, Soil Moisture and Ocean Salinity (SMOS), launched at the end of 2009, is another passive microwave radiometer, which will be used to derive ocean salinity.

20.2.1.2 Ocean Colour Remote Sensing

In this chapter the focus lies on ocean colour remote sensing, i.e. remote sensing in the visible-near-infrared (VIS/NIR) range, 400–900 nm, with passive satellite radiometers. VIS/NIR radiation, i.e. the sun light, is scattered and absorbed on its way through the atmosphere. As the radiant flux reaches the sea surface, some of it is reflected, and some of it is refracted as it enters the water body. Once in the water, the radiant flux is either absorbed or scattered by the optical components in the water body, which changes its spectral signature. The radiance that is scattered back into the atmosphere, the so-called water-leaving radiance, now contains information about the optical water constituents. It is changed, again, on its way through the atmosphere. The VIS/NIR signal measured remotely by a sensor placed on an aircraft or a satellite therefore carries information on both the optical in-water constituents and the atmosphere. The NIR channels of the radiometer are used for atmospheric correction, whereas the visible channels are used to derive information about water quality.

Widely used ocean colour radiometers are, e.g. NASA's[3] SeaWiFS (launched in 1997) and MODIS (1999 and 2002) as well as ESA's MERIS (2002). The common spatial resolution for ocean colour images is about 1 km, which is sufficient for open ocean applications. MODIS has also medium-resolution bands (250 and 500 m) which were designed for land applications. These can also give valuable information about specific features in coastal waters (Kutser et al. 2007). A method called pan-sharpening can be used to improve the resolution of MODIS multispectral images from 1 km to 250 m resolution. In this method the resolution of multispectral radiometer is increased on the basis of higher resolution bands from the same or another radiometer of similar acquisition terms (Carper et al. 1990, Chavez et al. 1991).

MERIS on ENVISAT offers currently the best spectral and radiometric resolution in operational ocean colour remote sensing radiometry (Doerffer et al. 1999). MERIS has 15 spectral bands with 10 nm bandwidth each. It also has an improved spatial resolution of 300 m and is therefore especially suitable for coastal applications. MERIS is suitable both for aquatic and for terrestrial remote sensing as it has a wide dynamic range, capable of detecting the low signals reflected from the dark water, as well as bright reflectance from sea ice, clouds, or land surfaces. Thus, MERIS is notably suitable to study land–ocean interactions and Earth system dynamics.

Figure 20.5 shows a true colour composite of Himmerfjärden, a fjord-like bay situated 60 km south of Stockholm. The left plate shows Himmerfjärden in full resolution (300 m), whereas the right plate shows the reduced resolution of MERIS (1,200 m). The figure demonstrates visually that full resolution MERIS data are suitable to analyse coastal bays, whereas the reduced resolution image only gives very few pixels from within the bay. These pixels are also clearly influenced by the strong reflection from land, which is called adjacency effect.

Ocean colour sensors fly on near-polar, sun-synchronous satellite orbits to obtain high temporal coverage. The length of the resonant orbit is 35 days for MERIS and 16 days for MODIS and SeaWiFS. The sensors have different swath widths and global coverage is provided every 72 h for MERIS, every 48 h for SeaWiFS, and every 24 h for MODIS.

20.2.1.3 Remote Sensing Products

Products derived from ocean colour remote sensing are commonly categorized into three levels (Bukata 2005). Level 1 products are calibrated and geo-located radiances, at sensor height, i.e. at the top of the atmosphere (TOA). Level 2 products are retrieved from TOA radiances and include the previously described in-water optical properties, i.e. the concentration of chlorophyll-*a* and TSM, as well as

---

[3]NASA: National Aeronautics and Space Administration; ESA: European Space Agency; MODIS: Moderate Resolution Imaging Spectroradiometer; SeaWiFS: Sea-viewing Wide Field-of-View Sensor; MERIS: Medium Resolution Imaging Spectrometer; ENVISAT: European Environmental Satellite.

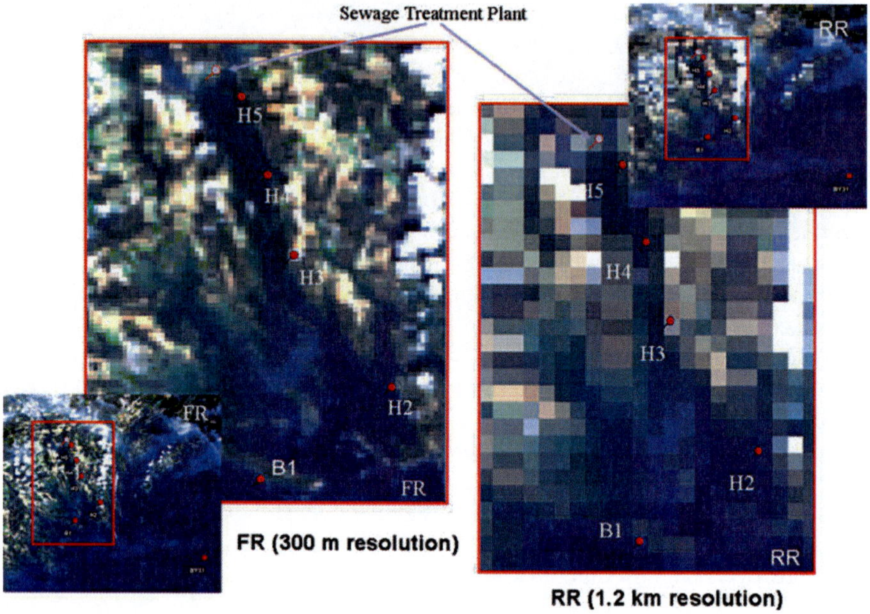

Fig. 20.5 RGB composite images from 19 August 2002 over Himmerfjärden, both in 300 m, full resolution (FR) and 1.2 km reduced resolution (RR). This comparison shows that the 300 m resolution of MERIS is more appropriate to view coastal bays than the common 1 km resolution of ocean colour sensors. Note the images have not been corrected for environmental effects (Kratzer and Vinterhav 2010)

CDOM absorption at 440 nm. Besides the optical in-water components, one can also derive so-called inherent optical properties describing the propagation of light in the water, such as absorption and scattering, as well as the diffuse attenuation coefficient, $K_{d,}$ characterizing the rate of light attenuation. Level 3 products are space and/or time binned data sets of level 2 products which are used to generate seasonal climatologies and to analyse long-term global trends.

20.2.1.4 Limitations and Challenges

The low sun elevation in the high-latitude regions of the Baltic basin limits the availability of satellite data from approximately early March to late October. Further limitations are caused by intermittent cloud cover. The extent of cloud cover in the Baltic Sea area is about 40–50% in summer and about 60–70% in winter (Karlsson 1996). Before the retrieval of the water quality parameters cloud masking techniques are applied to the satellite data so that only cloud-free areas are used in the processing. Due to high CDOM concentrations in the Baltic Sea, the absorption is very high, especially in the blue part of the spectrum. Thus, Baltic Sea water is relatively dark compared to other seas leading to especially low water signals. Therefore, a high signal-to-noise ratio is a crucial sensor requirement for Baltic Sea remote sensing.

The signal measured at TOA is influenced by atmospheric, oceanic, and coupled atmosphere–ocean effects. Within the VIS/NIR region of the electromagnetic spectrum, the water signal is only about 10% of the total TOA signal. About 90% of the signal thus originates from atmospheric processes, such as scattering by aerosols. For water quality monitoring, the atmospheric contribution of the detected signal is unwanted and needs to be removed. This process is referred to as atmospheric correction (Fischer and Fell 2001, Schroeder et al. 2007a). Accurate atmospheric correction is critical for the correct retrieval of water quality products.

Areas close to the coast are usually influenced by high reflectance from land. Satellite data from water areas close to the coastline have to be corrected for these so-called adjacency or environmental effects. Prototype algorithms to correct for adjacency effects have been developed lately, e.g. the Improved Contrast between Ocean and Land (ICOL) processor for MERIS data (Santer et al. 2007).

20.2.1.5 Baltic Sea Remote Sensing

In order to retrieve the in-water constituent concentrations in the Baltic Sea from visible, spectral remote sensing data, a number of algorithms can be applied. Darecki et al. (2003) suggested using reflectance ratios between spectral bands at 550 and 590 nm to derive chlorophyll-*a* concentration. The algorithm showed robust results for Baltic Sea regions and is only little influenced by seasonal variations in CDOM. Siegel and Gerth (2008) give a comprehensive overview on ocean remote sensing algorithms in the Baltic Sea. Yet, no regional ocean colour algorithm has been developed that is valid for the whole Baltic Sea basin. For the retrieval of the three independently varying in-water constituents, complex approaches are needed, taking into account the full measured TOA spectrum (Doerffer and Schiller 2006a, Schroeder et al. 2007b). A number of coastal processors based on multispectral regression techniques are available for MERIS data processing. Besides the official ESA MERIS ground segment (IPF, i.e. the MERIS standard algorithm), four further processors can be used in the Baltic area. For fresh water lakes, the Boreal Lakes Water Processor and the Eutrophic Lakes Water Processor are available (Doerffer and Schiller 2008a, b). For the Baltic Sea, the FUB/WeW Water Processor and the Coastal Case-2 Regional Water Processor (C2R) can be applied (Schroeder et al. 2007a, b, Doerffer and Schiller 2006b, 2008a). The algorithms are based on neural network inversion techniques to derive a number of bio-optical parameters simultaneously. Most of the in situ data used for the development of these processors are from the North Sea or other European seas. These data do not represent the high background CDOM absorption of 0.4/m^{-1} typical for the open Baltic Sea (Kratzer and Tett 2009), and level 2 products are underestimated in Baltic Sea waters, especially CDOM (Ohde et al. 2007, Kratzer et al. 2008, Vinterhav 2008). More work is needed on the atmospheric correction process over this optically complex water (Sørensen et al. 2007, Moore and Lavender 2010). Furthermore, the ICOL processor for correction of adjacency effects is also currently being improved. Another important issue is the estimation of cyanobacterial biomass in relation to the biomass of other phytoplankton species (Kutser et al. 2006, Reinart and Kutser 2006).

20.2.1.6 Operational Satellite Systems in the Baltic Sea

Frequent occurrences of massive cyanobacterial blooms in the open Baltic Sea require operational satellite data for adequate monitoring of their extent. Different environmental institutes make retrieved bio-geophysical satellite products available online to a wide range of end-users, such as environmental agencies, tourism, or fisheries. The Swedish Meteorological and Hydrological Institute (SMHI) and the Finnish Environmental Institute (SYKE) have developed online information systems for water quality monitoring of the Baltic Sea, which are operational since 2002 and 2003, respectively, and can be accessed by the user through the Internet.[4] Both open web systems use Advanced Very High Resolution Radiometer (AVHRR) for near real-time monitoring of sea surface temperature, cyanobacterial blooms, and clouds (Rud and Kahru 1995, Kahru et al. 1995, Kahru 1997). Recently both institutes have also started to include MERIS data in their operational monitoring systems.

The Water Quality Service System (WAQSS, http://www.waqss.de/) was developed by Brockman Consult, Germany within the ESA project MarCoast for Global Monitoring of Environment and Security GMES (Fig. 20.6). WAQSS is a prototype of a marine downstream service that adds value to the data and delivers customized

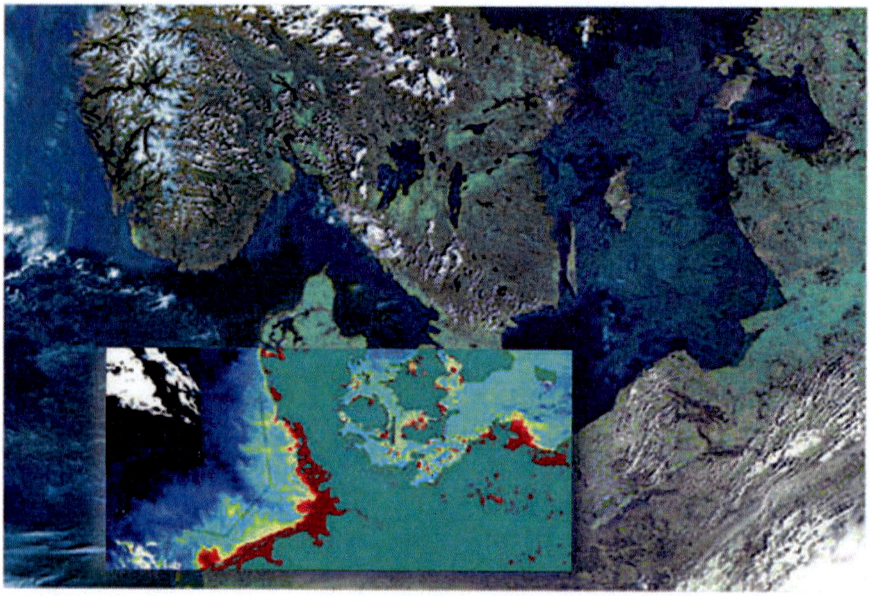

Fig. 20.6 WAQSS – The Water Quality Service System, a service for coastal management provided by Brockmann Consult, Germany; http://www.brockmann-consult.de/waqss/. The system is user friendly and provides both standard and customized products to the end-user

[4]http://www.smhi.se/oceanografi/oce_info_data/BAWS/algal_blooms_baltic_en.htm; http://www.ymparisto.fi/default.asp?contentid=87506&lan=en

products to end-users. Since 2006, the MarCoast service has been providing information on oil spills and water quality including information on chl-*a* concentration, SPM load, water transparency, and SST. The standard products of the WAQSS monitoring system are thematic maps of the North Sea and the Baltic Sea, including daily animations. The end-users can also request specific, custom-made products.

Another web-based information system for water quality products, specifically for the lakes Vänern, Vättern, and Mälaren, has been developed by Vattenfall Power Consultant Sweden (www.vattenkvalitet.se). This service has recently been extended to the coastal areas of the northwestern Baltic Sea, including Himmerfjärden bay and the Stockholm Archipelago and is currently extended to the Gulf of Bothnia. The water quality products include distribution of SPM, chl-*a*, and CDOM, derived from MODIS and MERIS data. Interactive maps are provided together with web-based user interface.

20.2.2 Autonomous Systems for Sea-Truthing of Satellite Data

20.2.2.1 The FerryBox System

The Baltic Sea is a productive ecosystem with strong nutrient input and frequent algal bloom development. In coastal areas the water mass exchange rates are high. For adequate monitoring of these highly dynamic areas it is necessary to develop monitoring systems that provide both a good temporal and a good spatial resolution. The European Union FerryBox program (Sørensen 2006) was developed to extend the two-dimensional coverage from satellite data through high frequency, one-dimensional transects from ships of opportunity and vertical measurements from oceanographic buoys at selected locations. The resulting operational system has a well-resolved temporal and spatial resolution. Figure 20.7 shows an example from the operational FerryBox website, where satellite products can be controlled with FerryBox data near real time.

A typical FerryBox system consists of automated sensors for measuring temperature, salinity, turbidity, and chlorophyll-*a* in vivo fluorescence. The data are transmitted to a station on land in real time via Internet, GPRS, or GMS connection. The Norwegian Institute for Water Research (NIVA) make their water quality products available on a web map server (www.ferrybox.no).

The FerryBox system is also equipped with an automatic, cooled water sampler that is used to take discrete water samples for calibrating the automatic sensors and also serves as sea-truthing data for satellite observations. The water samples are taken into harbour and subsequently analysed in the laboratory. Chlorophyll-*a* in vivo fluorescence can be used as a proxy for chlorophyll-*a* concentration, but it must be calibrated against measured chlorophyll-*a*. The turbidity sensor data can be used as a proxy for TSM after calibration of turbidity against TSM concentration. The measured water quality samples can also be used directly to validate satellite data when the sample meets the satellite match-up criteria of water sampling within half an hour of the satellite overpass.

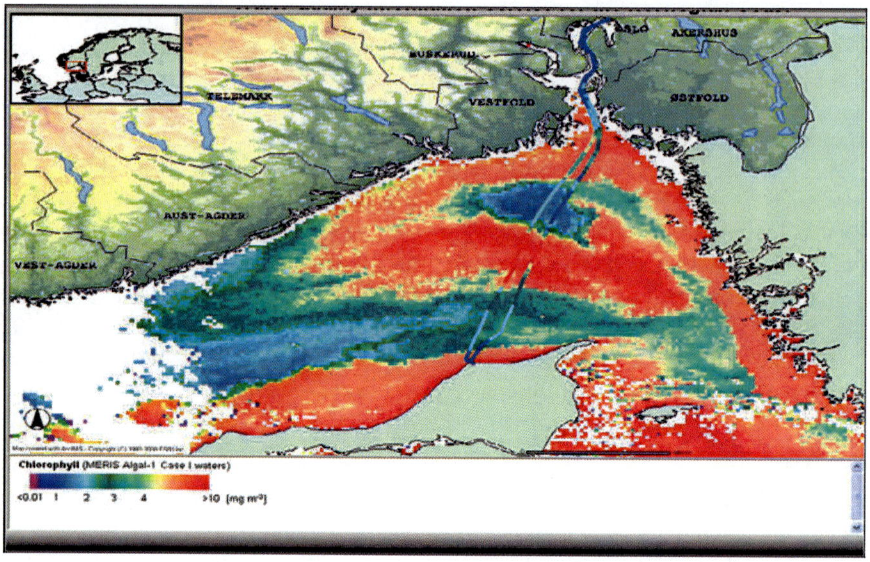

Fig. 20.7 A satellite Algal_1 (chlorophyll) image from MERIS in February 2008, overlaid with Ferrybox chlorophyll-*a* fluorescence data. Operational web site www.ferrybox.no

The development of seasonal stratification may lead to large discrepancies between chlorophyll-*a* derived from Ferrybox and from satellite data. It is therefore necessary to analyse Ferry Box data sets carefully. Some FerryBox lines are also equipped with fluorometers that are sensitive to CDOM and/or fluorometers that are sensitive to phycobilin pigments in order to detect cyanobacteria. Other FerryBox lines, e.g. the FerryBox from Oslo to Kiel, also have radiance sensors on deck to validate water-leaving radiance and remote-sensing reflectance derived from satellite TOA measurements. In the Baltic Sea, there are several FerryBox systems in operation (Ainsworth 2008), which are summarized in Fig. 20.8. The FerryBox data will be used in future monitoring and forecasting of the marine environment, e.g. within the EU FP7 project MyOcean.

20.2.2.2 The In Situ Autonomous NASA AERONET-Ocean Colour Stations

One way to improve truthing of satellite observations is to use autonomous validation stations placed on fixed platforms, such as oceanographic towers or light houses. This is done within the NASA AERONET-OC (AEROsol RObotic NETwork – Ocean Colour), which is the most advanced examples of an autonomous in situ station (http://aeronet.gsfc.nasa.gov). On an AERONET-OC tower, sun photometers are installed measuring atmospheric properties and water-leaving signals (Zibordi et al. 2006, 2009). The in situ data provided is of very high value as the oceanic and the atmospheric signal is measured simultaneously and continuously. Currently, there are eleven AERONET-OC stations worldwide, three of which are based in the Baltic Sea area: on Gustaf Dalén lighthouse at the Swedish coast in the

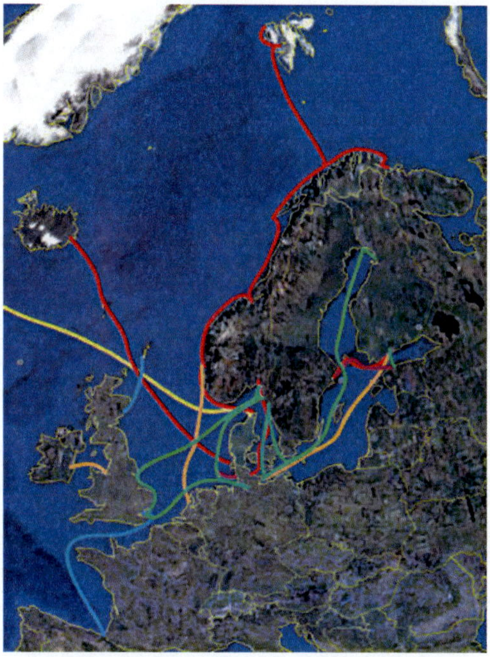

Fig. 20.8 Map of FerryBox systems in Europe operated by NIVA (NO), GKSS (GER), IMR (NO), BCCR/UoB (NO), NOCS (UK), POC (UK), Marlab/FRS (UK), NIOZ (NL) SYKE (FIN), SMHI (SWE), TTU (EST), LOMI (EST) (map from Durand et al. 2010)

northwestern Baltic Sea, on Pålgrunden lighthouse in the lake Vänern, and on the Helsinki lighthouse in the Gulf of Finland.

20.3 Recent Results and Developments

In the previous sections an overview of the vast variety of remote sensing and bio-optical techniques was presented, including various online operational and autonomous systems. In the following section some recent results are shown in order to illustrate the new knowledge that can be gained from combining remote sensing with bio-optical techniques.

20.3.1 Assessment of Eutrophication from Space

As mentioned before, Secchi depth is an important parameter that has historically been used to monitor eutrophication in the Baltic Sea, and it is inversely related to the diffuse attenuation coefficient, K_d. The diffuse attenuation coefficient can be derived from remote sensing data, and in ocean colour remote sensing it is common to derive the spectral diffuse attenuation coefficient, $K_d(490)$, from the reflectance ratio at 490 and 550 nm (Mueller 2000, Kratzer et al. 2003, Pierson et al. 2008). One uses the band at 490 nm as it contains information about all optical in-water constituents, i.e. phytoplankton pigments as well as CDOM and TSM, whereas the band at 550 nm contains least information about all in-water constituents. $K_d(490)$ is also of special interest from a monitoring point of view, because it has previously been shown to be the most reliable product that can be derived from remote sensing imagery over the Baltic Sea (Darecki and Stramski 2004, Kratzer et al. 2008). Figure 20.9 shows a Secchi depth map of the Baltic Sea derived with SeaWiFS with a 1 km resolution. Secchi depth and $K_d(490)$ correlations were modelled based on in situ data of the northwestern Baltic, and $K_d(490)$ was derived from SeaWiFS data. For validation purposes, in situ measured Secchi depth data from the open Baltic Sea and the Gulf of Riga were compared to satellite-derived Secchi depth and showed good agreement (Kratzer et al. 2003). Kratzer et al. (2008) derived $K_d(490)$ and Secchi depth from MERIS full resolution data of 300 m using MERIS channels 3 (490 nm) and 6 (620 nm). Figure 20.10 shows the spectral diffuse attenuation coefficient $K_d(490)$ for Himmerfjärden bay and adjacent areas, derived from MERIS full resolution data. The algorithm was validated with sea-truthing data using another MERIS scene. The example demonstrates that the data can be used to monitor water transparency also in coastal areas. This type of information with its synoptic view is very valid as it cannot be gained from any other technique.

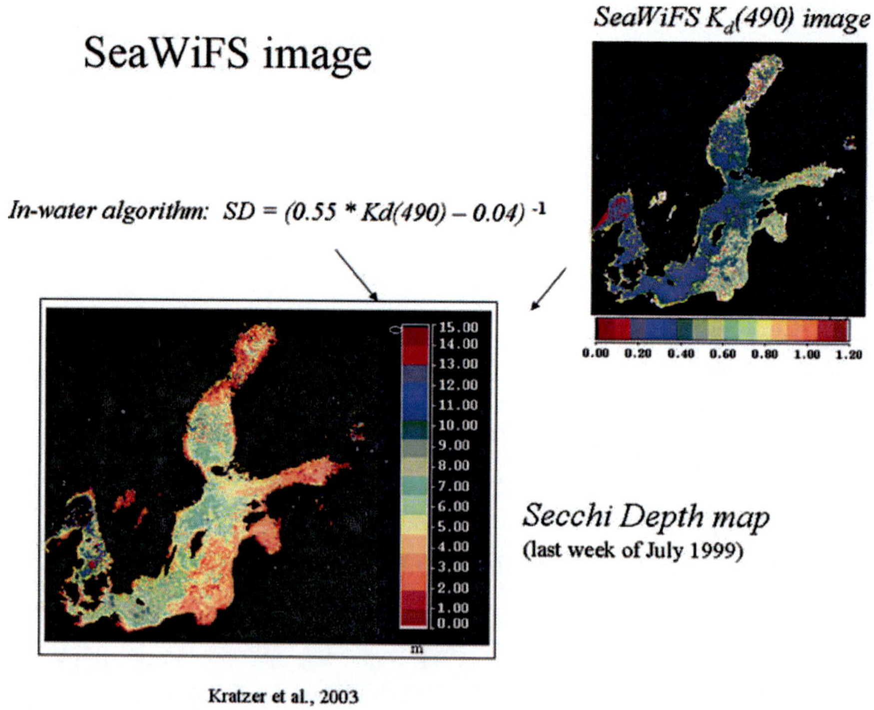

Fig. 20.9 Secchi depth map derived from SeaWiFS diffuse attenuation at 490 nm, $K_d(490)$, and a local in-water algorithm (Kratzer et al. 2003); composite image for the first week of August 1999. Strictly speaking, the map is only valid in the Baltic proper and/or areas with YS concentrations similar to those in the northwestern Baltic Sea (as YS is the dominant optical component)

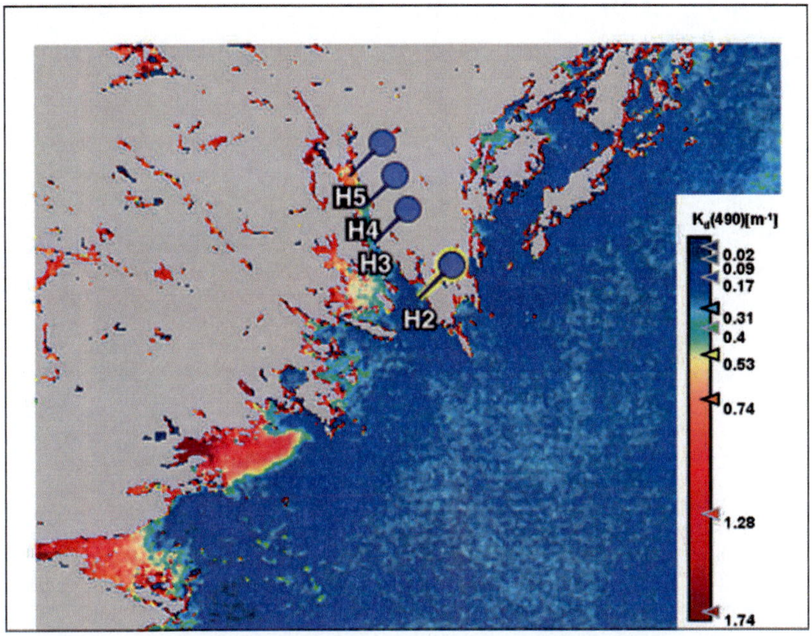

Fig. 20.10 Diffuse attenuation has been shown to be the most accurate product for the Baltic Sea. The image presented here shows that it is possible to monitor water quality in coastal areas of the northwestern Baltic Sea (from Kratzer et al. 2008). H2, H3, H4, and H5 are standard stations of the Swedish national monitoring program. H5 is situated about 0.7 km south of the outlet of Himmerfjärden sewage treatment plant

20.3.2 Optical Gradients of Inorganic Suspended Matter in Coastal Waters

Using bio-optical data, Kratzer and Tett (2009) have developed an attenuation model for the northwestern Baltic Sea that explains the contribution of each optical component to the diffuse attenuation of light. Figure 20.11 illustrates the changes in attenuation from source (coast) to sink (open sea, in this case Landsort Deep). CDOM is the dominant optical component in both the open sea and the coastal areas, with a steady increase towards the head of the fjord. Organic suspended particulate matter (organic SPM) did not show a spatial trend when comparing open sea to coastal data. However, inorganic suspended particulate matter (inorganic SPM) showed a clear spatial trend. The optical influence of inorganic SPM can be detected to approximately 15–20 km off the coast, which means that the coastal influence reaches much beyond the one nautical mile line (i.e. 1.85 km) from the coastal baseline as defined by the EU Water Framework Directive (WFD). This means that strictly speaking the breadth of the coastal zone should be in the range of tens of kilometres, which is also in the same dimension as the influence of coastal upwelling (5–20 km off the coast), which may bring up nutrient-rich bottom waters into the

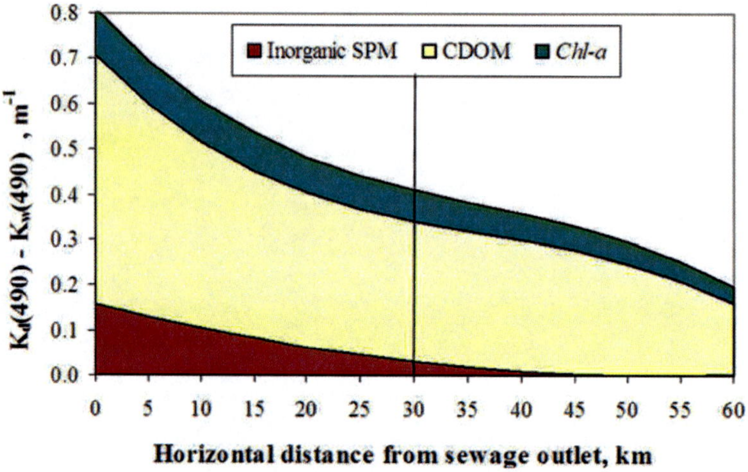

Fig. 20.11 Stacked contributions of the main optical component (CDOM, *chl-a* and inorganic SPM) to the diffuse attenuation coefficient, $K_d(490)$, along a transect from the outlet of Himmerfjärden sewage treatment plant to Landsort Deep, the deepest part of the Baltic Sea (459 m depth). Note that $K_d(490)$ was corrected for the attenuation of water itself, $K_w(490)$. The *black line* indicates the end of Himmerfjärden and the beginning of the open sea (after Kratzer and Tett 2009)

surface mixed layer, stimulating primary production (Fig. 20.2). From this point of view it would be therefore desirable to further extend the breadth of the coastal zone as defined in the WFD. As the northwestern Baltic Sea is characterized by relatively low terrestrial runoff compared to, e.g., the southern Baltic Sea, coastal waters extend even further offshore in the southern Baltic Sea. Chlorophyll is the optical component that is most variable over the year as it is so dependent on the nutrient status. However, both CDOM and TSM also have a strong seasonal cycle governed by seasonal changes in precipitation and runoff. In coastal areas of the Baltic Sea, there is also a significant contribution of inorganic SPM to the optical signal, and it also increases the reflective signal from the water close to the coast. Darecki et al. (2003) have also observed very large variability in reflectance in the Southern Baltic, which may be related to relatively high sediment loads compared to the northern Baltic Sea.

20.3.3 Synoptic Use of Remote Sensing and In Situ Techniques

Figure 20.12 shows the spatial distribution of TSM in the Baltic Sea on 23rd April 2008. The FUB/WeW[5] MERIS Water Processor was here applied to MERIS RR data (Schroeder et al. 2007a, b). The panels illustrate qualitatively and quantitatively

---

[5]FUB: Freie Universität Berlin, WeW: Institut für Weltraumwissenschaften (Institute for Space Science).

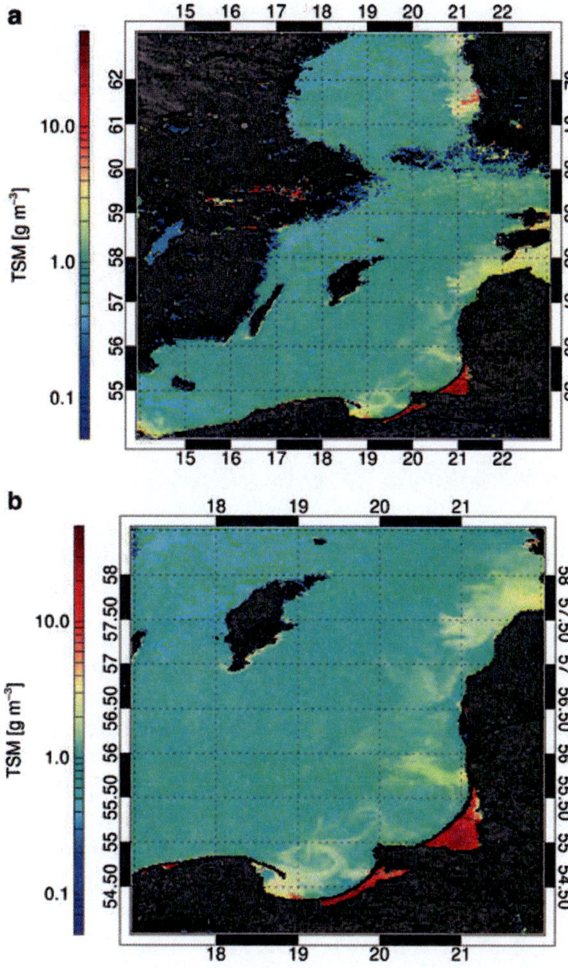

Fig. 20.12 (**a**) Total suspended matter (TSM) load in the Baltic Sea on 23 April 2008. RGB-composite (**b**) zoom of TSM load of southwestern Baltic with river runoff. Geo-physical products were derived from ENVISAT MERIS RR data (Algorithm description in Schroeder et al. 2007b). Courtesy of Institute for Space Science, Freie Universität Berlin. Longitude in °E, Latitude in °N

the gradients in SPM concentrations, and the clear differences between the southern and the northern Baltic Sea. The processor used has been especially developed for coastal areas and showed good retrieval of TSM and chlorophyll in the northwestern Baltic Sea (Kratzer et al. 2008). In coastal areas, the percent error for SPM was about 16% and in the open sea approximately 6%. For chlorophyll, the percent error was approximately 67% error in coastal areas and –35% in the open Baltic Sea.

The absorption coefficient of CDOM, g440, was underestimated by 37% in the coastal waters and by –74% in the open Baltic Sea. The FUB/WeW showed overall best results when taking into account all optical in-water constituents as well as $K_d(490)$ (Vinterhav 2008).

However, when first applying the adjacency correction using ICOL (Santer et al. 2007) to the level 1 data, and then reprocessing the level 2 data, both the MERIS standard processor and the FUB/WeW Water Processor showed rather good results (Kratzer and Vinterhav 2010). The MERIS standard processor retrieved chlorophyll

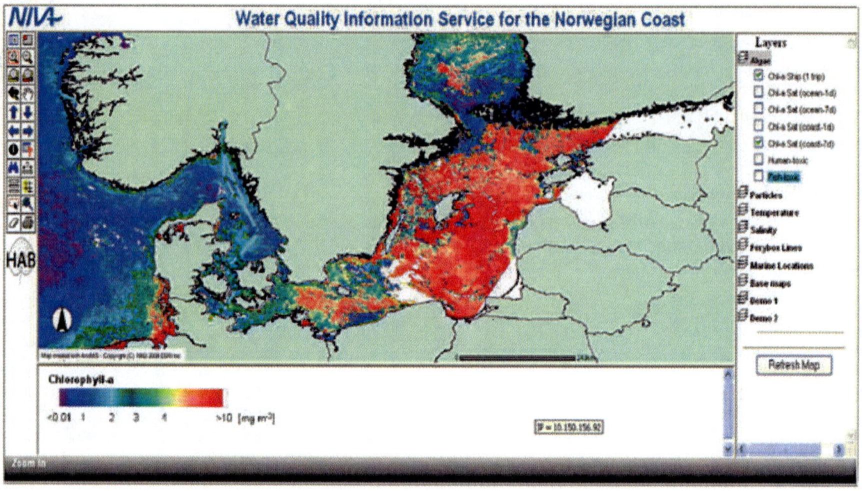

Fig. 20.13 NIVA Water Quality Information Service for southern Scandinavia and the Baltic Sea, here showing a 7-day binned Algal_2 product from July 2008. The data from FerryBox and MERIS were combined in real time using data from the Kattegat and Skagerrak

within 19% in coastal areas and the FUB/WeW within 25% at the open sea stations after ICOL correction.

This has great implications for Baltic Sea management, as chlorophyll is a better indicator for eutrophication than Secchi depth or $K_d(490)$. Chlorophyll is directly influenced by the nutrient status, whereas Secchi depth and $K_d(490)$ are influenced not only by the concentration of chlorophyll but also by TSM load and CDOM attenuation. By combining conventional monitoring with optical in situ techniques and remote sensing the number of observations as well the spatial coverage can be improved substantially.

Figure 20.13 shows how the NIVA Water Quality Information Service combines satellite data in near real time with information from the FerryBox system. The new FerryBox line in the southern Baltic has been operational since 2008. The data gathered by the automatic sampling system give information from 4 to 5 m depth, determined by the intake of the flow-through system. The combined use of different methods and several parameters gives a synoptic and consistent view into the physical drivers, bio-geochemical interactions, and dynamical processes of the ecosystem.

20.4 Conclusions and Outlook

Visible spectral remote sensing combined with sea-truthing and conventional monitoring provides information with high temporal resolution. The high spatial, spectral, and radiometric resolution of ESA's ocean colour sensor MERIS is of special interest for coastal applications. Remote sensing is a repetitive and stable method to provide relatively cost-effective synoptic observations over the whole large area of the Baltic Sea basin, including the Skagerrak and Kattegat. The strength of remote sensing data is foremost to display the geographical patterns of complex ecosystem processes. The use of remote sensing data in combination with in situ measurements has greatly advanced our understanding of the Baltic Sea ecosystem.

The water of the Baltic Sea is optically very complex. Remote sensing combined with a number of autonomous in situ monitoring techniques can be used to analyze the spatial dynamics and seasonal patterns of the Baltic Sea. In situ data are required to develop bio-optical retrieval algorithms, to validate them, and to assure the quality of the derived atmospheric and in-water products. In situ data are obtained from ships-of-opportunity, dedicated sea-truthing campaigns, and optical moorings. Remote sensing and bio-optical monitoring are relatively new disciplines, but have already significantly increased our understanding of the Baltic Sea ecosystem. Further development of ocean colour technology and retrieval algorithms in combination with advanced in situ instrumentation, and complex bio-geochemical coastal zones models, may lead to a revolution in knowledge and management of coastal ecosystem (IOCCG Report 2000).

There is a strong need to further link up remote sensing and operational in situ techniques with conventional monitoring techniques in order to secure quality-controlled synoptic monitoring of the Baltic Sea. This is also crucial for improved management of the Baltic Sea. The objective of the Helsinki Commission (HELCOM) for the protection of the Baltic Sea environment is to maintain its biological status and diversity as expressed in the Baltic Sea Action Plan (HELCOM 2007), which is based on the new European Marine Strategy Directive. In traditional in situ monitoring, Secchi depth and chlorophyll are used as biological indicators for water quality. Reliable water quality maps can now be derived from MERIS data and will be used in future to monitor and evaluate the HELCOM objective of restoring water transparency. Furthermore, they can be used to monitor the effects of eutrophication from space. The wide aerial coverage, the repetition and continuity of the satellite observations, the consistency of the measured data, and a relative cost-effectiveness clearly respond to the demands of a modern operational monitoring system. MERIS is now able to sense water quality parameters in the coastal zone and will in future be increasingly used in integrated coastal zone management.

More research is needed in order to realize the full potential of remote sensing in the Baltic Sea. This objective is addressed by the MERIS Date Quality Working Group in collaboration with the MERIS validation team. One major goal is to develop and validate atmospheric correction procedures over optically complex waters, as well as to improve and validate adjacency corrections, so that subsequently the retrieval of water quality parameters can be improved. Furthermore, ESA is going to continue its ocean colour mission. The Ocean and Land Colour Instrument (OLCI), an instrument optically similar to MERIS, will be launched on SENTINEL-3 in 2013, and is planned to be operational until 2023. This will allow for continuous and consistent long-term trend assessment of water quality and climate change induced effects in the Baltic Sea basin.

Short Glossary

AERONET-OC	AEROsol RObotic NETwork – Ocean Colour
AOPs	Apparent Optical Properties
AVHRR	Advanced Very High Resolution Radiometer
C2R	Coastal Case-2 Regional Water Processor
Chl-*a*	Chlorophyll-*a*
CDOM	Chromophoric or Coloured Dissolved Organic Matter
DIN	Dissolved Inorganic Nitrogen
DIP	Dissolved Inorganic Phosphorus
EC	European Commission
ENVISAT	European ENVIronmental SATellite
ESA	European Space Agency
EU	European Union
FUB	Freie Universität Berlin
GPRS	General Packet Radio Service
GMES	Global Monitoring of Environment and Security
GSM	Global System for Mobile communications
HELCOM	HELsinki COMmission
ICOL	Improved Contrast Between Ocean and Land processor
IOPs	Inherent Optical Properties
MERIS	MEdium Resolution Imaging Spectrometer
MODIS	MODerate Imaging Spectroradiometer
NASA	National Aeronautics and Space Administration
NIR	Near-InfraRed
NIVA	Norsk institutt for vannforskning, Norwegian Institute for Water Research
OLCI	Ocean and Land Colour Instrument
RGB	Red Green Blue
SeaWiFS	Sea-viewing Wide Field-of-view Sensor
SMHI	Swedish Meteorological and Hydrological Institute

SPM	Suspended Particulate Matter
SST	Sea Surface Temperature
SYKE	Finnish Environmental Institute
TOA	Top of Atmosphere
TSM	Total Suspended Matter
VIS	Visible
VSF	Volume Scattering Function
WAQSS	Water Quality Service System
WFD	Water Framework Directive of EC
WeW	Institut für Weltraumwissenschaften, Institut of Space Science
a	Absorption coefficient
b	Scattering coefficient
b_b	Backward Scattering coefficient
b_f	Forward Scattering coefficient
I	Radiance
E	Irradiance
E_d	Downwelling irradiance
K_d	Diffuse Vertical Attenuation Coefficient
R	Sea surface reflectance

References

Aiken J, Moore G, Trees CC, Hooker SB, Clark DK (1995) Volume 29, The SeaWiFS CZCS-type pigment algorithms. In: Hooker SB, Firestone ER, Acker JG (eds) SeaWiFS technical report series, NASA Technical Memorandum 104566, Maryland, 29:6–13

Ainsworth C (2008) FerryBoxes begins to make waves. Science 322:1627–1629

Askne J, Dierking W (2008) Sea ice monitoring in the Arctic and Baltic Sea using SAR. In: Barale V, Gade M (eds) Remote sensing of the European Seas. Springer, Netherlands, pp 383–398

Bernes (2006) Change beneath the surface – an in-depth look at Swedenś marine environment, Monitor 19, Swedish Environmental Protection Agency, ISBN 91-620-1246-0

Bianchi TS, Engelhaupt E, Westman P, Andrén T, Rolff C, Elmgren (2000) Cyanobacterial blooms in the Baltic Sea: natural or human-induced? Limnology and Oceanography 45:716–726

Brekke C, Solberg AHS (2005) Oil spill detection by satellite remote sensing. Remote Sensing of Environment 95:1–13

Bukata R (2005) Satellite monitoring of inland and coastal water quality: restrospection, introspection, future directions. CRC Press, Boca Raton, FL, 246 pp

Carper WJ, Lillesand TM, Kiefer RW (1990) The use of Intensity-Hue-Saturation transformations for merging SPOT panchromatic and multispectral image data. Photogrammetric Engineering and Remote Sensing 56(4):459–467

C.E.C. (2000) Directive/60/EC of the European Parliament and of the Council of 23 October 2000 establishing a framework for Community action in the field of water policy. Official Journal of the European Communities L 327:1–73

Chavez PS Jr, Sides SC, Andersen JA (1991) Comparison of three different methods to merge multiresolution and multispectral data: Landsat TM and SPOT panchromatic. Photogrammetric Engineering and Remote Sensing 57(3):295–303

Darecki M, Weeks A, Sagan S, Kowalczuk P, Kaczmarek S (2003) Optical characteristics of two contrasting Case 2 waters and their influence on remote sensing algorithms. Continental Shelf Research 23(3):237–250

Darecki M, Stramski D (2004) An evaluation of MODIS and SeaWiFS bio-optical algorithms in the Baltic Sea. Remote Sensing Environment 89:326–350

Diaz RJ, Rosenberg R (2008) Spreading dead zones and consequences for marine ecosystems. *Science* 321(5891):926–929

Doerffer R, Sorensen K, Aiken J (1999) MERIS potential for coastal zone application. International Journal of Remote Sensing 20(9):1809–1818

Doerffer R, Schiller H (2006a) The MERIS neural network algorithm, Chapter 6. In: Lee ZP (ed) Remote sensing of inherent optical properties: fundamentals, tests of algorithms, and applications. Reports of the International Ocean-Colour Coordinating Group, MacNab Print, Dartmouth, Canada

Doerffer R, Schiller H (2006b) MERIS advanced water algorithm, Version 1.0. GKSS

Doerffer R, Schiller H (2008a) MERIS regional coastal and lake Case 2 water project atmospheric correction ATBD, Version 1.0. GKSS

Doerffer R, Schiller H (2008b) MERIS lake water algorithm for BEAM ATBD, Version 1.0. GKSS

Durand D, Folkestad A, Sørensen K (2010) http://www.mercator-ocean.fr/documents/lettre/lettre_39_en.pdf

Fischer J, Fell F (2001) Numerical simulation of the light field in the atmosphere-ocean system using the matrix-operator method. Journal of Quantitative Spectroscopy & Radiative Transfer 69:351–388

Fleming V, Kaitala S (2007) Phytoplankton spring bloom biomass in the Gulf of Finland, Northern Baltic Proper and Arkona Basin in 2007. HELCOM Indicator Fact Sheet, last visited 25 February 2009

Gidhagen L (1987) Coastal upwelling in the Baltic Sea – satellite and in situ measurements of sea-surface temperatures indicating coastal upwelling. Estuarine, Coastal and Shelf Science 24:449–462

Gordon HR, Morel AY (1983) Remote assessment of ocean colour for interpretation of satellite visible imagery: a review. Springer, New York, NY

HELCOM (2007) The HELCOM Baltic Sea action plan, http://www.helcom.fi/press_office/news_helcom/en_GB/BSAP_full/, last visit: 2008-12-05

IOCCG (2000) Remote sensing of ocean colour in coastal, and other optically-complex, waters. In: Sathyendranath S (ed) IOCCG report 3. International Ocean-Colour Coordinating Group, MacNab Print, Dartmouth, Canada

Jorgensen PV (1999) Standard CZCS Case 1 algorithms in Danish coastal waters. International Journal of Remote Sensing 20(7):1289–1301

Kahru M, Håkansson B, Rud O (1995) Distribution of sea-surface temperature fronts in the Baltic Sea as derived from satellite imagery. Continental Shelf Research 15(6): 663–679

Kahru M (1997) Using satellites to monitor large-scale environmental change: a case study of cyanobacteria blooms in the Baltic Sea. In: Kahru M, Brown CW (eds) Monitoring algal blooms, New techniques for detecting large-scale environmental change. Springer, Berlin Heidelberg, New York, pp 43–61

Karlsson K-G (1996) Cloud classifications with the SCANDIA model. SMHI reports meteorology and climatology, no. 67

Kautsky N, Kautsky H, Kautsky U, Waern M (1986) Decreased depth penetration of *Fucus vesiculosus* (L.) since the 1940's indicate eutrophication of the Baltic Sea. Marine Ecology Progress Series 28:1–8

Kirk JTO (2010) Light and photosynthesis in aquatic ecosystems, 3rd edn. Cambridge University Press, Cambridge, 649pp

Kononen K, Leppänen J-M (1997) Patchiness, scales and controlling mechanisms of cyanobacterial blooms in the Baltic Sea: application of a multiscale research strategy. In: Kahru M, Brown CW (eds) Monitoring algal blooms. New techniques for detecting large-scale environmental change. Springer, Berlin, Heidelberg, New York, pp 63–84

Kowalczuk P, Stedmon CA, Markager S (2006) Modelling absorption by CDOM in the Baltic Sea from season, salinity and chlorophyll. Marine Chemistry 101:1–11

Kratzer S (2000) Bio-optical properties of coastal waters. PhD thesis, School of Ocean Sciences, University of Wales Bangor, UK

Kratzer S, Håkansson B, Sahlin C (2003) Assessing Secchi and photic zone depth in the Baltic Sea from Space. Ambio 32(8):577–585

Kratzer S, Brockmann C, Moore G (2008) Using MERIS full resolution data (300 m spatial resolution) to monitor coastal waters – a case study from Himmerfjärden, a fjord-like bay in the north-western Baltic Sea. Remote Sensing of Environment 112(5):2284–2300

Kratzer S, Tett P (2009) Using bio-optics to investigate the extent of coastal waters – a Swedish case study. Hydrobiologia 629:169–186

Kratzer K, Vinterhav C (2010) Improvement of MERIS data in Baltic Sea coastal areas by applying the Improved Contrast between Ocean and Land processor (ICOL). Oceanologia 52(2): 211–223

Kullenberg G (1981) Physical oceanography. In Voipio A (ed) The Baltic Sea. Elsevier Oceanography Series 30, 418 pp. Elsevier, New York, NY

Kutser T, Metsamaa L, Strombeck N, Vahtmae E (2006) Monitoring cyanobacterial blooms by satellite remote sensing. Estuarine, Coastal and Shelf Science 67(1–2):303–312

Kutser T, Metsamaa L, Vahtmäe E, Aps R (2007) Operative monitoring of the extent of dredging plumes in coastal ecosystems using MODIS satellite imagery. Journal of Coastal Research SI50:180–184

Lehmann A, Myrberg K (2008) Upwelling in the Baltic Sea – a review. Journal of Marine Systems 74:3–12

Leppänen J-M, Niemi A, Rinne I (1988) Nitrogen fixing of cyanobacteria (blue-green algae) and the nitrogen cycle of the Baltic Sea. Symbiosis 6(1–2):181–194

Meier HEM, Kauker F (2003) Sensitivity of the Baltic Sea salinity to the freshwater supply. Climate Research 24:231–242

Milliman JD (2001) Delivery and fate of fluvial water and sediment to the sea: a marine geologist's view of European rivers. Scientia Marina 65(2):121–132

Moore G, Lavender S (2010) MERIS ATBD 2.6 – Case II.S Bright pixel atmospheric correction. ESA Algorithm Theoretical Basis Documents, revision 6, Argans & Bio-Optika, UK

Morel A, Prieur L (1977) Analysis of variations in ocean colour. Limnology and Oceanography 22:709–722

Mueller JL (2000) SeaWiFS algorithm for the diffuse attenuation coefficient, K (490), using water-leaving radiances at 490 and 555 nm. In: Hooker SB, Firestone ER (eds) SeaWiFS postlaunch calibration and validation analyses, Part 3. NASA GSFC, Greenbelt, MD, pp 24–27

Ohde T, Siegel H, Gerth M (2007) Validation of MERIS Level-2 products in the Baltic Sea, the Namibian coastal area and the Atlantic Ocean. International Journal of Remote Sensing 28(3&4):609–624

Pierson D, Kratzer S, Strömbeck N, Håkansson B (2008) Relationship between the attenuation of downwelling irradiance at 490 nm with the attenuation of PAR (400 nm–700 nm) in the Baltic Sea. Remote Sensing of Environment 112:668–680

Preisendorfer RW (1986) Secchi disk science: visual optics of natural waters. Limnology and Oceanography 31:906–926

Reinart A, Kutser T (2006) Comparison of different satellite sensors in detecting cyanobacterial bloom events in the Baltic Sea. Remote Sensing of Environment 102(1–2):74–85

Robinson IS (2004) Measuring the oceans from space: the principles and methods of satellite oceanography. Springer Praxis Books/Geophysical Sciences. Springer, Berlin, Chapter 7, 669 pp

Rud O, Kahru M (1995) Monitoring of harmful algal blooms in the Baltic Sea. Remote Sensing 26:8–10

Sandén P, Håkansson B (1996) Long-term trends in the Secchi depth in the Baltic Sea. Limnology and Oceanography 41(2):346–351

Santer R, Zagolski F, Gilson M (2007) Université du Littoral, France, ICOL ATBD, Version 0.1, February 28, 2007

Sathyendranath S, Hoge FE, Platt T, Swift RN (1994) Detection of phytoplankton pigments from ocean color: improved algorithms. Applied Optics 33:1081–1089

Schroeder T, Behnert I, Schaale M, Fischer J, Doerffer R (2007a) Atmospheric correction for MERIS above Case-2 waters. International Journal of Remote Sensing 28(7): 1469–1486

Schroeder T, Schaale M, Fischer J (2007b) Retrieval of atmospheric and oceanic properties from MERIS measurements: a new Case-2 water processor for BEAM. International Journal of Remote Sensing 28(24):5627–5632

Schwarz JN, Kowalczuk P, Kaczmarek S, Cota GF, Mitchell BG, Kahru M, Chavez FP, Cunningham A, McKee D, Gege P, Kishino M, Phinney AD, Raine R (2002) Two models for absorption by coloured dissolved organic matter (CDOM). Oceanologia 44(2): 209–241

Siegel H, Gerth M, Neumann T, Doerffer R (1999) Case studies on phytoplankton blooms in coastal and open waters of the Baltic Sea using Coastal Zone Color Scanner data. International Journal of Remote Sensing 20(7):1249–1264

Siegel H, Gerth M, Tschersich G (2006) Sea surface temperature development of the Baltic Sea in the period 1990–2004. Oceanologia 48(S):119–131

Siegel H, Gerth M (2008) Optical remote sensing applications in the Baltic Sea. Barale V, Gade M (eds) Remote sensing of European Seas. Springer, New York, NY

Sørensen K (2006) Use of ferrybox data for validation purposes of satellite data. Ferrybox Work Package 5 Team. Deliverable D-5-4. EU-FerryBox. Contract No EVK2-2002-0014

Sørensen K, Aas E, Høkedal J (2007) Validation of MERIS water products and bio-optical relations in the Skagerrak. International Journal of Remote Sensing 28(3–4):555-568

Stigebrandt A (2001) Physical oceanography of the Baltic Sea, Chapter 2. In: Wulff F, Rahm L, Larsson P (eds) A system analysis of the Baltic Sea. Springer, Berlin, pp 19–74

Subramaniam A, Kratzer S, Carpenter JC, Söderbäck E (2000) Remote sensing and optical in-water measurements of a cyanobacteria bloom in the Baltic Sea. Proceedings of the 6th International conference on remote sensing for marine and coastal environments. Veridian ERIM International, Charleston, SC, I-57-64

Tett P, Gilpin L, Svendsen H, Erlandsson CP, Larsson U, Kratzer S, Fouilland E, Janzen C, Lee J-Y, Grenz C, Newton A, Ferreira JG, Fernandes T, Scory S (2003) Eutrophication and some European waters of restricted exchange. Continental Shelf Research 23:1635–1671

Vahtera E, Laanemets J, Pavelson J, Huttunen M, Kononen K (2005) Effect of upwelling on the pelagic environment and bloom-forming cyanobacteria in the western Gulf of Finland, Baltic Sea. Journal of Marine Systems 58(1–2):67–82

Victorov SV (1996) Regional satellite oceanography. Taylor & Francis, London, 306 pp

Vinterhav C (2008) Remote sensing of Baltic coastal waters using MERIS – a comparison of three Case-2 water processors. Examensarbete (undergraduate degree thesis, 30 ECTS), Department of Physical Geography and Quaternary Geology, Stockholm University

Voipio A (1981) The Baltic Sea. Elsevier Science Publishers, New York, NY

Wasmund N, Andrushaitis A, Lysiak-Pastuszak E, Muller-Karulis B, Nausch G, Neumann T, Ojaveer H, Olenina I, Postel L, Witek Z (2001) Trophic status of the south-eastern Baltic Sea: a comparison of coastal and open areas. Estuarine Coastal and Shelf Science 53(6):849–864

Williams T, Thouveny N, Creer KM, Punufimana AN, Seuseu NKF, Shawyer RJ, Gani ASB, Hojerslev NK, Holt N, Aarup T (1996) Optical measurements in the North Sea-Baltic Sea transition zone. I. On the origin of the deep water in the Kattegat. Continental Shelf Research 16:1329–1342

Zibordi G, Strömbeck N, Mélin F, Berthon J-F (2006) Tower-based radiometric observations at a coastal site in the Baltic Proper. Estuarine, Coastal and Shelf Science 69:649–654

Zibordi G, Holben B, Slutsker I, Giles D, D'Alimonte D, Mélin F, Berthon JF, Vandemark D, Feng H, Schuster G, Fabbri BE, Kaitala S, Seppälä J (2009) AERONET-OC: a network for the validation of ocean color primary products. Journal of Atmospheric and Oceanic Technology 26:1634–1651

Index

A
Absorption (a), 111, 408–409, 412–413, 418–419, 427
Acceleration due to gravity, 269
Accommodation, 233–248
Accumulation, 6–7, 31, 37–39, 54, 56–58, 61, 64–68, 100, 103, 105, 114, 123, 128, 153, 182, 203, 205, 207–208, 219, 234, 242–243, 245, 247–248, 259, 263, 282–284, 292, 294–295, 339, 341, 345, 348, 358, 367, 369, 382, 411, 416
  rates, 67
Acids, humic and fulvic, 408
Acoustic echoe, 106
Acoustic index, 111, 116
Actinocyclus octonarius Ehrenberg, 120
Adjacency effect (environmental effect), 417–419
Aeolian processes, 349, 357
Age model, 6, 107–110, 112–113, 122–123
Agriculture, 92, 192, 317, 325, 331, 341, 384, 386–387
Ahrensburgian culture, 311–312, 314
Algal, 374, 382, 387, 421–422, 428
Algorithms, 54–55, 67, 237, 284, 407, 413, 415, 419, 424, 427, 429
Allerød, 190, 214, 308, 311–312
Altdarss, 282
Alum shale, 21, 35, 39
AMS method, 113, 172, 183, 223, 237, 241, 248, 302, 321
Anaerobic, 369
Ancylus
  Lake, 86–88, 112, 115–118, 127, 135, 137, 167–170, 175, 180–185, 209, 213, 315–325
  transgression, 87, 207, 214
Animal husbandry, 317, 326, 331
Anoxia, 8, 125, 355, 366, 368–371, 410
Anthropogenic activities, 8, 338–339, 342, 356–358, 361
Anthropogenic changes, 270
Anthropogenic impact, 264, 341, 358, 360
Anthroposphere, 150, 152, 302, 307, 330
Aphanizomenon sp, 409, 412
Apparent optical properties (AOPs), 412–413
ArcGIS, 210
Archaeological record, 302, 304, 317, 330
Archaeological sites, 8, 168, 243, 302, 304, 306, 317, 320–321, 324–326, 329
Archipelago, 88, 151, 182, 257, 259, 312, 320–322, 331, 338, 421
Arctic Oscillation (AO), 101
Arkona Basin, 76, 82–83, 89, 101, 235
Assessment of eutrophication from space, 423–426
Atlantic Multi-decadal Oscillation (AMO), 101
Atlantic period, 220–221, 224–225, 230
Atmospheric correction, 416, 419, 429
Atmospheric temperature, 91, 113, 153
Aulacoseira granulata (Ehrenberg) Simonsen, 121, 141, 229
Aulacoseira islandica (O. Mũller) Simonsen, 119
Aulacoseira subarctica (Mũller) Krammer, 119–120
Axial lines, 208, 210–212, 215

B
Backscatter (bb), 412–413
Bacteria, 369–370
Baltica, 4, 26–27, 31, 34–35

J. Harff et al. (eds.), *The Baltic Sea Basin*, Central and Eastern European Development Studies (CEEDES), DOI 10.1007/978-3-642-17220-5,

Baltic basin, 3–6, 8, 14–20, 23–29, 31–34, 37–41, 43, 77, 79, 82–84, 87–88, 90–91, 100–101, 113–114, 123–124, 127–128, 151, 156, 209, 235, 241–242, 307–308, 310, 315–316, 418
Baltic Ice Lake (BIL)
  final drainage, 84, 310
  first drainage, 83, 198
Baltic Sea Action Plan, 373–374, 429
Barents Sea, 5, 55, 78
Barotropic inflow, 124
Basin-to-basin transport, 114, 124, 128
Bathymetry, 81, 175, 191–192, 204, 208, 210, 256, 259, 266–267, 285, 369
Beach
  almost equilibrium, 255–276
  bayhead, 262, 273, 275
  degradation, 264, 361
  embayed, 259
  high-energy, 212, 263, 267, 275
  low-energy, 263
  nourishment, 268
  re-nourishment, 161
  ridge, 242, 245–246, 248, 262, 325
  skären type, 258
  step-like development, 211, 275
  subaerial, 272
  uplifting, 257
  young, 43, 76
Bedload transport, 261
Bedrock, 5, 54–58, 62–68, 76–77, 80, 82, 151, 204–205, 257, 259, 338, 367
  surface, 5, 57–58, 63–66
Belgian coast, 238, 240
Berm, 271
Billingen drainage, 169, 180, 182, 192, 199–200
Bio-available, 378, 382, 384–385
Bio-optical data, 425
Biosphere, 3, 302
Black shales, 22, 39–40
Blooms, 88, 369, 374, 409–413, 415, 420–421
Bluff, 234, 261–262, 275
Bock Island, 282–283, 294
Bodden, 234, 283, 294–295
Bond Cycle, 123, 126
Bornholm, 16, 21, 27, 31–32, 34–35, 41, 76, 80, 82, 88, 101–102, 128
Bothnian Sea, 23, 59, 65–66, 338, 408–409
Bottom, 5–7, 13, 30, 65–66, 78, 91, 100–104, 106, 113–119, 124–128, 136–137, 143–144, 151, 199, 204, 206–209, 212, 220–221, 234–235, 245, 267–268, 274, 285, 317, 327, 337, 341, 344, 347–349, 352–358, 366–371, 374, 376, 379, 385, 397, 400, 408–410, 425
Bottom erosion, 347–348, 352–355, 357
Boulder(s), 135, 261–262, 347–348, 355
Boundary condition, 376, 395, 397–398, 400
Brackish freshwater diatom taxa, 88
Brackish sequence, 114
Brackish waters, 3, 89, 120, 303, 365, 373, 410
Breadth of the coastal zone, 425–426
Breaker line, 267
Breakwater, 266, 274
Brommian culture, 308
Bronze Age, 238, 321, 326–327, 329, 331
Bruun's Rule coastal engineering structure(s), 258, 271, 273
Budget, 8, 258, 284, 374, 382–384, 386, 392–394, 402–403
Bulwark, 266–267
Burial, 21, 39, 41–42, 167, 183, 185, 302, 367
Buried organic matter, 168–169, 183–185

C

Caledonian, 15, 20–22, 24, 28–29, 32–35
Caledonides, 3, 15, 20, 24, 32, 34–35, 101, 151
Cambrian, 4, 16–18, 21, 24–26, 30–31, 33, 35, 37–41, 43, 63–66
Ca-Mn-carbonate, 117
Carbon, 80, 89–90, 204, 369, 381, 415
Carboniferous, 4, 19, 22, 24, 26–27, 32, 35, 41
Catastrophe management, 150, 158
Catastrophic, 150, 161–162, 229, 340, 342, 357
Catchment, 8, 55, 77, 125, 366, 374, 376–377, 391–403
Cauchy (third kind boundary condition), 397
^{14}C dated, 88–89, 321–322
Cenozoic, 14, 19, 24, 27, 35, 43
Central Swedish moraine, 189
CERC formula, 270
Channels, 56, 88, 114, 241, 246, 294–295, 327, 345, 358, 416, 424
Chlorophyll *a*, 409, 412–414, 417, 419, 421–422, 429
Chronology, 136, 140–141, 185, 198, 204, 303, 306, 318, 320, 325–326, 330–331
Chronostratigraphic subdivision, 20, 112, 309
Cliff, 7, 136, 151, 203–216, 234–235, 237, 242–245, 247–248, 261, 264, 271, 283, 339, 346, 348, 355
Cliff retreat, 244, 248, 346

Climate
archives, 100
change, 6, 91, 100, 150–151, 292, 308, 311, 338, 361, 391–393, 402, 429
driving forces, 100
forcing, 92
Closure depth (also depth of closure), 268–269, 271–273, 276
CO_2-concentration, 289–290
Coastal, 65, 207, 344, 347–348
barrier, 235, 242–244
current, 118, 258, 261
defence, 150
erosion, 114, 124, 161, 248, 263, 283, 292, 326, 338–342, 345–347, 350–351, 355, 359–360
landforms, 166, 168–169, 175, 182–183, 185, 192, 198
landscape, 256, 275, 302–303, 316–317, 325–327, 361
lowlands, 7
mire, 237–238, 243
processes, 4, 7, 419
protection strategies, 7
upwelling, 411–412, 425
waters, 220, 230, 358, 374–376, 387, 392, 413, 415, 417, 425–427
zone, 8, 37, 65, 88, 150, 162, 165–185, 212, 220, 243, 261, 269, 304, 317, 325–326, 331, 337–361, 425–426, 429
Coast-to-basin transport, 114
Coastline, 6–7, 88, 113, 149–162, 165–185, 189–200, 202–216, 219–230, 233–248, 259, 270, 272–273, 275, 281–296, 303, 311–312, 316, 318, 325, 344, 347, 352, 354, 392–393, 395, 400, 419
Coastline scenarios, 6, 150, 282
Cobbles, 262, 347
Coloured dissolved organic matter (CDOM), 408–409, 412–414, 418–419, 421–423, 425–428
Communication route, 308
Compaction, 108, 233, 284
Compression modulus, 111
Contaminants, 392
Core-to-core correlation, 114–115
Correlation coefficient, 105
Cover peat, 242, 247
Cretaceous, 5, 13, 19, 34, 135–136, 204
Curonian Lagoon, 134, 347, 349, 357, 361
Curonian Spit, 39, 134, 137, 209, 346–349, 361
Cyanobacteria, 88, 91, 369, 376, 409–412, 416, 419–420, 422
Cymatopleura elliptica (Brebisson) Smith, 120

D

Danish basin, 15
Danish Islands, 157, 304
Darss
Sill area, 76, 87, 317, 326
-Zingst Peninsula, 7, 154, 242, 247, 281–296
Data management system, 150
Dating, 6, 15, 20, 80, 89, 104–105, 113, 122, 124, 136–137, 139–141, 144, 168, 170, 172, 174, 180, 198, 204, 207, 215, 219, 224, 237, 242, 302, 304–306, 312, 314, 323–324, 326
Dean's equilibrium profile, 258
Defence water level, 158–159
Deflation, 345
Deglaciation, 6, 67, 76–77, 80, 82–84, 86, 126–127, 167, 169, 175, 180, 182, 198, 302–303, 307–308, 311–312, 316, 328, 338
Delta, 22, 103, 214, 221, 229, 244, 259, 265, 349, 371
Dendrological analyses, 223, 229
Denmark, ice advances, 5, 80
Density, 41, 63–65, 68, 105–106, 108–112, 114, 123, 125, 143, 288, 342, 408
Depositional sequence, 260
Depression, 14–16, 19, 23, 30, 35–37, 43, 56, 64, 67, 78, 143, 151, 205, 207, 241–242, 294, 358
Depth-to-time transformation, 105, 122
Desertification, 8, 365–371
Detergents, 385
Devonian, 4, 19, 22, 24, 26–27, 29–33, 35, 37, 41
Diabase, 19, 24–25, 27, 30, 33
Diatom
analysis, 89, 100, 111, 123, 128, 136, 141, 323
assemblage zone, 119
flora composition, 121
Diatomological analyses, 223, 225
Diatoms, brackish marine, 5, 89, 100, 113, 120–121, 123–124, 126–128, 152, 228
Diffuse attenuation coefficient, Kd, 414, 418, 423–424, 426

Digital elevation model (DEM), 6, 105, 151, 153, 237, 282, 284–286, 295
Digital terrain model (DTM), 6, 101, 165–167, 175, 182, 185, 191–192
Dinantian, 22
DIN:DIP ratio, 410
Diploneis dombilitensis (Ehrenberg) Cleve, 120
Dirichlet (first kind boundary condition), 397
Discharge (river), 102, 115, 258, 273–275, 408
Drainage system, 194, 245, 248, 311, 386
Drill core, 259
Drowned forest, 219–230
Dugout canoe, 326
Dunes
  bluff, 261
  building, 263
  coastal, 182, 351
  face, 263
  forest, 261
  toe, 263, 265, 270
Dyke, 22, 24–25, 27, 31, 161–162
Dynamic landscape, 305

E
Earthquake, 338–340, 345, 356
East Avalonia, 26–27, 31, 35
Eastern European Platform, 157, 345
Eastern Gotland Basin, 5, 88, 92, 100, 102–103, 108, 110, 112–121, 124, 126–128
East European Craton, 4, 13–14, 35
ECHAM/LSG, 289
Echo-sounder, 204, 210–212, 215
Ecological hazards, 8
Economic strategy, 302, 307, 317
Ecosystem, 4, 7–9, 303, 338, 342, 358–359, 366, 374–375, 385–386, 407–429
EC Water Framework Directive (WFD), 374, 415, 425–426
Ediacaran, 4, 16, 20, 24–25, 28, 30–31, 43
Eemian, 5, 37, 66, 77–78, 143
Eemian Baltic Sea, 78
Eemian interglacial, 5, 37
Endmoraine, 192, 194
Endogenic processes, 338, 340, 344–346, 349–350
ENVISAT (European ENVIronmental SATellite), 407, 410, 417, 427
Equilibrium beach profile (EBP), 7, 265, 267–268, 271–272, 276
ERGOM, 375, 377–382, 386
Erosion
  accumulation modeling, 56
  pattern, 57
  rate, 54–58, 67, 350
Erosional, 5–6, 14, 35–36, 43, 55, 57–58, 63, 259, 267, 284, 346, 352, 359
Ertebølle culture, 320
Esker, 61, 262
Estonia, 6, 14, 16, 30, 32–33, 88, 167–168, 172, 175, 182–183, 185, 189–190, 192–199, 257, 259, 262, 265, 269, 275, 308, 324–325, 338
Estonian beaches, 7, 259, 263, 269
Estuarine current system, 101
Estuary, 8, 185, 317, 373–387, 408
European Caledonides, 101
Eustatic change, 6, 154, 159, 162, 213
Eustatic curve, 154–155, 209, 213
Eustatic rise, 162, 236, 302, 320
Eustatic sea level rise, 7, 82, 90, 157, 235, 288, 324, 392
Eutrophication, 8, 75, 92, 366, 369–370, 373–374, 381–384, 386–387, 409–410, 413–414, 423–424, 428
Evapotranspiration, 394–397, 402–403
Evolution, 4, 13–45, 53–69, 92, 114, 203–204, 212–215, 233–248, 257–258, 264, 283, 392, 429
Exogenic processes, 340, 346, 350–355
Extreme sea level scenario, 153, 158–161

F
Facies zone, 100, 104
Fairway, 265
Falster, 20, 234, 241
Fast ferry, 263
Fault, 5, 15, 27–35, 64, 135, 207, 338, 345, 349, 355–356
Federmesser culture, 308
Feeder cliff, 235, 237, 244–245
Fe/Mn concretions, 370
Fennoscandian Shield, 63, 101, 151, 153, 156–157
Fennoscandian uplift, 234
FerryBox System, 421–423, 428
Fertilizer, 92, 386–387
Fe-sulphide, 112, 118
Fetch, 259
Filamentous nitrogen-fixing cyanobacteria, 410
Finite-difference, 395, 399–401

Finland, 5, 15, 32, 43, 56, 59, 61, 63–64, 66–67, 80, 91, 113, 151, 172, 190–193, 195–198, 255–276, 304, 308, 337–361, 365–371, 409, 423
Fischland, 235–236, 239–240, 242, 247–248, 282–283, 294–295
Fishermen, 312, 315, 320
Fishing fence, 302, 304, 306, 318, 326
Fish trap, 306, 318
Flexural rigidity, 67
Flood, 22, 86–87, 89, 159–161, 167, 225, 229, 245–246, 257, 265, 275, 293, 302, 306, 310, 312, 317, 320, 324, 331, 339–341, 345, 349, 351–352, 355, 357–358
Fluorometers, 422
Fluvial process, 284
Fluvioglacial, 53, 56, 63–64, 67, 259
Forcing, 92, 103–104, 122, 126, 256, 258–264, 270, 276
  anthropogenic, 92
  conditions, 276
Forebulge, 197, 234
Foreklint, 259
Fosna-Hensbacka group, 312, 314
Fossil, 84, 152, 204, 207–212, 214–215, 306, 308
Fourier transform, 105
Fragilariopsis cylindrus (Grunow) Krieger, 121
Freshwater input, 90, 408
Funnel Beaker culture, 320

G

Gas, 5, 8, 20, 32, 37–38, 41–42, 105, 338–340, 342, 345, 356–357, 392, 410–411
  vacuoles, 410–411
Gauge, 6, 150, 157, 160–162, 236, 240, 288
Geodetic relative sea-level data, 184
Geoinformation, 204
Geological hazard potential, 8, 337–361
  classification, 356–357
Geological hazards, 8, 337–361
Geological history, 4, 76, 162, 259, 392
Geological processes, 8, 149–162, 282, 339, 345, 356–357, 360–361
Geological and tectonic evolution, 4, 13–45, 53–69
Geomorphic (features), 270, 273
Geophysical survey, 104, 234, 247
Geosphere, 3, 150, 152, 302, 317, 331, 392
Geostatistical correlation, 166
Geo-system, 150, 317
Geotectonic setting, 151
Geothermal, 5, 14, 37
German Baltic coast, 7, 234, 238, 240, 285, 339
GIS, 6, 100, 109, 151, 165, 175, 191, 210, 376–377, 394–395
GIS-based spatial calculation, 165
Glacial
  cycle, 5, 54–56, 58, 61–62, 66–68, 76–77, 100, 135
  erosion, 5, 13–14, 36–37, 43, 53–68, 101, 207
  formations, 259
  and glaciofluvial erosion, 77
  isostatic adjustment (GIA), 6, 44, 153–154
  Lake Peipsi, 192–193, 199
  Lake Võrtsjärv, 191–192
  sedimentation, 61–62, 67, 85
  varved clay, 82
  varves, 82, 113
Glaciation, 44–45, 55–56, 64, 67, 76, 80, 136–137, 143–144, 197, 262, 301
Glaciodislocations, 137, 143
Glacio-isostatic adjustment, 154
Glacio-isostatic rebound, 209, 213, 304
Global warming, 289, 291, 340, 366, 370
Golf of Bothnia, 101
Gondwana, 22, 34
Gotland
  Basin, 5, 88, 92, 100–103, 106, 108, 110, 112–116, 119, 124–128
  Island, 39, 115
Grain size
  class, 287
  mean, 267
Granites, 20, 22–25, 265, 350
Granitoids, 15, 43
Granulometry, 270
Graptolitic shales, 17, 19
Gravity corer, 107, 109, 127
Great Belt, 88, 90, 235
Greifswald, 35, 159, 326
Grenvillian Sea, 25
Groin, 271
Ground penetrating radar (GPR), 237, 241–243
Groundwater
  discharge, 8–9, 391–403
  flow model, 9, 393, 395, 398
  quality, 392
  recharge, 9, 392–403

Groyne, 161
Gulf of Finland, 15, 43, 56, 59, 61, 63–64, 67, 91, 151, 255–277, 337–361, 365–371, 409, 423
Gulf of Gdańsk, 7, 157, 207, 209, 219–230
Gyrosigma attenuatum (Kützing) Cleve, 120
Gyttja, 135, 137, 167, 172, 175, 182, 185, 223–225, 243, 367, 370

H
Hadley Centre (HC), 9, 41, 393
Haff, 220, 283, 381
Halocline, 5, 8, 78, 84, 86, 91, 101–102, 104, 113–114, 127–128, 367–368, 371, 408
Hamburgian group, 38, 310
Hamming windowing, 105
Harbour, 8, 135, 160, 221, 260, 264–266, 271, 302, 306, 312, 326–330, 421
Harbour facilities, 306, 327
Hazard potential, 8, 337–361
Headland, 234–235, 245, 259, 275, 339
Heat flow, 37, 41
Heavy metals, 370
Helsinki Commission (HELCOM), 373–374, 409, 413, 429
Hiddensee, 236, 239–241, 243–245
Highest shoreline, 6, 82, 190
Hindcasting, 150
Holocene, 5–7, 61, 64, 78, 84, 90, 100, 103–104, 106–107, 112–114, 116–121, 123, 126, 135, 150–152, 167–168, 175, 180, 182, 184–185, 192, 203–216, 219–230, 233–248
Horizontal surface current field, 282–283, 305–306, 308, 312, 317
H_2S diffusion, 86
Human
  activity, 183, 366
  intervention, 264, 276
  life, 8, 337, 340
  occupation, 182–185
Hunter-gatherers, 315–325, 331
Hydraulic head, 397–398, 400–401
Hydraulic model, 274–275
Hydraulic parameters, 274
Hydrocarbon, 37–39, 41
Hydrodynamic load, 257
Hydrogeological modelling, 8
Hydrothermal, 32
Hypoxia, 91–92, 368
Hypsometric, 208–209

I
Ice
  cover, 58, 126, 269, 366
  impact, 355
  season, 261, 276
  sheet advance, 5, 80, 83
  stream, 56–58, 60–61, 64–65
Industrial Revolution, 92
Inflow event, 119, 125
Inherent optical properties (IOPs), 412, 418
Initial Littorina Sea, 88
Integrated coastal zone management, 359–360, 429
Intergovernmental Panel on Climate Change (IPCC), 150–151, 159–162, 289–290, 392–393
Internal eutrophication, 381–384, 386
Interpolation, 104–105, 127, 153, 162, 166, 168, 174–175, 191–192, 210, 242
Intra-continental sea, 100
Inundation, 7, 248, 306, 316
Inversion, 5, 13, 28, 33–34, 258, 419
Iron Age, 326
Iron-phosphates, 376, 379, 381–382
IR-OSL dating, 136–137, 139–140
Irradiance (E), 412, 414
Isostasy, 68, 204, 209, 213, 233–248, 282
Isostatic
  rebound, 54, 82, 85, 167, 209, 213, 302, 304, 320, 325
  response, 67
  uplift, 6, 67–68, 82, 88, 152, 155, 184, 247, 306, 310, 315, 320, 327, 338, 392

J
Jetty, 260, 265
Jõekalda settlement, 167
Jotnian, 15–16, 24, 63–66
Jurassic, 19, 34, 135–136

K
Kalbådagrund, 269
Kaliningrad, 19, 30, 33–34, 37–38, 129, 207, 220, 337–361
Karelia, 5, 76, 78, 184, 194–196
Karelian Isthmus, 184, 194–196
Klaipėda Strait, 135–136
Kleines Haff, 381
Klint, 259
Kriging, 168, 191, 237

L
Lacustrine, 6, 19, 78, 80, 82, 113, 123–124, 127, 135–136, 192, 241–244, 248

Lake Ladoga, 24, 56, 59, 64–66, 193–196
Lamination, 100, 113, 116
Land
uplift, 7, 166, 174, 182, 197, 257, 271, 302–306, 311–312, 324–326, 330
use, 92, 376, 392, 394–397, 402
Landscape degradation, 360
Landslide, 339–341, 345–346, 348, 355, 357, 402
Last Glacial Cycle (LGC), 5, 54–55, 67, 100
Last interglacial/glacial cycle, 4–5, 54–56, 58, 61–62, 66–68, 76–78, 100, 135
Late Glacial, 65, 135, 166–168, 189–191, 197, 221, 235, 241–242, 244–245, 248, 309, 331
Late Holocene, 7, 103–104, 113, 184, 213
Late Subatlantic transgression, 238, 247–248
Latvia, 6, 15–16, 19, 25, 30–31, 38–39, 44, 87, 144, 190–193, 195–197, 199, 207, 257, 308, 311, 325, 338
Laurentia, 4, 26, 31
Leba ridge, 32–33, 39, 41
Leister prongs, 318
Lemmetsa settlements, 167, 180, 184–185
Levees and lakes, 88
Level 1, level 2, level 3 products, 417–419
Levene moraine, 198
Liepaja-Saldus ridge, 28–33, 39
Limestone, 19, 39, 65, 194
Limnaea Sea, 325–330
Lithology, 43, 56, 58, 60–61, 64–65, 80, 83–84, 89, 111, 135, 139, 143, 208, 348, 397
Lithosphere, 4, 26–27, 31, 33, 43–44, 67, 77, 340
Lithuania, 15, 19, 25, 28, 30–31, 34, 36–39, 41–42, 44, 134–136, 141–142, 144, 192–193, 199, 207, 220, 308, 312, 324–325, 338
Lithuanian Coastal Area, 135–136, 142, 144
Little Ice Age (LIA), 113, 116, 124, 238, 248
Littoral drift, 257–259, 265, 270–271, 275–276
Littorina littorea, 152
Littorina Sea, 7, 88–92, 113, 122, 127, 135, 137, 167–169, 171–172, 175, 180–185, 209, 213, 305, 315–325
Littorina transgression, 6–7, 89, 114, 116, 123–124, 127–128, 152, 154–155, 157, 159, 207, 214, 216, 235, 247–248, 283, 306, 317, 319, 323–324
Long-term prognosis, 340
Longterm simulation, 7
Lowland coast, 151

M

Macro remain, 237
Magmatic, 4, 13, 22, 24
Magnetic susceptibility, 108, 110, 112, 114
Malda settlement, 167, 180, 184–185
Managed retreat, 258
Management, 4, 8–9, 150, 158, 162, 257, 276, 281, 359–360, 373–387, 392, 408, 420, 428–429
Mapping, 8, 115, 135–137, 142, 208–209, 340, 342, 344, 360–361
Marine isotope stage (MIS), 5–6, 77–80, 136–137, 143–144
Marine resource, 303, 326
Marine wind, 269
Maritime zone, 304, 306, 312, 314, 327
Marked lithological chance, 89
Mass- balance, 66–68, 282, 400
Master stations, 104–106, 108, 110, 112–113, 117–119, 126–127
Maturity, 41
Mazury High, 27
Mecklenburgian Bight, 101, 316–317, 325
Mecklenburg-Vorpommern, 9, 234, 282, 285, 292, 392
Medieval Climate Anomaly (MCA), 113, 124–125
Medieval Warm Period, 91
Meiendorf interstadial, 308, 311
MERIS, 407, 410, 417–422, 424, 426–429
Mesolithic, 184, 306–307, 315, 317–318, 320–321, 324–325, 328, 331
Mesolithic settlements, 306, 320
Microtidal, 261
Middle Age, 303, 326, 330
Middle-late Littorina, 91–92
Miiduranna Port, 265
Mineral resources, 3, 14, 341–342
MIS, 5–6, 77–80, 136–137, 143–144
Modeling, 56, 67, 100, 109, 391–403
Model simulation, 378–379, 384
Modern warm period, 5, 124, 127
MODIS, 417, 421
Molluscs, 136–137, 141, 241
MOM3 code, 103–104
Moneris, 375–378, 384
Monitoring, 9, 347, 375, 379, 407–431
Moraine cliffs, 151
Morphodynamic processes, 284, 350

Morphostructural, 206
Moscow basin, 16–17, 27
Mud, 19, 104, 106, 111–112, 114, 117, 235, 242, 244–245, 286–287, 340, 345, 348, 355, 358, 367, 381
Multi-corer (MUC), 113, 124
Multi-scale approach to monitoring, 415–416
Multi-sensor core logger (MSCL), 105, 108–109, 113, 115–116, 125
Multispectral radiometers, 417
Mussel farm, 303, 387

N
Namurian, 22
Narva Bay, 261–262, 268–269
Narva-Jõesuu, 258, 261–267, 273–275
Narva River, 255, 258, 262, 265–266, 274
NASA AERONET-OC (AEROsol RObotic NETwork – Ocean Color), 422–423
Natural processes, 276, 339, 342, 358
Nautical chart, 6, 210
Navigation, 106, 160, 204–205, 265–266, 292, 354, 360
Near-bottom currents, 102–104, 354, 376
Nearshore, 239, 242–243, 261–264, 267, 269, 275, 339, 342, 344, 347–348, 352, 354, 358, 360
Neolithic period, 317, 321, 326, 331
Neolithic settlements, 184, 324
Neotectonic, 36, 150, 154, 160–162, 209, 213, 216, 234
  modelling, 150
Neumann (second kind boundary condition), 397–398
Neural networks, 419
Neva, 180, 190–192, 194–196, 198, 350–351, 355, 358–359
Nitrogen, 8, 88, 374, 376–377, 382, 385, 387, 409–410, 414
Nodularia spumigena, 409, 411
No-flow, 398
North Atlantic Oscillation (NAO), 6, 91–92, 101, 124–126, 128
Northeast-German Depression, 151
North German Basin, 15, 20, 22–23, 33, 35
North Sea, 3, 92, 101, 124–125, 128, 151, 159, 235, 238, 240, 242, 290, 292, 308–309, 311, 315–316, 325, 329, 368, 376, 408, 413, 419, 421
  coast, 238, 240
Numerical modelling, 7, 103
Nutrient cycle, 386–387
Nutrients, 8, 86, 91–92, 225, 228, 355, 369–370, 374–377, 380, 385–387, 392, 409, 412, 414, 421, 425–426, 428
NW Russia, 6, 184, 190–197, 199

O
Ocean colour, 407, 410, 412, 415–419, 422–423, 428–429
Ocean colour remote sensing, 415–419, 423
Ocean waters, 127, 412
Oder Lagoon, 374–375, 377, 379, 381–385
Oder palaeo-valley, 235, 241
Odra, 8, 374
Oil, 5, 8, 20, 37–39, 41–42, 105, 339–340, 342, 345, 356–358, 416, 421
Older Dryas, 311
Old Red, 26
OMNIDIA, 111
One-layer flow, 275
Operational monitoring, 408, 420, 429
Optical Case-1 waters, 412
Optical Case-2 waters, 413
Optical gradients, 425–426
Orbital velocity, 270
Ordovician, 19, 21–22, 25–26, 31, 33–35, 38–41, 65
Öresund Strait, 76, 80, 82, 89, 114
OSL (optical-stimulated luminescence), 6, 89, 135–144
Oxygen, 77, 86, 91, 100, 102, 112–113, 116–117, 119, 125–128, 228–229, 368–369, 374, 376, 379, 381, 385–386, 410
Oxygenation, 125

P
Pakri, 273
Palaeocoastline, 165, 167–175, 180, 182
Palaeo-ecological proxies, 7
Palaeo-environmental reconstructions, 6
Palaeogeographic model, 165–185
Palaeogeography, 189, 230
Palaeolithic, 308–313, 317, 331
Paleoclimatic records, 79
Paleogeographic, 83, 86–87, 90, 135–136
Paleosalinity, 90–91, 119, 124
Palivere stade (or ice-marginal zone), 191, 193, 198–199
Palynological analyses, 223
Pandivere/Neva stage (or ice-marginal zone), 190
Parametric sediment echosounder, 106

Pärnu area, 165, 169–170, 173, 175, 180, 182–185
Passive continental margin, 25–26, 34
Pauliella taeniata (Grunow) Round & Basson, 121
Peak period, 268–269
Peat, 7, 87, 123, 135, 137, 165, 167–168, 170, 175, 182–185, 192, 199, 219–220, 223–225, 229, 237–238, 241–245, 247–248
Pebbles, 262, 348
Pelagic deposition, 103, 118, 124
Periodicity analysis, 105, 121, 216
Permocarboniferous, 13, 22, 27, 29–30, 32–33, 43
Phosphorus, 8, 91, 365, 369–371, 373–387, 410, 412
Physico-stratigraphic facies zones, 100
Physico-stratigraphic unit, 105
Physico-stratigraphic zonation, 112, 117–119, 123, 127
Phytoplankton abundance, 91
Pirita, 259–260, 262–263, 265, 268–269, 271–273
Pirita Beach, 7, 255, 258–265, 267–268, 270, 272–273
Plain, 61, 205, 207–208
Platform, 3, 25, 27, 56, 65, 101, 157, 204, 208, 210, 212, 242, 345, 422
Pleistocene, 5, 37, 43, 53–54, 58, 64–65, 99–100, 107, 112–113, 126–127, 133, 135–137, 141, 143–144, 151, 203, 208, 213, 234–235, 237, 241, 245, 312, 317, 324
Pleniglacial, 133, 143–144
Pockmark, 338, 345, 350, 356
Poland, 3, 5, 13, 15, 17, 33, 38, 144, 199, 220, 257, 311, 339, 385
Polish basin, 13, 15, 22
Polish coast, 37, 199, 220, 229, 238, 240, 244, 408
Pollen, 7, 78, 124, 136, 141, 143, 213, 219, 225, 227, 229, 237
Pollution, 8, 337–338, 341, 355–357, 366, 377–379
Polygenic, 212, 214
Pomeranian Bight, 234–235, 242, 339
Porosity, 39, 67, 288, 397–398
Porosity coefficient, 269
Portlandia (Yoldia) arctica, 84
Positive buoyancy, 410
Postglacial, 53–68, 82, 85, 90, 99–101, 106, 112, 157, 167, 169, 182, 185, 205, 207, 255, 257, 264, 275, 365, 370
Post-glacial sedimentation, 67, 85
Post-glacial uplift, 5, 53–68, 213
Post-Littorina Sea, 325
Potential (immersed weight) transport rate, 269–270
Preboreal, 84–85, 220, 243, 311–312, 316
Precipitation, 9, 77, 91, 101–102, 113, 118, 125, 275, 382, 391, 393–400, 402–403, 409, 426
Primary production, 89, 91, 374, 376, 384–385, 426
Principle Component Analysis (PCA), 117
Probability, 225, 226–228, 268–269, 339–340, 344
Processor, 419, 426–427, 430
Proglacial lake, 189–191, 194, 198–199
Progradation, 214, 233, 236, 242, 245–248
Projective scenarios, 150
Proterozoic, 14–15, 64, 151
  crystalline bedrock, 64, 151
Pseudosolenia calcar-avis, 120
Pseudosolenia calcar-avis (Schultze) Sundstrom, 120
Pseudostaurosira brevistriata (Grunow in Van Heurck) Williams and Round, 121
Puck Lagoon, 220
Pulli settlement, 180, 182–184
p-wave velocity, 108–109, 111–112, 116, 122–123

Q
Quaternary, 4–5, 13–45, 53–57, 62, 64–66, 88, 99–128, 133–145, 203, 205, 233, 349, 355, 391
Quaternary glaciations, 76, 101, 151
Quay, 265

R
Radiance (I), 412
Radon, 349
Raised beach, 87
Rapakivi, 16, 23–25, 43–44
Reconstruction, 5–6, 36, 41, 56, 100, 119, 122, 124, 144, 150, 159–162, 165, 167–168, 175, 180, 184–185, 189, 191, 193, 197–199, 209, 258, 292, 302, 305, 321
Recreational value, 265, 268
Reed belt, 387
Refill, 264, 270

Regression, 6, 38, 84, 86–88, 152–153, 156, 161, 165, 169, 172, 180, 182–183, 207, 214, 303, 309, 315, 321, 324, 329, 394, 419
Reindeer, 308, 311–312, 317, 331
Relative sea level curves, 6, 149–150, 155, 162, 213, 234, 237, 247, 283, 305
Reservoir, 5, 14, 37–39, 89, 395
effect, 89
Resuspension, 104, 382
Retention, 374, 377, 384–386
Revetment, 265
Rift, 23–27, 33
Risk, 160, 337, 339–340, 342, 344–345, 348–349, 356
Risk prevention, 357–360
River
channels, 88
discharge, 102, 258, 273–275, 408
load, 377, 383
plume, 274
Rock art, 327
Rodinia, 25, 30
Roman Climate Optimum (RCO), 157
Rostock, 159, 234–235, 283
Rügen, 21–22, 27, 41, 234–243, 247
Rügen Island, 5, 13, 20, 34–35, 154, 157, 317, 339
Rule of thumb, 306
Runoff, 58, 118, 125, 394, 399–403, 407, 409, 426–427
Runup, 263
Russian Plate, 101
R/V "Petr Kotsov", 106
R/V "Poseidon", 100, 106, 128

S
Saalian ice sheet, 78
Saint (Sankt) Petersburg, 261
Salinity
maximum postglacial, 90
stratification, 78, 91, 367, 412
Salpausselkä, 192–194, 196, 198
Salzhaff, 160
Sambian peninsula, 6, 205, 209, 211–212, 214, 220, 344, 346–348
Sand
bar, 258, 261, 265, 267, 270, 273–274
coarse(-grained), 260, 267, 354
fine, 89, 139, 222, 234, 241–243, 260, 263, 268, 286–287
loss, 257, 265, 271–272, 276
supply, 245
well-sorted, 259, 268, 352
Sandstone, 15–17, 19, 21, 25, 31, 34, 37–39, 41, 63, 204, 262
Sandy Holocene spit, 151
Satellite, 4, 351, 359, 411, 415–422, 424, 428–430
Scandinavian Ice Sheet (SIS), 6, 77, 80, 82, 84, 86, 88, 182, 189
Scarp, 56, 65, 206, 265, 271
Scattering, 412, 418–419, 431
Scenario
emission, 392–393
palaeogeographic, 150, 154–157, 162
regional, 282
sea level, 9, 150, 158–161, 393
Scuba divers investigation, 223
Sea bottom, 206
Seafloor, 8, 210–212, 215, 355, 365–371
Sea level
change, 6, 67, 77, 82, 84, 150, 153–157, 162, 182, 203, 209, 213–214, 236–238, 284, 289–291, 295, 329–330
change, 182, 203, 209, 213–214, 236–238, 329–330, 340, 352
curve, 213, 234, 236–240, 247, 304–306, 312, 314, 320–321, 323, 327, 330–331
index points, 8, 306–307, 317, 327, 331
relative, 6, 79, 82, 84, 150, 153–155, 157, 159, 162, 169, 203, 213–215, 229, 234, 236–237, 239, 247, 283, 289, 305–306, 321, 330
rise, 6–7, 82, 84, 89–90, 127, 150, 152, 155, 157, 159–160, 235–237, 245, 247–248, 283, 288, 290, 294–295, 316, 324, 338, 392–393, 395, 399–402
eustatic, 7, 82, 90, 157, 235, 288, 324, 392
Seasonal variation, 101, 258, 261, 267, 273, 275, 395, 409, 419
Sea surface reflectance, R, 412, 431
Seawall, 265
SeaWiFS, 417, 424, 430
Secchi depth, 89, 413–414, 423–424, 428–429
Sedimentary compartment, 259
Sediment echosounding (SES), 114, 237, 241, 243–244
Sediments
accumulation, 6, 58, 61, 66–68, 103, 105, 114, 234, 259, 295, 341
deficit, 263, 265, 275, 347–348

dynamics, 4, 7, 248, 344
flow, 345, 352–355
fluvial, 115, 229, 262, 265, 275
flux, 350, 352, 360
homogeneous, 111–113, 116, 119, 128, 368
interstadial lacustrine organic, 80
inter-till, 136–139, 141, 143–144
laminated, 91, 111, 113, 116, 123–128
loss, 271–273, 354
map, 286–288
redistribution, 54–55, 58, 63, 67
sequence, 104, 106, 126, 242
source, 114–115, 283, 288, 294
terrestrial organic, 80
texture, 121, 124–125, 267
transgressive, 166
transport
cross-shore, 282
eolian, 263–264, 275
longshore, 7, 265, 269, 282, 348
volume, 234, 243, 248
SEDSIM, 7, 281–295
Seismic cross-section, 105
Seismicity, 338
Selective denudation, 57, 64
Senegal type, 265
SENTINEL-3, 407, 429
Settlement development, 301–331
Shifting coastline, 6, 185
Ship carving, 327
Shipping barrier, 327
Shore displacement
curve, 86, 166–167, 170, 185, 198, 314
models, 8, 304–306, 324, 326, 331
Shoreline
changes, 219–230, 325
database, 167–168, 191
recession, 245, 343, 346, 351
tilting, 169, 180, 184
Side-scan sonar, 344, 349–350, 353
Signal-to-noise enhancement, 105
Signal-to-noise ratio, 418
Significant wave height, 259, 261–262, 268–269, 291–292
Sill, 33, 76, 87, 258, 261, 265, 273–275, 317, 326
Sill height, 261, 273–275
Silt, 80, 108, 135, 137, 139, 234, 241–243, 260, 283, 287–288
Silting, 345, 360
Silurian, 4, 19, 26, 29, 31–34, 38–41
SINCOS, 150, 152, 156, 221, 302, 317, 320, 331
Sindi-Lodja settlements, 183
Slides, 341, 348, 355
Slope, 6, 58, 63–65, 67, 204–208, 210–211, 214–215, 220–221, 245, 265, 269, 271–272, 284, 341, 344, 347–349, 352, 355, 360, 367, 392, 413
Socio-economic development, 403
Socio-economy, 403
Soil nutrient leakage, 92
Sorting, 267
Sorting, well-sorted, 259, 268, 352
Source rock, 32, 39–41
Spatial interpolation, 153, 162
Spatial resolution of satellite data, 411
Specific weight, 269
Stage, 5, 17, 24–28, 32, 34, 35, 39, 43, 62, 64, 84, 113, 144, 181, 182, 190–200, 208–209, 213–214, 310
Stakeholders, 4
Staurosira construens var. *binodis* Ehrenberg, 121
Steady-state, 9, 393, 395, 397, 400
Stephanodiscus alpinus Hustedt, 120
Stolpe Channel, 102–103, 116, 128
Stolpe Foredelta, 103, 113–116, 128
Stone Age, 6, 167, 172, 175, 182, 185–186, 305, 313, 322, 324–325
settlements, 167, 172, 175, 305
Storm
events, 292, 295, 339, 342, 347, 361
surges, 150, 161, 229, 284, 306, 338–339, 346, 352
St. Petersburg, 344–345, 349, 351–352, 355, 358, 360, 366
Stralsund, 159, 243, 283
Stratification, 78, 90–92, 258, 274–275, 367, 369, 408, 410–412, 422
Stratigraphic zonation, 107, 112, 117–119, 123, 127
Stratigraphy, 58, 80, 82, 136, 185
Structure-dependent, 204, 212, 214–215
Sub-Jotnian, 15
Submarine groundwater discharge, 9, 392–395, 399–402
Submarine terrace, 351–355
Submerged, 67, 87, 167, 183–184, 203–216, 221, 265, 306, 317–318, 329
Subsidence, 4–5, 14–15, 19, 23–28, 41, 43, 67–68, 150, 152–154, 161, 166, 236, 240, 247, 345, 347, 349
Subsistence strategy, 317

Sunken forest, 219
Surface runoff, 394, 399–403
Surf zone, 261, 263, 267–270
Surge, 150, 159–161, 183, 229, 265, 282, 284, 292, 295, 305, 338–340, 346, 351–352
Suspended matter, 103, 114, 262, 358–359, 409, 413, 425–427, 431
Sweden, 31–32, 64, 80, 82, 84, 87–88, 151, 172, 183, 198, 203, 257, 303–306, 308–312, 315, 320, 322, 327, 421
  coast, 301–331
Swedish Meteorological and Hydrological Institute (SMHI), 9, 393, 420, 430
Swiderian culture, 311–312
System shift, 126
Szczecin, 159, 374–375
  Lagoon, 375

T
Tallinn
  Bay, 258–260, 262, 269
  Port of, 261
Technogenic hazards, 340
Tectonic
  Early Ediacaran, 30–31, 43
  movements, 36, 208–209, 213, 342, 345, 349, 356
  processes, 25, 32, 34–35
Thalassionema nitzschioides, 120
Thalassionema nitzschioides decrease, 120
Thalassionema nitzschioides (Grunow) Grunow, 120
Thalassiosira baltica (Grunow) Ostenfeld, 121
Thalassiosira oestrupii, 120
Thalassiosira oestrupii (Ostenfeld) Hasle, 120
Thermal stratification, 411
Thickness analysis, 103, 114–116
Thickness map, 6, 17, 115, 127
Threshold, 77, 80, 82, 105, 192, 194–196, 199, 203, 213, 235, 268–269, 385
Tide, tidal range, 157, 261, 276, 366, 408
Till, 6, 64, 66–67, 135–144, 234, 241–243, 264–265, 275, 283, 286–287, 347, 350, 367, 392–393, 396
Time series, 6, 100, 105, 122, 124, 126, 128, 269, 289–291, 431
Time/space modeling, 100, 109
Topographical map, 273
Tornquist Zone, 15, 20, 28, 31–32, 34
Total suspended matter (TSM), 409, 412–414, 417, 423, 426–428, 431
Trading centre, 327–329, 331
Transgression
  Ancylus, 87, 207, 214
  Holocene, 152, 161
  Littorina, 6, 124, 214, 317
Transient, 393, 395, 399–401
Transport
  bulk, 270
  net, 271
Transportation system, 303
Tree stumps, 7, 220–221, 223, 237, 318, 329
Triassic, 19, 22, 27, 34
Tundra, 308, 311
Two-layer flow, 275

U
Underwater archaeology, 307
Up-dammed lakes, 167
Uplift
  postglacial, 53–68, 257, 264, 275
  and subsidence pattern, 67
Usedom, 20, 22, 234–235, 237, 241, 248, 287, 339, 392

V
Valley, 54, 63–65, 80, 82, 84, 182–183, 194–196
  erosional, 259
Variscan, 22, 27, 32, 35
Ventilation, 100, 116, 119
Vertical crustal movement, 150, 157, 160–162, 213, 216, 284
Viimsi Peninsula, 259, 265
Visean, 22
Vistula Delta, 221, 229–230
Volcanic, 15, 22–23, 33, 35, 339–340

W
WAM model, 268
Warnow River, 288
Warthanian, 137, 144
Water budget, 374, 392–393
Water Framework Directive (WFD), 374, 415, 425–426, 431
Water level
  changes, 82
  reconstruction, 168, 191, 197, 199
  rise, 182, 184, 271
  change model, 6, 166–167, 174, 183, 185, 289
Waterline, 267–268
Water transparency, 414–415, 429
Wave action, activity, 7, 259, 261–262, 265, 267–268, 273, 275–276

Waves
approach angle, 270
atlas, 268
breaking, 268, 270
climate, 7, 259, 261–263, 268, 275–276
-cut, 203–216
direction, 284
energy flux, 270
height, 259, 261–262, 268–270, 273, 276, 284, 291–292, 416
hindcast, 268
load, 261, 269
measurements, 268
period, 261, 268, 273, 275
power, 246, 270
refraction, 284, 291
Web-based information system, 421
Weichselian
glacial deposits, 151
ice sheet, 5, 78–79, 157
Western European Platform, 3
Westphalian, 22
Wet bulk density, 106, 108, 112
Wind field, 159, 262
Wind flat, 243, 245
Wisconsinan ice sheet, 79
Wisla River, 115
Wismar, 9, 160, 235–236, 238–240, 242, 247, 317–321, 328–329, 395–396, 402
Wismar Bight, 152–153, 160–161, 247, 317–321, 328–329, 331
Written sources, 302
Wustrow Peninsula, 160

X
X-ray fluorescence (XRF), 111, 117–118, 122–123, 125
XRF Core Scanner, 111

Y
Yoldia Phase, 112
Yoldia Sea, 7, 84, 86, 113, 135, 169, 182, 189, 207, 209, 214, 308–314, 310–311, 314, 320
Younger Dryas, 83–85, 190, 198, 214, 309, 311–312
– Preboreal transition, 84

Z
Zeros padding technique, 105
Zingst, 7, 154, 234–235, 237, 242–248, 281–295
Zura depression, 30